Handbook of Environmental Engineering

Volume 18

More information about this series at http://www.springer.com/series/7645

Zhao Youcai • Zhang Chenglong

Pollution Control and Resource Reuse for Alkaline Hydrometallurgy of Amphoteric Metal Hazardous Wastes

Springer

Zhao Youcai
State Key Laboratory of Pollution Control and Resource Reuse
School of Environmental Science and Engineering
Tongji University, Shanghai, China

Zhang Chenglong
Shanghai Cooperative Center for WEEE Recycling
Shanghai Polytechnic University
Shanghai, China

Handbook of Environmental Engineering
ISBN 978-3-319-55157-9 ISBN 978-3-319-55158-6 (eBook)
DOI 10.1007/978-3-319-55158-6

Library of Congress Control Number: 2017933684

Printed on acid-free paper

This Springer imprint is published by Springer Nature
The registered company is Springer International Publishing AG
The registered company address is: Gewerbestrasse 11, 6330 Cham, Switzerland

Preface

Amphoteric metal is a metal susceptible to leaching in both acid and alkaline media, especially in aqueous solutions, generally in the form of oxides, such as aluminum, zinc, lead, etc. Zinc and lead are two of the commonest amphoteric heavy metals used in the world. As a result, a large quantity of hazardous wastes containing zinc and lead are being generated or stored at landfills and factories. The treatment and recycling of zinc and lead hazardous wastes have thus received great concern. Zinc and lead in these hazardous wastes may be generally extracted by leaching with acidic or alkaline solutions. For the acidic leaching process, though zinc and lead will be dissolved to an acceptable high level, the bulk materials, iron, calcium, etc., will also be dissolved completely, and the dissolved iron and other elements have to be precipitated from the leaching solutions. Moreover, a big fraction of zinc exists as zinc ferrites in the dust, which cannot be attacked effectively by acidic leaching processes. Therefore, the acidic leaching process seems not to be economically viable for the treatment of these wastes.

In contrast, considering that only the oxides of lead and zinc as well as part of aluminum will be dissolved in alkaline solution, it may be a cost-effective method to extract zinc and lead from the wastes by alkaline leaching processes. The thermodynamics and kinetics of alkaline leaching of zinc and lead hazardous wastes show that ZnO, $ZnCO_3$, and Zn_2SiO_4 can be dissolved by strong alkaline solution. Compared with ZnO and $ZnCO_3$, a higher concentration of OH^- is required to dissolve Zn_2SiO_4. ZnS cannot be leached by alkaline solution directly under atmospheric pressure. PbO, $PbSO_4$, and $PbCO_3$ can be dissolved in concentrated $NaOH$ solutions, while the dissolution of PbS may be negligible. For the ZnS in solid wastes, the leaching rate of zinc in alkaline solution is greatly improved via chemical conversion with $PbCO_3$. In this book, the alkaline hydrometallurgy of zinc and lead hazardous wastes is fully described. The alkaline leaching process for the selective leaching of zinc and lead, selective separation between zinc and lead in the leaching solution using sodium sulfide, electrowinning of high-purity zinc powders from the purified leach solution, operational costs and mass balance analysis for all possible processes, flow sheets, and chemical reactions that take place in the processes are provided in detail. The industrial application process and engineering design is also given. The main contents include zinc and lead hazardous wastes and hydrometallurgical processes, leaching of zinc and lead hazardous wastes, purification of leach solution of zinc and lead, electrowinning of zinc from purified alkaline solutions, chemical reactions taking place in the processes and proposed flow sheets, thermodynamic and spent electrolyte regeneration, alkaline hydrometallurgy of low-grade smithsonite ores, recovery of associated valuables from lean leach solutions, and industrial-scale production of 1500–2000 t/a zinc powder using alkaline leaching-electrowinning processes. The process is cost-effective and generates little secondary pollutants and has been applied widely in China.

The readers include solid waste engineers, managers, technicians, recycling coordinators and government officials, undergraduates and graduates, and researchers.

Shanghai, China

Zhao Youcai
Zhang Chenglong

Contents

List of Figures

List of Tables

Contributors

Guo Cuixiang State Key Laboratory of Pollution Control and Resource Reuse, School of Environmental Science and Engineering, Tongji University, Shanghai, China

Xia Fafa State Key Laboratory of Pollution Control and Resource Reuse, School of Environmental Science and Engineering, Tongji University, Shanghai, China

Jiang Jiachao School of Environmental Science and Engineering, Tongji University, Shanghai, China

China Mining and Metallurgy University, Xuzhou, Jiangsu, China

Li Qiang School of Environmental Science and Engineering, Tongji University, Shanghai, China

Liu Qing The Hunan Key Laboratory of Pollution Control and Resource Reuse, University of South China, Hengyang, China

School of Environmental Science and Engineering, Tongji University, Shanghai, China

Zhao Youcai State Key Laboratory of Pollution Control and Resource Reuse, School of Environmental Science and Engineering, Tongji University, Shanghai, China

Deng Yue State Key Laboratory of Pollution Control and Resource Reuse, School of Environmental Science and Engineering, Tongji University, Shanghai, China

Zhang Chenglong Shanghai Cooperative Center for WEEE Recycling, Shanghai Polytechnic University, Shanghai, China

School of Environmental Science and Engineering, Tongji University, Shanghai, China

About the Authors

Zhao Youcai is currently a professor of environmental engineering at the School of Environmental Science and Engineering, Tongji University, China. He got his bachelor's degree from Sichuan University (1984) and Ph.D. from the Institute of Chemical Metallurgy (now the Institute of Process Engineering), Chinese Academy of Sciences, Beijing (1989). After finishing his postdoctoral research work at Fudan University, Shanghai, China, he joined Tongji University in 1991. Meanwhile, he had worked at Aristotle University, Greece; National University of Singapore; Tulane University, USA; and Paul Scherrer Institute, Switzerland, for 4 years as research fellow or visiting professor. He had authored or coauthored 138 publications published in peer-reviewed internationally recognized journals, 420 publications in Chinese journals, and 75 books (as an author or editor in chief). Currently, his research interests include treatment of municipal solid wastes, sewage sludge, hazardous wastes, polluted construction wastes, industrial wastes, and agricultural and forest wastes.

Zhang Chenglong is currently a professor of environmental engineering at Shanghai Collaborative Innovation Centre for WEEE Recycling, Shanghai Polytechnic University, China. He got his bachelor's degree from Wuhan University, China (1996), master's degree from the National Engineering Research Center of Waste Resource Recovery (2003), and Ph.D. from East China University of Science and Technology, China (2008). He conducted a series of research work on the recovery of metals and materials from industry wastes, WEEE, tailings, and so on. He had authored or coauthored 69 publications published in peer-reviewed internationally recognized journals or Chinese journals and 5 books. Currently, his research interests include cleaner production and treatment of WEEE, industrial wastes, and hazardous wastes.

Summary

Zinc and lead are two of the commonest amphoteric heavy metals used in the world, and their wastes are all classified as hazardous. In this book, pollution control and resource reuse processes for zinc and lead hazardous wastes are involved, including zinc and lead hazardous waste characterization and hydrometallurgical processes, leaching of zinc and lead hazardous wastes, purification of leach solution of zinc and lead, electrowinning of zinc and lead from purified alkaline solutions, chemical reactions taking place in the processes and proposed flow sheets, thermodynamics and spent electrolyte regeneration, alkaline hydrometallurgy of low-grade smithsonite ores, recovery of associated valuables from lean leach solution, and industrial-scale production of 1500–2000 t/a zinc powder using alkaline leaching-electrowinning processes. The process is cost-effective and generates little secondary pollutants and has been applied widely in China.

Chapter 1
Amphoteric Metal Hazardous Wastes and Hydrometallurgical Processes of Zinc and Lead

Abstract Various wastes containing zinc and lead are generated in industries such as galvanizing, casting, scrap recycling, smelting, and pyrometallurgical and hydrometallurgical process, as forms of dust, tailings, residues, sludge, and lower-grade lean ores. The disposals of these wastes are now becoming expensive due to the need for the treatment to render the wastes nonhazardous. The options currently available can be comprehensively classified as security landfilling and pyrometallurgical and hydrometallurgical processes. As the wastes containing zinc and lead have been classified as hazardous, conventional landfilling processes should be modified to security landfilling in order to meet the environmental constraints required for the hazardous. The pyrometallurgical processes to treat hazardous wastes containing zinc and lead include Inred, Wala Kiln, Plasma, etc., in which zinc and lead can be easily extracted due to the high temperature used, though a large amount of energy will be consumed and serious secondary pollutions arise. The hydrometallurgical treatment method for wastes containing zinc and lead is by dissolution in mineral acids and alkaline solutions. For acidic leaching process, though zinc and lead will be dissolved to an acceptable high level, the bulk materials, iron, calcium, etc., will also be dissolved completely, and the dissolved iron and other elements have to be precipitated from the leach solutions, possibly leading to the generation of new hazardous wastes and wastewaters. Moreover, a big fraction of zinc exists as zinc ferrites in the dust, which cannot be attacked effectively by acidic leaching processes. Therefore, the acidic leaching process seems not to be economically viable for the treatment of dust. The alkaline process is considered to be a cleaner technology for extracting zinc and lead from the hazardous waste bearing zinc and lead and will be fully introduced in this book.

Keywords Hazardous wastes containing zinc and lead • Landfilling • Pyrometallurgical processes • Hydrometallurgical processes • Acidic process • Alkaline process

Y. Zhao, C. Zhang, *Pollution Control and Resource Reuse for Alkaline Hydrometallurgy of Amphoteric Metal Hazardous Wastes*, Handbook of Environmental Engineering 18, DOI 10.1007/978-3-319-55158-6_1

1.1 Amphoteric Metal Hazardous Wastes

Amphoteric metal has the characteristics of an acid and a base and capable of reacting chemically either as an acid or a base. Zinc and lead are typical amphoteric heavy metals of highest interest as they are commonly used in the world. This book focuses mainly on the pollution control and resource reuse of amphoteric metal hazardous wastes bearing zinc and lead.

The metal melting industry is one of the largest industries worldwide, including iron and steel foundries and mills, as well as brass and bronze foundries. During the metal melting process, the electric arc furnace (EAF) can reach temperatures of 1600 °C or higher. Under these conditions, many of the components of the charges, including zinc, cadmium, and lead, are volatilized and enter the vapor phase. In addition, due to the inherent turbulence of the melt, some fine particular matter is also captured by the baghouse. As the fume is cooled, the vapor tends to oxidize and condense on the fugitive matter, and this results in a chemically and physically complex dust. This dust contains up to 32% mass percent zinc. Zinc and lead hazardous wastes are also being generated at light chemical industry and hydrometallurgical and pyrometallurgical processes of zinc and lead, as forms of dust, tailings, residues, sludge, and lower-grade lean ores.

However, the average zinc content is approximately 19% mass percent with somewhat smaller content of lead and trace levels of cadmium and other heavy metals. The bulk waste is generally iron oxide. The fumes from a brass melting furnace can have up to 65% or higher zinc. Zinc ferrite spinel accounts for 20 to 50 mass percent of the zinc in the dust. The quantity of such dust is enormous. About 600,000 t/a of EAF dust are generated in North America alone. The weight of dust collected in a typical carbon steel electric furnace steel making shop is about 10–15 kg/t steel. Therefore, the quantity of dust produced in any one location is comparatively small, about 10,000–30,000 t/a. Because of their metal leaching potential, the wastes are frequently classified as hazardous wastes under the US EPA classification and China regulations.

Moreover, zinc and lead hazardous wastes may come from chemical and metallurgical industries, including fly ash, slags, residues, construction and demolition wastes, explosion and fire disasters, etc. These wastes may be collected and landfilled or may be discarded at dumping sites. Before 1990s, these wastes, including the steel dust, were actually dumped as supporting materials for road construction.

Classification as hazardous wastes greatly increases the cost of disposal, due to the need for treatment to render the wastes nonhazardous, as well as greater transportation and disposal costs and increases the paperwork requirements. A number of treatment processes are in operation, particularly high-temperature reduction processes which reduce and fume off the volatile metals. Most of these installations are based on rotary kiln technology, which is sensitive to economies of scale, and therefore the dust must be collected from numerous sources and transported to the relatively large processing plant. Plasma-based treatment processes are currently being developed, which are custom designed for the capacity of a special steelmaking location. All pyrometallurgical operations have high thermal energy requirements, elaborate dust collecting systems, and require additional processing to separate the lead, zinc, and halides as intermediate products. Moreover, the quantity of dust treated by this process is also very limited.

The hydrometallurgical approach is aimed at eliminating the above problems. In the 1970s, Amax operated a pilot plant in New Jersey based on caustic leaching, purification of solutions with zinc dust, and electrowinning. This development was eventually abandoned due to a combination of technical and economic problems. A similar plant, based on direct caustic leaching, was later built in France, but this also has been abandoned. So far, most of these wastes are disposed of in the industrial landfills, with the exception of steel mill electric arc furnace (EAF) dust. As abovementioned, the dust and fumes have been classified as hazardous, conventional landfilling processes should be modified in order to meet the need of the environmental constraints required for the hazardous. Consequently, the treatment costs increase remarkably, and the treatment processes become more complex. Although the landfills are still the most important places for the wastes to go, it is necessary that an economically viable process is developed to treat these wastes.

Metals, including zinc and lead, have been consumed at an exponential rate as a result of population and capital growth. The resources of original mineral ores are quite limited and are being consumed rapidly. Thus, reclamation from low-grade oxides ores, industrial solid wastes, and wastewaters is becoming more and more attractive. In the case of zinc and lead, such solid wastes may include the dust generated in brass and steelmaking foundries and the wastes generated in chemical industries with varied contents of the valuables. For the dust, it is known that millions of tons of the toxic wastes have been dumped in the landfill all over the world, resulting to the great loss of lead and zinc as well as the heavy environmental pollution in the vicinity of the landfills.

It has been confirmed that zinc and lead and other metals in the dust and some industrial solid wastes usually exist as oxides, which could not be selectively separated by conventional sulfide flotation and acidic leaching processes. In addition, up to 20–40% of iron oxide may coexist with the zinc and lead oxides in the dust. Many researches have been done on the reclamation of dust and industrial solid wastes containing zinc, lead, iron, and copper oxides, using both acidic and alkaline leaching processes and pyrometallurgical reduction. Nevertheless, a lot of problems remain unsolved in the leaching process.

Conventional landfills used for the disposal of municipal solid wastes and nonhazardous and nontoxic solid wastes could not be used for the treatment of dust and fumes. However, the dust and fumes may be filled in the landfills specifically designed for the filling of industrial hazardous solid wastes and solvents. If the space is available, it is feasible and practical and cost-effective to dispose of the dust and fumes in the industrial hazardous landfills. Nevertheless, in this case, all the valuables may be lost and cannot be recovered, though currently nearly all the dust and fumes are dumped in the landfills.

1.2 Pyrometallurgical Treatment Processes for Zinc and Lead

The zinc and lead in the dust and fumes, especially the EAF dust, can be reduced to gaseous state and collected as another fume, which has been practiced in Japan and the USA for years. After the zinc, lead, and cadmium are volatilized, the remaining residues become nonhazardous and can be treated more easily. But this process will consume a large amount of energy and seems to give rise to the secondary pollution. The basic chemical reactions occurred in the pyrometallurgical processes may be expressed as follows:

$$ZnO(s) + C(s) = Zn(g) + CO(g) \tag{1.1}$$

$$ZnO(s) + CO(g) = Zn(g) + CO_2(g) \tag{1.2}$$

The major problem with the current industrial processes is that during the zinc condensation stage, the carbon dioxide off-gas back-reacts with the zinc vapors resulting in zinc oxide. One possible alternative to carbothermic reduction would utilize either solid or liquid iron as a reducing agent according to the following reaction:

$$Zn(FeO_2)(s) + 2Fe(s\ \text{or}\ l) = Zn(g) + 4FeO(s) \tag{1.3}$$

In this case, the off-gas would consist mainly of zinc, which would facilitate the condensation of a purer zinc product. The iron addition to the dust also dilutes impurities, such as copper and sulfur, in the EAF dust, and, hence, the iron-rich residue could be recycled to the iron or steelmaking process. In addition, in the current industrial processes, the iron oxide may be preferentially reduced by carbon, and this iron may play a role in the reduction of the zinc-containing oxides in the EAF dust.

Pyrometallurgical treatment processes have been used widely in the world for the recovery of zinc, lead, and other metals in forms of oxides from lower-grade wastes and ores, and the resultant-rich oxides can be used as the raw materials for the production of metallic zinc and lead using alkaline or acidic processes as described in this book.

1.3 Stabilization of Heavy Metals for Hazardous Wastes

Most generators treat and dispose their waste materials. The standard treatment process uses cement and chelating agents to stabilize the heavy metals in the dust, although other competing stabilization process using additives such as phosphate are also used. The EAF dust seems not particularly suitable for cement treatment; because of very small particle size and because of high metals contents and salts, all of them tend to disrupt the cement settings. Phosphate stabilization is very effective for wastes with low metal contents but may become quite expensive for the higher metal content wastes, since the amount of additive required is roughly proportional to the amount of heavy metals present. Also, there are no valuables to be recovered in this process.

1.4 Acidic Leaching Process for Zinc and Lead Ores and Wastes

The composition of the zinc and lead hazardous wastes is very complex, which may include oxides of Zn, Pb, Al, Fe, Cu, Si, Cd, Mn, Mg, etc. When the wastes are digested in strong mineral acids such as hot HCl acid, nitric acid, and sulfuric acid, most of Zn, Pb, Al, Cu, Fe, etc. may be dissolved. In general, sulfuric acid is used, if zinc is recovered as electrolytic metal zinc in the later process. The solutions are neutralized to pH 4–5 to precipitate out the iron and then filtered. The filtrates are purified by the addition of zinc metal powder to remove most of the other metals and used for electrolysis.

This process has not been used widely in industries, majorly because the bulk materials in the wastes are the oxides of iron and the solubility of zinc ferrite is quite limited. When sulfuric acid is used for the digestion of dust, all the iron will be dissolved, which consumes a lot of acid and thus increases the treatment costs. In addition, when the acidic digestion solutions are neutralized, a certain part of zinc and lead usually coprecipitates with the voluminous iron hydroxide, which may lead to the loss of the valuables and make the disposal of the precipitates more difficult.

1.5 Alkaline Leaching Process for Zinc and Lead Ores and Wastes

The oxides of Zn, Pb, and Al can be dissolved nearly completely in strong NaOH solution. The solubility of Cu and Cr oxides in alkaline solution are much lower than those of Zn, Pb, and Al, especially when these oxides coexist in the solid wastes. Unfortunately, nearly half of zinc exists as zinc ferrite in the EAF dust, which cannot be dissolved in both acidic and alkaline solutions. Therefore, conventional alkaline leaching process cannot be used for the effective and selective extraction of the amphoteric metals—zinc, lead, etc., from the ferrite wastes. It was reported that a treatment work was established for the alkaline leaching and extraction of zinc from dust decades ago but closed in a short term. The reason was unclear. From previous experiences, zinc ferrite cannot be attacked and broken by alkaline solution. However, nearly half of zinc in the dust exists as zinc ferrite.

Obviously, the extraction and leaching efficiencies of zinc from ferrite dust by alkaline leaching process would be quite low and become technically and economically unpractical and unfeasible, when it is used in industrial scale.

Obviously, the destruction of zinc ferrite is the key to improve the extraction efficiencies of zinc from dust, using both acidic and alkaline leaching processes. Several methods have been developed for the acidic leaching process, such as using acetic acid as leaching agent or heating the dust in sulfuric acid solution at relatively high temperature. Pressure leaching technique using acidic solution has also been used for the destruction of zinc ferrite in the leaching residue of roasting product of zinc concentrates in zinc hydrometallurgical industry. For alkaline leaching process, it seems to have no much space to be remained for the improvement of the leaching efficiencies.

Scientists in Belgium and France developed a direct alkaline leaching—electrowinning process for the treatment of EAF dust containing 21.2% Zn, 21.8% Fe, 3.6% Pb, and 2.5% Mn. The dust was preliminarily separated roughly by magnetic separator. The magnetic fractions, majorly zinc ferrites, were leached by 11 M NaOH solution for 4 h at 95 °C, and the nonmagnetic fractions, supposedly to be ZnO, were leached by 6 M NaOH solution for 1.5 h at 95 °C. Around 80 to 85% of Zn and Pb can be leached out of the dust by this process used.

The lead in the leach solution was cemented by the addition of Zn powder. The residual concentration in solution was 50–100 mg/L Pb and 0.5–1 mg/L Cu which is fully convenient for the electrowinning. The zinc metallic power obtained in electrowinning was quite fine and can be sold directly. The current efficiency was 95% and the power consumption 2.7 kWh/kg Zn.

Anyway, considering that the hydroxides of Fe, Ca, Cd, and Mg cannot be leached out in alkaline medium, it seems to be attractive to develop an innovative and cost-effective alkaline leaching process for the reclamation of the dust and some industrial wastes.

1.6 Hydrometallurgical Production of Zinc

The production and uses of zinc and lead in industrial scale have at least over 100 years. However, the zinc extraction techniques are developed gradually, from traditional pyrometallurgy to modern hydrometallurgy. Currently, most of the zinc used in the world is produced hydrometallurgically. Nevertheless, lead is produced majorly pyrometallurgically.

Although the extraction and recycling of zinc from wastes have become attractive in recent years, the zinc used in the modern world is in fact produced predominantly from mineral ores. The most important zinc ores applied for the extraction of zinc in industry are sphalerite, ZnS, with specific gravity 4.0 and marmatite, $ZnS{\cdot}x FeS$, with specific gravity around 3.9. The zinc ores usually coexist with the lead and copper. The ores are pretreated by crushing, grinding, and classifying. The lead and copper in the ores are concentrated firstly by the flotation process using xanthates as collectors; lime, soda ash, etc. as pH modifiers; and sodium cyanide, zinc sulfate, sulfur dioxide or sodium bisulfite, zinc hydrosulfite, etc. as zinc and iron depressants. Dowfroth, cresylic acid, pine oil, methyl isobutyl carbinol, coal tar, triethyloxybutanel, sodium silicate, etc. are usually used as frothers. The pH values should be controlled at 4.5–6.0, depending on the depressants used. After the lead and copper concentrates are separated out of the ores, the flotation of zinc is followed, using copper sulfate as activator, xanthates, aerofloates, etc. as collectors and methyl isobutyl carbinol, Dowfroth, cresylic acid, pine oil, etc. as frothers, with pH values from 8–10.5. As in lead and copper flotation, lime is generally used as pH regulator because calcium ions depress pyrite and often have an effect over and above that created by high pH alone. The grades of zinc concentrates strongly affect subsequent reduction costs. The contents of zinc oxide in the commercial zinc concentrates may be over 60–65%.

The flotation and concentration of non-sulfides zinc and lead ores seem to be much more difficult. For oxides ores, it should be converted into sulfides before the flotation operation is conducted, which may be realized by the addition of sodium sulfide into the ores slurry. Non-sulfide zinc ores, such as oxidized zinc ores (zincite, hydrozincite, hemimorphite, smithsonite, and willemite), have long been an important source of zinc; however, their concentration was difficult, and until relatively recently, only rich ores were exploited, using limited concentration by washing and gravity methods.

If the zinc sulfide concentrates are directly dissolved in sulfuric acid solution, the dissolution is found to be slow, and the resultant zinc sulfate solutions are not suitable for the subsequent purification and electrolysis. Therefore, the concentrates should be roasted firstly at 600–800 °C with fluidized roasters and multi-hearth roasters, followed by the leaching using sulfuric acid solution as leaching agent. The dissolution of the roasted product is quite easy. The roasting process may be expressed as follows:

$$ZnS + \frac{3}{2}O_2 = ZnO + SO_2, \quad \Delta H^0 = -111,000 \text{ cal/mol} \tag{1.4}$$

Meanwhile, zinc ferrites will also form at high temperature:

$$ZnO + Fe_2O_3 = Zn(FeO_2)_2 \tag{1.5}$$

The zinc oxide calcines, containing a few percent of zinc sulfate from the roasting of zinc sulfide concentrates, are leached with spent electrolyte solution from the electrolytic precipitation cells which follow. The acidity (sulfuric acid) of the spent solution has been increased as a consequence of zinc deposition from it and therefore is a much more effective solvent of zinc oxide at the leaching stage. The leaching may be conducted in two different ways, i.e., continuous process leaching and batch process leaching. The continuous process leaching is much more commonly used of the two and is in general a two- or three-stage, cyclic procedure.

In this process, the zinc oxide calcine, sometimes first finely ground to minus 325 mesh in a wet, close-circuit ball mill-centrifugal classifier arrangement, is added to an agitation-type leach tank. The zinc oxide calcines, added in excess to use up the free acid, raise the pH to a neutral 5–5.2 as acidity decreases and the spent electrolyte which contained 10–11.5% H_2SO_4 drops to 0.5% H_2SO_4. From 50–75% of the soluble, zinc present in the calcine is dissolved in this first step.

The final high pH of 5.2 also causes the precipitation as oxides and hydroxides of various impurities which have dissolved into solution along with the zinc oxide. To assist in this precipitation, air is injected under the impellers in those leach tanks using mechanical agitators, or oxidizing agent such as MnO_2 is added. A good elimination of impurities is achieved, with ferric hydroxide precipitation removing most of the iron and insoluble oxides and hydroxide compounds forming to take out silica, alumina, arsenic, and antimony, as well as a considerable portion of the copper. Cadmium is not affected and is soluble with the zinc.

The major problem encountered in the continuous process leaching is the low extraction efficiencies of non-leachable zinc ferrites. The leaching residues include about 16–25% zinc and 27–28% iron, which exist mainly as zinc ferrite, so the next leaching step for dissolving zinc ferrite should be necessary. Various processes for treating these leaching residues are applied in practical hydrometallurgical plants, majorly by acidic dissolution at relatively high temperature (100–106 °C). Pyrometallurgical processes such as lead blast furnace treatment, rotary kiln process, and others are also applied for treatment of leaching residues.

Batch process leaching of zinc calcine is not used commonly, but it has certain advantages. One of these is that zinc ferrite ($ZnO{\cdot}Fe_2O_3$) which is formed during roasting and is insoluble during normal leaching can be processed. In batch process leaching, high acid-spent electrolyte of at least 60 g/L is heated to 60 °C and combined with a charge of ferrites in an agitated tank for 1 h. The zinc ferrites decompose and MnO_2 is added to oxidize the iron as it is freed. The pH of the solution is raised by adding zinc oxide free of ferrites to use up the free acid by taking the soluble zinc into solution as $ZnSO_4$, and as the pH increases, the oxidized iron and other metal impurities will precipitate. The final remaining acid can be neutralized to pH 5–5.3 by adding more zinc oxide calcine or sometimes lime. The heat of decomposition reaction of zinc ferrites raises the solution temperature to 106 °C, and a considerable amount of water is evaporated as steam. Stacks are provided over the leach tanks to aid in the removal of this evolution. Most of zinc ferrite can be dissolved in this stage.

The leach solutions are filtered. The cakes are water washed, repulped with water, and thickened, and the thickener underflow is filtered, dried, and finally shipped to the smelter to recover the valuable metallic contents, which are principally lead and zinc, with some silver.

The filtrate of leach solutions contain 215 g/L zinc, 600 mg/L cadmium, 300 mg/L copper, and 300 mg/L cobalt and are purified by the addition of zinc scrap or dust and antimony dust. Finally, the purified solution can be used for the electrolytic precipitation of zinc.

The formation of zinc ferrite in the zinc oxide calcines renders the leaching efficiencies of zinc from the calcines to great difficulty. Many alternatives improving the decomposition and leaching of zinc ferrites have been proposed. Nevertheless, the problems have not been solved completely in terms of dissolution rates and costs. The loss of zinc due to the poor leaching efficiencies of zinc ferrites is considered to be of great significance, which is also encountered in the treatment and recovery of zinc from EAF dust using both acidic and alkaline leaching processes.

As the problems are encountered in the leaching processes of zinc calcines, the zinc ferrites in the dust cannot be leached effectively by common acidic and alkaline leaching processes. Around 20% to 50% of zinc in the dust exists as zinc ferrites. Zinc ferrites will be easily formed when the mixtures of metallic Zn and Fe are heated at temperature over 650 °C at which the melting and roasting is operated in EAF and roasters.

Sulfuric acid is the most important leaching agent for zinc ferrites in the dust. However, zinc ferrites cannot be attacked effectively by sulfuric acid solution. Some modifications have been proposed, such as the pretreatment by acetic acid solution, followed by the precipitation of sulfide and dissolution by sulfuric acid solution. Another method is to use hot sulfuric acid solution to attack zinc ferrites, as used in the leaching processes of the leaching residues generated in the zinc hydrometallurgical processes. Nevertheless, the bulk materials in the dust are iron oxide, which will be also dissolved to leach solution. The neutralization and precipitation of iron from the acidic solution becomes an obstacle encountered in this leaching process, which hinders the application of this process in industry.

Similarly, zinc ferrites cannot be decomposed by alkaline solution regardless of the leaching temperature and concentration of NaOH used, although it is well known that zinc oxide can be dissolved readily in alkaline solution.

1.7 Metallurgy of Lead

Currently, metallic lead is produced pyrometallurgically at high temperature. In the term of metallurgy principle, the modern lead metallurgy can be classified into the following categories.

1.7.1 Oxidation-Reduction Smelting

In the process, the lead sulfide and other sulfides in the lead sulfide concentrate are oxidized to produce oxide (perhaps metal) firstly, and then the oxides are reduced to obtain metal. The reactions of lead sulfide are shown as follow:

$$Pb + \frac{3}{2}O_2 = PbO + SO_2 \tag{1.6}$$

$$PbO + CO(C) = Pb + CO_2(CO) \tag{1.7}$$

$$PbS + O_2 = Pb + SO_2 \tag{1.8}$$

90% metallic lead are produced by the method of sintering-blast furnace reduction. The oxidative desulfurization of lead sulfide concentrate is carried out in the sintering machine, and the feedstock is made to sinter lump, mixed with coke, and reduced to obtain crude metallic lead in the blast furnace. The method is mainly applied for producing lead nowadays for its good adaptability, stable production, and large output. However, it consumes a lot of energy and results in serious environmental pollution.

1.7.2 Reaction-Melting

The PbS in the lead sulfide concentrate is oxidized to PbO and $PbSO_4$ and then reacted with the oxidized PbS immediately to produce metallic lead. Metallic lead may be also produced for some PbO reacting with coke. The possible reactions in the process are shown as follow:

$$2PbS + 3O_2 = 2PbO + 2SO_2 \tag{1.9}$$

$$PbS + 2O_2 = PbSO_4 \tag{1.10}$$

$$2PbO + PbS = 3Pb + SO_2 \tag{1.11}$$

$$PbSO_4 + PbS = 2Pb + 2SO_2 \tag{1.12}$$

$$PbO + CO(C) = Pb + CO_2(CO) \tag{1.13}$$

High recovery efficiency could be obtained for treating high-grade lead ore ($w_{Pb} \geq 65\%$). There existed a strict restriction for the content of impurities, especially SiO_2. The fusible silicate is formed since the binding between PbO and SiO_2 and sunk into the ash. Therefore, the content of SiO_2 should be limited below 2–3%. In order to raise the melting temperature of feedstock, $CaCO_3$ ($w_{CaCO_3} < 2\%$) is usually mixed into the feedstock.

1.7.3 Precipitation Melting

The metal, whose sulfur affinity is stronger than that of lead, could replace the lead in lead sulfide. Iron is the common replacement agent, as follows:

$$PbS + Fe = Pb + FeS \tag{1.14}$$

The reaction cannot finish completely for the form of $PbS{\cdot}3FeS$ from the combination of PbS and FeS. Hence, the addition of iron can extract 72–79% lead from PbS. The actual reaction occurred in the precipitation-melting process at the high temperature (>1000 °C) is given as follows:

$$4PbS + 4Fe = 3Pb + PbS \cdot 3FeS + Fe \tag{1.15}$$

The addition of iron is determined commonly by production experience, sometime reached to 30–40% of feedstock, and the waste iron scrap is usually used. The process of the precipitation-melting method is pretty simple and of easily operation. The crude lead of good quality can be obtained with little lead evaporation and low investment. However, the consumption of iron scrap and fuel is huge with low lead recovery efficiency, labor productivity, and bad working environment. Hence, the method is not adapted to manufacture on a large scale.

1.7.4 Alkali Melting

The sodium carbonate (or sodium hydrate) and carbonaceous fuel (coal or coke) are mixed, then the mixture is smelt to produce lead and matte residue, as shown below.

$$2PbS + 2Na_2CO_3{+}C = 2Pb + 2Na_2S + 3CO_2 \tag{1.16}$$

$$PbS + Na_2CO_3{+}C = Pb + Na_2S + CO + CO_2 \tag{1.17}$$

$$PbS + Na_2CO_3{+}CO = Pb + Na_2S + 2CO_2 \tag{1.18}$$

The melting temperature is 1000–1100 °C, the valuable metals (Au, Ag, Bi, etc.) are concentrated in the crude lead, and the quality of lead reaches to 98–98.5% with the lead direct recovery efficiency of 98.4%. The content of copper in the crude lead is 0.25–0.35%, fume dust is 3.8%. The matte residue concentrates 94.5% copper; 71.2% zinc, a majority of Se, Te, and Mo (w = 90–96%); and 0.1–0.3% lead. The fume dust concentrates 21% Zn and nearly all Cd and can be used as the raw material for Cd extraction.

Chapter 2
Thermodynamics of Alkaline Leaching of Zinc and Lead Hazardous Wastes

Abstract The thermodynamics of alkaline leaching of solid wastes bearing zinc or lead is carried out for the hydrometallurgy of zinc or lead hazardous wastes. The E-pH equilibrium diagrams of leaching systems of zinc or lead in different forms are drawn. In strong alkaline solution, most of zinc ions exist in the form of $Zn(OH)_4^{2-}$. ZnO, $ZnCO_3$, and Zn_2SiO_4 are the predominant mineralogical phases in solid wastes bearing zinc. ZnO and $ZnCO_3$ can be dissolved directly by strong alkaline solution, while higher concentration of OH^- is required to dissolve Zn_2SiO_4. PbO, PbS, $PbCO_3$, and $PbSO_4$ are the predominant mineralogical phases in solid wastes bearing lead. PbO, $PbSO_4$, and $PbCO_3$ can be dissolved in concentrated NaOH solutions, while the dissolution of PbS may be negligible. The equilibrium concentrations of Cd in leach solutions at 180 ~ 240 g/L NaOH is about 0.45 ~ 0.79 g/L. The dissolution of other impurities such as Fe, Cu, Co, Ni, Mg, Ca, etc. can be negligible.

Keywords Thermodynamics • Alkaline leaching • Zinc hazardous waste • Lead hazardous waste • Impurities

2.1 Thermodynamics of Alkaline Leaching of Zinc Hazardous Wastes

The thermodynamics of alkaline leaching of solid wastes bearing zinc and lead is the basis for the hydrometallurgy of zinc and lead hazardous wastes, facilitating the calculation of all chemicals used and the secondary pollutants generated in the process.

2.1.1 Morphology Distribution of Zinc in Alkaline Solution

Morphology distribution of zinc in alkaline solution includes Zn^{2+}, ZnO_2^{2-}, $Zn(OH)^+$, $Zn(OH)_2$, $Zn(OH)_3^-$, $HZnO_2^-$, $Zn(OH)_4^{2-}$, H^+, and OH^-, in the Zn(II)–NaOH–H_2O or Zn(II)–H_2O systems. The thermodynamic data are presented in Table 2.1.

Y. Zhao, C. Zhang, *Pollution Control and Resource Reuse for Alkaline Hydrometallurgy of Amphoteric Metal Hazardous Wastes*, Handbook of Environmental Engineering 18, DOI 10.1007/978-3-319-55158-6_2

Table 2.1 Gibbs free energy of related species at $T = 298$ K (kJ/mol)

Species	$\Delta_f G_m^\theta$	Species	$\Delta_f G_m^\theta$	Species	$\Delta_f G_m^\theta$
ZnO(s)	−323.131	$Zn(OH)_2$(aq)	−537.398	$Zn(OH)^+$	−330.540
$HZnO_2^-$	−465.780	H_2O	−238.098	$Zn(OH)_3^-$	−702.912
ZnO_2^{2-}	−390.729	Zn^{2+}	−147.773	$Zn(OH)_4^{2-}$	−868.031
OH^-	−157.899	Na^+	−261.890		

On the base of the simultaneous equilibrium principle, every zinc complexes are in equilibrium with zinc oxide in the presence of zinc oxide in the alkaline system.

$$ZnO(s) + (j-1)H_2O = Zn(OH)_j^{2-j} + (j-2)H^+ \quad (j = 2\sim4) \tag{2.1}$$

$$ZnO(s) + OH^- = HZnO_2^- \tag{2.2}$$

$$ZnO(s) + 2OH^- = ZnO_2^{2-} + H_2O \tag{2.3}$$

According to the exponential computation method, the concentration of these species can be expressed as

$$[R] = \exp(A + B\,\text{pH}) \tag{2.4}$$

where $[R]$ is every species' mole concentration, A is the constants calculated from equilibrium constants or thermodynamic data, and B is the multiplication of ln10 and gained and lost proton number. There exist eight balance equations in Zn(II)–H_2O system as follows. The concentration expressions of related species can be thus determined:

$$ZnO(s) + H_2O = Zn^{2+} + 2OH^- \tag{2.5}$$

$$[Zn^{2+}] = \exp(25.324 - 4.606\,\text{pH})$$

$$ZnO(s) + H_2O = HZnO_2^- + H^+ \tag{2.6}$$

$$[HZnO_2^-] = \exp(-38.525 + 2.303\,\text{pH})$$

$$ZnO(s) + H_2O = ZnO_2^{2-} + 2H^+ \tag{2.7}$$

$$[ZnO_2^{2-}] = \exp(-68.819 + 4.606\,\text{pH})$$

$$ZnO(s) + H_2O = ZnOH^+ + OH^- \tag{2.8}$$

$$[ZnOH^+] = \exp(2.990 - 2.303\text{pH})$$

$$ZnO(s) + H_2O = Zn(OH)_2(aq) \tag{2.9}$$

$$[Zn(OH)_2(aq)] = \exp(-9.619)$$

$$ZnO(s) + 2H_2O = Zn(OH)_3^- + H^+ \tag{2.10}$$

$$\left[Zn(OH)_3^-\right] = \exp(-38.915 + 2.303\ pH)$$

$$ZnO(s) + 3H_2O = Zn(OH)_4^{2-} + 2H^+ \quad (2.11)$$

$$\left[Zn(OH)_4^{2-}\right] = \exp(-68.371 + 4.606\ pH)$$

$$H_2O = H^+ + OH^- \quad (2.12)$$

$$[OH^-] = \exp(-32.370 + 2.303\ pH)$$

According to the principle of conservation of mass, the sum concentration of zinc can be expressed as Eq. (2.13):

$$[Zn]_T = \sum_{j=1}^{4}\left[Zn(OH)_j^{2-j}\right] + \left[Zn^{2+}\right] + \left[HZnO_2^-\right] + \left[ZnO_2^{2-}\right] \quad (2.13)$$

where $[Zn]_T$ is the total mole concentrations of zinc and j is the complex numbers of hydroxide.

The distribution coefficient of zinc can be calculated by Eq. (2.14):

$$\delta_{species} = \frac{[species]}{[Zn]_T} \quad (2.14)$$

where $\delta_{species}$ is the distribution coefficient of species and [species] is the concentration of species.

The relationship among the three variables of $[Zn]_T$, $\delta_{species}$, and pH is confined by Eqs. (2.13 and 2.14). If pH is given, $[Zn]_T$ and $\delta_{species}$ may be obtained from the abovementioned simultaneous equations. The calculated results have been plotted into Figs. 2.1 and 2.2. As shown in Fig. 2.1a, at the temperature of 25 °C, when pH is less than 6, the main present shape of zinc in the system of Zn(II)–H_2O is Zn^{2+} as given in Eq. (2.5). When pH is between 6 and 9, the solubility of zinc reduces rapidly, and Zn^{2+} has been hydrolyzed and hydrolyzate $Zn(OH)_2$(s) formed, and the species of zinc in the solution has changed into the shape of $Zn(OH)_2$(aq).

When pH varies between 9 and 11, the present shape of zinc in the aqueous solution is $Zn(OH)_2$(aq). The concentration of zinc is only 6.7×10^{-5} mol/L, so zinc is insoluble in the weak alkaline solution.

When pH is between 11 and 12, the shape of $Zn(OH)_2$(aq) has gradually transformed into $Zn(OH)_3^-$ and $HZnO_2^-$. $Zn(OH)_4^{2-}$ and ZnO_2^{2-} have appeared at pH = 12. The equilibrium concentration of zinc in the solution has increased slowly with the increase of pH. It indicates that a portion of the precipitation of Zn $(OH)_2$(s) has dissolved into the solution again. When pH is between 13 and 14, the main species in the system are $Zn(OH)_4^{2-}$ and ZnO_2^{2-}. $Zn(OH)_3^-$, $HZnO_2^-$, and $Zn(OH)_2$(aq) have transformed to the two former species completely at pH = 14.

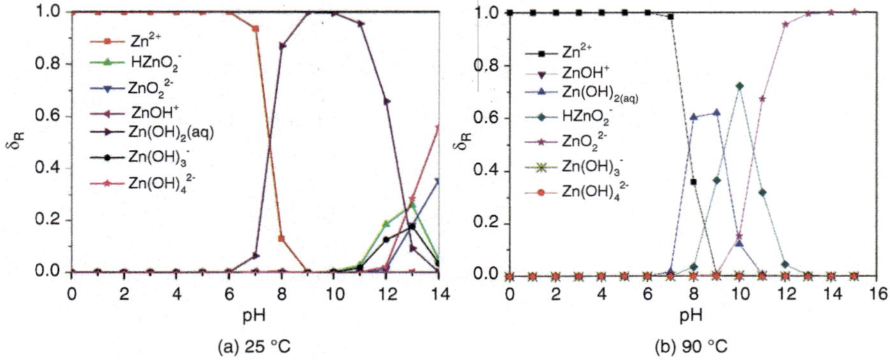

Fig. 2.1 Distribution curves of zinc in the system of Zn (II)–H_2O

Fig. 2.2 Zinc equilibrium concentrations with various pH

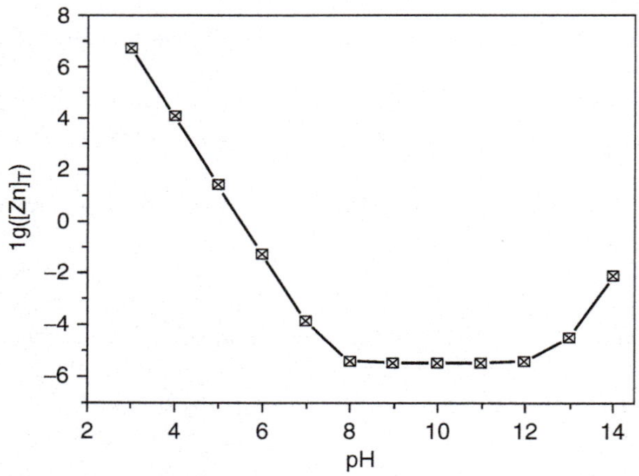

As shown in Fig. 2.1b, the form of zinc exists as Zn^{2+}, mostly with distribution coefficient of 99.9% in Zn(II)–H_2O system at 90 °C and pH $\leq$ 6, while its relationship with pH shows significant differences compared with normal temperature (25 °C) at pH > 6. When pH is between 6 and 9, Zn^{2+} in the solution would transform to $Zn(OH)_2$(s) and partly to $HZnO_2^-$ at pH > 7. $Zn(OH)_2$(s) represents the largest portion of its total zinc content $[Zn]_T$ in the solution at pH = 9. As pH grows up, $Zn(OH)_2$(aq) would be able to transform to $HZnO_2^-$ and ZnO_2^{2-} gradually, and the content of $HZnO_2^-$ peaks as the biggest part of $[Zn]_T$ at pH = 10. After that, the content of ZnO_2^{2-} is increasing due to the transformation from $HZnO_2^-$ and reaches the top at pH = 13 with distribution coefficient of 99.5%.

2.1.2 Experimental Verification

Zinc experimental and theoretical values under a balance state in strong alkali solution are presented in Fig. 2.3, which contains two different temperatures, 25 °C and 90 °C, respectively. From Fig. 2.3a, it can be seen that zinc equilibrium concentration in Zn(II)–NaOH–H_2O system keeps increasing as the NaOH concentration increases at 25 °C, with average relative error less than 5%, which testifies the correctness of this thermodynamic model within the scope of experiment.

Figure 2.3b shows that under 90 °C Zn(II)–NaOH–H_2O system, zinc equilibrium concentration increases when the NaOH concentration goes up. However, zinc experimental values agree well with theoretical values with NaOH concentration below 160 g/L; after that, experimental values are bigger than theoretical values as NaOH concentration keeps increasing. The data fitting for the results at 90 °C has been completed which shows a good coherence with the coefficient of determination r^2 of 0.98676:

$$[Zn]_T = 18.51803 - 0.37962[NaOH] + 0.00267[NaOH]^2 \tag{2.15}$$

Alkali quantity in leaching section and the concentration of free NaOH which needed to be controlled during the whole process can be determined from dissolution and equilibrium concentration of zinc calculated from thermodynamic model. Zinc concentration is able to reach the top in the leaching part during the whole alkaline leaching-electrowinning process. So the concentration of free NaOH should be adjusted on the basis of zinc concentration in the leaching process. However, one important thing to note is that high NaOH concentration could lead to a viscous solution which would harm the process of leaching and solid-liquid separation and then impacts residue washing section. Generally speaking, zinc concentration would be controlled between 35 and 40 g/L in the leaching process, so that the need of free NaOH should be 180 g/L. Timing of sampling for leaching solution ought to be adopted in real production to make sure that NaOH concentration maintains around 180 g/L.

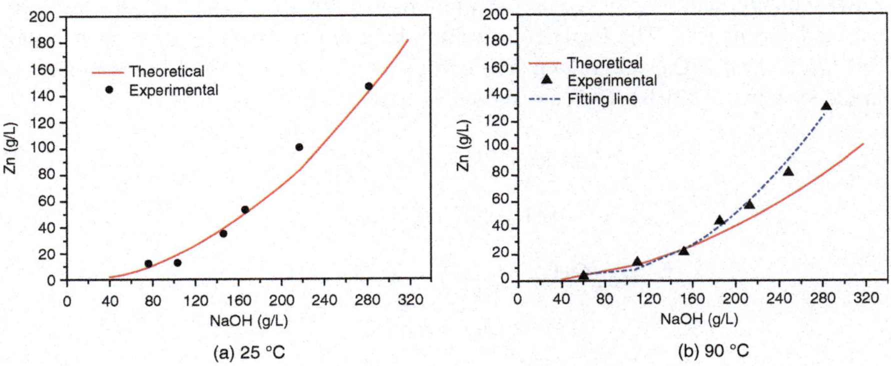

Fig. 2.3 Relationship of Zn solubility with NaOH concentration at different temperatures

Table 2.2 Relationship between K_C and NaOH concentration in the reaction of ZnO dissolved in NaOH solution

NaOH concentration (mol/L)	3.152	3.921	4.152	4.690	5.382	5.920
Zn concentration (mol/L)	0.348	0.552	0.708	0.895	1.147	1.351
K_C (mol/L)	0.0350	0.0359	0.0411	0.0407	0.0396	0.0386

2.1.3 Apparent Equilibrium Constant for Zinc Dissolved in NaOH Solution

In strong alkaline solution, most of zinc ions exist in the form of $Zn(OH)_4^{2-}$ at normal temperature. In order to investigate the apparent equilibrium constant $\left(K_C = \left[Zn(OH)_4^{2-}\right]/[OH^-]^2\right)$ for reaction of ZnO dissolved in NaOH solution, leaching experiments were carried out in a flask placed on a thermostatically controlled magnetic stirrer. To a 500 mL NaOH solution of required concentration, excess ZnO was added and then leached at a constant temperature. About 5 mL solution was removed at time intervals of 5 min and analyzed zinc and OH^-.

The relationship between K_C and NaOH concentration in the reaction of ZnO dissolved in NaOH solution is shown in Table 2.2. It was found that K_C was increased with the NaOH concentration up to 4.2 mol/L. The K_C dropped down when sodium hydroxide was over 5 mol/L because the high sodium hydroxide concentrations increased the viscosity of the solution, which would tend to reduce the diffusion rate of the ions. Thus, a NaOH solution of 4–5 mol/L(160–200 g/L) was considered suitable for leaching reaction.

2.1.4 E-pH Equilibrium Diagrams of Leaching Systems of Zinc

ZnO, $ZnCO_3$, and Zn_2SiO_4 are the predominant mineralogical phases in solid wastes bearing zinc. The E-pH equilibrium diagrams of leaching systems of ZnO, $ZnCO_3$, and Zn_2SiO_4 were drawn. The complex reactions and equilibrium constants in the systems of Zn(II)–H_2O are shown as follows:

$$ZnO(s) + 2H^+ = Zn^{2+} + H_2O \tag{2.16}$$

$$pH = 5.5 - \frac{1}{2}\lg a_{Zn^{2+}}$$

$$Zn(OH)_4^{2-} + 2H^+ = ZnO + 3H_2O \tag{2.17}$$

$$pH = 14.84 + \frac{1}{2}\lg a_{Zn(OH)_4^{2-}}$$

$$Zn^{2+} + 2e = Zn \tag{2.18}$$

$$E = -0.76 + \frac{0.0591}{2} \lg a_{Zn^{2+}}$$

$$ZnO(s) + 2H^{+} + 2e = Zn + H_2O \tag{2.19}$$

$$E = -0.44 - 0.0591pH$$

$$Zn(OH)_4^{2-} + 4H^{+} + 2e = Zn + 4H_2O \tag{2.20}$$

$$E = 0.44 + \frac{0.0591}{2} \lg a_{Zn(OH)_4^{2-}} - 0.1182pH$$

Figure 2.4 shows the E-pH equilibrium diagram of Zn(II)–H_2O at zinc ion activity of 0.1. Dissolved by strong alkaline solution, ZnO can be converted into $Zn(OH)_4{}^{2-}$. The solid wastes bearing ZnO can be leached by NaOH solution directly.

The complex reactions and equilibrium constants in the systems of Zn(II)–$CO_3{}^{2-}$–H_2O are shown as follows:

$$ZnCO_3(s) + 2H^{+} = Zn^{2+} + H_2CO_3(aq) \tag{2.21}$$

$$pH = 3.69 - \frac{1}{2} \lg a_{H_2CO_3(aq)} - \frac{1}{2} \lg a_{Zn^{2+}}$$

$$Zn(OH)_2(s) + CO_3^{2-} + 2H^{+} = ZnCO_3(s) + H_2O \tag{2.22}$$

$$pH = 11.08 + \frac{1}{2} \lg a_{CO_3^{2-}}$$

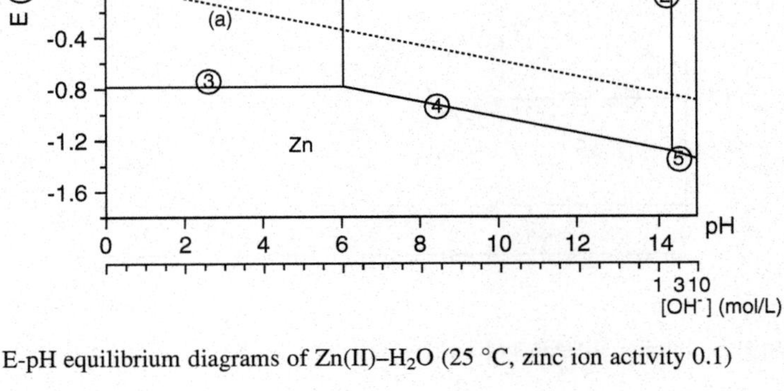

Fig. 2.4 E-pH equilibrium diagrams of Zn(II)–H_2O (25 °C, zinc ion activity 0.1)

$$Zn(OH)_4^{2-} + 2H^+ = Zn(OH)_2 + 2H_2O \quad (2.23)$$

$$pH = 13.959 + \frac{1}{2}\lg a_{Zn(OH)_4^{2-}}$$

$$Zn^{2+} + 2e = Zn \quad (2.24)$$

$$E = -0.76 + \frac{0.0591}{2}\lg a_{Zn^{2+}}$$

$$ZnCO_3(s) + 2H^+ + 2e = Zn + H_2CO_3(aq) \quad (2.25)$$

$$E = -0.55 - 0.0591pH - 0.02955\lg a_{H_2CO_3(aq)}$$

$$ZnCO_3 + 2e = Zn + CO_3^{2-} \quad (2.26)$$

$$E = -1.04 - 0.02955\lg a_{CO_3^{2-}}$$

$$Zn(OH)_2(s) + 2H^+ + 2e = Zn + 2H_2O \quad (2.27)$$

$$E = -0.44 - 0.0591pH$$

$$Zn(OH)_4^{2-} + 4H^+ + 2e = Zn + 4H_2O \quad (2.28)$$

$$E = 0.44 + 0.01182\lg a_{Zn(OH)_4^{2-}} - 0.11821pH$$

Figure 2.5 shows the E-pH equilibrium diagram of $Zn(II) - CO_3^{2-} - H_2O$ at zinc ion activity of 0.1. $ZnCO_3$ can be converted into $Zn(OH)_4^{2-}$ when $pH > 13.459$. The solid wastes bearing $ZnCO_3$ can be leached by NaOH solution directly.

The complex reactions and equilibrium constants in the systems of $Zn(II) - SiO_3^{2-} - H_2O$ are shown as follows:

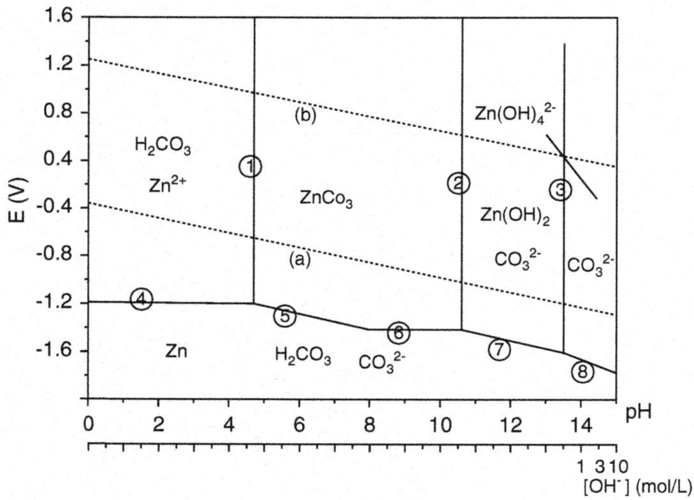

Fig. 2.5 E-pH equilibrium diagrams of $Zn(II)–CO_3^{2-}–H_2O$ (25 °C, zinc ion activity 0.1)

$$ZnSiO_3(s) + 2H^+ = Zn^{2+} + H_2O + SiO_2(aq) \tag{2.29}$$

$$pH = 3.43 - \frac{1}{2}\lg a_{Zn^{2+}}$$

$$2Zn(OH)_4^{2-} + SiO_3^{2-} + 2H^+ = Zn_2SiO_4(s) + 7H_2O \tag{2.30}$$

$$pH = 15.23 + \frac{1}{3}\lg a_{Zn(OH)_4^{2-}} + \frac{1}{6}\lg a_{SiO_3^{2-}}$$

$$Zn^{2+} + 2e = Zn \tag{2.31}$$

$$E = -0.76 + \frac{0.0591}{2}\lg a_{Zn^{2+}}$$

$$ZnSiO_3(s) + 2H^+ + 2e = Zn + H_2SiO_3(aq) \tag{2.32}$$

$$E = -0.57 - 0.0591pH - 0.02955\lg a_{H_2SiO_3(aq)}$$

$$ZnSiO_3(s) + 2e = Zn + SiO_3^{2-} \tag{2.33}$$

$$E = -1.16 - 0.02955\lg a_{SO_3^{2-}}$$

$$Zn_2SiO_4(s) + 4H^+ + 4e = 2Zn + H_4SiO_4(aq) \tag{2.34}$$

$$E = -0.42 - 0.0591pH - 0.0148\lg a_{H_4SiO_4(aq)}$$

$$Zn(OH)_4^{2-} + 4H^+ + 2e = Zn + 4H_2O \tag{2.35}$$

$$E = 0.44 + 0.01182\lg a_{Zn(OH)_4^{2-}} - 0.11821pH$$

Figure 2.6 shows the E-pH equilibrium diagram of $Zn(II) - SiO_3^{2-} - H_2O$ at zinc ion activity of 0.1. Compared with ZnO and $ZnCO_3$, higher concentration of OH^- was required to dissolve Zn_2SiO_4. $ZnSiO_3$ can be converted into $Zn(OH)_4^{2-}$

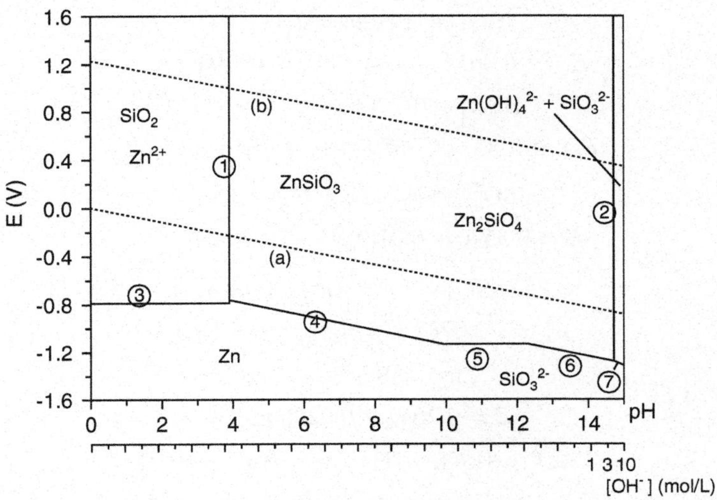

Fig. 2.6 E-pH equilibrium diagrams of $Zn(II)–SiO_3^{2-}–H_2O$ (25 °C, zinc ion activity 0.1)

when the concentration of OH^- was over 5.37 mol/L. The solid wastes bearing $ZnSiO_3$ can be leached by NaOH solution directly.

The complex reactions and equilibrium constants in the systems of ZnS–H_2O are shown as follows:

$$H_2S = H^+ + HS^- \tag{2.36}$$

$$pH = 7.0016 + \lg a_{HS^-} - \lg a_{H_2S}$$

$$HS^- = H^+ + S^{2-} \tag{2.37}$$

$$pH = 13.9997 - \lg a_{HS^-} + \lg a_{S^{2-}}$$

$$SO_4^{2-} + H^+ = HSO_4^- \tag{2.38}$$

$$pH = 1.9051 + \lg a_{SO_4^{2-}} - \lg a_{HSO_4^-}$$

$$HSO_4^- + 9H^+ + 8e = H_2S + 4H_2O \tag{2.39}$$

$$E = 0.2890 - 0.0665pH + 0.00739\ \lg a_{HSO_4^-} - 0.00739\ \lg a_{H_2S}$$

$$SO_4^{2-} + 10H^+ + 8e = H_2S + 4H_2O \tag{2.40}$$

$$E = 0.3033 - 0.0739pH + 0.00739\lg a_{SO_4^{2-}} - 0.00739\lg a_{H_2S}$$

$$SO_4^{2-} + 9H^+ + 8e = HS^- + 4H_2O \tag{2.41}$$

$$E = 0.2515 - 0.0665pH + 0.00739\lg a_{SO_4^{2-}} - 0.00739\lg a_{HS^-}$$

$$SO_4^{2-} + 8H^+ + 8e = S^{2-} + 4H_2O \tag{2.42}$$

$$E = 0.1480 - 0.0591pH + 0.00739\lg a_{SO_4^{2-}} - 0.00739\lg a_{S^{2-}}$$

$$S + 2H^+ + 2e = H_2S \tag{2.43}$$

$$E = 0.1418 - 0.0591pH - 0.0296\lg a_{H_2S}$$

$$S + H^+ + 2e = 2HS^- \tag{2.44}$$

$$E = -0.0652 - 0.0296pH - 0.0296\lg a_{HS^-}$$

$$HSO_4^- + 7H^+ + 6e = S + 4H_2O \tag{2.45}$$

$$E = 0.3383 - 0.06898pH + 0.00985\lg a_{HSO_4^-}$$

$$SO_4^{2-} + 8H^+ + 6e = S + 4H_2O \tag{2.46}$$

$$E = 0.3570 - 0.0788pH + 0.0098\lg a_{SO_4^{2-}}$$

$$ZnS(s) + 2H^+ = Zn^{2+} + H_2S(g) \tag{2.47}$$

$$pH = -2.0787 - 0.5\lg a_{Zn^{2+}} - 0.5\lg a_{H_2S}$$

$$Zn^{2+} + S + 2e = ZnS \tag{2.48}$$

$$E = 0.2647 + 0.0296\lg a_{Zn^{2+}}$$

$$Zn^{2+} + HSO_4^- + 7H^+ + 8e = ZnS + 4H_2O \tag{2.49}$$

$$E = 0.3199 - 0.0517pH + 0.00739\lg a_{HSO_4^-} + 0.00739\lg a_{Zn^{2+}}$$

$$Zn^{2+} + SO_4^{2-} + 8H^+ + 8e = ZnS + 4H_2O \tag{2.50}$$

$$E = 0.3340 - 0.0591\text{pH} + 0.00739\lg a_{SO_4^{2-}} + 0.00739\lg a_{Zn^{2+}}$$

$$2Zn^{2+} + SO_4^{2-} + 2H_2O = ZnSO_4 \cdot Zn(OH)_2 + 2H^+ \tag{2.51}$$

$$\text{pH} = 3.6519 - 0.5\lg a_{SO_4^{2-}} - \lg a_{Zn^{2+}}$$

$$ZnSO_4 \cdot Zn(OH)_2 + SO_4^{2-} + 18H^+ + 16e = 2ZnS + 10H_2O \tag{2.52}$$

$$E = 0.3610 - 0.0665\text{pH} + 0.00370\lg a_{SO_4^{2-}}$$

$$ZnSO_4 \cdot Zn(OH)_2 + 2H_2O = 2Zn(OH)_2 + SO_4^{2-} + 2H^+ \tag{2.53}$$

$$\text{pH} = 8.6686 + 0.5\lg a_{SO_4^{2-}}$$

$$Zn(OH)_2 + SO_4^{2-} + 10H^+ + 8e = ZnS + 6H_2O \tag{2.54}$$

$$E = 0.4251 - 0.0739\text{pH} + 0.00370\lg a_{SO_4^{2-}}$$

$$Zn(OH)_4^{2-} + SO_4^{2-} + 12H^+ + 8e = ZnS + 8H_2O \tag{2.55}$$

$$E = 0.6290 - 0.0887\text{pH} + 0.00739\lg a_{SO_4^{2-}} + 0.00739\lg a_{Zn(OH)_4^{2-}}$$

$$Zn(OH)_4^{2-} + 2H^+ = Zn(OH)_2 + 2H_2O \tag{2.56}$$

$$\text{pH} = 13.959 + \frac{1}{2}\lg a_{Zn(OH)_4^{2-}}$$

$$Zn^{2+} + 2e = Zn \tag{2.57}$$

$$E = -0.76 + \frac{0.0591}{2}\lg a_{Zn^{2+}}$$

$$ZnS + 2e = Zn + S^{2-} \tag{2.58}$$

$$E = -1.5057 - \frac{0.0591}{2}\lg a_{S^{2-}}$$

$$ZnS + H^+ + 2e = Zn + HS^- \tag{2.59}$$

$$E = -1.0918 - \frac{0.0591}{2}\lg a_{HS^-} - \frac{0.0591}{2}\text{pH}$$

$$ZnS + 2H^+ + 2e = Zn + H_2S \tag{2.60}$$

$$E = -0.8848 - \frac{0.0591}{2}\lg a_{H_2S} - 0.0591\text{pH}$$

$$Zn(OH)_4^{2-} + S^{2-} + 4H^+ = ZnS + 4H_2O \tag{2.61}$$

$$\text{pH} = 16.27 + \frac{1}{4}\lg a_{Zn(OH)_4^{2-}} + \frac{1}{4}\lg a_{S^{2-}}$$

E-pH equilibrium diagram in ZnS–H_2O system shown in Fig. 2.7 is finally determined based on the results above.

Although direct acid leaching for ZnS could be achieved through the reaction $ZnS(s)+2H^+=Zn^{2+}+H_2S(g)$, it only happens when H^+ concentration is above

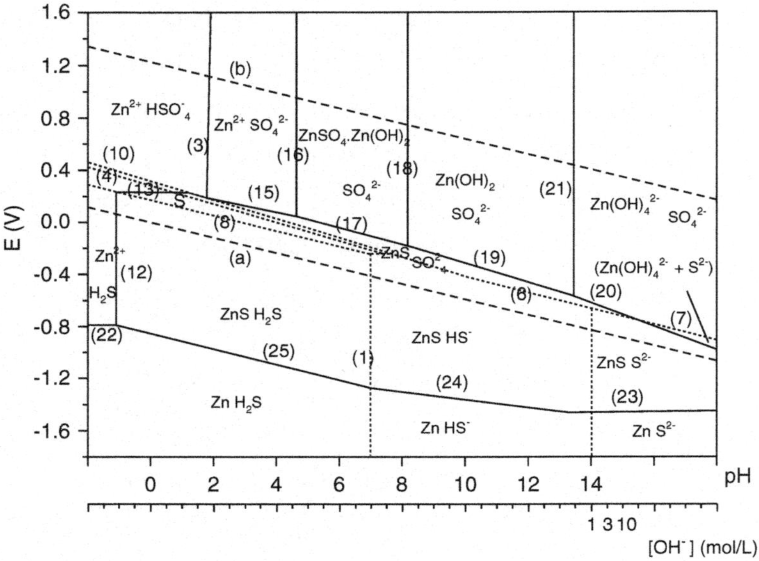

Fig. 2.7 E-pH equilibrium diagrams for the systems of ZnS–H_2O

11.99 mol/L. High H^+ concentration is hard to implement in the industry, meanwhile, the recycle of waste electrolyte is painfully difficult under such situation. Therefore, direct acid leaching for ZnS is unreasonable and infeasible.

Theoretically, when the Zn^{2+} activity is 0.1, a stable region between $Zn(OH)_4^{2-}$ and SO_4^{2-} is able to exist when pH is > 13.459, while there is a stable domain between $Zn(OH)_4^{2-}$ and S^{2-} when pH is > 14, $[OH^-] > 58.88$ mol/L. Hence, there remain two ways for dissolution of ZnS in alkali solution. One can be achieved through the reaction of $ZnS + 4OH^- + 2O_2 = Zn(OH)_4^{2-} + SO_4^{2-}$ under an oxidizing condition. However, it has two questions needed to be solved. First, the solubility of oxygen in strong alkali solution is quite low, so a high pressure condition is necessary to improve oxygen solubility, in order to enhance oxidation rate for ZnS. Second, closed circulation of leaching solvent in the system is hard to be completed due to an easy transformation from NaOH to Na_2SO_4. Based on those problems, this method has been blocked on the way to industrial implementation. The other way relies on the reaction $ZnS + 4OH^- = Zn(OH)_4^{2-} + S^{2-}$. But it also has its own problems. The reaction could happen only when $[OH^-] > 58.88$ mol/L. Nevertheless, solubility product for ZnS of 2×10^{-23} is much lower than instability constant for $Zn(OH)_4^{2-}$ of 3.6×10^{-16}, and $\Delta_r G_m^\theta$ remains 55.986 kJ/mol under that condition, which shows that the reaction can't go on spontaneously. It is unrealistic to dissolve ZnS with strong alkali solution alone.

2.2 Thermodynamics of Alkaline Leaching of Solid Wastes Bearing Lead

2.2.1 Morphology Distribution of Lead in Alkaline Solution

The main species of lead-bearing waste are PbO, $PbCO_3$, and $PbSO_4$ which all can dissolve in the alkaline solution and dissociate into $Pb(OH)_2$, $CO_3{}^{2-}$, and $SO_4{}^{2-}$. The dissociation process and total concentration of lead in the solution can be described as follows:

$$Pb^{2+} + OH^- = PbOH^+ \quad K_1 = \frac{[PbOH^+]}{[Pb^{2+}][OH^-]} \tag{2.62}$$

$$Pb^{2+} + 2OH^- = Pb(OH)_2 \quad K_2 = \frac{[Pb(OH)_2]}{[Pb^{2+}][OH^-]^2} \tag{2.63}$$

$$Pb^{2+} + 3OH^- = Pb(OH)_3^- \quad K_3 = \frac{[Pb(OH)_3^-]}{[Pb^{2+}][OH^-]^3} \tag{2.64}$$

where K_i is the cumulative stability constant and could be obtained from the literature. Here, the total concentration of lead is the sum of Pb^{2+}, $PbOH^+$, Pb $(OH)_2$, and $Pb(OH)_3^-$, and δ_0, δ_1, δ_2, and δ describe the distribution coefficient of Pb $^{2+}$, $PbOH^+$, $Pb(OH)_2$, and $Pb(OH)_3^-$, respectively, as calculated below. In alkaline solution, especially when pH is greater than 12, $Pb(OH)_3^-$ is the main part of the dissolved lead (Fig. 2.8).

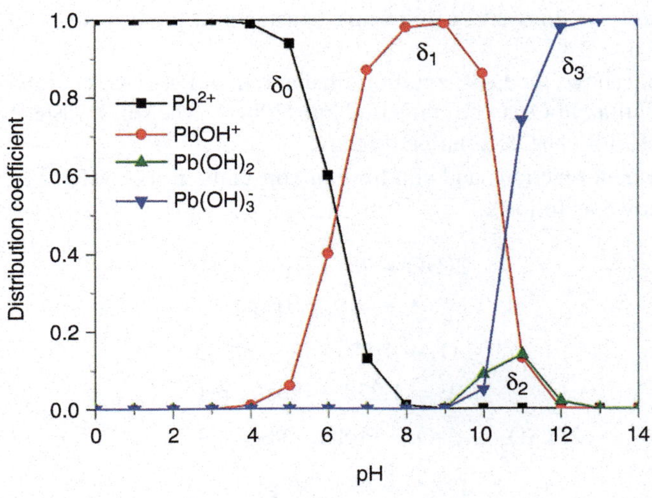

Fig. 2.8 Distribution coefficients of lead at different pH

Table 2.3 Solubility of lead in NaOH solution with different concentrations at 25 °C

$C(NaOH)$ ($mol{\cdot}L^{-1}$)	1.00	2.06	2.99	4.00	5.00
$C(Pb)$ ($g{\cdot}L^{-1}$)	9.22	18.90	23.62	24.49	25.56

Leaching experiments were carried out to investigate the solubility of lead in NaOH solution (Table 2.3). The solubility of leaching increased with the increase of NaOH concentration, the solubility of lead in 5 mol/L NaOH solution at 25 °C was observed to be 25.56 g/L.

$$\delta_0 = \frac{[Pb^{2+}]}{[Pb]_T} = \frac{1}{1 + K_1[OH^-] + K_2[OH^-]^2 + K_3[OH^-]^3} \tag{2.65}$$

$$\delta_1 = \frac{[PbOH^+]}{[Pb]_T} = \frac{K_1[OH^-]}{1 + K_1[OH^-] + K_2[OH^-]^2 + K_3[OH^-]^3} \tag{2.66}$$

$$\delta_2 = \frac{[Pb(OH)_2]}{[Pb]_T} = \frac{K_2[OH^-]^2}{1 + K_1[OH^-] + K_2[OH^-]^2 + K_3[OH^-]^3} \tag{2.67}$$

$$\delta_3 = \frac{[Pb(OH)_3^-]}{[Pb]_T} = \frac{K_3[OH^-]^3}{1 + K_1[OH^-] + K_2[OH^-]^2 + K_3[OH^-]^3} \tag{2.68}$$

2.2.2 *E-pH Equilibrium Diagrams of Leaching Systems of Lead*

The complex reactions and equilibrium constants in the systems of $Pb–H_2O$ are shown in Table 2.4.

Figure 2.9 shows the E-pH equilibrium diagram of $Pb–H_2O$. Dissolved by strong alkaline solution, PbO can be converted into $HPbO_2^-$. The solid wastes bearing PbO can be leached by NaOH solution directly.

The complex reactions and equilibrium constants in the systems of $Pb–SO_4^{2-}–H_2O$ are shown as follows:

$$2H_2O + 2e = H_2 + 2OH^- \tag{2.69}$$

$$E = -0.0591\ pH$$

$$O_2 + 4H^+ + 4e = 2H_2O \tag{2.70}$$

$$E = 1.229 - 0.0591\ pH$$

$$2PbSO_4 + H_2O = PbSO_4 \cdot PbO + SO_4^{2-} + 2H^+ \tag{2.71}$$

$$pH = 7.59 + 0.5lg[SO_4^{2-}]$$

Table 2.4 Chemical reactions and equilibrium constants in the systems of Pb–H$_2$O system

No.	Chemical reactions	Equilibrium constants
1	$Pb^{2+} + H_2O = PbO + 2H^+$	$pH = 6.36–0.5\ lg[Pb^{2+}]$
2	$Pb^{2+} + 2e = Pb$	$E = -0.125 + 0.0296\ lg[Pb^{2+}]$
3	$3PbO_2 + 4H^+ + 4e = Pb_3O_4 + 2H_2O$	$E = 1.12–0.0591pH$
4	$PbO_2 + 4H^+ +2e = Pb^{2+} + 2H_2O$	$E = 1.\ 47–0.1182pH -0.296\ lg[Pb^{2+}]$
5	$Pb_3O_4 + 2H^+ + 2e = 3PbO + H_2O$	$E = 1.05–0.0591pH$
6	$Pb_3O_4 + 8H^+ + 2e = 3Pb^{2+} + 4H_2O$	$E = 2.18–0.2364pH -0.0886\ lg[Pb^{2+}]$
7	$PbO + 2H^+ + 2e = Pb + H_2O$	$E = 0.\ 25–0.0591pH$
8	$PbO + H_2O = HPbO_2^- + H^+$	$pH = 15.33 + lg[HPbO_2^-]$
9	$HPbO_2^- + 3H^+ + 2e = Pb + 2H_2O$	$E = 0.\ 71–0.0886pH + 0.0296\ lg[HPbO_2^-]$
10	$2H_2O+ 2e = H_2 + 2OH^-$	$E = -0.0591pH$
11	$O_2 + 4H^+ + 4e = 2H_2O$	$E = 1.229–0.0591pH$

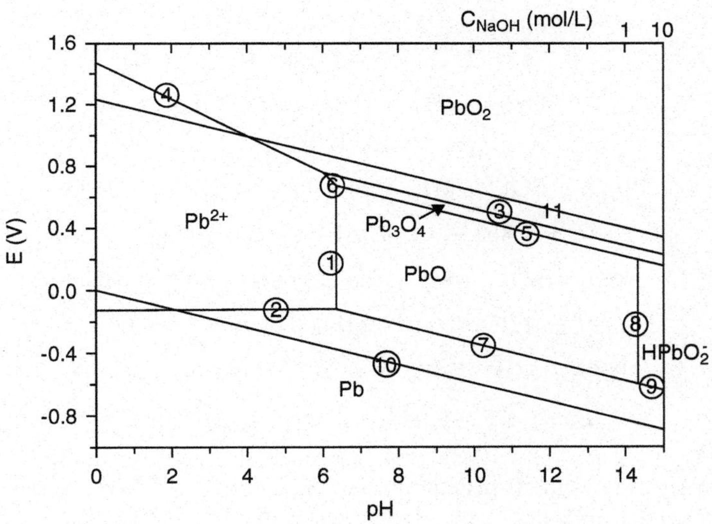

Fig. 2.9 E-pH equilibrium diagrams of Pb–H$_2$O (25 °C)

$$3PbSO_4 \cdot PbO + H_2O = 2PbSO_4 \cdot 2PbO + SO_4^{2-} + 2H^+ \tag{2.72}$$
$$pH = 11.31 + 0.5lg\left[SO_4^{2-}\right]$$
$$PbSO_4 \cdot 2PbO + H_2O = 3PbO + SO_4^{2-} + 2H^+ \tag{2.73}$$
$$pH = 13.72 + 0.5lg\left[SO_4^{2-}\right]$$
$$PbO + H_2O = HPbO_2^- + H^+ \tag{2.74}$$
$$pH = 15.33 + lg\left[HPbO_2^-\right]$$

$$PbSO_4 + 2e = Pb + SO_4^{2-} \quad (2.75)$$

$$E = -0.355 - 0.0296\ \lg\left[SO_4^{2-}\right]$$

$$PbSO_4 \cdot PbO + 2H^+ + 4e = 2Pb + SO_4^{2-} + H_2O \quad (2.76)$$

$$E = -0.13 - 0.01478\lg\left[SO_4^{2-}\right] - 0.0296pH$$

$$PbSO_4 \cdot 2PbO + 4H^+ + 6e = 3Pb + SO_4^{2-} + 2H_2O \quad (2.77)$$

$$E = -0.0188 - 0.0099\lg\left[SO_4^{2-}\right] - 0.0394pH$$

$$PbO + 2H^+ + 2e = Pb + H_2O \quad (2.78)$$

$$E = 0.25 - 0.0591pH$$

$$HPbO_2^- + 3H^+ + 2e = Pb + 2H_2O \quad (2.79)$$

$$E = 0.71 - 0.0886pH + 0.0296\lg\left[HPbO_2^-\right]$$

$$Pb_3O_4 + 2H^+ + 2e = 3PbO + H_2O \quad (2.80)$$

$$E = 1.05 - 0.0591pH$$

$$3PbO_2 + 4H^+ + 4e = Pb_3O_4 + 2H_2O \quad (2.81)$$

$$E = 1.12 - 0.0591pH$$

$$Pb_3O_4 + SO_4^{2-} + 4H^+ + 2e = PbSO_4 \cdot 2PbO + 2H_2O \quad (2.82)$$

$$E = 1.862 - 0.1183pH + 0.0296\lg\left[SO_4^{2-}\right]$$

$$3PbO_2 + SO_4^{2-} + 8H^+ + 6e = PbSO_4 \cdot 2PbO + 4H_2O \quad (2.83)$$

$$E = 1.366 - 0.0788pH + 0.0099\lg\left[SO_4^{2-}\right]$$

$$2PbO_2 + SO_4^{2-} + 6H^+ + 4e = PbSO_4 \cdot PbO + 3H_2O \quad (2.84)$$

$$E = 1.477 - 0.0887pH + 0.01478\lg\left[SO_4^{2-}\right]$$

Figure 2.10 shows the E-pH equilibrium diagram of $Pb - SO_4^{2-} - H_2O$. $PbSO_4$ can be converted into PbO and $HPbO_2^-$ when pH is > 13.72. The solid wastes bearing $PbSO_4$ can be leached by NaOH solution directly.

The complex reactions and equilibrium constants in the systems of $Pb - CO_3^{2-} - H_2O$ are shown as follows:

$$H_2CO_3 = H^+ + HCO_3^- \quad (2.85)$$

$$pH = 6.38 + \lg\left(\left[HCO_3^-\right]/[H_2CO_3]\right)$$

$$HCO_3^- = H^+ + CO_3^{2-} \quad (2.86)$$

$$pH = 10.25 + \lg\left(\left[CO_3{}^{2-}\right]/\left[HCO_3^-\right]\right)$$

$$PbCO_3 + 2H^+ = Pb^{2+} + H_2CO_3(aq) \quad (2.87)$$

$$pH = 1.89 - 0.5\left(\lg\left[Pb^{2+}\right] + \lg[H_2CO_3(aq)]\right)$$

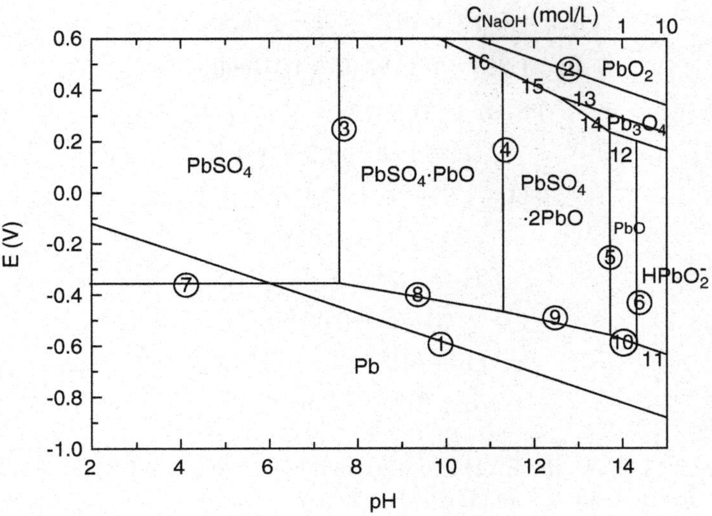

Fig. 2.10 E-pH equilibrium diagrams of Pb–SO_4^{2-}–H_2O (25 °C)

$$PbCO_3 + H_2O = PbO + CO_3^{2-} + 2H^+ \tag{2.88}$$

$$pH = 12.82 + 0.5lg[CO_3^{2-}]$$

$$PbO + H_2O = HPbO_2^- + H^+ \tag{2.89}$$

$$pH = 15.33 + lg[HPbO_2^-]$$

$$Pb^{2+} + 2e = Pb \tag{2.90}$$

$$E = -0.125 + 0.0296\ lg[Pb^{2+}]$$

$$PbCO_3 + 2H^+ + 2e = Pb + H_2CO_3(aq) \tag{2.91}$$

$$E = -0.013 - 0.0591pH - 0.0296\ \ lg[H_2CO_3(aq)]$$

$$PbCO_3 + H^+ + 2e = Pb + HCO_3^- \tag{2.92}$$

$$E = -0.2 - 0.0296pH - 0.0296lg[HCO_3^-]$$

$$PbCO_3 + 2e = Pb + CO_3^{2-} \tag{2.93}$$

$$E = -0.506 - 0.0296lg[CO_3^{2-}]$$

$$PbO + 2H^+ + 2e = Pb + H_2O \tag{2.94}$$

$$E = 0.25 - 0.0591pH$$

$$HPbO_2^- + 3H^+ + 2e = Pb + 2H_2O \tag{2.95}$$

$$E = 0.71 - 0.0886pH + 0.0296lg[HPbO_2^-]$$

$$PbO_2 + CO_3^{2-} + 4H^+ + 2e = PbCO_3 + 2H_2O \tag{2.96}$$
$$E = 1.853 - 0.1182pH + 0.0296lg[CO_3^{2-}]$$

$$Pb_3O_4 + 2H^+ + 2e = 3PbO + H_2O \tag{2.97}$$
$$E = 1.05 - 0.0591pH$$

$$3PbO_2 + 4H^+ + 4e = Pb_3O_4 + 2H_2O \tag{2.98}$$
$$E = 1.12 - 0.0591pH$$

$$2H_2O + 2e = H_2 + 2OH^- \tag{2.99}$$
$$E = -0.0591pH$$

$$O_2 + 4H^+ + 4e = 2H_2O \tag{2.100}$$
$$E = 1.229 - 0.0591pH$$

Figure 2.11 shows the E-pH equilibrium diagram of $Pb - CO_3^{2-} - H_2O$. $PbCO_3$ can be converted into PbO and $HPbO_2^-$ when pH > 12.82. The solid wastes bearing $PbCO_3$ can be leached by NaOH solution directly.

The complex reactions and equilibrium constants in the systems of $PbS–H_2O$ are shown as follows:

$$2H_2O + 2e = H_2 + 2OH^- \tag{2.101}$$
$$E = -0.0591pH$$

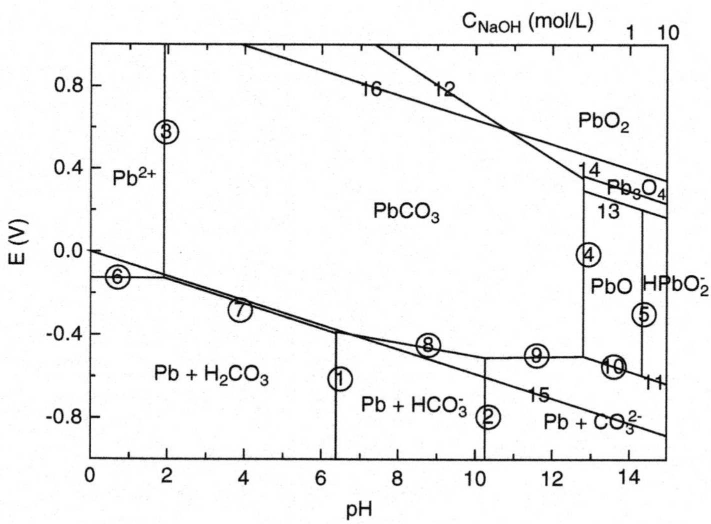

Fig. 2.11 E-pH equilibrium diagrams of $Pb–CO_3^{2-}– H_2O$ (25 °C)

$$2PbSO_4 + H_2O = PbSO_4 \cdot PbO + SO_4^{2-} + 2H^+ \quad (2.102)$$

$$pH = 7.59 + 0.5\lg\left[SO_4^{2-}\right]$$

$$3PbSO_4 \cdot PbO + H_2O = 2PbSO_4 \cdot 2PbO + SO_4^{2-} + 2H^+ \quad (2.103)$$

$$pH = 11.31 + 0.5\lg\left[SO_4^{2-}\right]$$

$$PbSO_4 \cdot 2PbO + H_2O = 3PbO + SO_4^{2-} + 2H^+ \quad (2.104)$$

$$pH = 13.72 + 0.5\lg\left[SO_4^{2-}\right]$$

$$PbO + H_2O = HPbO_2^- + H^+ \quad (2.105)$$

$$pH = 15.33 + \lg\left[HPbO_2^-\right]$$

$$H_2S_{(aq)} = H^+ + HS^- \quad (2.106)$$

$$pH = 6.97 + \lg([HS^-]/[\,H_2S])$$

$$HS^- = H^+ + S^{2-} \quad (2.107)$$

$$pH = 12.9 + \lg(\left[S^{2-}\right]/[HS^-])$$

$$PbS + 2H^+ + 2e = Pb + H_2S_{(aq)} \quad (2.108)$$

$$E = -0.358 - 0.0591pH - 0.0296\ \lg\left[H_2S_{(aq)}\right]$$

$$PbS + H^+ + 2e = Pb + HS^- \quad (2.109)$$

$$E = -0.564 - 0.0296pH - 0.0296\lg[HS^-]$$

$$PbS + 2e = Pb + S^{2-} \quad (2.110)$$

$$E = -0.949 - 0.0296\ \lg\left[S^{2-}\right]$$

$$PbSO_4 + 8H^+ + 8e = PbS + 4H_2O \quad (2.111)$$

$$E = 0.302 - 0.0591pH$$

$$HPbO_2^- + SO_4^2 + 11H^+ + 8e = PbS + 6H_2O \quad (2.112)$$

$$E = 0.567 - 0.0813pH + 0.0074\{\lg\left[HPbO_2^-\right] + \lg\left[SO_4^{2-}\right]\}$$

$$PbO + SO_4^{2-} + 10H^+ + 8e = PbS + 5H_2O \quad (2.113)$$

$$E = 0.454 - 0.074pH + 0.0074\ \lg\left[SO_4^{2-}\right]$$

$$PbSO_4 \cdot 2PbO + 2SO_4^{2-} + 28H^+ + 24e = 3PbS + 14H_2O \quad (2.114)$$

$$E = 0.386 - 0.069pH + 0.0049\ \lg\left[SO_4^{2-}\right]$$

$$PbSO_4 \cdot PbO + SO_4^{2-} + 18H^+ + 16e = 2PbS + 9H_2O \quad (2.115)$$

$$E = 0.358 - 0.0665pH + 0.0037\ \lg\left[SO_4^{2-}\right]$$

Figure 2.12 shows the E-pH equilibrium diagram of $PbS–H_2O$. The dissolution of PbS may be negligible in concentrated NaOH solutions. The solid wastes bearing PbS cannot be leached by NaOH solution directly.

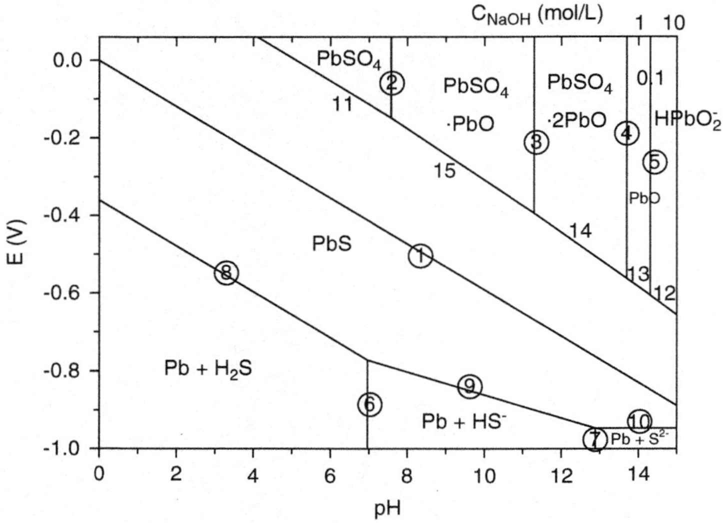

Fig. 2.12 E-pH equilibrium diagrams of PbS (25 °C)

2.3 Thermodynamics of Alkaline Leaching of Impurity Ions

Studies are carried out to explore the behavior of metal impurities in the leaching agent. The equilibrium concentrations of Cd in leaching solutions at 180 ~ 240 g/L NaOH is about 0.45 ~ 0.79 g/L. The dissolution of other impurities such as Fe, Cu, Co, Ni, Mg, Ca, etc. can be negligible.

2.3.1 *Thermodynamic Behavior of Cu(II) in Alkaline Solution*

According to the following calculation, the concentration of metal impurity of Cu (II) with respect to the pH was determined (Fig. 2.13).

$$CuO(s) + H_2O = Cu(OH)_2(aq) \tag{2.116}$$

$$\Delta_r G^{\theta}_{298.15} = 118.963\ \mathrm{kJ/mol}, \quad K^{\theta}_{298.15} = [Cu(OH)_2(aq)]$$

$$[Cu(OH)_2(aq)] = \exp(-0.048)$$

$$CuO(s) + H_2O = Cu^{2+} + 2OH^- \tag{2.117}$$

$$\Delta_r G^{\theta}_{298.15} = 117.520\ \mathrm{kJ/mol}, \quad K^{\theta}_{298.15} = [Cu^{2+}][OH^-]^2$$

$$[Cu^{2+}] = \exp(17.063 - 4.606\,\mathrm{pH}) \tag{2.118}$$

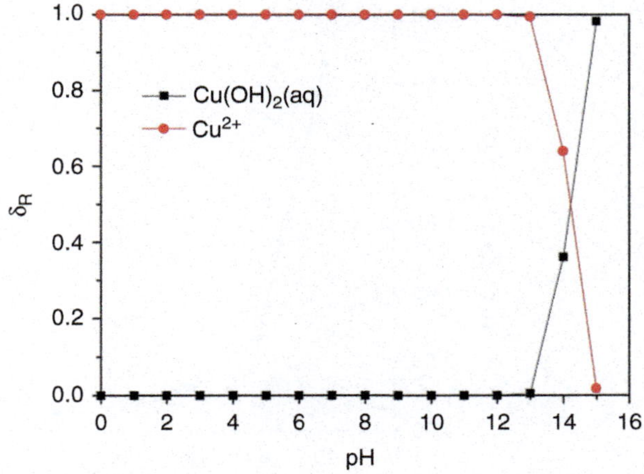

Fig. 2.13 Relationship between pH value and δ_R of different Cu-containing species (25 °C)

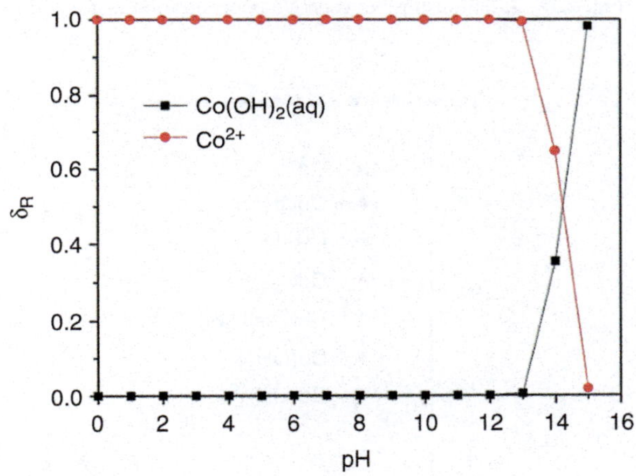

Fig. 2.14 Relationship between pH value and δ_R of different Co-containing species (25 °C)

$Cu(OH)_2(aq)$ may be the main form in alkaline solution when pH is high enough. The equilibrium concentrations of Cu in leaching solutions at 180 ~ 240 g/L NaOH are about $8.07 \times 10^{-21} \sim 4.54 \times 10^{-21}$ g/L. The dissolution of Cu can be negligible.

2.3.2 *Thermodynamic Behavior of Co(II) in Alkaline Solution*

According to calculation, the concentration of metal impurity of Co(II) with respect to the pH was determined as in Figs. 2.14 and 2.15.

$Co(OH)_2(aq)$ may be the main form in alkaline solution when pH is high enough. The equilibrium concentrations of Co in leaching solutions at 180 ~ 240 g/L NaOH are about $13.03 \times 10^{-15} \sim 7.33 \times 10^{-15}$ g/L. The dissolution of Co can be negligible.

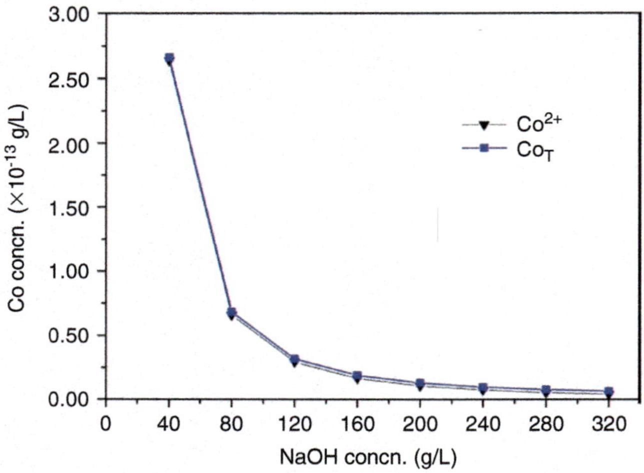

Fig. 2.15 Relationship between Co concentration and NaOH concentration (25 °C)

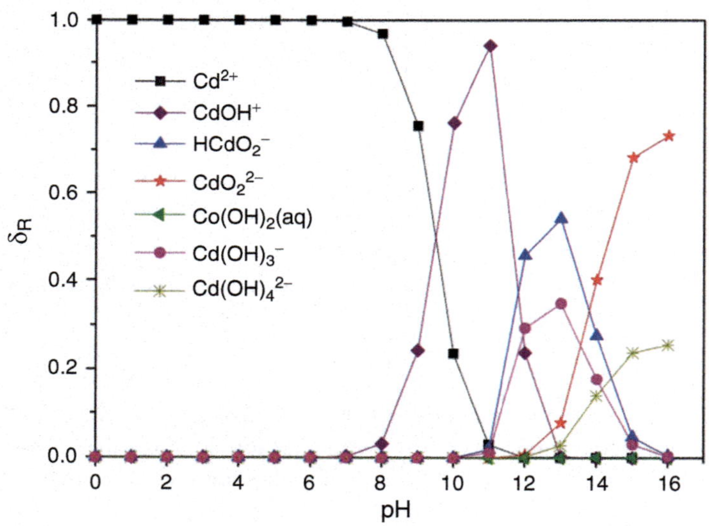

Fig. 2.16 Relationship between pH value and δ_R of different Cd-containing species (25 °C)

2.3.3 Thermodynamic Behavior of Cd(II) in Alkaline Solution

The concentration of metal impurity of Cd(II) with respect to the pH was determined as in Fig. 2.16.

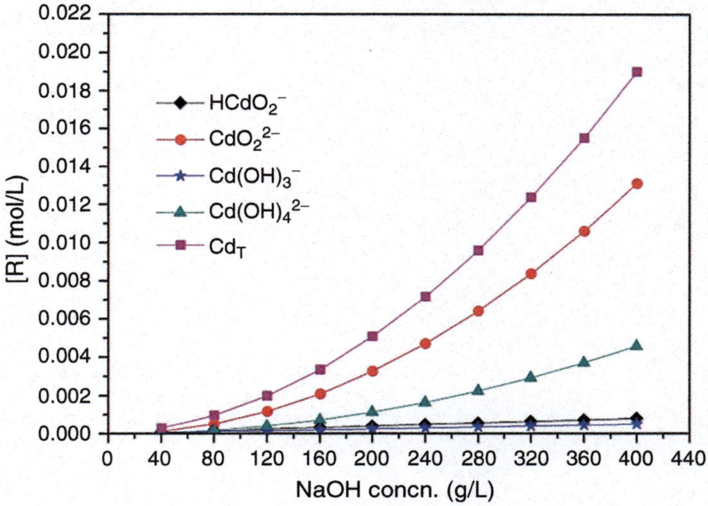

Fig. 2.17 Relationship between Cd concentration and NaOH concentration (25 °C)

$HCdO_2^-$, CdO_2^{2-}, $Cd(OH)_3^-$, and $Cd(OH)_4^{2-}$ are the main parts of dissolved Cd in alkaline solution when pH is greater than 14. Figure 2.17 shows the relationship between Cd concentration and NaOH concentration. The equilibrium concentrations of Cd in leaching solutions at 180 ~ 240 g/L NaOH are about 0.45 ~ 0.79 g/L, which should be removed.

$$CdO(s) + OH^- = HCdO_2^- \tag{2.119}$$

$$\Delta_r G_{298.15}^\theta = 23.357\ \mathrm{kJ/mol}, \quad K_{298.15}^\theta = \frac{[HCdO_2^-]}{[OH^-]}$$

$$CdO(s) + 2OH^- = Cd(OH)_2^{2-} + H_2O \tag{2.120}$$

$$\Delta_r G_{298.15}^\theta = 22.160\ \mathrm{kJ/mol}, \quad K_{298.15}^\theta = \frac{[CdO_2^{2-}]}{[OH^-]^2}$$

$$CdO(s) + OH^- + H_2O = Cd(OH)_3^- \tag{2.121}$$

$$\Delta_r G_{298.15}^\theta = 24.449\ \mathrm{kJ/mol}, \quad K_{298.15}^\theta = \frac{[Cd(OH)_3^-]}{[OH^-]}$$

$$CdO(s) + 2OH^- + H_2O = Cd(OH)_4^{2-} \tag{2.122}$$

$$\Delta_r G_{298.15}^\theta = 24.762\ \mathrm{kJ/mol}, \quad K_{298.15}^\theta = \frac{[Cd(OH)_4^{2-}]}{[OH^-]^2}$$

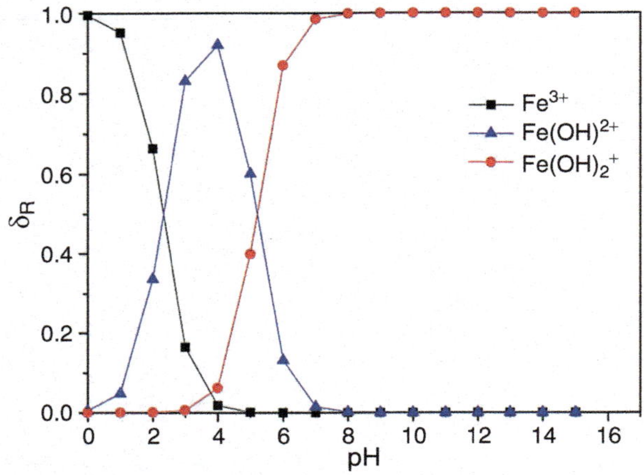

Fig. 2.18 Relationship between pH value and δ_R of different Fe-containing species (25 °C)

2.3.4 *Thermodynamic Behavior of Fe(III) in Alkaline Solution*

According to calculation, the concentration of metal impurity of Fe(III) with respect to the pH is determined in Fig. 2.18. $Fe(OH)_2^+$ is the main part of dissolved Fe(III) in alkaline solution when pH is greater than 14. The equilibrium concentrations of Fe(III) in leaching solutions at 180 ~ 240 g/L NaOH are about 6.98×10^{-23} ~ 5.25×10^{-23} g/L; the dissolution of Fe can be negligible.

2.3.5 *Thermodynamic Behavior of Ni(II) in Alkaline Solution*

According to the calculation, the concentration of metal impurity of Ni(II) with respect to the pH was determined as in Fig. 2.19. $Ni(OH)^+$ is the main part of dissolved Ni(II) in alkaline solution when pH is greater than 14. The equilibrium concentrations of Ni(II) in leaching solutions at 180 ~ 240 g/L NaOH are about 7.10×10^{-11} ~ 5.28×10^{-11} g/L; the dissolution of Ni can be negligible.

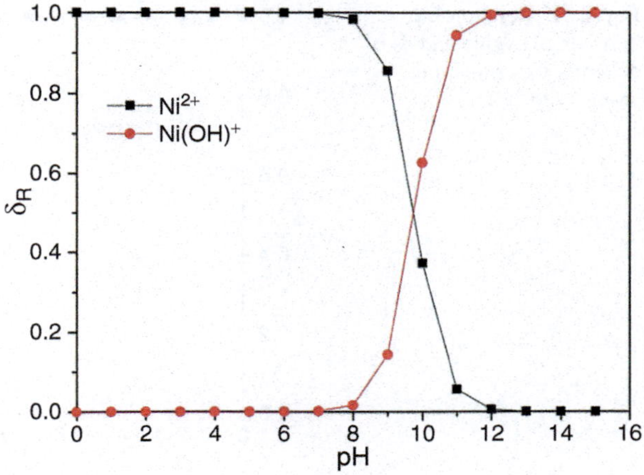

Fig. 2.19 Relationship between pH value and δ_R of different Ni-containing species (25 °C)

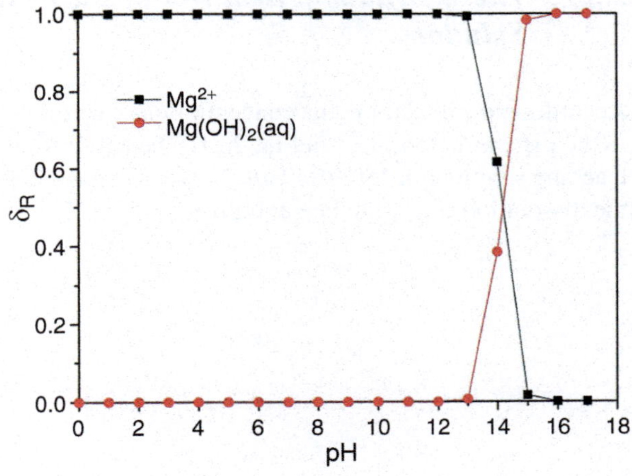

Fig. 2.20 Relationship between pH value and δ_R of different Mg-containing species (25 °C)

2.3.6 Thermodynamic Behavior of Mg(II) in Alkaline Solution

According to calculation, the concentration of metal impurity of Mg(II) with respect to the pH was determined as in Fig. 2.20. The equilibrium concentrations of Mg(II) in leaching solutions at 180 ~ 240 g/L NaOH are about 8.05×10^{-6} ~ 7.78×10^{-6} g/L; the dissolution of Mg can be negligible.

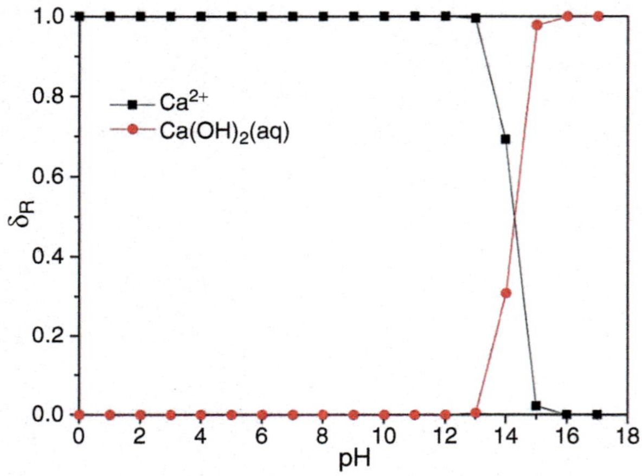

Fig. 2.21 Relationship between pH value and δ_R of different Ca-containing species (25 °C)

2.3.7 *Thermodynamic Behavior of Ca(II) in Alkaline Solution*

According to calculation, the concentration of metal impurity of Ca(II) with respect to the pH was determined in Fig. 2.21. The equilibrium concentrations of Ca(II) in leaching solutions at 180 ~ 240 g/L NaOH are about $2.28 \times 10^{-4} \sim 2.18 \times 10^{-4}$ g/L; the dissolution of Ca can be negligible.

Chapter 3
Kinetics of Alkaline Leaching of Solid Wastes Bearing Zinc and Lead

Abstract The leaching reaction kinetics of waste bearing zinc and lead are studied. For waste bearing zinc, plot of $1-(1-\eta)^{1/3}$ versus leaching time giving linear relationship indicates that the velocity of leaching is controlled by the chemical reaction step; the apparent activation energy is obtained to be 49.22 kJ/mol from an Arrhenius plot. For smithsonite and willemite, during the early stage of leaching reaction, $1-(2/3)\eta-(1-\eta)^{2/3}$ is linear with leaching time, with the apparent activation energy of 19.95 kJ/mol and 19.32 kJ/mol, respectively. In the later stage, $1-(1-\eta)^{1/3}$ is linear with leaching time, with the apparent activation energy of 46.54 kJ/mol and 32.64 kJ/mol, respectively. Thus, it indicates that the leaching mechanism of smithsonite changes from inner diffusion control to chemical reaction control and that of willemite changes from inner diffusion control to mixed control, during the leaching process in alkaline solution. The leaching of lead oxide may be controlled by diffusion of NaOH solution passing through the solid-phase layer at lower temperature and less NaOH concentration and by surface chemical reaction at high temperature and high NaOH concentration.

Keywords Kinetic analysis • Alkaline leaching • Zinc hazardous waste • Lead hazardous waste

3.1 Kinetic Model of Leaching in Alkaline Solution

The leaching is a complex heterogeneous reaction process between the solvent and the solid phase. There is a hypothesis that the reaction between the liquid and solid is developed from the surface to the center. Before the leaching is complete, there is an unreaction core in the center of the granule, and it is wrapped by the solid product layer or the material layer which is not leached. The leaching process of ore particles is shown in Fig. 3.1, which is generally composed of the following steps.

(1) The leaching agent in the liquid phase is diffused through the liquid film layer on the solid phase to the solid surface; this step is called external diffusion.
(2) The leaching agent is diffused to the reaction interface through the solid product layer/the material layer which is not leached; this step is called internal diffusion.

Y. Zhao, C. Zhang, *Pollution Control and Resource Reuse for Alkaline Hydrometallurgy of Amphoteric Metal Hazardous Wastes*, Handbook of Environmental Engineering 18, DOI 10.1007/978-3-319-55158-6_3

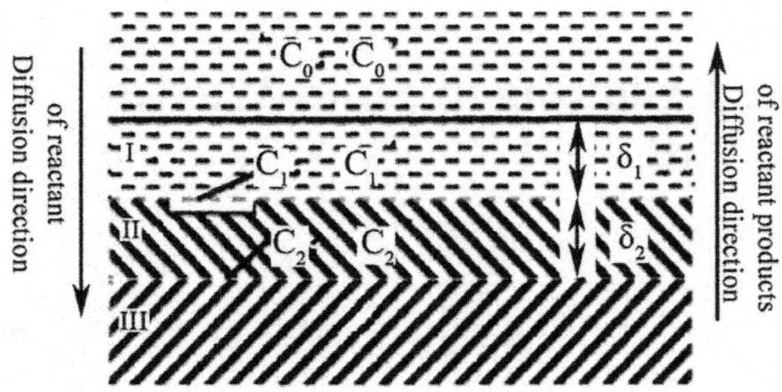

Fig. 3.1 Sketch of the leaching process for ores or solid wastes

(3) Chemical reactions occur at the reaction interface between the leaching agent and the ore particle.
(4) The soluble product passes through the solid product layer/the material layer which is not leached and spreads out from the reaction surface to the surface of the ore particle.
(5) The product is diffused through the liquid film layer covering the ore particle into the solution.

Steps 2 and 4 do not exist if there is no solid-phase product generation, but when the extracted solid-phase reactant is dispersed in an inert material (e.g., silica), even if no new solid product is formed, the solubility product formed by the interaction reaction must be diffused through the inert material layer, so the process still includes the above 5 steps.

The ore is broken and refined before leaching process, and the particle size is very fine, which are generally dense particles. The reaction process is shown in Fig. 3.2.

The main task of the study is to find out the control steps of the leaching process, so as to take measures to strengthen this leaching process. The characteristics of various control steps are as follows.

(1) Controlled by chemical reactions

The reaction rate is controlled by the chemical reaction if the diffusion layer of the liquid film and the resistance of the solid film are very small; at this time, the number of ore mole per unit of time is

$$-\frac{dN}{dt} = kS_1C^n \tag{3.1}$$

where:
N – the mole number of solid particles at the time of t, mol
k – chemical reaction rate constant, cm/s
S_1 – the area of the nuclear surface area of the solid ore, cm^2
C – concentration of leaching agent, mol/mL
n – reaction order

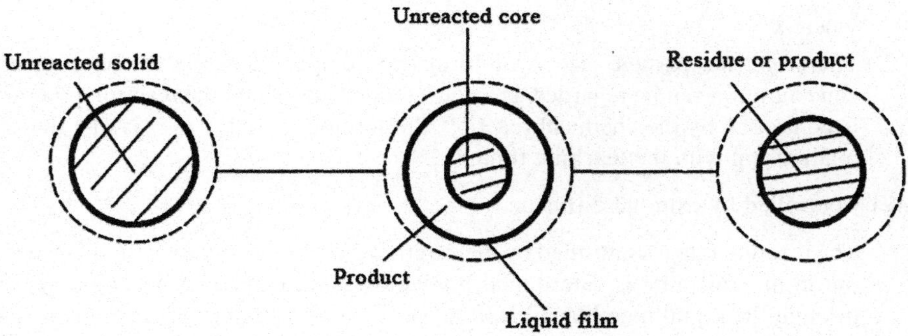

Fig. 3.2 Sketch of shrinking-core reaction model

During the reaction, the surface area of S_1 of the ore particles will change. It is assumed that the particle is spherical, dense, and nonporous, and its radius is r, the density is ρ, and M is the molar mass of ore and then

$$N = \frac{4}{3}\pi r^3 \rho / M \tag{3.2}$$

$$-\frac{dN}{dt} = -\frac{4}{M}\pi r^2 \rho \times \frac{dr}{dt} = 4\pi r^2 k C^n \tag{3.3}$$

$$\text{So,} \ -\frac{dr}{dt} = \frac{kMC^n}{\rho} \tag{3.4}$$

It is assumed that the leaching agent is too much in the leaching process; the concentration C of the leaching process can be regarded as the same, which is kept as C_0, and is then integral to the Eq. 3.4:

$$r_0 - r = \frac{kMC_0^n}{\rho} t \tag{3.5}$$

where:

r_0 – the original radius of the solid ore, cm
r – the radius of the solid ore particles at the time of t, cm

Because the particle radius is not easy to measure and *the reaction rate*η is usually used to indicate, so

$$\eta = \frac{\frac{4}{3} \times \frac{\pi r_0^3 \rho}{M} - \frac{4}{3} \times \frac{\pi r^3 \rho}{M}}{\frac{4}{3} \times \frac{\pi r_0^3 \rho}{M}} = 1 - \frac{r^3}{r_0^3} \tag{3.6}$$

$$r = r_0 (1 - \eta)^{1/3} \tag{3.7}$$

$$1 - (1 - \eta)^{1/3} = k' t \tag{3.8}$$

where $k^{'} = \frac{kMC_0^n}{r_0\rho}$.

Therefore, the particle size is uniform and compact, and the leaching agent concentration can be regarded as a constant process; when the reaction process is controlled by the chemical reaction, the formula $1-(1-\eta)^{1/3}$ is in a linear relationship with the leaching time.

(2) Controlled by external diffusion

The reaction rate is controlled by the external diffusion if the amount of leached ore in the unit time is determined by the diffusion rate of the leaching agent through the liquid membrane layer; at this time, the number of leaching agent which through the diffusion layer per unit of time is

$$L = \frac{C_0 D_1 S}{\delta_1} \tag{3.9}$$

It is assumed that per mol of ore extraction is required to consume α mol of leaching agent; the mole number of ore leached per unit of time is

$$-\frac{dN}{dt} = \frac{C_0 D_1 S}{\alpha\delta_1} \tag{3.10}$$

where S (cm^2) is the ore surface area that includes solid film.

① When the formation of the solid film or the existence of inert residual layer, and the size of the particle size, including the solid film, is essentially unchanged, the *S* is a constant, which is the surface area of the initial particle size:

$$-\frac{\mathrm{d}N}{\mathrm{d}t} = \frac{C_0 D_1 4\pi r_0^2}{\alpha\delta_1} = \text{constant} \tag{3.11}$$

That means leaching rate formula is in a linear relationship with the leaching time.

② When the leaching process does not produce a solid film, *S* is gradually reduced with constant leaching:

$$-\frac{dN}{dt} = \frac{C_0 D_1 4\pi r^2}{\alpha\delta_1} \tag{3.12}$$

According to Eqs. 3.9 and 3.12, it is calculated:

$$1-(1-\eta)^{1/3} = k''t \tag{3.13}$$

where $k'' = \frac{D_1 M C_0}{\alpha\delta_1 r_0 \rho}$

When there is no solid film, the relationship between the leaching rate and time is similar to that of chemical reaction control, so it is necessary to determine the control step by combining the apparent activation energy.

Under normal circumstances, the activation energy of chemical reaction control is larger, about 40 ~ 300 kJ/mol; by external diffusion control, the activation energy is smaller, about 8 ~ 20 kJ/mol.

(3) Controlled by internal diffusion

It is assumed that there is a dense solid product layer or an inert residual layer in the leaching process, and the diffusion resistance of them to the leaching agent is much greater than that of the outer diffusion. At the same time, the chemical reaction rate is very fast, and the leaching process is controlled by the diffusion of the inner membrane, the number of leaching agent which through the solid film per unit of time is:

$$L = D_2 S \frac{dC}{dr} \tag{3.14}$$

after integration:

$$L = 4\pi D_2 \left(\frac{r_0 r}{r_0 - r} \right) C_0 \tag{3.15}$$

It is assumed that per mol of ore extraction is required to consume α mol of leaching agent; the number of ore mole per unit of time is

$$-\frac{dN}{dt} = -\frac{4}{M}\pi r^2 \rho \times \frac{dr}{dt} = 4\pi D_2 \left(\frac{r_0 r}{r_0 - r} \right) C_0/\alpha \tag{3.16}$$

comprehensive above formulas:

$$1 - (2/3)\eta - (1 - \eta)^{2/3} = k'''t \tag{3.17}$$

where $k''' = \frac{2MD_2C_0}{\alpha r_0^2 \rho}$

Therefore, when the reaction process is controlled by internal diffusion, the formula $1 - (2/3)\eta - (1 - \eta)^{2/3}$ is in a linear relationship with the leaching time.

3.2 Kinetics of Alkaline Leaching of Zinc Hazardous Wastes

Kinetic experiments are carried out in the 2000 mL three-necked flask. The particle size of the leached material ranged from 200 to 160 mesh, and the average particle size was 0.0825 mm. In order to keep the concentration of the leaching agent basically unchanged in the leaching process, the liquid-solid ratio was set to 50:1. The concentration of zinc was analyzed by ICP-OES.

3.2.1 *Alkaline Leaching Kinetic Analysis of Waste Bearing Zinc*

The relationship between the leaching rate of waste bearing zinc (zinc oxide) and the leaching time in alkali solution is shown in Fig. 3.3. The relationship between the leaching time and $1-(1-\eta)^{1/3}$ is shown in Fig. 3.4. As show in Fig. 3.4, the $1-(1-\eta)^{1/3}$ is in a linear relationship with the leaching time.

According to Arrhenius theorem:

$$\frac{d(\ln k)}{dT}=\frac{E_a}{RT^2} \tag{3.18}$$

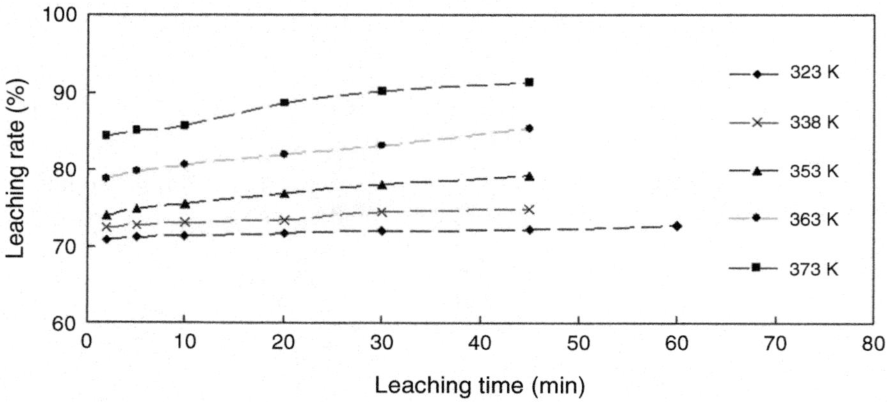

Fig. 3.3 Relationship between leaching rate and time for waste bearing zinc

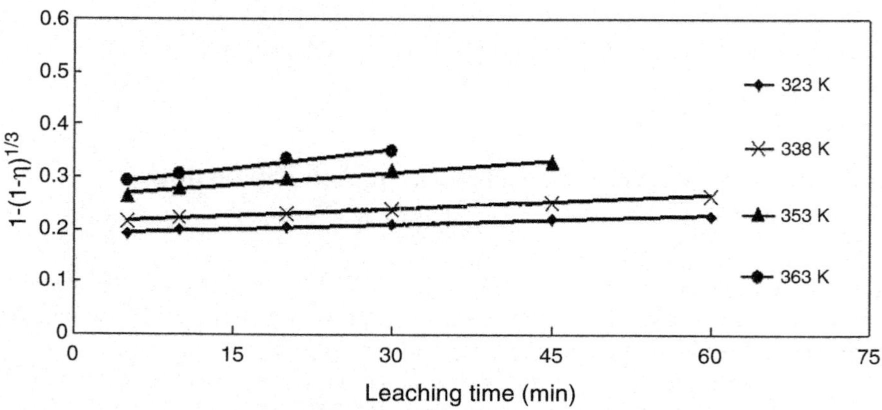

Fig. 3.4 Relationship between $1-(1-\eta)^{1/3}$ and time for waste bearing zinc

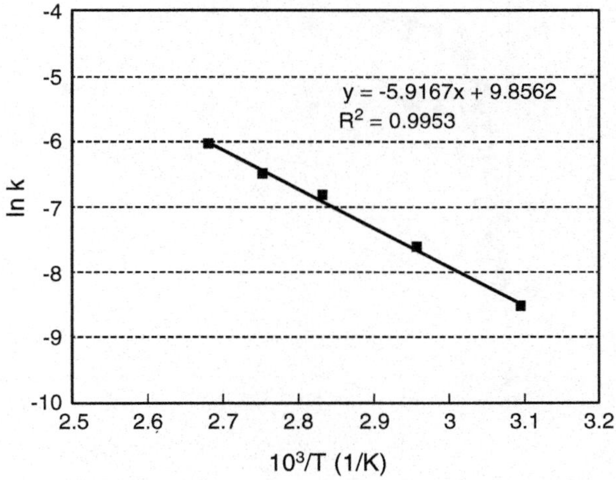

Fig. 3.5 Arrhenius plot for leaching experiments of waste bearing zinc

where:
k – reaction rate constant at the reaction temperature T, K
R – general constant of ideal gas, J/(mol·K)
E_a – apparent activation energy, J/mol

The result of integrating the above formula is:

$$\ln k = -\frac{E_a}{RT} + B \tag{3.19}$$

According to Fig. 3.4, a series of ln$k$ at different temperatures and 1/T are plotted as Arrhenius linear graph (Fig. 3.5). The activation energy is estimated to be 49.22 kJ/mol by using the slope of the Arrhenius linear graph.

It can be judged that the ash containing zinc leaching process is controlled by chemical reaction, because the formula $1-(1-\eta)^{1/3}$ and leaching time showed a linear relationship and the activation energy is 49.22 kJ/mol, which indicated that the main way to increase the leaching rate is to increase the reaction temperature and the concentration of the NaOH solution.

3.2.2 Alkaline Leaching Kinetic Analysis of Zinc Carbonate

The relationship between the leaching rate of zinc carbonate ($ZnCO_3$) and the leaching time in alkali solution is shown in Fig. 3.6. According to the analysis of experimental data of leaching process, $ZnCO_3$ in alkali solution leaching process can be divided into two stages. In the first period of 0–6 min, the formula $1-(2/3)\eta-(1-\eta)^{2/3}$ and leaching time show a linear relationship (Fig. 3.7). According to the Arrhenius theory, using the linear Arrhenius diagram as shown

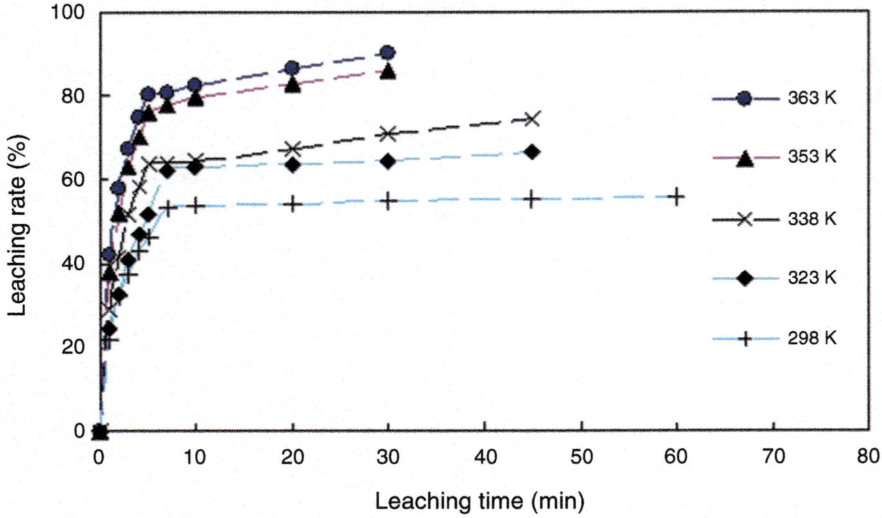

Fig. 3.6 Relationship between leaching rate and time for $ZnCO_3$

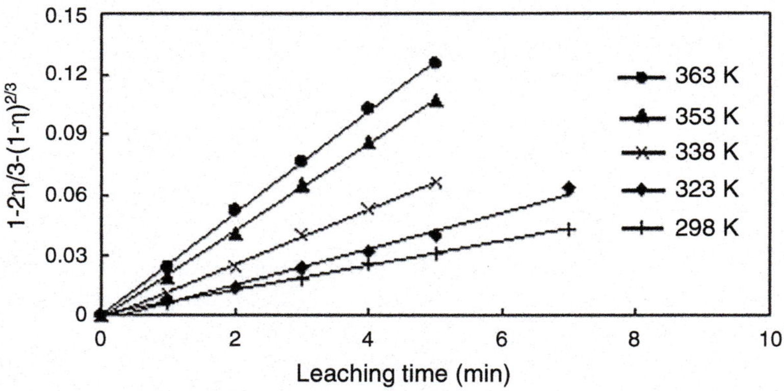

Fig. 3.7 Relationship between $1-(2/3)\eta-(1-\eta)^{2/3}$ and time for $ZnCO_3$

Fig. 3.8 Arrhenius plot for leaching experiments of $ZnCO_3$ during the early stage

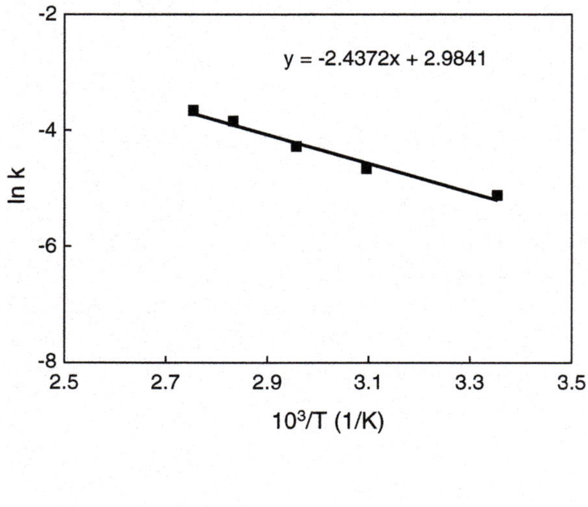

in Fig. 3.8, the apparent activation energy can be calculated to be 19.95 kJ/mol. After 6 min at the second stage, the formula$1-(1-\eta)^{1/3}$ is in a linear relationship with the leaching time (Fig. 3.9), and the apparent activation energy of the Arrhenius diagram is 46.54 kJ/mol (Fig. 3.10).

The leaching of $ZnCO_3$ ore in alkaline solution is controlled by internal diffusion firstly, and subsequent leaching process is controlled by chemical reaction. The selected $ZnCO_3$ sample was low-grade ore (13.5% Zn) and contained a lot of alkali-insoluble inert ingredients; hence, the diffusion rate of the alkali solution in the inert layer limited the reaction rate of leaching at the beginning. For this leaching process, the main way to improve the leaching rate is to increase the reaction

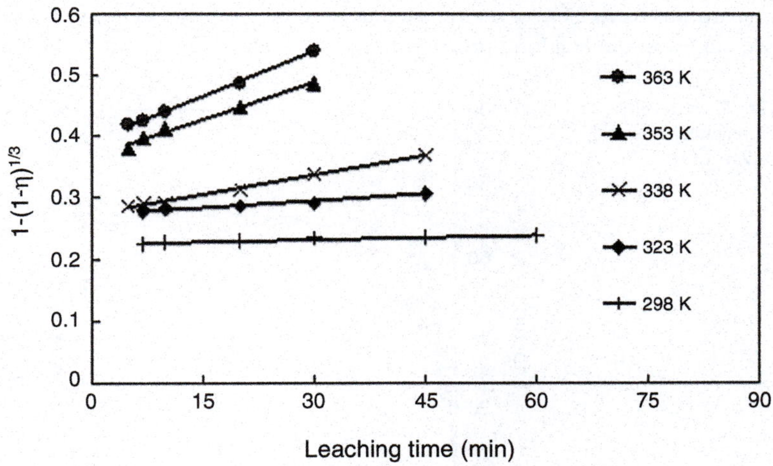

Fig. 3.9 Relationship between $1-(1-\eta)^{1/3}$ and time for $ZnCO_3$

Fig. 3.10 Arrhenius plot for leaching experiments of $ZnCO_3$ during the later stage

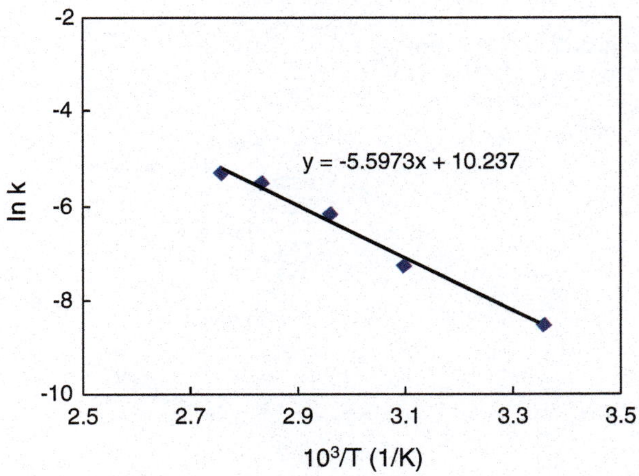

temperature and the concentration of the NaOH solution. In addition, reducing the particle size of the sample can also increase the leaching rate, but if ore particles is too fine, a large amount of fine particles of mud may be formed and hinder the dissolution of solid particles. When the particle size is finer than 100 mesh, the leaching rate changes little.

3.2.3 Alkaline Leaching Kinetic Analysis of Zinc Silicate

The relationship between the leaching rate of zinc silicate and the leaching time in alkali solution is shown in Fig. 3.11. According to the leaching experimental data analysis, similar to $ZnCO_3$ ore, zinc silicate in alkali solution can also be divided into two stages in the leaching process. In the first period of 0–5 min, the formula

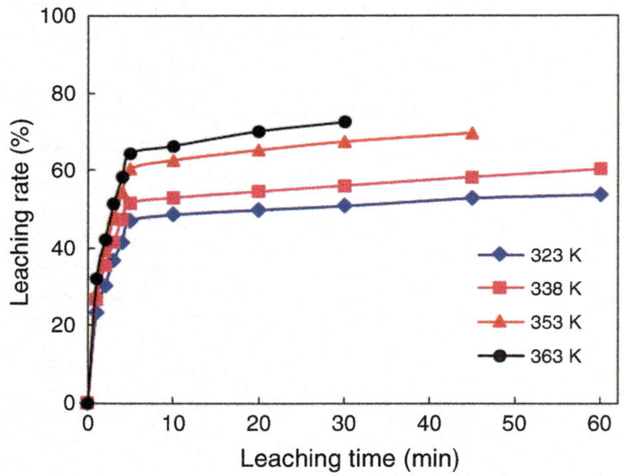

Fig. 3.11 Relationship between leaching rate and time for Zn_2SiO_4

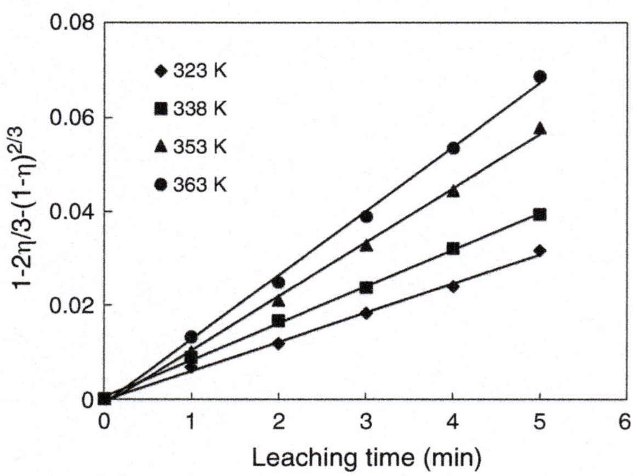

Fig. 3.12 Relationship between $1-(2/3)\eta-(1-\eta)^{2/3}$ and time for Zn_2SiO_4

$1-(2/3)\eta-(1-\eta)^{2/3}$ and leaching time were in linear relationship (Fig. 3.12). According to the Arrhenius theory, the Arrhenius linear graph (Fig. 3.13) was obtained, and the apparent activation energy was calculated to be 19.32 kJ/mol. After 5 min, the formula $1-(1-\eta)^{1/3}$ and leaching time showed a linear

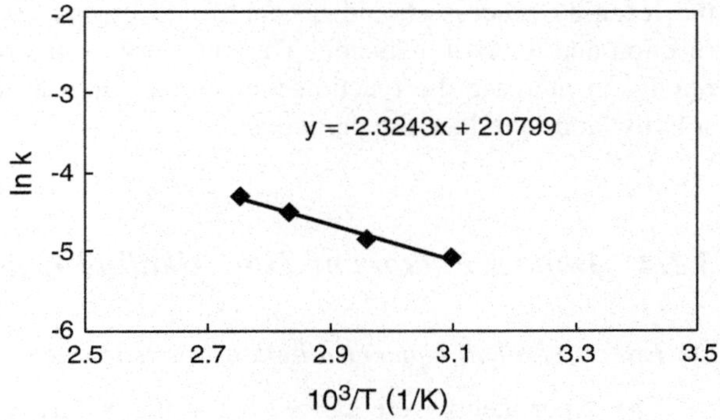

Fig. 3.13 Arrhenius plot for leaching experiments of Zn_2SiO_4 during the early stage

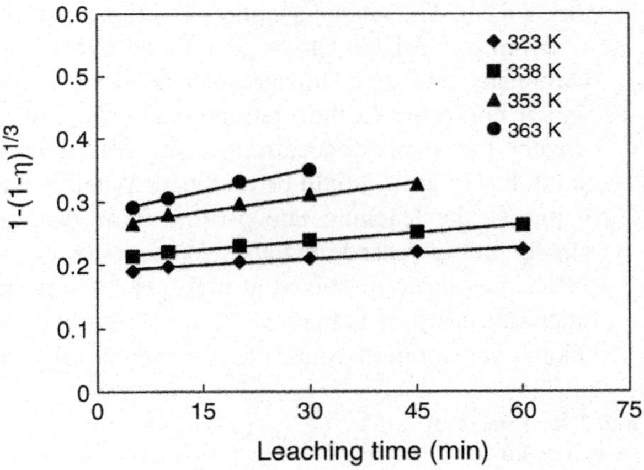

Fig. 3.14 Relationship between $1-(1-\eta)^{1/3}$ and time for Zn_2SiO_4

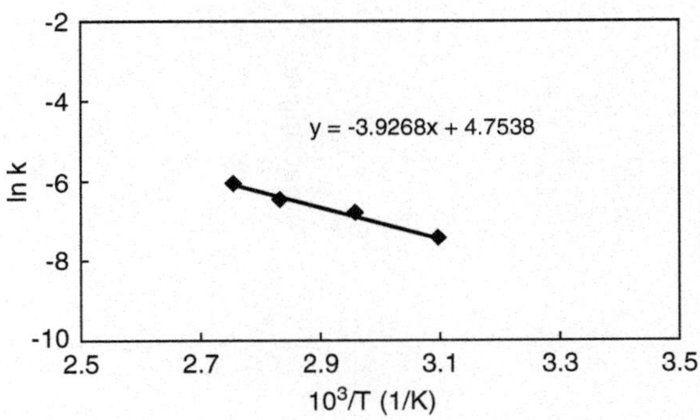

Fig. 3.15 Arrhenius plot for leaching experiments of Zn_2SiO_4 during the later stage

relationship (Fig. 3.14), as calculated from the Arrhenius graph, the apparent activation energy for 32.64 kJ/mol (Fig. 3.15).

Therefore, when the zinc silicate ore is leached in the alkali solution, it is controlled by the internal diffusion at the beginning, and the subsequent leaching process is controlled by the mixing extent. Since the formula $1-(1-\eta)^{1/3}$ is linear with the leaching time, but the activation energy is lower, which is 32.64 kJ/mol, so the leaching process should be controlled by the combined effects of chemical reaction and internal diffusion. The main way to improve the leaching rate is still mainly to increase the reaction temperature and the concentration of the NaOH solution and to reduce the particle size.

3.2.4 Impact Factors of Zinc Alkaline Leaching Process

(1) *Effects of alkali concentration on zinc leaching*

The experiments were carried out at the NaOH concentration of 2–10 mol/L, the temperature of 90 °C, liquid-solid ratio of 10:1 (mass ratio), and leaching time of 2 h. The results are shown in Fig. 3.16.

From Fig. 3.16, it can be seen that the leaching rate of zinc fume dust, zinc carbonate, and zinc silicate increase with increasing alkali concentrations, which conforms to the thermodynamic equilibrium calculation results. The greater the alkali concentration, the better for leaching, and the huger the solubility of zinc, would be obtained. When the alkali concentration exceeded 6 mol/L, the leaching rate of zinc fume dust and zinc carbonate increased slowly, but increased obviously for zinc silicate, which conformed to the result that zinc silicate dissolved at high pH. However, the leaching rate of all raw materials decreased after the alkali concentration up to 8 mol/L with increasing alkali concentration. Since the viscosity of alkali solution is proportional to the

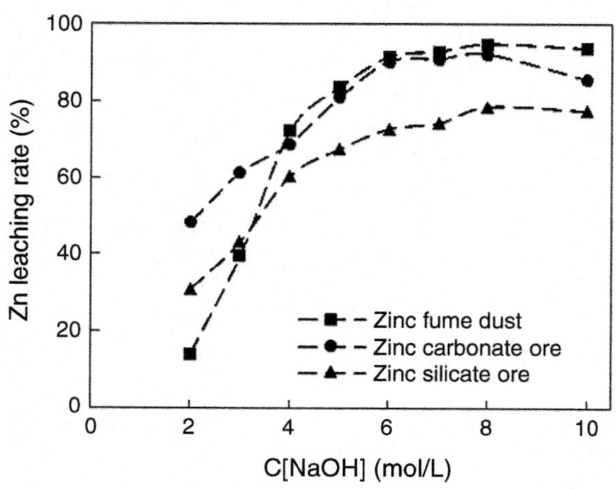

Fig. 3.16 Effects of sodium hydroxide concentrations on the zinc leaching rate

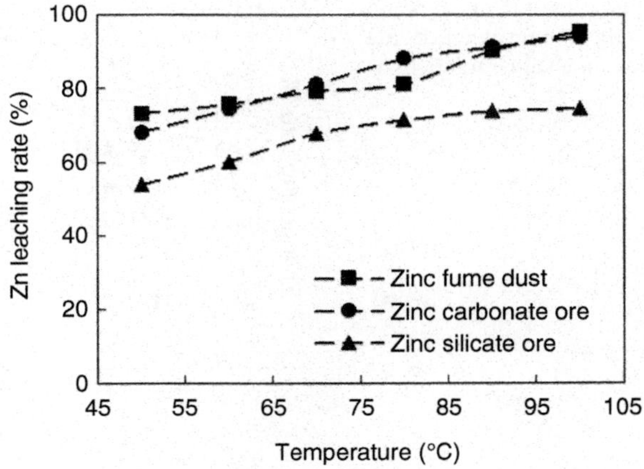

Fig. 3.17 Effects of temperature on zinc leaching rate in alkaline solutions

alkali concentration, and is inversely proportional to the temperature, the rising viscosity of alkali solution is unfavorable to the reaction between alkali and mineral grain. Moreover, compared with zinc carbonate, the leaching rate of zinc silicate is at low level.

(2) *Effects of temperature on zinc leaching*

At a liquid-solid ratio of 10:1, alkali concentration of 6 mol/L, and leaching time of 2 h, the effects of temperature on zinc leaching rate of three kinds of raw materials were investigated. The results are shown in Fig. 3.17. Leaching rate increased with temperature increasing up to the boiling point of the solution, which was costly and hard for industrial operation. Hence, the optimum leaching temperature was determined at 90 °C.

(3) *Effects of liquid-solid ratio on zinc leaching*

At the alkali concentration of 6 mol/L, temperature of 90 °C, and leaching time of 2 h, the effects of liquid-solid ratio on zinc leaching rate of three kinds of raw materials are shown in Fig. 3.18. Leaching rate increased with liquid-solid ratio increasing and went slowly when it reached to a certain degree. Since increasing liquid-solid ratio is beneficial to diffusion, and hydroxy zinc ion in solution has certain solubility, low liquid-solid ratio would limit the leaching rate of zinc. Exorbitant liquid-solid ratio can result in low zinc concentration in leaching solution and influence efficiency of the subsequent electrolytic process. Hence, the optimum liquid-solid ratio of zinc fume dust was ten, around five for zinc carbonate and zinc silicate.

(4) *Effects of leaching time on zinc leaching*

At an alkali concentration of 6 mol/L, temperature of 90 °C, and liquid-solid ratio of 10, the effects of leaching time on zinc leaching rate of three kinds of raw materials are shown in Fig. 3.19. Leaching rate increased slowly with leaching time increasing, with the optimum leaching time determined at 60–90 min.

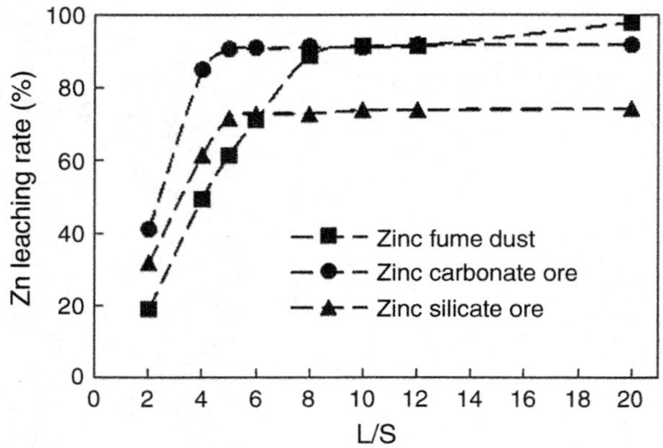

Fig. 3.18 Effects of liquid-solid ratio on zinc leaching rate in alkaline solutions

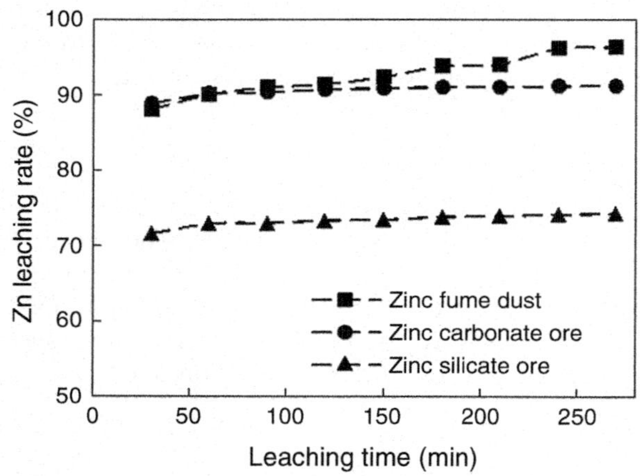

Fig. 3.19 Effects of leaching time on zinc leaching rate in alkaline solutions

(5) *Effects of particle size on zinc leaching*

At the alkali concentration of 6 mol/L, temperature of 90 °C, liquid-solid ratio of 10, and leaching time of 120 min, the effects of particle size on zinc leaching rate of three kinds of raw materials were investigated. The results are shown in Fig. 3.20. The particle size of zinc fume dust was nearly finer than 100 mesh. With the decrease of the particle size, the zinc leaching rate increased slightly. While the leaching rate of zinc carbonate and silicate zinc was significantly low when the particle size was coarser than 100 meshes, the zinc raw materials should be crushed to pass through 100 meshes sieve for a high leaching rate in the actual production.

(6) *Effects of stirring speed on zinc leaching*

During alkali leaching, the effects of the stirring speed on zinc leaching rate for zinc fume dust, zinc carbonate, and zinc silicate are shown in Fig. 3.21. The

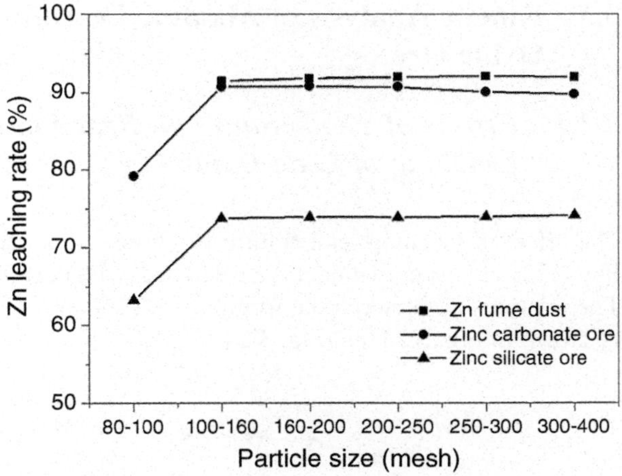

Fig. 3.20 Effects of particle size on zinc leaching rate in alkaline solutions

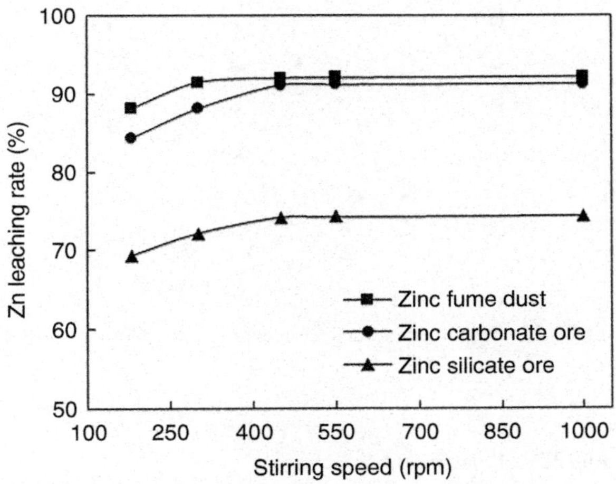

Fig. 3.21 Effects of stirring speed on zinc leaching rate in alkaline solutions

leaching rate increased with the increasing stirring speed which can make solid and liquid mixed adequately and improve the mass transfer efficiency so as to enhance the reaction rate. However, when stirring intensity makes the solid particles in a fully suspended state (solid particles stay for no more than 1 s at the bottom of the container), the stirring speed has little effect on increasing zinc leaching rate. Since zinc fume dust is a fine-particle size and low in density, the stirring speed is set to 300 r/min to maintain particles fully suspended, while the stirring speed for that of zinc carbonate and zinc silicate particles should be 450 r/min.

3.3 Kinetic Analysis of Alkaline Leaching of Lead Oxide Ore

3.3.1 *Effects of Temperature on Reaction Rate of Alkaline Leaching of Lead Oxide Ore*

The effect of leaching temperature and time on the lead leaching rate is shown in Fig. 3.22, and the core model is used to deal with the experimental data in the graph. The relationship between the formula $1-(2/3)\eta-(1-\eta)^{2/3}$ and the leaching time is shown in Fig. 3.23 and Fig. 3.24.

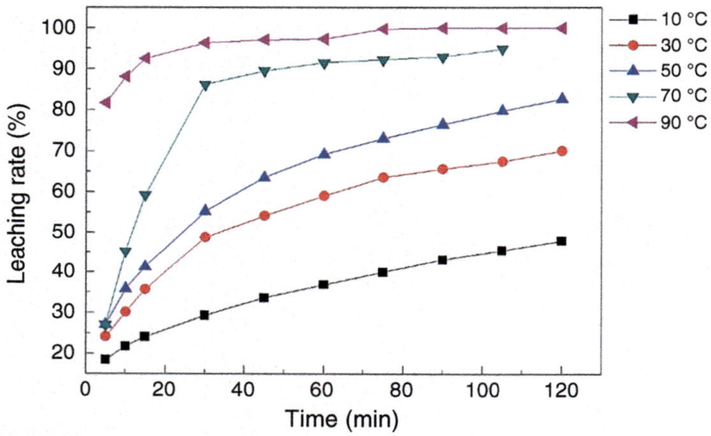

Fig. 3.22 Relationship between leaching rate and time for lead oxide ore

Fig. 3.23 Relationship between $1-(2/3)\eta-(1-\eta)^{2/3}$ and time for lead oxide ore in alkaline solutions

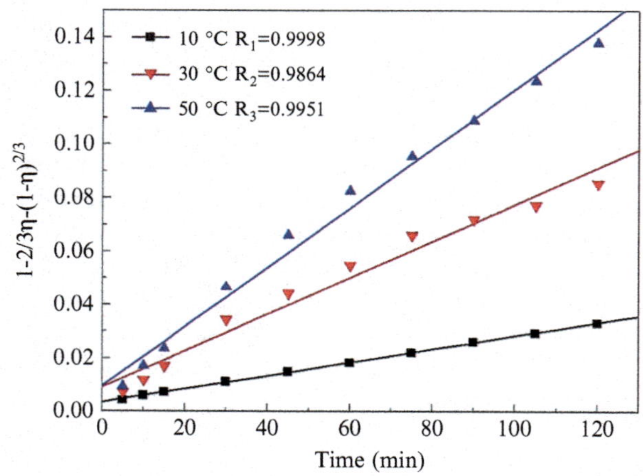

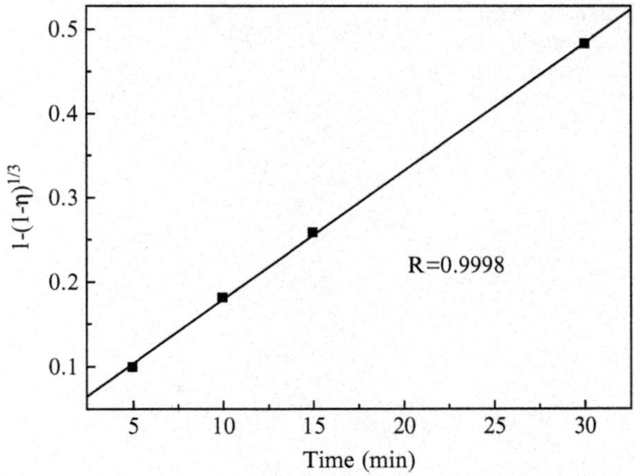

Fig. 3.24 Relationship between $1-(1-\eta)^{1/3}$ and time for lead oxide ore at 70 °C in alkaline solutions

As shown in Figs. 3.23 and 3.24, when the temperature is less than 50 °C, the formula $1-(2/3)\eta-(1-\eta)^{2/3}$ is linear with the leaching time. This shows that the leaching process of lead is a solid film diffusion control. At this time, the leaching rate of lead is low, and the diffusion of NaOH in the inert residue layer is the controlling step of the leaching reaction.

The leaching process of lead was in accord with the control law of surface reaction, and the leaching rate of lead was higher, when the temperature was 70 °C, $1-(1-\eta)^{1/3}$ and time t had a good linear relationship as lead oxide ore contained higher alkali-soluble components and can form larger pore channels in inert residue layer. As a result, the diffusion process did not affect the rate of leaching reaction, and the leaching process was controlled by surface reactions of lead phase and NaOH solution. When the temperature was 90 °C, the leaching rate of 5 min can reach more than 80%; at this time, the leaching reaction rate of lead is also great.

For the irreversible leaching reaction, the leaching rate under different temperature can be expressed as

$$\left(\frac{\mathrm{d}x}{\mathrm{d}t}\right)_{T_1} / \left(\frac{\mathrm{d}x}{\mathrm{d}t}\right)_{T_2} = k_{T_1}/k_{T_2} \tag{3.20}$$

According to the Arrhenius equation

$$k = A\exp\frac{-E_a}{RT} \tag{3.21}$$

where:

A – frequency factor, s^{-1}
E_a – the activation energy, $\mathrm{kJ \cdot mol^{-1}}$
R – the ideal gas constant, $\mathrm{J \cdot mol^{-1} \cdot K^{-1}}$
T – temperature, K

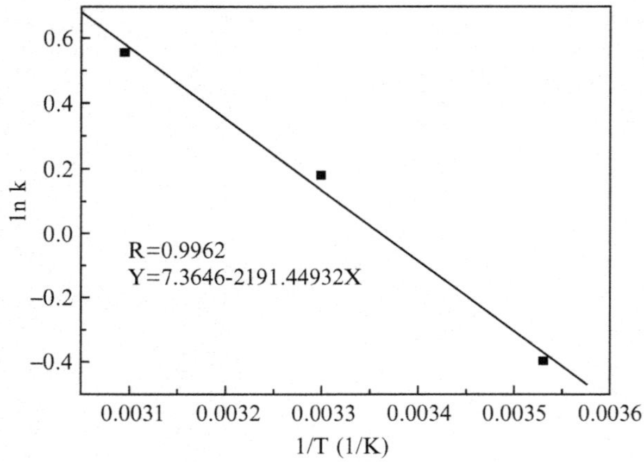

Fig. 3.25 Arrhenius diagram when the temperature was below 50 °C in alkaline solution

Using $\ln\frac{dx}{dt}$ to 1/T plotting can get a straight line in Fig. 3.25, and the slope of the straight line is $-E_a/R$. The apparent activation energy E_a derived from the linear equation is 18.22 kJ/mol. It confirmed that the leaching process of lead in lead oxide ore complies with the solid membrane diffusion process at temperature below 50 °C.

Similarly, apparent activation energy of lead leaching reaction E_a was 57.44 kJ/mol when the temperature was greater than 70 °C, meaning the lead leaching process was controlled by surface chemical reaction. Higher lead leaching rate indicated that lead-containing substance in the lead oxide ore was easier dissolved in NaOH solution, with a higher reaction speed.

3.3.2 *Effects of NaOH Concentration on Reaction Rate of Alkaline Leaching of Lead Oxide Ore*

The core model is used to deal with the experimental data in the Fig. 3.26, and the relationship between $1-(2/3)\eta-(1-\eta)^{2/3}$, $1-(1-\eta)^{1/3}$, and the leaching time is shown in Fig. 3.27 and Fig. 3.28. When the concentration of NaOH was 3 mol/L, the $1-(2/3)\eta-(1-\eta)^{2/3}$ was linear with the leaching time. It shows that the leaching process of lead is a solid film diffusion control. At this time, the leaching rate of lead is low, and the diffusion of NaOH in the inert residue layer is the controlling step of the leaching reaction. When the concentration of NaOH was 5 mol/L, the $1-(1-\eta)^{1/3}$ was linear with the leaching time. It shows that the leaching process of lead was controlled by surface reaction. At this time, the leaching rate of lead was higher, and the diffusion of NaOH in the inert residue layer was not the controlling step of the leaching reaction, but controlled by lead phase and NaOH solution surface reaction.

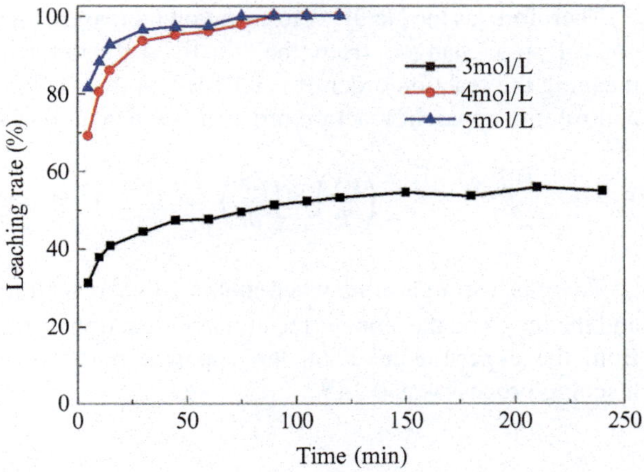

Fig. 3.26 Relationship between leaching rate and concentration of NaOH for lead oxide ore in alkaline solutions

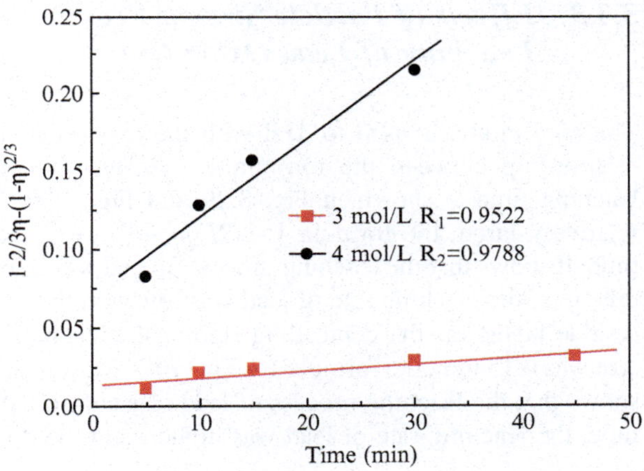

Fig. 3.27 Relationship between $1-(2/3)\eta-(1-\eta)^{2/3}$ and time for lead oxide ore at different concentrations of NaOH

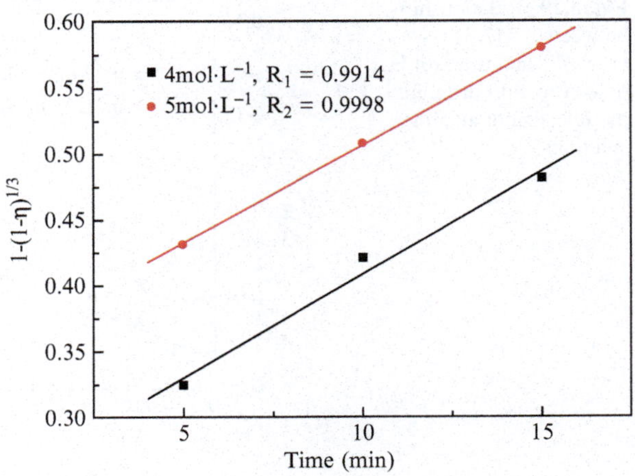

Fig. 3.28 Relationship between $1-(1-\eta)^{1/3}$ and time for lead oxide ore at different concentrations of NaOH

Therefore, as the NaOH concentration increased, the lead leaching process of the control step changed from the kinetic diffusion region to the kinetic region, meaning the reaction order of NaOH is less than 1. In the surface reaction kinetics control, the apparent reaction order of NaOH can be calculated by Eq. 3.22:

$$\left(\frac{dx}{dt}\right)_1 / \left(\frac{dx}{dt}\right)_2 = \left(\frac{c_{01}}{c_{02}}\right)^n = \text{constant} \tag{3.22}$$

There is a straight line, when making a chart with the relationship between $\ln \frac{dx}{dt}$ and the $\ln c_0$, and the slope is the apparent reaction order of the reactants. Calculated from the experimental data, the apparent reaction order of NaOH in the lead leaching process was 0.745.

3.3.3 *Effects of Particle Size on Reaction Rate of Alkaline Leaching of Lead Oxide Ore*

The core model is used to deal with the experimental data in Fig. 3.29, and the relationship between the formula $1-(2/3)\eta-(1-\eta)^{2/3}$, $1-(1-\eta)^{1/3}$, and the leaching time is shown in Fig. 3.29 and Fig. 3.30. When the particle size was relatively large, the formula $1-(2/3)\eta-(1-\eta)^{2/3}$ was linear with the leaching time. It shows that the leaching process of lead was a solid film diffusion control. At this time, the leaching rate of lead was low, and the diffusion of NaOH in the inert residue layer was the controlling step of the leaching reaction. When the particle size was 0.15 mm, the formula $1-(1-\eta)^{1/3}$ was linear with the leaching time. This shows that the leaching process of lead was controlled by surface reaction. At this time, the leaching rate of lead was higher, and the diffusion of NaOH in the inert

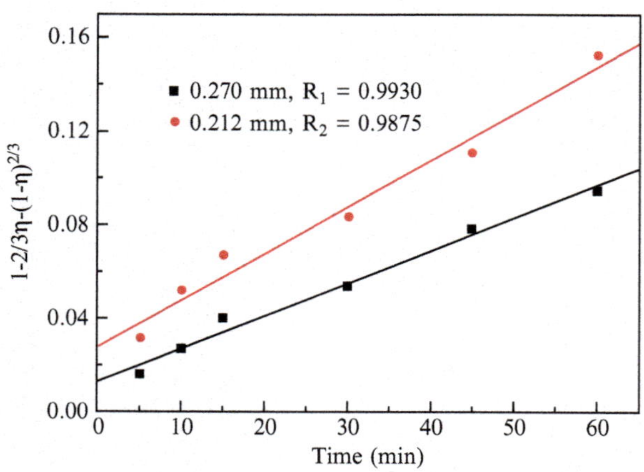

Fig. 3.29 Relationship between $1-(2/3)\eta-(1-\eta)^{2/3}$ and time for lead oxide ore with different particle size in alkaline solutions

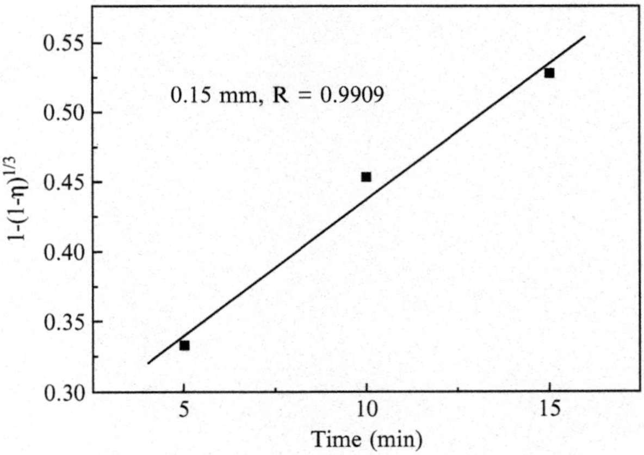

Fig. 3.30 Relationship between $1-(1-\eta)^{1/3}$ and time for lead oxide ore with particle size of 0.15 mm in alkaline solutions

residue layer was not the controlling step of the leaching reaction, but controlled by lead phase and NaOH solution surface reaction.

The reaction rate and the leaching rate of lead are interrelated. When the leaching rate is low, because of residue layer, the leaching rate of lead is controlled by the diffusion of NaOH solution in the solid residue layer. When the temperature and NaOH concentration are higher and the particle size is lower, the leaching rate of lead could reach a high level in a short time. At this time, because most of the lead phase dissolves in alkali solution, the residue layer diffusion process can no longer affect the rate of reaction, and part of lead phase and NaOH solution reaction activation energy is relatively high, making the lead leaching process controlled by surface chemical reaction.

Chapter 4
Leaching of Zinc and Lead Hazardous Wastes in Alkaline Solutions

Abstract ZnO, $ZnCO_3$, and Zn_2SiO_4 are the predominant mineralogical phases in solid wastes bearing zinc. ZnO, $ZnCO_3$, PbO, $PbCO_3$, and $PbSO_4$ can be easily dissolved directly by strong alkaline solutions. However, quite a big portion of zinc in the dust exists as zinc ferrites or zinc sulfide and lead as lead sulfide which cannot be dissolved completely in alkaline solution. Integrated hydrometallurgical processes for leaching the zinc and lead from zinc ferrites, zinc sulfide, and lead as lead sulfide in alkaline medium are developed. For zinc ferrites in EAF dust, the zinc is extracted by direct leaching-melting-leaching of melt or hydrolysis-melting-leaching processes. For the zinc sulfide in solid wastes, the leaching rate of zinc in alkaline solution is improved via chemical conversion with $PbCO_3$ so that the zinc in zinc sulfide can be extracted in NaOH solution with lead carbonate as additive, and the lead sulfide in the leach residues can be converted to lead carbonate by reacting with sodium carbonate solution using air as the oxidizing agent, and then the lead can be recycled in the whole process by dissolving the lead carbonate in NaOH solution. Over 90% of zinc can be extracted from the zinc sulfide, and over 95% of lead sulfide in leach residues can be converted. The Pb in leaded glass can be extracted in NaOH solution after mechanochemical reduction with metallic iron as additive. Over 90% of Pb can be extracted from the mechanochemical-reduced leaded glass, compared with less than 5% of Pb extraction for alkaline leaching of nonactivated samples, and 67% Zn extraction for alkaline leaching of activated samples for 2 h by planetary ball mill. Alkaline leaching process of smoke dust and lead oxide ore is conducted, which found that the leaching rate was over 90% after 30 min or longer. The resultant leaching residue contains lower than 1–2% of zinc, 0.5% of lead, 0.3% of copper, and 0.1% of cadmium, and over 35% of Fe, and may be classified as nonhazardous waste according to the results of leaching tests on the residues.

Keywords Alkaline leaching • Zinc ferrites • Zinc sulfides • Leaded glass • Chemical conversion • Mechanochemical reduction

Y. Zhao, C. Zhang, *Pollution Control and Resource Reuse for Alkaline Hydrometallurgy of Amphoteric Metal Hazardous Wastes*, Handbook of Environmental Engineering 18, DOI 10.1007/978-3-319-55158-6_4

4.1 Leaching of Zinc and Lead Dust from Steelmaking Plants with Lower Iron Contents in Alkaline Solutions

It is well known that ZnO, PbO, and Al_2O_3 can be easily dissolved in strong alkaline solution. However, quite a big portion of zinc in the dust exists as zinc ferrites or zinc sulfide and lead as lead sulfide which cannot be dissolved completely in alkaline solution. The dissolution of zinc carbonates and silicates is also quite poor.

5 g dust was digested in 30 mL of concentrated nitric acid, heated to nearly boiling point (90 °C) for 10–20 min, and then added water to dilute the acidity, continued to heat for 60 min. A small portion of the solid in the samples cannot be dissolved. The supernatant was analyzed by ICP-OES. The results are shown in Table 4.1. The particles of dust are very fine. No pretreatment was conducted in the study.

4.1.1 Effects of Leaching Time for Leaching of Zinc and Lead Dust in Alkaline Solutions

The results are shown in Fig. 4.1. It took about 72 h for the leaching process to reach an equilibrium of the dissolution of zinc and lead from dust. Nevertheless, when the dust were contacted for 42 h, the extraction of zinc and lead increased slightly. Therefore, a 42 h leaching time was used in the following experiments.

4.1.2 Effects of Liquid-Solid Ratio on Leaching of Zinc and Lead Dust in Alkaline Solutions

For zinc extraction and leaching, a liquid-solid ratio (v/w) of 3 seemed enough to reach a maximum leaching and extraction of zinc from dust under the conditions shown in Fig. 4.2. The extraction rate of zinc was found to be around 35%.

Table 4.1 Composition of the dust of test

Elements	Content (%) of first portion of dust sampled
Zn	24.80
Fe	32.00
Pb	1.84
Mn	3.31
Ca	4.08
Al	1.03
Cd	0.03
Cu	0.02
Water	0.01

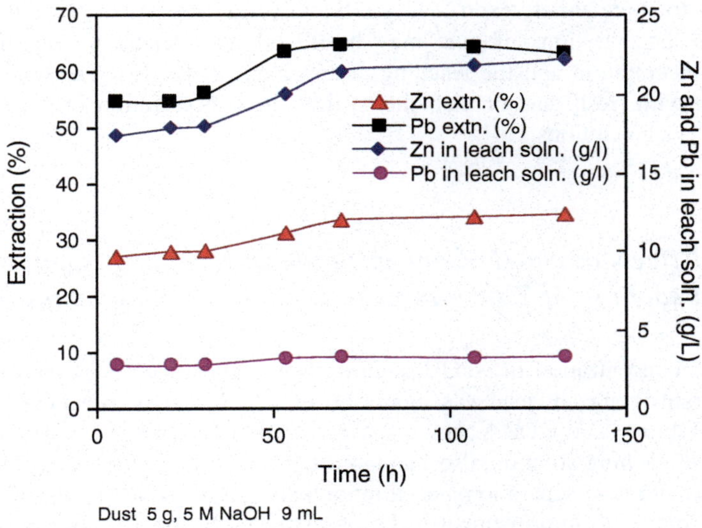

Fig. 4.1 Effects of leaching time on the extraction of Zn and Pb in alkaline solutions

Fig. 4.2 Relationship between the Zn extraction from dust and liquid-solid ratios (v/w) in alkaline solutions

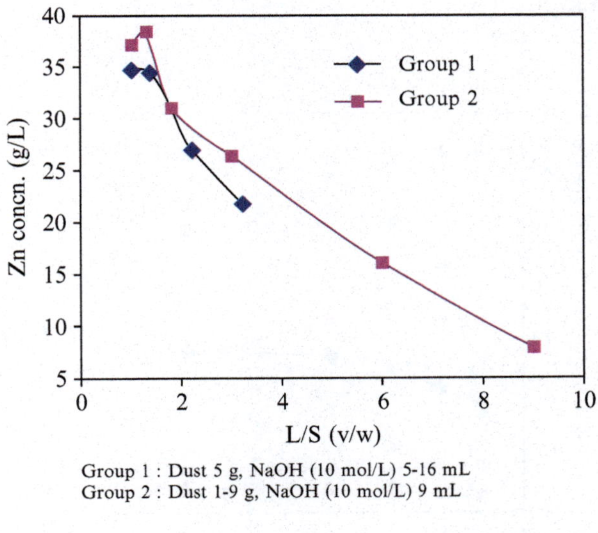

However, the extraction and leaching of lead increased as the liquid-solid ratio increased. When phase ratio was kept at 3, the extraction rate of lead in dust was found to be about 55–60%. Obviously, the higher the liquid-solid ratio was used, the lower the elemental concentrations would be obtained.

The maximum extraction can be obtained when the dust or their melts were leached for several times with fresh NaOH solution or unsaturated leach solutions, e.g., two to three times, even a lower liquid-solid ratio (v/w) were used. The saturated concentrations of zinc may be 40–60 g/L, depending on the initial NaOH concentration and the leaching arrangement. If the leaching solution was contacted with fresh dust or their melts, the zinc concentration increased in the resultant leach solutions.

4.1.3 *Effects of NaOH Concentration in Leaching Agent on Leaching of Zinc and Lead Dust in Alkaline Solutions*

When the concentration of NaOH solution was higher than 5 M, the maximum leaching rate of zinc and lead was obtained (Fig. 4.3). In most experiments, a 5 M of NaOH solution was used as leaching agent. Additionally, higher NaOH concentration in the leaching solution also increased the viscosity considerably, leading to more difficult phase separation and filtration between the solid and leach liquids.

There was little improvement to be observed when the leaching of dust was conducted in an environment of 50 °C, in comparison with that in ambient temperature. It was suggested that the leaching texts were carried out at room temperature (25 °C).

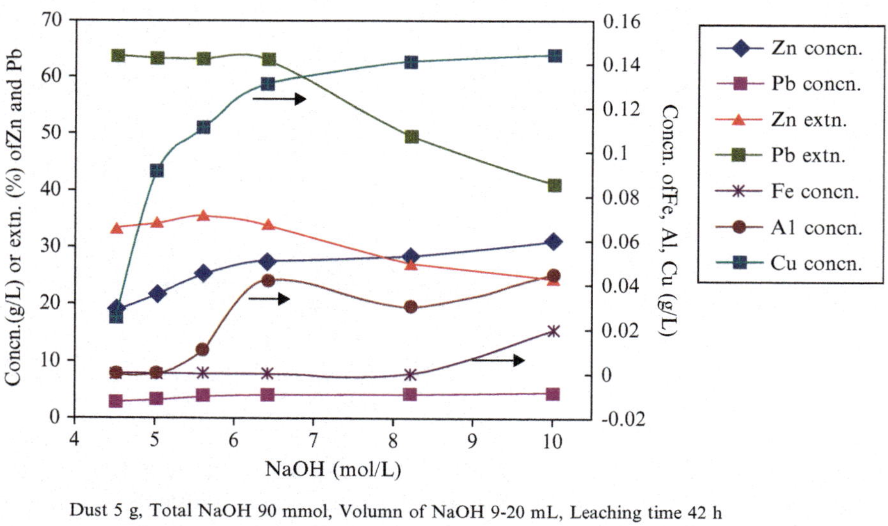

Fig. 4.3 Effects of NaOH concentration on the leaching of elements from dust

Table 4.2 Sequential extraction of zinc and lead from dust

Mixing sequences	Zn in leach solution (g/L)	Pb in leach solution (g/L)	Fe in leach solution (g/L)
Paralleling test I			
First extraction	20.13	2.80	0.10
Second extraction	9.82	0.82	0.09
Third extraction	3.85	0.15	0.08
Fourth extraction	0.49	0.00	0.08
Total extraction	53.67%	73.66%	0.13%
Paralleling test II			
First extraction	21.11	2.88	0.11
Second extraction	9.80	0.85	0.08
Third extraction	2.63	0.14	0.08
Fourth extraction	0.23	0.00	0.08
Total extraction	52.86%	75.72	0.13%
Paralleling test III			
First extraction	22.31	2.89	0.10
Second extraction	8.75	0.72	0.09
Third extraction	2.73	0.21	0.09
Fourth extraction	0.00	0.00	0.08
Total extraction	52.86%	74.74	0.13%
Average extraction	53.13%	74.71%	0.13%

Dust 50 g, NaOH (5 M) 180 mL, leaching time 42 h for each extraction

4.1.4 Sequential and Multistage Leaching of Dust in Alkaline

One-stage leaching rate of zinc was quite low (<40% for zinc). In order to improve the leaching efficiencies, multistage leaching process was conducted as follows. 50 g of dust were contacted with 180 mL of 5 M NaOH solution for 42 h, then filtered, the residue was contacted with another 180 mL of fresh NaOH solution for 42 h, and so on. The results are shown in Table 4.2. It can be seen that the maximum extraction rate for zinc was around 53%, and two-stage leaching was found to be enough to get such leaching rate. It may consider that the portion of leachable zinc in the dust was around 53%, and the remaining 47% of zinc may exist as zinc ferrites, which cannot be attacked further by alkaline solution.

When the leaching solution was contacted, the fresh dust multiply, and the elemental concentrations increased as shown in Table 4.3. It can be seen that the highest concentration of zinc in leaching solution was around 33 g/L and lead around 9 g/L.

Table 4.3 Multistage extraction of zinc and lead from dust

Extraction stages	Zn in leaching solution (g/L)	Pb in leaching solution (g/L)	Fe in leaching solution (g/L)
First stage	21.13	2.89	0.10
Second stage	25.58	6.40	0.11
Third stage	28.63	8.20	0.10
Fourth stage	33.08	9.01	0.12

For each extraction, dust 50 g, volume of filtrate 180 mL, leaching time 48 h

Table 4.4 Typical composition of the leaching residues after three-stage leaching

Elements	Zn	Fe	Pb	Mn	Ca	Al	Cu
Content (%)	12.15	39.24	0.32	3.31	3.97	0.01	0.09

Table 4.5 Effects of the addition of chemicals on the leaching of residues given in Table 4.2

No	Chemicals added (g)	Zn extraction (%)
1	0.32 Fe powder	0.0
2	1.16 Fe powder	0.0
3	2.48 Fe powder	0.0
4	2.00 NaCl	0.0
5	1.90 NaAc	0.0
6	2.67 NH_4Ac	0.0
7	2.70 Citric acid	0.0

Residue 2.5 g, 5 M NaOH 9 mL, leaching time 42 h

4.1.5 Tests on the Leaching Enhancement for Leaching Residues

After three-stage leaching of dust, the zinc content in the residues decreased to around 12% (Table 4.4). Fe powder, NaCl, NaAc, NH_4Ac, and citric acid were added to the leaching system of residues, and the results are shown in Table 4.5. It can be seen that the zinc in residue cannot be extracted further in the presence of the chemicals used.

About half of zinc in dust was possible to be leached and transferred into alkaline solution by contacting the dust with NaOH solution. The remaining zinc in leaching residue cannot be extracted further by the addition of Fe powder, NaCl, NaAc, NH_4Ac, and citric acid.

4.2 Extraction of Zinc from Dust by Direct Melting with Solid NaOH

Zinc in dust can be extracted quantitatively, if the dust are leached firstly by NaOH solution, followed by melting the leaching residues with solid NaOH at 318 °C for 1 h and then leaching the melts with NaOH solution. It was named such a process as

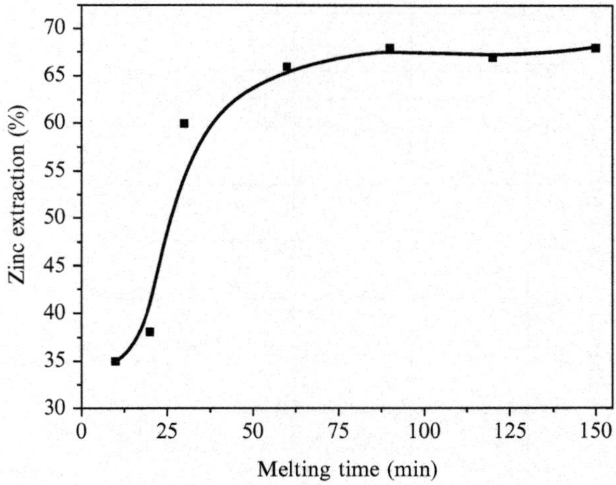

Fig 4.4 Effects of melting time on the extraction of zinc from the dust melted at 318 °C in NaOH

two-step process, leaching and melting-leaching processes. It was thought that the two-step process may merge into one-step process, melting the dust directly, followed by the leaching of resulting melts.

4.2.1 Effects of Melting Time for Extraction of Zinc from Dust by Direct Melting with Solid NaOH

The results are shown in Fig. 4.4. One hour of melting seemed to be enough to reach the maximum leaching and extraction from dust (around 68% extraction of zinc).

4.2.2 Effects of Leaching Time on Extraction of Zinc from Dust by Direct Melting with Solid NaOH

At least 24 h was needed for the melt to reach the maximum extraction rates (Fig. 4.5). In general, a leaching time of 42 h was used in the following experiments. Extraction rates were found to be little affected by increasing the leaching temperature. Also, it should be pointed out that the extraction of lead decreased to around 30% when the dust were melted with NaOH, in comparison with about 65% when the dust were leached directly with NaOH solution.

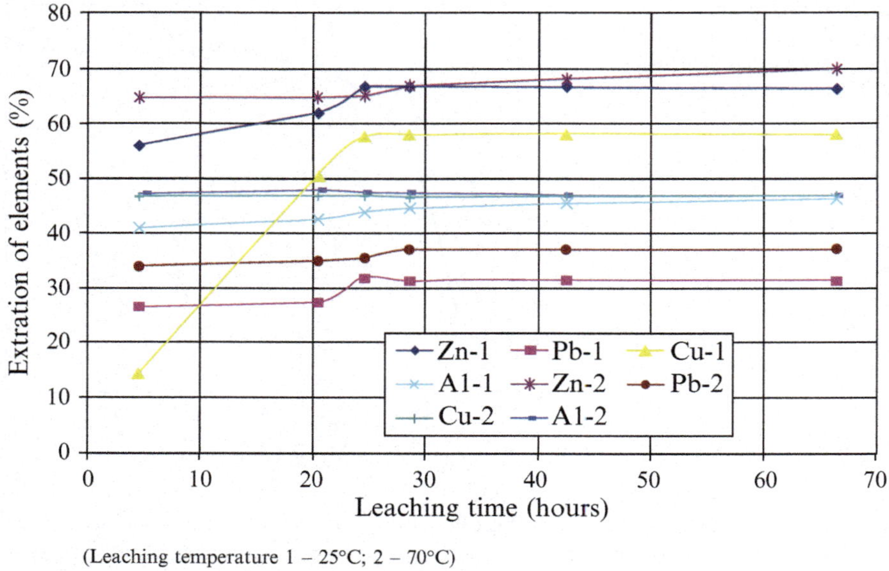

Fig. 4.5 Effects of leaching time on the extraction of elements from the dust melted by 5 M NaOH (37 mL)

Table 4.6 Relationship between the mass ratios of dust to solid NaOH in the melts and the leaching efficiencies

No.	Dust/NaOH mass ratios	Zinc extraction (%)
1	0.00	38.19
2	0.38	43.98
3	1.14	65.89
4	1.27	67.88
5	1.67	70.98

Dust 1.5 g, leaching time 42 h, melting time 1 h

4.2.3 Effects of Mass Ratios of Dust to Solid NaOH in the melts on leaching

The extraction of zinc from melts of dust increased as the mass ratios of NaOH to dust increased (Table 4.6). A mass ratio of 1.27 was used in this experiment. High mass ratios seemed unreasonable in industry.

4.2.4 Effects of NaOH Concentration on Leaching

ZnO can only be dissolved in strong alkaline solution. From Fig. 4.6 it can be seen that the leaching efficiencies of zinc from the molten product increased as NaOH

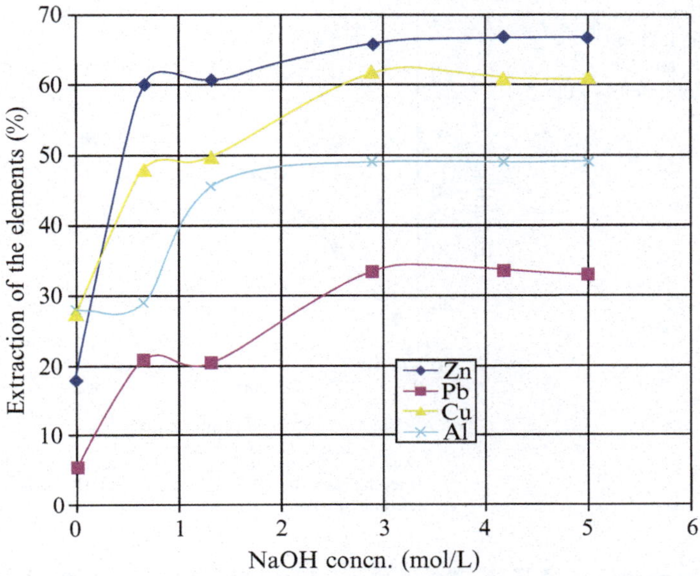

Fig. 4.6 Effects of NaOH concentration on the extraction of the elements from the dust melted at 318 °C

concentrations in leaching agent increased. At least of 1 M NaOH solution should be used as leaching agent. The leaching efficiencies of zinc were increased to around 65%, while those of lead decreased to around 30%, when the dust were melted with solid NaOH at the melting point of NaOH (318 °C) for 1 h, in comparison with the results that around 38% of zinc and 65% of lead can be extracted when the dust were leached directly with NaOH solution.

4.2.5 Effects of Temperature on Extraction of Zinc from Dust by Direct Melting with Solid NaOH

The relationship between the melting temperature and the extraction of zinc for dust are shown in Fig. 4.7. It can be seen that the extraction increased as the melting temperature increased. From the viewpoint of energy-saving and cost reduction, the melting temperature should be selected at or slightly above the melting point of 318 °C. In the following experiments, a melting temperature of 318 °C was used. It can be seen from Fig. 4.7 that at 318 °C, around 65% of zinc can be extracted from the dust, which was higher than that obtained by direct contact of dust with NaOH solution. But a large fraction of zinc was still left in the leaching residues.

Fig. 4.7 Relationship between the melting temperature and the extraction of zinc from the dust melted

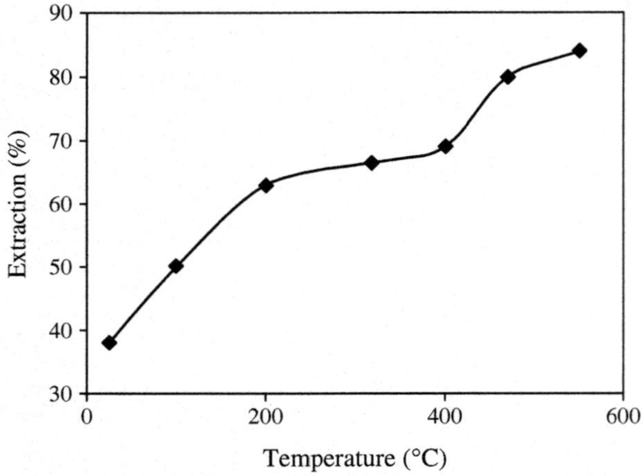

Dust 1.5 g, NaOH/Dust 1.27, melting time 1 hour, leaching agent for melts 37 mL 5 M NaOH solution, leaching time 42 hours

4.3 Extraction of Zinc from Leaching Residues by Melting the Residues with Solid NaOH

4.3.1 Effects of Melting Time on the Extraction of Zinc from the Leaching Residues

Both leaching time and melting time had significant effects on the extraction of metals from the leaching residues. A longer melting time and leaching time was found to be favorable to the release and extraction of zinc from the residues, as shown in Fig. 4.8. Obviously, over 40 min of melting time and 42 h of leaching time should be used in order to get the maximum extraction of zinc from the residues.

4.3.2 Effects of Melting Temperature on the Extraction of Zinc from Leaching Residues by Melting the Residues with Solid NaOH

As shown in Table 4.7, the zinc in the zinc ferrites can be released quantitatively when the residues were melted with solid NaOH at or above the melting point of NaOH (318 °C).

Fig. 4.8 Effects of melting time on the extraction of zinc from residues by melting at 318 °C

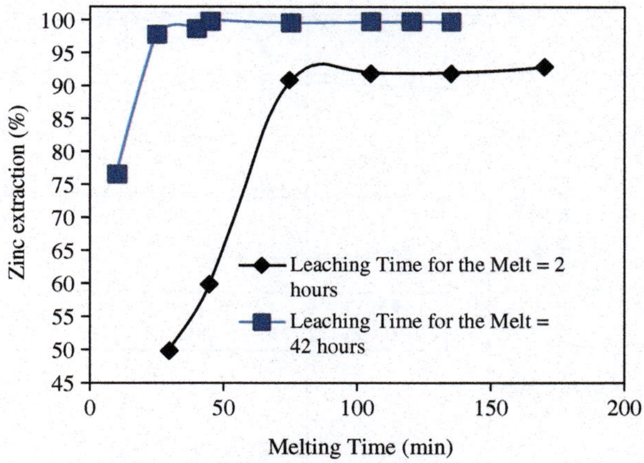

Residues 1.5 g, NaOH/Residues 0.91

Table 4.7 Effects of melting temperature on the extraction of zinc from leaching residue

Melting temperature (°C)	200	318	400	450	550
Zn extraction (%)	80.09	99.98	99.99	100	100

Leaching agent 20 mL 5 M NaOH solution, residues 1.5 g, leaching time 42 h

4.3.3 Effects of Leaching Time on the Extraction of Zinc from Leaching Residues Melted with Solid NaOH

Around 30 h of leaching time was found to be enough to get the quantitative extraction and release of zinc from the molten products of the residues (Fig. 4.9).

4.3.4 Effects of NaOH Concentration on the Extraction of Zinc from Leaching Residues Melted with Solid NaOH

The zinc can be released from the melts of the residues completely when an over 5 M NaOH solution was used as leaching agent under the conditions shown in Fig. 4.10.

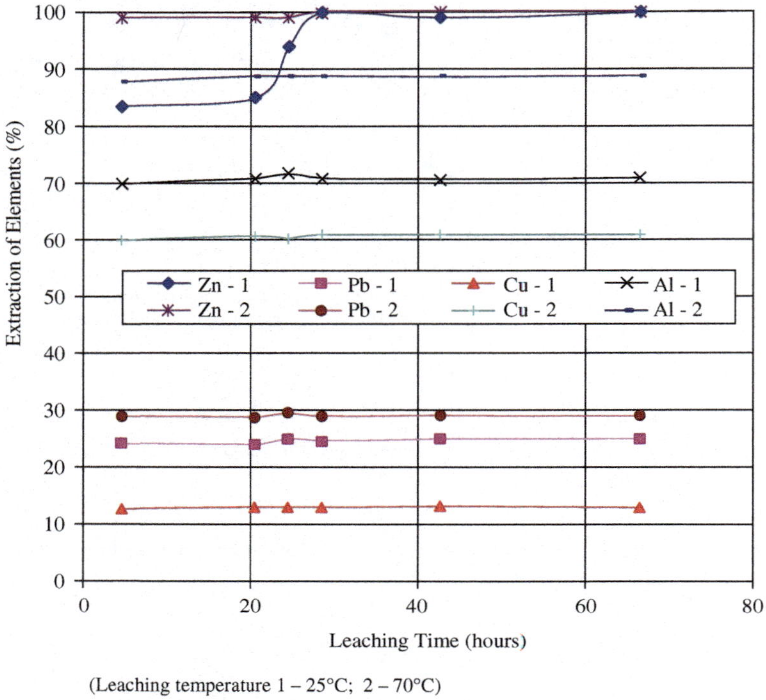

Fig. 4.9 Effects of leaching time on the extraction of zinc from the dust by melting at 318 °C

Fig. 4.10 Effects of NaOH concentration on the extraction of the elements from the residues melted with solid NaOH

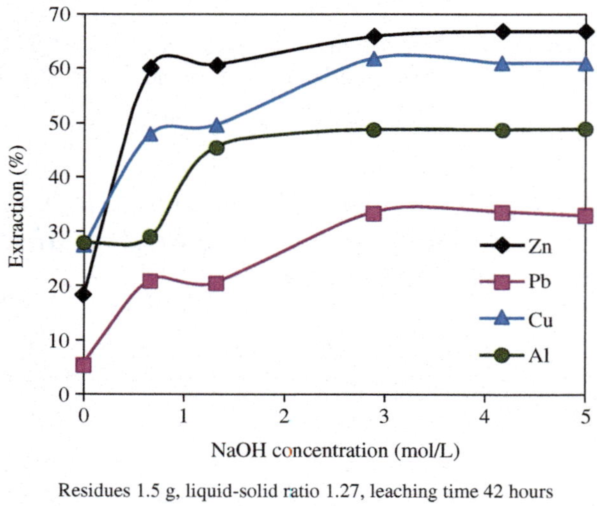

4.3.5 *Melting with Sodium Phosphate Instead of Sodium Hydroxide*

It was attempted to lower the melting temperature of melts. Thus, sodium phosphate was selected. It was found that the leaching and extraction of zinc was considerably lower than that using NaOH as melting agents. It was also found that zinc phosphate can be easily dissolved in 5 M NaOH solution. Therefore, the poor effects of sodium phosphate on the melting and leaching of leaching residues of dust may be contributed to its incapability of attacking the zinc ferrites.

4.3.6 *Composition of Leaching Solution of the Melts and the Resultant Zn-Free Residues*

The concentrations of elements in leaching solutions are dependent on the liquid-solid ratio used in the leaching processes. Table 4.8 gave the results under the conditions used.

The composition of resultant residues obtained from Table 4.8 was determined by digested the samples in nitric acid and the results are shown in Table 4.9. It can be seen that the content of iron was as high as around 37% (53% as Fe_2O_3), which may be used as the feed for steelmaking plants.

The zinc in the leaching residues of dust can be leached and extracted quantitatively when the residues were melted with NaOH solid at 318 °C for 1 h and a mass ratio of residue/NaOH over 0.75 in the melt. The solid wastes thus obtained contained a very low content of hazardous and toxic metals and may be classified as nontoxic and nonhazardous wastes. It can be used as the filling materials for sea-filling or the feed materials for steelmaking.

Table 4.8 Typical composition of the leaching solution of the molten product of leaching residue

Element	Liquid-solid ratio	Zn	Pb	Fe	Cu	Al	Mn	Cd
Concentration (g/L)	13.6	8.64	0.12	0.008	0.40	0.44	0.00	0.00
Extraction (%)		99.98						
Concentration (g/L)	3.6	32.19	0.10	0.02	0.12	0.78	0.00	0.00
Extraction (%)		99.23						

[a]Liquid-solid ratio: NaOH solution/residues, v/w

Table 4.9 Typical composition of the resultant residues after leaching and extraction from the melts of the leaching residue of dust

Element	Zn	Pb	Fe	Cu	Al	Mn	Cd	Nitric acid insoluble
Contents (%)	0.00–0.10	0.11–0.20	35.39–39.90	0.15–0.30	0.10–0.20	1.25–2.00	0.055–0.090	5.00–10.19

Liquid-solid ratio 13.6, the residues digested by nitric acid solution

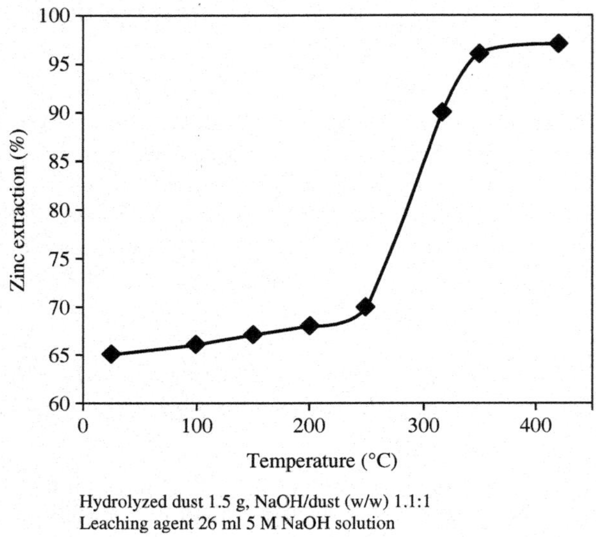

Fig. 4.11 Effects of melting temperature on the extraction from the melts of hydrolyzed dust

4.4 Extraction of Zinc from Dust via Hydrolysis-Melting-Leaching Process

It has been proven that the zinc in dust cannot be extracted effectively, when dust are melted directly with solid NaOH and then leached with NaOH solution. However, it can be seen that the zinc can be extracted quantitatively by two-step process, i.e., direct leaching of dust followed by melting-leaching of the resultant leaching residues. Obviously, in the two-step processes, chemical and physical properties of dust may change when it is leached directly with NaOH solution, which facilitates the destruction of zinc ferrites structures when the leaching residues are melted with NaOH pellets above melting point of NaOH, leading to a more complete release of zinc from dust. Based on the above considerations, it was decided to "wash" or "hydrolyze" the dust and then melt the hydrolyzed dust. The experimental results show this method does work. All the original dust should be hydrolyzed for at least 4 h with water/dust mass ratio of 1:1 followed by the washing of cakes by another half amount of freshwater before melting with solid NaOH.

4.4.1 Effects of Melting Temperature on Extraction of Zinc from Dust via Hydrolysis-Melting-Leaching Process

From Fig. 4.11, the extraction efficiencies of zinc increased slowly from ambient temperature to 250 °C but increased rapidly from 250 to 350 °C and then kept

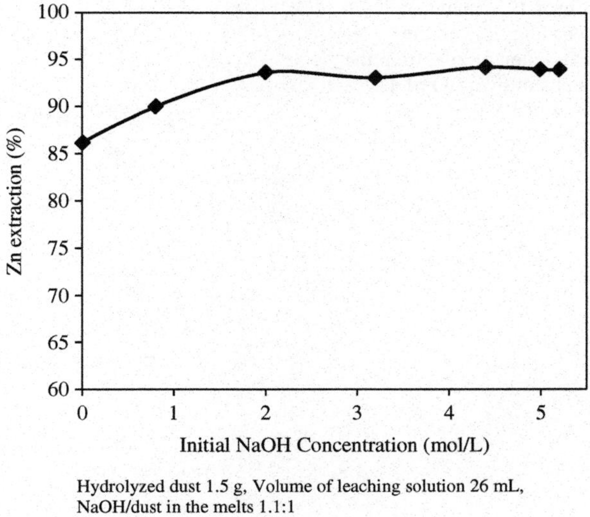

Fig. 4.12 Effects of NaOH concentration in leaching agent on the extraction of the melts of hydrolyzed dust by melting at 350 °C

nearly unchanged after 350 °C. Therefore, a melting temperature of 350 °C was used in the following experiments.

The zinc cannot be extracted quantitatively, even melted the hydrolyzed dust at 350 °C, in comparison with the complete release of zinc from dust in the two-step process using a lower melting temperature of 318 °C for leaching residues.

4.4.2 Effects of NaOH Concentration in Leaching Agent on the Extraction of Zinc from Dust via Hydrolysis-Melting-Leaching Process

Similar to the previous experimental results, 4 M NaOH solution should be used as leaching agent for the maximum extraction of zinc from the melts of hydrolyzed dust (Fig. 4.12). In the experiment, 5 M NaOH was used as leaching solutions as used previously.

4.4.3 Effects of Leaching Time on Extraction of Zinc from Dust via Hydrolysis-Melting-Leaching Process

Obviously, a longer leaching time was favorable to the extraction of zinc from the melts of hydrolyzed dust (Fig. 4.13). In general, a 42 h leaching time was used in the experiments, unless indicated otherwise.

Fig. 4.13 Effects of leaching time on the extraction of zinc from the melts of hydrolyzed dust obtained at 350 °C

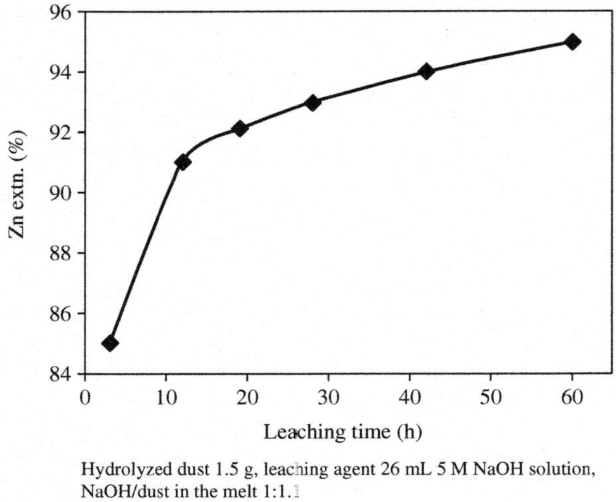

Fig. 4.14 Effects of NaOH/dust mass ratios in the melts on the extraction of zinc from the melts of hydrolyzed dust melted at 350 °C

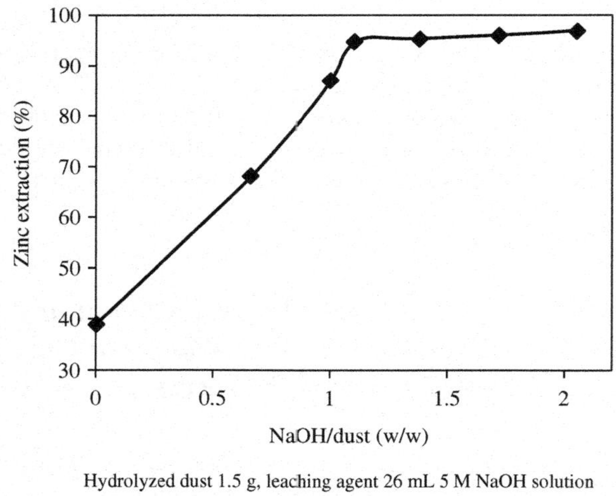

4.4.4 Effects of NaOH/Dust Mass Ratios in the Melt on the Extraction of Zinc from Dust via Hydrolysis-Melting-Alkaline Leaching Process

A NaOH/dust mass ratio of at least 1.1:1 should be used in order to achieve the maximum extraction from the dust (Fig. 4.14). The mass ratio of 1.1:1 was used. So far, the fundamental conditions for an acceptably high extraction level of zinc from hydrolyzed dust has been obtained, i.e., melting at 350 °C, leaching the melt with 5 M NaOH solution for 42 h. As a result, around 95% Zn and over 80% Pb can be extracted from the dust.

4.4.5 *Effects of Water-Dust Ratio and Hydrolysis Time in the Hydrolysis Step via Hydrolysis-Melting-Alkaline Leaching Process*

The results are shown in Tables 4.10 and 4.11. The water-dust mass ratio of 1:1 and the hydrolysis time of at least 4 h should be utilized, in order to obtain a maximum extraction of zinc. In addition, it was found that the cake must be washed by freshwater after the hydrolyzed dust was filtered (Table 4.12), in order to get a higher zinc extraction from the melts.

4.4.6 *Composition of the Supernatant in the Hydrolysis of Dust in the Hydrolysis Step via Hydrolysis-Melting-Alkaline Leaching Process*

Dilute NaOH solution, nitric acid solution, and water were used as hydrolysis agents for the dust. The composition of the supernatant (filtrates) was determined after hydrolysis processes, and results were given in Tables 4.13 and 4.14. It indicated that the concentrations of Pb and Zn were quite high, especially when NaOH or nitric acid solutions were used (Table 4.13). When water was used as

Table 4.10 Effects of water-dust ratio (v/w) on the melting and leaching of hydrolyzed dust

No.	1	2	3	4	5
Water/dust ratios (v/w)	0.0	0.5	1.0	2.0	3.3
Zn extraction from the melts of hydrolyzed dust (%)	68.91	95.19	96.02	95.62	94.56

Dust 1.5 g, leaching solution 26 mL, washing time 24 h, leaching time 42 h

Table 4.11 Effects of hydrolysis time on the melting and leaching of hydrolyzed dust

No.	1	2	3	4	5	6	7	8
Hydrolysis time for the dust (h)	0	2	3	4	9	23	28	48
Zn extraction from the melt of hydrolyzed (%)	66.98	85.79	92.10	96.01	93.09	95.23	94.33	94.89

Water/dust (v/w) 1, dust 1.5 g, leaching solution 26 mL, leaching time 42 h

Table 4.12 Comparison of washing or non-washing of the hydrolyzed cake of No. 4 in Table 4.11

Treatment ways for the cake	Washing with 3 mL freshwater	Non-washing
Zn extraction from the melt of hydrolyzed dust (%)	95.01	83.98

Table 4.13 Composition of the aqueous solution after contacting dust with various hydrolysis agents for 42 h

Hydrolysis agents	0.5 M NaOH	0.2 M nitric acid	water
Volume of hydrolysis agents (mL)	500	500	500
Weight of the dust (g)	150	150	150
Filtrate obtained (mL)	444	444	444
Composition of the filtrates (mg/L)			
Zn	58.3	1359.0	1.8
Pb	1000.0	8.41	197.0
Fe	0.06	0.06	0.2
Cu	3.6	0.05	0.2
Al	2.1	1.3	1.4
Mn	0.0	65.7	0.0
Cd	0.0	55.8	0.0

Table 4.14 Composition of the aqueous solution after contacting dust with water for 4 h

Purposes of water used	Hydrolysis	Washing of hydrolyzed cake
Volume of water used (mL)	400	200
Weight of the dust (g)	400	400
Filtrate obtained (mL)	250	200
Color of the filtrates	Yellow	Nearly no color
pH of the filtrate	10.54	10.79
Composition (mg/L)		
Zn	1.7	1.0
Pb	1.0	0.9
Fe	0.2	0.3
Cu	0.2	0.1
Al	1.4	0.4
Mn	0.0	0.0
Cd	0.0	0.0

hydrolysis agent, the concentrations of elements in the resultant supernatant in both hydrolysis step and washing step of the hydrolyzed cake were found to be much lower (Table 4.14).

4.4.7 Recycling of the Filtrate for the Hydrolysis of Dust in the Hydrolysis Step via Hydrolysis-Melting-Alkaline Leaching Process

The filtrate (first filtrate) obtained from the first hydrolysis step was used as the solvent for the hydrolysis of the next fresh dust; filtered, the resultant filtrate (second filtrate) was used again for the hydrolysis of the third fresh dust, and so

Table 4.15 Effects of the recycling of hydrolysis filtrate on the extraction of zinc from the melts of hydrolyzed dust

Hydrolysis agent	Hydrolysis time resulting to the filtrate	Zinc extraction from the melts of hydrolyzed dust(%)
Water		95.78
First filtrate	42	75.09
Second filtrate	42	74.26
Third filtrate	42	70.12
First filtrate	4	94.78
Second filtrate	4	93.45
Third filtrate	4	91.09

Filtrate/dust (v/w): 1:1

forth, in order to observe the effects of such recycling on the extraction of zinc from the melts of the corresponding hydrolyzed dust. The results are shown in Table 4.15. It can be seen that the effects of recycling the filtrates were dependent on the hydrolysis time from which the filtrate was obtained. Minor effect was observed for the recycling of filtrate resulting from a 4 h hydrolysis process. Nevertheless, the extraction efficiency of zinc from the dust was decreased to a great extent for recycling of filtrate resulting from a 42 h hydrolysis process.

4.4.8 *Effects of the Addition on the Melting and Extraction of Zinc from Dust via Hydrolysis-Melting-Leaching Process*

NaF, Na_2HPO_4, and ZnO were added to hydrolyzed dust, melted, and then leached. The results are shown in Table 4.16. The addition of NaF and ZnO had little effect on the extraction of zinc, as it was thought that the silica and ZnO in the dust might be one of reasons to be responsible for relatively low extraction efficiencies of zinc by direct melting of dust (without previous hydrolysis). In addition, NaOH cannot be substituted by the chemical with lower melting point, Na_2HPO_4.

4.4.9 *Effects of Liquid-Solid Ratio on the Extraction of Zinc from Melted Dust via Hydrolysis-Melting-Leaching Process*

The leaching experiments were carried out. The melted dust were obtained by 1.5 g hydrolyzed dust melting with 2.0 g solid NaOH at the temperature of 350 °C for 1 h.

Table 4.16 Effects of the addition on the melting and leaching of hydrolyzed dust

Chemicals	NaOH (g)	Melting temperature (°C)	Melting time (h)	Zn extraction (%)
1.0 g NaF	1.0	350	1	72.71
2.0 g NaF	2.0	350	1	97.10
0.5 g Na_2HPO_4	0	110	2	41.84
1.0 g Na_2HPO_4	0	110	2	46.40
1.5 g Na_2HPO_4	0	110	2	46.98
2.0 g Na_2HPO_4	0	110	2	46.40
2.0 g Na_2HPO_4	0.8	110	2	46.32
2.0 g Na_2HPO_4	1.4	110	2	56.32
27.7 mg ZnO	2.0	350	1	96.77
73.7 mg ZnO	2.0	350	1	97.10
120.2 mg ZnO	2.0	350	1	97.99

Hydrolyzed dust 1.5 g, leaching agent 26 mL 5 M NaOH solution

Table 4.17 Effects of liquid-solid ratio in the leaching process on leaching melted dust

No.	1	2	3	4
Volume of the leaching solution (mL)	7.5	13.5	21.0	25
Liquid/solid (v/w)	5	9	14	17
Zn extraction (%)	83.15	89.66	92.11	93.45
Pb extraction (%)	82.50	89.00	80.89	86.11
Composition of the leaching solutions (g/L)				
Zn	37.18	22.91	14.67	12.00
Pb	2.97	1.78	1.04	0.93
Fe	0.050	0.034	0.030	0.030
Cu	0.56	0.26	0.17	0.12
Al	1.25	1.22	0.75	0.70
Mn	0.00	0.00	0.00	0.00
Cd	0.00	0.00	0.00	0.00
Volume of the supernatant after centrifugation (mL)	6.0	12.0	19.5	23.5

In the leaching process, the 5 M NaOH solution was used, and the liquid-solid ratio was calculated by the volume of leaching agent and the hydrolyzed dust. And the leaching maintained for 42 h with weak stirring. The total extraction efficiency of zinc from melts was found to be dependent on the liquid-solid ratio in the leaching process (Table 4.17). Moreover, there existed a nearly linear relationship between the zinc concentration in leach solution and liquid-solid ratio. The lower liquid-solid ratio was used. the higher concentration of zinc in the leaching solution was obtained.

In order to find out the saturation concentration of zinc in the leaching solution, each leach solution in Table 4.17 was contacted with the fresh melts according to the abovementioned method of Table 4.17, respectively. The results are shown in Table 4.18. The highest concentration of zinc in the leaching solutions seemed to be

Table 4.18 Leaching of the melts of hydrolyzed dust using the corresponding supernatant (leaching solutions of the same No.) shown in Table 4.17

No. (same as in Table 4.17)	1	2	3	4
Supernatant added to the melt for leaching of the elements (mL)	5.0	11.0	18.5	23.0
5 M NaOH solution added (mL)	2.5	2.5	2.5	2.5
Liquid/solid (v/w)	5	9	14	17
Zn extraction from the fresh melts (%)	30.02	78.22	95.89	94.29
Pb extraction from the fresh melts (%)	30.55	66.48	80.01	86.90
Composition of the leaching solutions (g/L)				
Zn	39.80	39.79	29.63	24.32
Pb	3.08	2.78	1.94	1.76
Fe	0.053	0.055	0.035	0.036
Cu	0.53	0.42	0.31	0.24
Al	1.30	1.23	0.89	0.99
Mn	0.00	0.00	0.00	0.00
Cd	0.00	0.00	0.00	0.00
Volume of the supernatant after centrifugation (mL)	6.0	11.0	18.5	23.0

Table 4.19 Leaching rates of the corresponding leaching residues obtained at Table 4.18 using 5 mL 5 M NaOH solution

No.	1	2	3	4
Total Zn extraction for the same melt after two contacts with 26 and 5 mL of NaOH solution (%), respectively	96.10	96.95	96.45	95.80
Composition of the second leaching solution				
Zn	9.56	5.01	3.09	2.99
Pb	0.56	0.47	0.31	0.35
Fe	0.015	0.007	0.007	0.008
Cu	0.095	0.040	0.030	0.029
Al	0.27	0.13	0.07	0.07
Mn	0.00	0.00	0.00	0.00
Cd	0.00	0.00	0.00	0.00
Volume of the supernatant after centrifugation (mL)	5.0	5.0	5.0	5.0

around 40 g/L under the conditions of tests. From Tables 4.17 and 4.18, the extraction of zinc was over 95% if the zinc concentration in final leach solution was lower than the saturation level.

It was obvious that the zinc remaining in the leaching residues shown in Table 4.18 may be further extracted by contacting the residues with fresh NaOH solution. The experimental results were given in Table 4.19. It can be seen that the highest zinc extraction was about 96%. The experimental results implied that the melts can be leached with relatively low NaOH-melted dust ratios, and the zinc can be extracted quite completely by multi-leaching processes. As a result, the quantity

Table 4.20 Miscellaneous tests for the extraction of melts of hydrolyzed dust

No.	Experimental conditions	Remarks	Zn extraction (%)
1	The melt was inserted into 26 mL 5 M NaOH solution, and analyzed the leaching solution after all the melts were transferred into the solution (2 h)	Around half of the melt could be transferred into the solution when the melts were quenched	70.70
2	The melts were transferred into NaOH solution, stirred with magnetic stirrer for 2 h	With and without heating (25 and 85 °C, respectively)	87.85 at 85 °C 84.52 at 25 °C
3	Both original and hydrolyzed dust was leached in 26 mL of 5 M NaOH solution in the environment of 121 °C steam for 1 h	Stirring unavailable	37.86 for original dust 36.51 for hydrolyzed dust

of NaOH solution required will be the lowest, in comparison with the leaching process in which a high liquid-solid ratio was used and all the leachable zinc was extracted in the primary leaching. For convenient analysis, in most experiments in this work, the liquid-solid ratios were usually kept over 25 (mL)/1.5 g, in order to get the possible highest zinc extraction in the primary leaching.

The wetting of melts was found to be very slow. In order to accelerate the wetting and dispersing of melts in NaOH solution, the hot melt was inserted into NaOH solution after it was taken out of the furnace. The melt cannot be dispersed completely in a quench, and remaining melt should be scraped further. After all the melt was scraped into NaOH solution which took about 2 h, the supernatant was analyzed. It was found that only 70.70% of zinc in the dust was extracted.

In another test, the melt was transferred into NaOH solution by hand, stirred with magnetic stirrer for 2 h with and without heating. It was found that around 87% and 84% of zinc in the dust can be extracted at temperature of 85 and 25 °C, respectively.

Finally, the original and hydrolyzed dust were leached at 121 °C in a pressurized steam tank. The extraction of zinc was found to be very low. The experimental results are summarized in Table 4.20.

4.4.10 Relationships Between Zinc Extractability and the Zinc and Iron Contents in the Dust by Direct Leaching Process

Simple calculations will show that the relationship between the zinc extraction rates and zinc or iron contents are linear, while those of zinc extraction rates and Fe/Zn

Table 4.21 Relationships among the zinc extraction, zinc, and iron contents, by direct alkaline leaching using 5 M NaOH solution

No.	1	2	3
Composition of the EAF dust (%)	Zn 21.2 Fe 21.8	Zn 24.8 Fe 33.0	Zn 16.8 Fe 39.9
Fe/Zn weight ratios in the dust	1.07	1.33	2.38
Zn extraction by direct alkaline leaching (%)	80	53	33

ratios seem more complex, but surely, the zinc extractions always decrease as the Fe/Zn weight ratios increase. It may be confirmed, for steelmaking dust, that the content of iron will affect the zinc extraction considerably from dust by direct alkaline leaching.

Obviously, the zinc in zinc ferrites cannot be leached directly by alkaline solution. Three dust with different zinc and iron contents were tested (Table 4.21).

4.4.11 Chemical Reactions in the Melting and Alkaline Leaching Processes

The chemical reactions are proposed according to the following data based on 1000 kg of dust treated: 53% of zinc exists as ZnO and 47% of zinc as $Zn(FeO_2)_2$; 95% of zinc in the dust can be extracted in which all the ZnO is supposedly extracted completely, i.e., 132.5 kg Zn as ZnO can be leached and reacted with NaOH; the Zn as zinc ferrites that is leached and reacted with NaOH is

$$250 \times 0.95 - 132.5 = 105\,\text{kg Zn} = 1607.5\ \text{mol Zn ferrites}\left(\text{Zn}(\text{FeO}_2)_2\right)$$

The consumption of NaOH in the melting and leaching processes is 163 kg (4075 mol). The molar ratio of NaOH reacted to the zinc ferrites extracted is

$$4075/1607.5 = 2.53$$

Therefore, the following reactions may be proposed in the melting and leaching processes:

Leaching of ZnO:

$$\text{ZnO(s, in dust)} + 2\text{NaOH(aq)} + \text{H}_2\text{O(l)} = \text{Na}_2\text{Zn(OH)}_4\text{(aq)} \tag{4.1}$$

Leaching of PbO:

$$\text{PbO(s, in dust)} + 2\text{NaOH(aq)} + \text{H}_2\text{O(l)} = \text{Na}_2\text{Pb(OH)}_4\text{(aq)} \tag{4.2}$$

This reaction will be taken place at ambient temperature, and the NaOH combined in the leaching products will be released and titrated by standard HCl titration method. So it may be proposed that there is negligible loss of NaOH in this step.

Melting:

$$Zn(FeO_2)_2(s, \text{in dust}) + 8NaOH(s) = Na_2ZnO_2(s) + 2Na_3FeO_3(s) + 4H_2O \quad (4.3)$$

$$ZnO(s, \text{in dust}) + 2NaOH(s) = Na_2ZnO_2(s) + H_2O \quad (4.4)$$

$$PbO(s, \text{in dust}) + 2NaOH(s) = Na_2PbO_2(s) + H_2O \quad (4.5)$$

$$SiO_2(s, \text{in dust}) + 2NaOH(s) = Na_2SiO_3(s) + H_2O \quad (4.6)$$

Leaching:

$$Na_2ZnO_2(s) + 2Na_3FeO_3(s) + 5H_2O = Na_2Zn(OH)_4(aq) + 2Fe(OH)_3(s) + 6NaOH \quad (4.7)$$

$$Na_2PbO_2(s) + 2H_2O = Na_2Pb(OH)_4(aq) \quad (4.8)$$

$$Na_2SiO_3(s) + H_2O = Na_2SiO_3(aq) + H_2O \quad (4.9)$$

Overall reactions in the melting and leaching processes:

$$ZnO(s, \text{in dust}) + Zn(FeO_2)_2(s, \text{in dust}) + PbO(s) + 6NaOH(s) + 6H_2O = 2Na_2Zn(OH)_4(aq) + 2Fe(OH)_3 + Na_2Pb(OH)_4(aq) \quad (4.10)$$

For zinc ferrites alone:

$$Zn(FeO_2)_2(s, \text{in dust}) + 2NaOH(s) + 4H_2O = Na_2Zn(OH)_4(aq) + 2Fe(OH)_3 \quad (4.11)$$

The molar ratio of NaOH needed to the zinc ferrites reacted is 2:1, which is approached to the actual estimation value of 2.53 mentioned above. Some of NaOH may be consumed in the formation of sodium silicates and iron hydroxides. It is sure that 163 kg NaOH will be lost as long as the dust is treated by melting followed by leaching. However, much less NaOH may be lost if the dust is treated only by alkaline leaching process because the forms of most of elements in the dust may not react with NaOH and keep unchanged in the leaching process. The consumption of electricity used for the melting of dust-NaOH mixture has not been determined in this work. 100 kW·h may be enough for the melting of 1000 kg dust for 1 h at 350 °C.

4.5 Scale-Up Experiments on Extraction of Zinc from Dust via Hydrolysis-Melting-Alkaline Leaching Process

Only 1.5 g dust were used in the above experiments. It was necessary to justify the experimental results by using a larger quantity of sample in the leaching and melting processes. Table 4.22 gives the experimental results using 400 g dust.

Table 4.22 Scale-up experiment and counter-contact alkaline leaching of the melts

(a) First batch leaching

Testing parameters	First leaching	Second leaching
Leaching agent	Fresh 5 M NaOH solution	Fresh 5 M NaOH solution
Total volume of the leaching solution	1520 mL	1318 mL
Liquid-solid ratio (v/w) in the leaching process	3.8	3.4
Volume of the filtrate	1120 mL	950 mL
Volume of the cake	400 mL	418 mL
Unit volume of the dust	1 mL/g dust	1.045 mL/g dust
Filterability	Very difficult	Very difficult
Composition of the filtrate (g/L)	Zn 49.60, Pb 2.89, Fe 0.071, Al 1.23, Cu 0.010, Mn 0.02, Cd 0.011	Zn 14.01, Pb 0.49, Fe 0.028, Cu 0.010, Al 0.17, Mn 0.03, Cd 0.001
Composition of the leaching residues (%)	Zn 4.80, Pb 0.60, Fe 37.75, Cu 0.29, Al 0.21, Mn 2.13, Cd 0.01	Zn 1.01, Pb 0.46, Fe 38.01, Cu 0.33, Al 0.27, Mn 2.17, Cd 0.01
Accumulated total zinc extraction based on the filtrate (%)	57.04	70.30
Accumulated total zinc extraction based on the total volume of the leaching system (%)	75.39	94.54
Accumulated total lead extraction based on the total volume of the leaching system (%)	61.01	78.95
Total zinc extraction based on the residues composition (%)	80.08	95.23
Total lead extraction based on the residues composition (%)	66.67	74.44

Dust 400 g, NaOH 440 g, melting at 350 C for 1 h, leaching time for 42 h

(b) Second batch leaching

Testing parameters	First leaching	Second leaching	Third leaching
Leaching agent	Leaching solution of the second leaching in Table 13-A	Fresh 5 M NaOH solution	Fresh 5 M NaOH solution
Liquid-solid ratio (v/w) in the leaching system	4.3	3.9	3.9
Volume of the filtrate	1310 mL	1180	1150
Volume of the cake	400 mL	408 mL	411 mL
Unit volume of the dust	1 mL/g dust	1.02 mL/g dust	1.03 mL/g dust
Filterability	Very difficult	Very difficult	Very difficult
Composition of the filtrate (g/L)	Zn 50.85, Pb 2.84, Fe 0.01, Al 0.96, Cu 0.00, Mn 0.00, Cd 0.010	Zn 15.91, Pb 0.15, Fe 0.00, Cu 0.00, Al 0.00, Mn 0.00, Cd 0.00	Zn 4.53, Pb 0.65, Fe 0.00, Cu 0.00, Al 0.00, Mn 0.00, Cd 0.00
Composition of the leaching residues (%)			Zn 1.05, Pb 0.48, Fe 39.01, Cu 0.31, Al 0.20, Mn 1.30, Cd 0.067

(continued)

Table 4.22 (continued)

(b) Second batch leaching			
Testing parameters	First leaching	Second leaching	Third leaching
Accumulated total zinc extraction based on the total volume of the leaching system (%)	63.36	86.28	92.39
Accumulated total lead extraction based on the total volume of the leaching system (%)	58.5	61.83	75.91
Total zinc extraction based on the residues composition (%)			94.60
Total lead extraction based on the residues composition (%)			73.33

Dust 400 g, NaOH 440 g, melting at 350 C for 1 h, leaching of the melt for 42 h

The total volume/weight of dust ratios in leaching step was kept at 3.4–4.3. It should be noted that the total volume was the actual volume of leaching system after all the melt was dispersed in NaOH solution. It was approximate to the volume sum of cake and filtrate. From Table 4.22, it can be seen that the saturation concentration of zinc in the leaching solutions was as high as around 50 g/L. After two sequential leaching of the same melt, around 94% of the zinc can be extracted and remained 1–2% of zinc in the leaching residue.

Using the unsaturated leaching solution (second leaching solution in Table 4.22a) as initial leaching agent for the fresh melt, the zinc in the resultant leaching solution will also be saturated (Table 4.22b). The remaining zinc in the cake can be further extracted by unsaturated leaching solution or fresh NaOH solution. The extraction rates based on the concentrations of zinc in the leaching solution and the contents determined in the leaching residues were compared. It can be seen that the extraction rates should be calculated based on the total volume of the leaching system instead of the volume of filtrate. It was obvious that the experimental results were in good agreement with each other using both 1.5 g and 400 g of samples.

4.6 Extraction of Lead and Other Metals from Zinc and Lead Hazardous Wastes in Alkaline Solution

The main targets were focused on the leaching and extraction of zinc, as the zinc content was as high as around 25% in the EAF dust. However, it was also very important to take the disposal of the resultant leaching residues into account after the zinc was taken out of dust. Actually, only Zn, Pb, Al, Cu, and Cr(III) may be dissolved in strong alkaline solution, and Fe, Cd, Ca, Mg, etc. cannot be dissolved. The previous experiments showed that the solubility of these amphoteric elements decreased in the following sequences:

$$Zn > Pb > Al > Cr(III) > Cu$$

In fact, the solubilities of Cr(III) and Cu were found to be negligible in the presence of zinc and lead. Meanwhile, the solubility of lead in alkaline solution was also depressed greatly in the presence of zinc, especially when the content of zinc was relatively high. As a result, the extraction percentages and concentrations in leaching solutions of lead became relatively low, as the zinc contents in dust were much higher than those of lead and the concentrations of zinc in the leaching solutions were thus much higher than those of lead.

In addition, the extraction was also some related to the treatment approaches by which the metals were leached. 70–75% of lead in the dust can be extracted when the direct leaching process was used. Nevertheless, slight higher lead extraction (about 75–85%) was obtained, when the dust was hydrolyzed and melted at 350 °C with 1.5 g of dust to be treated. In scale-up experiments, the lead extraction was found to be 70–80%. Therefore, it may be concluded that, in any case, around 70–80% of lead in EAF dust can be extracted, regardless of the extraction approaches used. It should be noted that the original lead content in dust was about 1.8%. After 70–80% of lead was removed, the lead content in the resultant leaching solutions was reduced to 0.4–0.5%. This was an acceptable lead level for the residue to be classified as nonhazardous wastes.

It indicated that the concentration of lead in leach solution was not always higher than about 3 g/L. Nevertheless, the lead concentration in leaching solutions may be over 9 g/L, when the dust was leached directly and the corresponding zinc concentration was relatively low (about 30 g/L), though the lead extraction percentages were kept at a range of 70–80%. The extraction of Al from dust was about 65–75%, regardless of the extraction approaches used. The typical concentration of Al in leaching solutions was around 1–1.5 g/L. Al was not a harmful element to the subsequent treatment of leaching solutions (purification and electrolysis). Because Al cannot be removed with lead when sodium sulfide was added to leaching solution, it may be accumulated if the leaching solutions are recycled after zinc was precipitated as zinc sulfide. In this case, part of Al should be removed in certain period of recycling.

The concentrations of other elements (Fe, Ca, Cu, Cd, Mn) in leaching solutions were all lower than 0.5 g/L and can be reduced further to lower than 0.1 g/L after sodium sulfide was added to remove selectively the lead. In hydrometallurgy of zinc using sulfuric acid as leaching agent, the presence of Cd always exerts a lot of problems in the subsequent purification and electrolysis steps for leaching solution. However, the Cd concentration in alkaline leaching solution was found to be negligible (<0.01 g/L). But the Cd in the residues was a major heavy metal that increased considerably the treatment costs, as Cd must be stabilized before the residue was disposed safely in landfills. Fortunately, the Cd content in the original dust was very low, and the resultant leaching residues may be treated easily.

4.7 Typical Composition and Supposed Treatment of the Alkaline Solution Leaching Residues

After the zinc was extracted as completely as possible, the leaching residue was washed with 5 M NaOH solution and then water, and the cake was dried at 110 °C. Table 4.23 shows the analysis results of the resultant cake (residue). It can be seen that the zinc and lead contents had been reduced from the original around 25 and 1.8% to lower 1 and 0.5%, respectively. The other heavy metals such as Cd and Cu also existed at a very low level in the residues. Such residue may be supposedly classified as nonhazardous wastes, as the leaching experiment results for the residues using water as leaching agent for 18 h at L/S ratio of 20:1 showed negligible release of the heavy metals from the residue (Table 4.24), but the pH values were quite high, which may be contributed to the entrained NaOH and the hydroxides in the residues.

The leaching residue can be use as materials for bricks, road substrates or as the stock feed for steelmaking plants, or just to be filled in landfill if available, depending on the economic and technical feasibility study case by case.

Around 94% of zinc can be extracted and recovered from EAF dust, when the dust was treated by hydrolyzing for over 4 h and then washed with freshwater, melting at 350 °C for 1 h, and leaching with 5 M NaOH solution for over 42 h, regardless of the quantity of dust samples used. In the saturated leaching solutions, the zinc, lead, and Al concentrations can reach 50 g/L, 3 /L, and 1.2 g/L,

Table 4.23 Typical composition of the alkaline leaching residue of the melts of hydrolyzed dust (nitric acid digestion method)

Elements	Zn	Pb	Fe	Cu	Al	Mn	Cd
Contents(%)	0.59–1.00	0.20–0.50	34.30–37.01	0.15–0.28	0.21–0.28	1.23–1.94	0.054–0.088

Table 4.24 Leaching tests on the alkaline leaching residues from the hydrolysis-melting-unsaturated leaching

Concentration of elements in the leaching solutions (mg/L)	Parallel test I	Parallel test II	Parallel test III
Zn	0.25	0.27	0.28
Pb	0.09	0.07	0.07
Fe	2.40	3.12	3.79
Cu	0.07	0.05	0.05
Al	3.81	5.26	13.0
Mn	0.26	0.30	0.40
Cd	0.005	0.008	0.006
Mg	0.19	0.23	0.24
pH	12.38	12.32	12.47

Liquid-solid ratio 20:1, leaching time 18 h, leaching agent water respectively. The concentrations of other elements such as Fe, Cu, and Cd in the solutions were lower than 0.5, 0.4, and 0.1 g/L, respectively. The resultant leaching residue contained lower than 1–2% of zinc, 0.5% of lead, 0.3% of copper, and 0.1% of cadmium and over 35% of Fe. It may be classified as nonhazardous waste according to the results of leaching tests on the residues.

4.8 Alkaline Treatment of Low-Leachable Zinc and Lead Hazardous Wastes with High Iron Contents

ZnO and PbO can be leached easily in alkaline solutions. However, the leachability may be reduced when Fe, Si, and S are present in the wastes. The zinc and lead contents in the dust may depend on the sources the dust are collected. The composition of dust may also vary with the melting temperature and the original components of the feed used in the foundries and steelmaking plants. In this work, another dust samples were collected from the different process in the same steelmaking plant, and composition of the dust are shown in Table 4.25.

4.8.1 Direct Alkaline Leaching of Low-Leachable Zinc and Lead Hazardous Wastes with High Iron Contents

The dust were contacted with NaOH solutions, and the results are shown in Table 4.26. It can be seen that the extraction rates of zinc and lead in one-stage leaching were around 32% and 44%, respectively. When the dust were contacted with fresh NaOH solution multiply for two and three times, the total extraction rates of zinc increased to around 35 and 38%, respectively. The zinc left in the leaching residues, which may be supposed to exist as zinc ferrites, cannot be leached out further by conventional alkaline leaching process.

Table 4.25 Composition of the dust of test

Elements	Content (%) of second portion of dust sampled
Zn	16.81
Fe	39.91
Pb	1.12
Mn	3.02
Ca	
Al	1.01
Cd	0.01
Cu	0.47
Water	0.02
Molar ratio of Fe/Zn	2.77

Table 4.26 Direct alkaline leaching of the second dust (duplicate test results)

Dust (g)	225
Leaching time (h)	42
Leaching agent	5 M NaOH solution
Filtrate obtained (ml)	835
Total volume of leaching solution (ml)	1050
Composition of leaching solution (g/L)	Zn 11.22, Pb 1.06, Fe 0.27, Cu 0.08, Al 0.12, Mn 0.10, Cd 0.00
Extraction of elements from dust (%)	Zn 31.96, Pb 44.44
Composition of the leaching residue (%)	Zn 11.92, Pb 0.34, Fe 40.92, Cu 0.20, Al 0.42, Mn 2.45, Cd 0.0010
Total zinc extraction after two leaching of the same dust (%)	35.00
Total zinc extraction after three leaching of the same dust (%)	37.10

Table 4.27 Effects of weight ratio of NaOH to residue on the melting and alkaline leaching of the direct leaching residue

No.	1	2	3
Leaching residues (g)	1.5	1.5	1.5
NaOH in the melt (g)	1.0	1.5	2.0
NaOH/residue in the melt	0.7	1.0	1.3
Melting time at 350 C (h)	1	1	1
Leaching time for the melts (h)	42	42	42
Total volume of leaching system (mL)	25	25	25
Extraction of zinc from the residue (%)	52.60	65.21	77.86
Total extraction of zinc from dust by direct leaching-melting-leaching process (%)	57.88	70.49	84.09
Zn content in the resultant residue after leaching-melting-leaching process (%)	5.65	4.15	2.63

4.8.2 *Melting and then Alkaline Leaching of the Leaching Residues Shown in Table 4.26*

The residues resulted from one-step leaching of the dust by 5 M NaOH solution were melted with solid NaOH at 350 °C for 1 h. The results are shown in Table 4.27. It can be seen that the maximum extraction rate of zinc was around 84% when a weight ratio of NaOH to the residue is 1.3. As a result, the zinc content in the ultimate residue after melting and leaching was found to be around 2.6%, which was slightly higher than the value of 1.0–2.0% for the first dust samples. The results are shown in Table 4.28. It seemed that the maximum total extraction of zinc appeared at 350 °C. The increase of melting temperature seemed unfavorable to the extraction of zinc when it was over 350 °C. It was found that a melting time of

Table 4.28 Effects of temperature on the extraction of zinc from the direct alkaline leaching residue

No.	Temperature (°C)	Zinc extraction from the melt (%)	Total zinc extraction after direct leaching and melting-leaching processes (%)
1	290	72.78	80.74
2	350	77.86	83.91
3	400	68.01	77.36
4	450	68.81	78.10
5	500	66.04	75.97

Mass ratio of NaOH to residue 1.3, melting time 1 h

Table 4.29 Effects of melting time on the extraction of zinc from the melt of direct alkaline leaching residue

No.	Time (h)	Zinc extraction from the melt (%)	Total zinc extraction after direct leaching and melting-leaching processes (%)
1	0.5	67.45	76.96
2	1.0	77.19	83.35
3	1.5	71.10	79.55
4	2.0	68.22	77.73
5	2.5	68.01	77.41

Mass ratio of NaOH to residue 1.3, temperature 350 °C

around 1 h was enough for the maximum extraction of zinc from the residue (Table 4.29).

It may be concluded that the maximum extraction of zinc can be obtained when the leaching residue was melted at 350 °C for 1 h with a weight ratio of NaOH solid to the residue of over 1.3. An extraction of 34–38% for Zn can be obtained when the dust was leached directly with NaOH solution. The total extraction of Zn may increase to around 84% when the Zn in the leaching residues was further recovered by melting and then leaching of the residues.

4.8.3 Melting and Alkaline Leaching of the Original and Hydrolyzed Dust with Higher Iron Contents

The dust was hydrolyzed and the conditions are given in Table 4.30. The results for melting and leaching of the original and hydrolyzed dust are shown in Table 4.31. It can be seen that the maximum extraction of zinc for the original dust was 76% under the conditions of Table 4.30. The corresponding extraction will be increased to 83% when the dust were hydrolyzed, which nearly reached the total extraction obtained by direct leaching-melting-leaching process shown in Table 4.30. The chemical formula of zinc ferrites may be expressed as $Zn(FeO_2)_2$, i.e., the molar ratio of Fe/Zn was 2:1. In the first sample, the Fe/Zn ratio was 1.50. According to

Table 4.30 Hydrolysis of the dust

Dust (g)	200
Water used for hydrolysis (mL)	200
Hydrolysis time (h)	4
Water used for washing of hydrolyzed dust (mL)	100

Table 4.31 Melting and alkaline leaching of the original and hydrolyzed dust

No.	1	2	3	4	5	6
Original dust (g)	1.5	1.5	1.5			
Hydrolyzed dust (g)				1.5	1.5	1.5
NaOH in the melt (g)	1.2	1.6	2.3	1.0	1.5	2.3
Melting time at 350 °C (h)	1	1	1	1	1	1
Volume of leaching system (5 M NaOH)	24	24	24	24	24	24
Leaching time for the melts (h)	42	42	42	42	42	42
Zn extraction (%)	57.40	67.42	76.19	60.51	74.03	82.75

the chemical formula of zinc ferrites, there was 32 g of Fe (0.571 mol Fe) per 100 g sample, which will combine with 18.67 g of Zn (0.286 mol Zn), supposing that all the Fe may exist as zinc ferrites. There was 24.80 g of Zn in the 100 g of the first sample. Hence, the content of ZnO was around

$$24.80 - 18.67 = 6.13 \text{ g, i.e., } (6.13/24.8) \times 100\% = 24.71\%,$$

which may be proposed to exist in the form of ZnO. The ZnO in the dust should be dissolved completely into NaOH solution.

The corresponding values of the second sample can be calculated as follows in 100 g of sample: Fe 39.91 g (0.712 mol), Zn 16.81 g (0.257 mol), Fe/Zn molar ratio 2.77.

This molar ratio was over 2. Hence, all the zinc in the second sample would be combined with Fe into zinc ferrites, and there would be no ZnO in the dust. But, obviously, it would not be true to propose that all the iron was combined into zinc ferrites. Anyway, the big difference of ZnO contents in the two dust samples based on the proposal may be used for the explanations that the zinc extraction of the first sample was more effective than that of the second sample.

4.9 Alkaline Leaching of Zinc Sulfide in Alkaline Solution via Chemical Conversion with Lead Carbonates

The oxidized zinc ores are usually coexisted with sphalerite which cannot be leached by NaOH solution under atmospheric pressure. In the conventional method of extracting zinc from sphalerite, the ore was roasted at 850–950 °C, leached the resultant zinc oxides with sulfuric acid solution, purified the leaching solution by a complex flowsheet, and then electrowon to produce the zinc flake in the purified sulfuric acid solution. The obstacle for this process was the generation and emission of SO_2 in the roasting process.

Hydrometallurgical treatment of metal sulfide minerals by chemical conversion process has been proposed. The S originally present in ZnS can be converted into PbS, while Zn can be converted into $Na_2Zn(OH)_4$ in the alkaline solution in the presence of lead compounds.

4.9.1 Effects of Pb/ZnS Mole Ratio on Leaching of Zinc Sulfide in Alkaline Solution via Chemical Conversion with Lead Carbonates

The leaching tests were carried out at 90 °C, NaOH concentration of 5 mol/L, liquid-solid ratio (v/w) of 25:1, and leaching time of 120 min. Only 0.1% zinc can be leached in the absence of lead compounds (PbO), as shown in Fig. 4.15. However, the leaching efficiency increased considerably as the lead compounds was added. The maximum leaching efficiency was obtained when the Pb/ZnS (molar ratio) ratio was over 0.9:1. XRD pattern of the leaching residue was given in Fig. 4.16. PbS phase was the main phase in the leaching residue, which showed that most of the S in the ZnS has been converted into PbS.

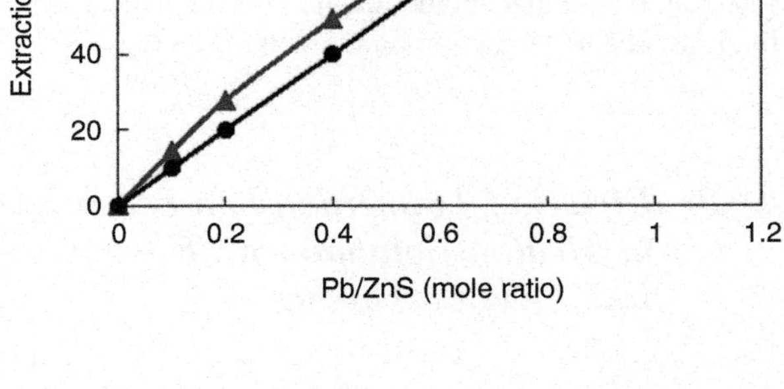

Fig. 4.15 Effects of the Pb/ZnS ratios on the zinc extraction (NaOH concentration 5 M, temperature 90 °C, liquid-solid ratio (v/w) 25, leaching time 120 min)

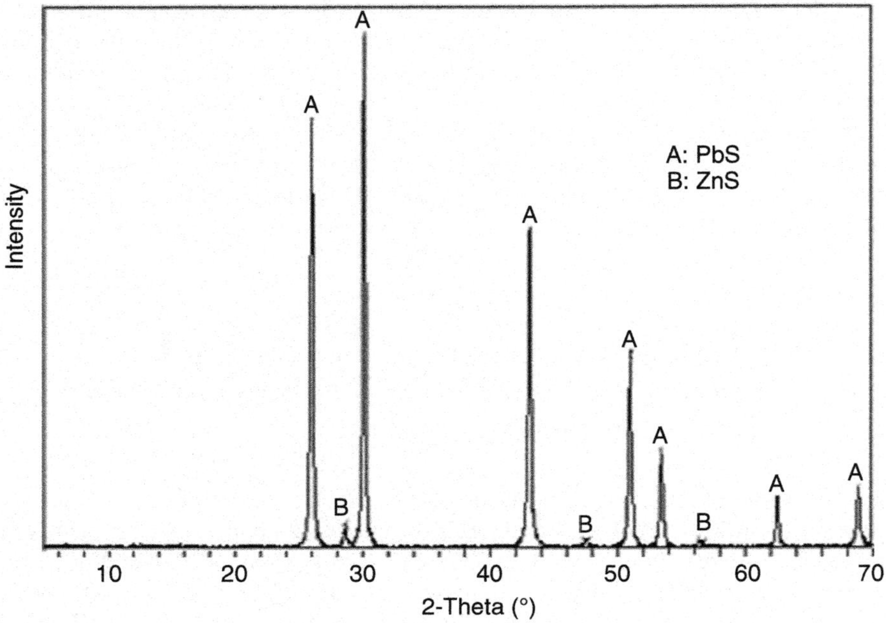

Fig. 4.16 XRD pattern of the leaching residue (temperature 90 °C, NaOH concentration 6 M, liquid-solid ratio (v/w) 25, leaching time 120 min, Pb/ZnS 0.9)

4.9.2 Effects of NaOH Concentrations on Leaching of Zinc Sulfide in Alkaline Solution via Chemical Conversion with Lead Carbonates

The effect of sodium hydroxide concentrations on the zinc extraction at 90 °C, liquid-solid ratio (v/w) of 25:1, Pb/ZnS of 0.9:1, and leaching time of 120 min is shown in Fig. 4.17. The zinc extraction increased with increase of sodium hydroxide concentrations, and 94% of zinc leaching was obtained at over 8 mol/L NaOH solution in 120 min. In practical application, NaOH solution of 5–6 mol/L may be selected. It should be noted that no zinc sulfide can be dissolved in water medium in the presence of lead compounds without the presence of NaOH.

4.9.3 Effects of Temperature on Leaching of Zinc Sulfide in Alkaline Solution via Chemical Conversion with Lead Carbonates

The effect of temperature on the zinc extraction at NaOH concentration of 6 mol/L, liquid-solid ratio (v/w) of 25:1, Pb/ZnS of 0.9:1, and leaching time of 120 min is

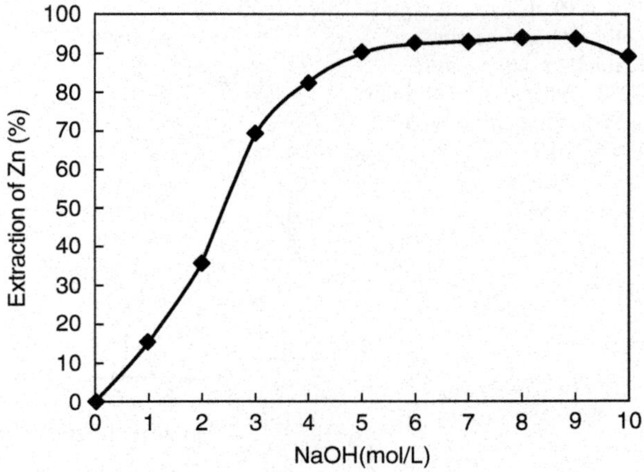

Fig. 4.17 Effects of the sodium hydroxide concentration on the zinc extraction (temperature 90 °C, liquid-solid ratio (v/w) 25, leaching time 120 min, Pb/ZnS 0.9)

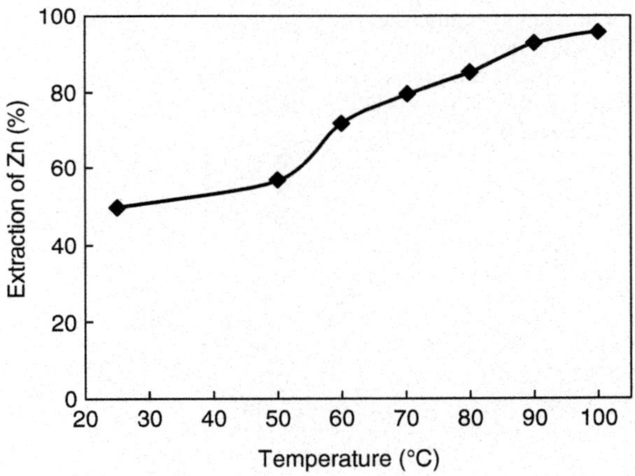

Fig. 4.18 Effects of temperature on the zinc extraction (NaOH concentration 6 M, liquid-solid ratio (v/w) 25, leaching time 120 min, Pb/ZnS 0.9)

shown in Fig. 4.18. The leaching efficiencies increased with increase of leaching temperatures from 25 to 100 °C. At 100 °C, 96.92% zinc extraction can be obtained.

4.9.4 Effects of Liquid-Solid Ratio on Leaching of Zinc Sulfide in Alkaline Solution via Chemical Conversion with Lead Carbonates

The effects of liquid-solid ratio on the leaching at 90 °C, 6 mol/L NaOH, Pb/ZnS of 0.9:1, and leaching time of 120 min are shown in Fig. 4.19. Higher liquid-solid ratio of NaOH solution to ZnS facilitated dissolution of Zn and Pb and improved surface contact between solution and ZnS particles thus increasing the extraction

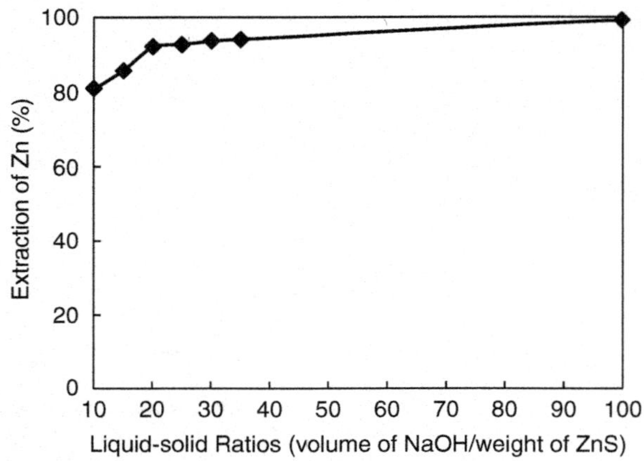

Fig. 4.19 Effects of liquid-solid ratio on the zinc extraction (temperature 90 °C, NaOH concentration 6 M, leaching time 120 min, Pb/ZnS 0.9)

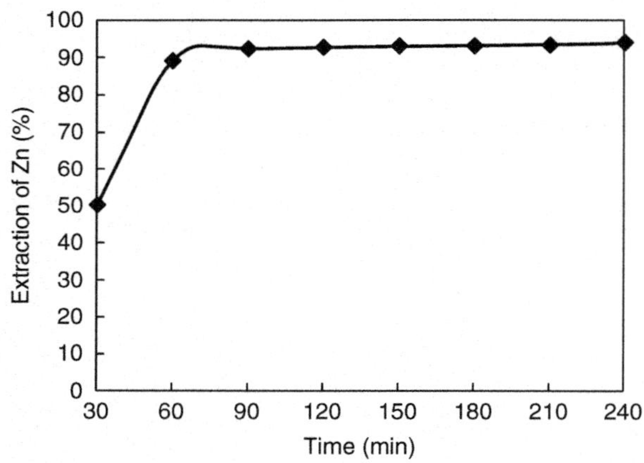

Fig. 4.20 Effects of leaching time on the zinc extraction (temperature 90 °C, NaOH concentration 6 M, liquid-solid ratio (v/w) 20, Pb/ZnS 0.9)

efficiencies of Zn. But higher liquid-solid ratio would increase the amount of NaOH required and reduced the Zn concentrations in the resultant leach solutions. Therefore, a phase ratio of 20 can be employed in order to optimize the leaching process.

4.9.5 Effects of Leaching Time on Leaching of Zinc Sulfide in Alkaline Solution via Chemical Conversion with Lead Carbonates

The effects of leaching time on the zinc extraction at 90 °C, 6 mol/L NaOH, liquid-solid ratio (v/w) of 20:1, and Pb/ZnS of 0.9:1 are shown in Fig. 4.20. As expected, the rate of zinc leaching increased with increase of leaching time. Around 90 min was sufficient to reach the maximum extraction and leaching equilibrium.

Table 4.32 Effects of the type of initial lead content on the leaching of Zn in alkaline solution

Type of initial lead content[a]	Zn extraction (%)
PbO	92.81
$Pb(NO_3)_2$	92.90
$PbSO_4$	90.52
PbO_2	95.21
$PbCO_3$	92.67
Pb_3O_4	93.25

Temperature 90 °C, NaOH concentration 6 M, liquid-solid ratio (v/w)25, leaching time 120 min
[a]The stoichiometric requirement of Pb was used in the experiments

4.9.6 *Effects of Type of Initial Lead Content for Leaching of Zinc Sulfide in Alkaline Solution via Chemical Conversion with Lead Carbonates*

Leaching tests were also carried out with $Pb(NO_3)_2$, $PbSO_4$, PbO_2, $PbCO_3$, and Pb_3O_4 as additives to compare with PbO. The results are shown in Table 4.32. It was found that the effects of Pb compounds on the leaching of Zn from ZnS were comparable, with an exception of PbO_2 which was more effective for zinc recovery because of its strong oxidation efficiency. However, in order to recycle the lead content cost-effectively, $PbCO_3$ was considered suitable as additive.

4.9.7 *Conversion of Leach Residue for Leaching of Zinc Sulfide in Alkaline Solution via Chemical Conversion with Lead Carbonates*

The lead sulfide formed thus in leaching residue should be recycled. Much work has been done on the conversion of galena to lead carbonate in ammonium or sodium carbonate solution. It has been shown that the conversion of lead was effective and cost-effective. In this work, the conversion of lead sulfide into lead carbonate was also tested. Experimental results are shown in Table 4.33. Over 95% of lead in leaching residue can be converted again into $PbCO_3$, which can dissolve completely in NaOH solution and be recycled for leaching reaction.

The solubility of ZnS and PbS is 2×10^{-23} and 1×10^{-28}, respectively. PbS is much more insoluble than ZnS. This relative insolubility accounts for one reason that zinc sulfide can be dissolved preferably in alkaline solution with lead compound as additive. The non-dissolution for zinc sulfide in water showed that the solid-solid reaction cannot take place between ZnS and $PbCO_3$. The main forms of lead and zinc in strongly alkaline solution may be $Na_2Pb(OH)_4$ and $Na_2Zn(OH)_4$. The dissolution of zinc sulfide by reacting with $PbCO_3$ in alkaline solution can be represented as:

Table 4.33 Typical conditions for the conversion Pb from alkaline leaching residues

Na_2CO_3/Pb (molar ratio)	Phase ratio (V/W)	Temperature (°C)	Flow rate of air (L/min)	Time (h)	Concentration of Cupric ions (M)	Pb conversion (%)
1–3	4–5	60–80	0.5	3–5	0.001–0.01	95.0–99.0

$$PbCO_3(s) + 2NaOH(aq) + H_2O(l) = Na_2Pb(OH)_4(aq) + Na_2CO_3(aq) \quad (4.12)$$

$$Na_2Pb(OH)_4(aq) + ZnS(s) = Na_2Zn(OH)_4(aq) + PbS(s) \quad (4.13)$$

The experimental results of leaching of Zn from ZnS via conversion by lead compound were in agreement with the theoretical results calculated by reaction stoichiometry shown in Fig. 4.15. And the transformation of S originally present in the ZnS can be evidenced in the XRD patterns of leaching residue shown in Fig. 4.16. The conversion of PbS into $PbCO_3$ by reacting with Na_2CO_3 catalyzed by cupric ion in the presence of air can be summarized as follow:

$$PbS(s) + Na_2CO_3(aq) + 2O_2(g) = PbCO_3(s) + Na_2SO_4(aq) \quad (4.14)$$

The ΔG_{298} and ΔH_{298} for the conservation reaction are −752.9 kJ/mol and −829.29 kJ/mol, which show that the reaction can take place in standard reference states.

The leaching solution can be used to produce metallic zinc powder by electrowinning after chemical separation of impurities. The NaOH solution was recycled to the next leaching operation after electrowinning of most zinc. The high-purity zinc powder with metallic zinc over 96% can be obtained. The conversion solution of Na_2CO_3 for lead can be recycled into the leaching step or the next conversion step by reacting with the addition of $Ca(OH)_2$ to remove sulfate.

NaOH losses in the whole process were estimated to be around 100 g per kg of Zn from titration analysis. The loss may be beyond this figure in practical application as the loss of NaOH was dependent on the type and composition of the ores. The presence and leaching of silicates, carbonates, and phosphates in the ores would increase the loss of NaOH. The main cost item of the process may be the energy consumed by electrowinning, heating of leaching system, and lead conversion system. Around 2.5–2.6 kW·h of electricity was needed to electrowin 1 kg of metallic zinc from the alkaline leaching solution. And the energy needed for heating was estimated as 0.9–1 kW·h/kg Zn on average.

Table 4.34 compares the alkaline process developed in this work with the classical acidic process for the production of metallic Zn from ZnS and oxidized Zn ores. Around 20% of energy for electrowinning in the alkaline solution is reduced in comparison to that in the acidic solution. The purification process in the alkaline process is simpler than the acidic process, as the undesired elements, such as Fe, Mg, Cu, Cd, Co, Ni, etc., can hardly be leached out in the alkaline solution. Moreover, the alkaline process is cleaner for sulfur in the zinc sulfide ores

Table 4.34 Comparisons for different processes for the production of Zn from ZnS ores (based on the production of 1 kg Zn)

Processes selected	Alkaline process developed in this work	Acidic process
Energy for electrowinning	2.5–2.6 kW·h	3.3–3.4 kW·h
Energy for heating leaching system	0.9–1 kW·h	1 kW·h
Purification of leaching solution	Simple	Complex
Chemical consumption	Caustic soda (small quantity)	Sulphuric acid (large quantity)
Loss of Zn powder for impurity removal	None	Necessary
Transformation of S present in the ZnS ores	S is converted to sulfide and sulphate in the leaching residue	S is converted sulfur dioxide in the gaseous effluent

which is converted to sulfide and sulfate, which remains in the leaching residue, rather than sulfur dioxide in the gaseous effluent to pollute the atmosphere in acidic process. The alkaline leaching via chemical conversion-electrowinning process can be considered as a cost-effective and environmentally friendly technology for extracting zinc from zinc sulfide ores or oxidized zinc ores coexisted with zinc sulfide.

Hence, the zinc in zinc sulfide can be extracted in NaOH solution with lead carbonate as additive, and the lead sulfide in the leach residues can be converted to lead carbonate by reacting with sodium carbonate solution using air as the oxidizing agent, and then the lead can be recycled in the whole process by dissolving the lead carbonate in NaOH solution. Over 90% of zinc can be extracted from the zinc sulfide, and over 95% of lead sulfide in leach residues can be converted.

4.10 Mechanochemical Leaching of Sphalerite in Alkaline Solution via Chemical Conversion with Lead Carbonates

The mechanical activation of minerals represents nowadays an important contribution to different fields of solid processing technology. In extractive metallurgy, activation by intensive grinding decreases the reaction temperature in pyrometallurgy and increases leaching kinetics of several sulfide and oxide minerals in hydrometallurgy.

The investigation was carried out using a sample of sphalerite from Gejiu City of China, and the average composition of the sphalerite was given in Table 4.35. X-ray diffraction analysis indicated that the major phase present in the sample was sphalerite (Fig. 4.21). The diffraction peak of sphalerite was narrower than that of

Table 4.35 Chemical composition of the sphalerite

Elements	Zn	Pb	Cd	Fe	As	S	Mg	Cu	Al
wt. %	29.57	0.08	0.08	23.68	0.04	27.82	0.57	0.72	1.55

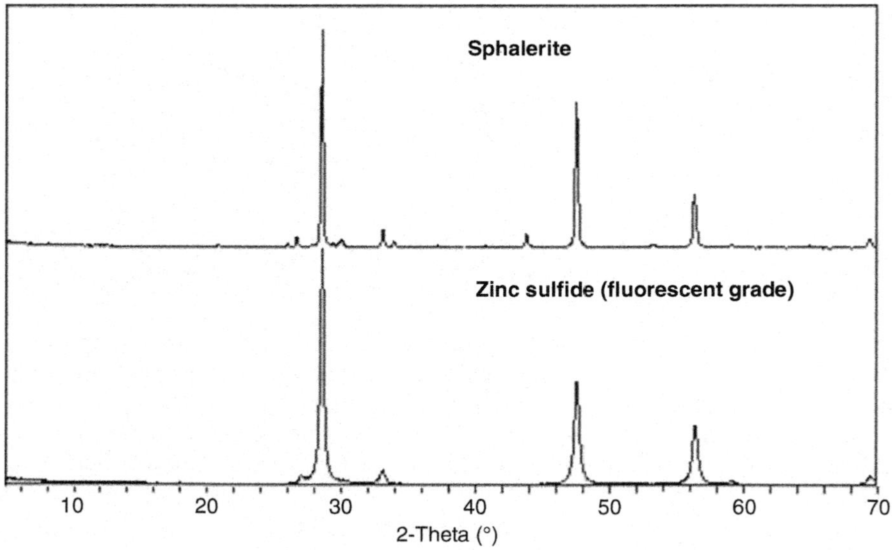

Fig. 4.21 XRD of ZnS (fluorescent grade) and sphalerite

zinc sulfide (fluorescent grade), which proved that sphalerite was more stable than zinc sulfide of fluorescent grade. Two modes of leaching were studied, i.e., the separate mechanical activation and subsequent chemical leaching were applied in the first mode and compared with the combined mechanical activation and chemical leaching (mechanochemical leaching) in the second mode.

Mechanical activation tests of sphalerite were carried out using three mills: vibratory, rolling mill, and stirring ball mill (Fig. 4.22). Stainless steel ball or corundum ball was used as grinding bodies with dimensions between 2 mm and 6 mm. Leaching experiments of sphalerite mechanically activated were carried out in a flask placed on a thermostatically controlled magnetic stirrer. To a 200 mL sodium hydroxide solution, zinc sulfide and lead compound was added and then leached at constant temperature of 90 °C. The volume was kept constant by adding water. Since vibratory and rolling mill were difficult to be closed and heated, mechanochemical leaching was performed in a stirring ball mill under the following conditions: Φ5 mm stainless steel ball or corundum ball as activation medium, the mass ratio of ball to raw materials of 30:1, Pb/ZnS of 0.9:1, NaOH concentration of 6 mol/L, and temperature 90 °C.

Fig. 4.22 Effects of types of mills for mechanical activation on alkaline leaching rate of sphalerite (NaOH concentration 6 M, temperature 90 °C, leaching time 120 min)

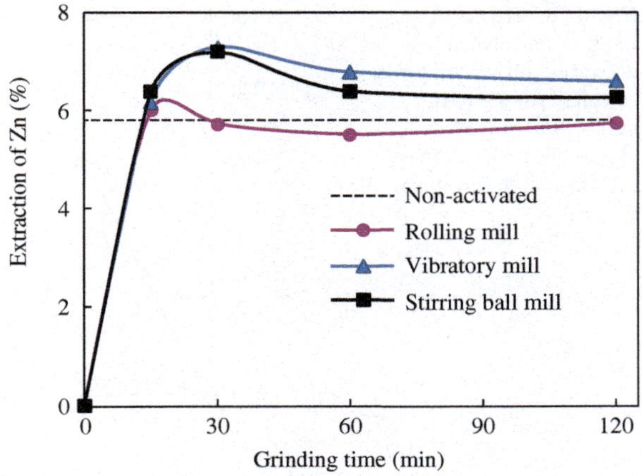

4.10.1 *Effects of Activation and Leaching Modes on Mechanochemical Leaching of Sphalerite in Alkaline Solution via Chemical Conversion with Lead Carbonates*

The results of leaching of sphalerite mechanically activated by different mills are shown in Fig. 4.23. Compared with the nonactivated sample, the leaching rate of sample mechanically activated had little increase. Comparative studies were performed with mechanochemical leaching and chemical leaching of nonactivated samples (Fig. 4.23). The results confirmed the favorable effects of mechanochemical leaching on the recovery of zinc from sphalerite. 86.8% Zn extraction can be achieved by mechanochemical leaching, compared with 5.8% Zn extraction for chemical leaching of nonactivated samples and 6.2% Zn extraction for chemical leaching of activated samples for 120 min.

The enhancement of leaching reaction due to mechanical activation was attributed to increased specific surface area, enhanced surface reactivity, and the changes in crystalline structure during the mechanical activation of sulfide minerals. Mechanochemical leaching has significant beneficial effect on the leaching rate of sphalerite, while leaching of sphalerite mechanically activated has little effect, which suggests that the increase in surface area of the finely ground particles is not the major factor that influences the rate of leaching.

During mechanical activation, the crystal structure of a mineral is usually disordered, and generation of defects or other metastable forms can be registered. These effects are not stable and have different relaxation times. If the process of mechanical activation is separated from leaching in time, then a number of highly

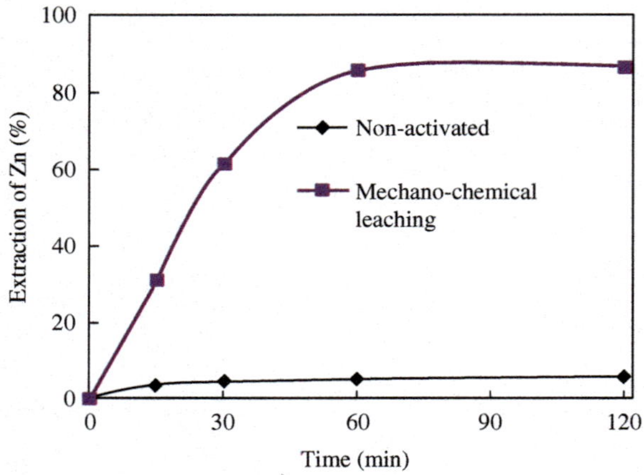

Fig. 4.23 Effects of methods of mechanical activation on leaching rate of sphalerite (NaOH concentration 6 M, temperature 90 °C, mass ratio of ball to raw materials 30:1)

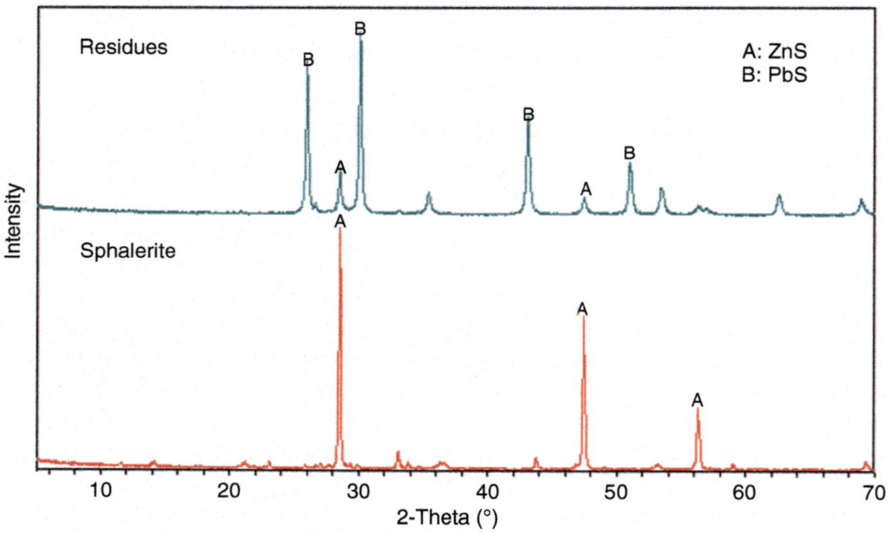

Fig. 4.24 XRD spectra of sphalerite and alkaline leaching residues

excited states may form and decay before leaching. If the mechanical activation and chemical leaching are integrated into a common step, all the excitation states can be utilized. Thus, mechanochemical leaching contributes to operational benefits and economics of the leaching of sphalerite. XRD pattern of the leaching residue was given in Fig. 4.24. PbS phase was the main phase in the leaching residue, which showed that most of the S in the sphalerite had converted into PbS.

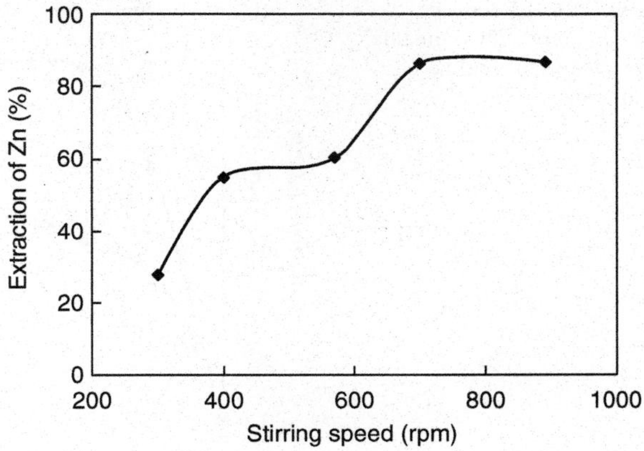

Fig. 4.25 Effects of stirring speed on zinc extraction in alkaline solution

4.10.2 Effects of Stirring Speed on Mechanochemical Leaching of Sphalerite in Alkaline Solution via Chemical Conversion with Lead Carbonates

At the NaOH concentration of 6 mol/L, temperature of 90 °C, mass ratio of ball to raw materials of 30, and leaching time of 120 min, the effect of speed on the zinc extraction is shown in Fig. 4.25. The zinc extraction increased with increasing speed, and 86% of zinc leaching was obtained at over 700 r/min.

4.10.3 Effects of Activation Medium on Mechanochemical Leaching of Sphalerite in Alkaline Solution via Chemical Conversion with Lead Carbonates

The effect of activation medium of stainless steel ball and corundum ball on the zinc extraction of sphalerite is shown in Fig. 4.26. It was clear that the effect of activation medium of high-density materials was better. Mechanical activation had a positive effect on the alkaline leaching of the sphalerite under study. By comparison with the separate mechanical activation and subsequent chemical leaching, the combined mechanical activation and chemical leaching (mechanochemical leaching) was more effective.

The zinc in sphalerite can be thus extracted by mechanochemical leaching in NaOH solution with lead carbonate as additive, and the lead sulfide in the leach residues can be converted to lead carbonate by reacting with sodium carbonate solution using air as the oxidizing agent, and then the lead can be recycled in the whole process by dissolving the lead carbonate in NaOH solution. Over 86% of zinc can be extracted from the sphalerite.

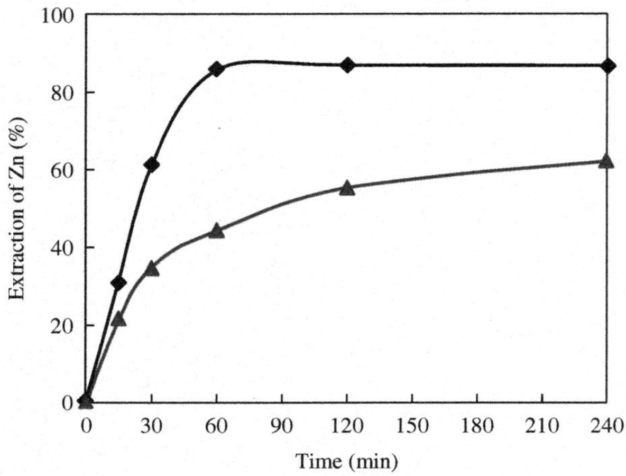

Fig. 4.26 Effects of activation medium on zinc extraction in alkaline solution

4.10.4 Process of Mechanochemical Leaching of Low-Grade Zinc Oxide Ore Containing Sphalerite

Mechanical activation enhances leaching transformation of low-grade zinc oxide ore containing sphalerite, which increased the leaching from 60.1% without activation to 81.5% using stainless steel ball with a diameter of 5 mm as activation medium, 30:1 ball-feedstock mass ratio, 90 °C, 0.9 Pb/ZnS ratio, and 700 r/min stirring rate in stirring mill summarized in Fig. 4.27. Firstly, zinc concentrates or low-grade mixed ore of zinc sulfate and zinc oxide was ground to 0.5–1 mm and then added to the stirring mill reactor alkaline solution, and lead carbonate at the same time so that leaching occurred with grinding activation at 5–6 mol/L NaOH, 80–90 °C, and 60–90 min stirring time. Zinc in zinc concentrates or low-grade mixed ore of zinc sulfate and zinc oxide was dissolved and transformed. After filtration and washing of filtration residue, lead in the washed filtration residue was then transformed and recovered by sodium carbonate solution. After lead transformation, the final residue was used for brick-making or landfill. Washing water was recycled for next process. Leaching filtrate that contained zinc was electrolyzed and zinc can be obtained in cathode. After electrolysis, residual waste electrolyte was directly recycled for next alkaline leaching process.

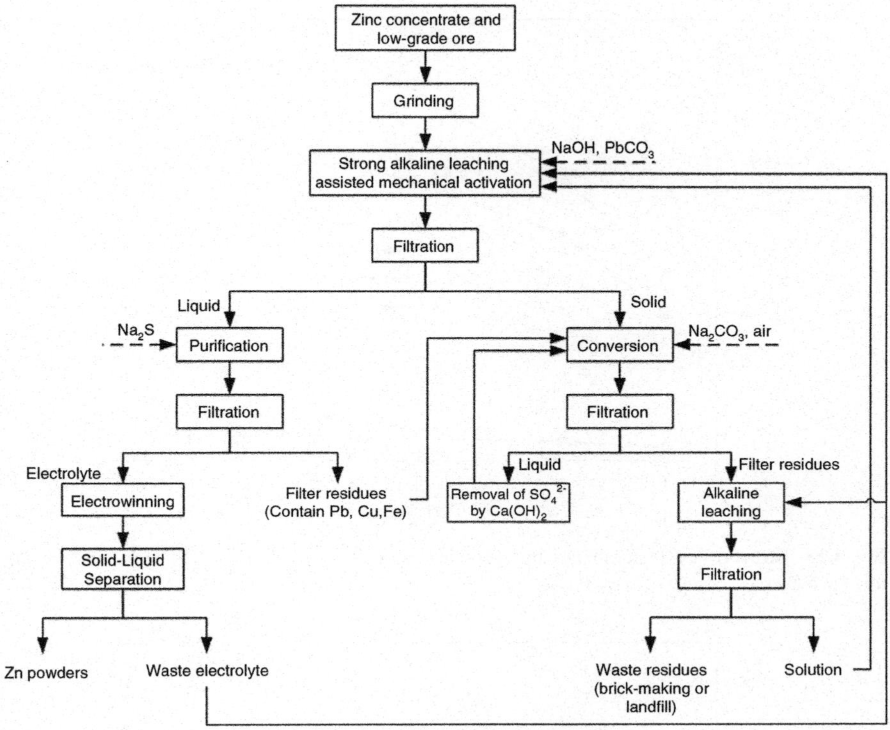

Fig. 4.27 Schematic flow sheet of hydrometallurgical process for zinc sulfide ores by alkaline leaching and chemical conversion

4.11 Other Enhanced Leaching Methods of Zinc Hazardous Wastes in Alkaline Solution

4.11.1 Leaching Process Enhanced by Microwave

Microwave is a kind of electromagnetic wave with a wavelength between 1 mm and 1 m and a frequency between 300 MHz and 300 GHz. It can selectively heat mineral inside ore so that the heating process can be controlled. Thus, application of microwave heating on some leaching process is one of the important ways to handle low leaching rate and long leaching time.

At NaOH solution concentration of 5 mol/L and liquid-solid ratio of 10:1, effects of radiant time (times of cycle operation) on metal leaching on leaching rate of zinc and lead are given in Fig. 4.28. A relatively low leaching rate of 30–40% with a single-time microwave irradiation was obtained. When increasing cycle operation times, recovery rate of zinc and lead gradually increased. When times of cycle operation were 4, leaching rate of zinc and lead was 82% and 60%, respectively.

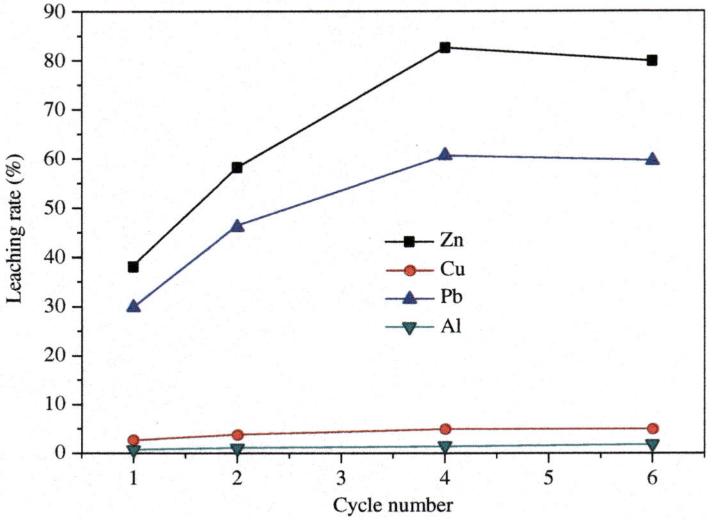

Fig. 4.28 Effects of cycle number on the metal leaching (5 M NaOH, 10:1 L/S) from wastes using microwave

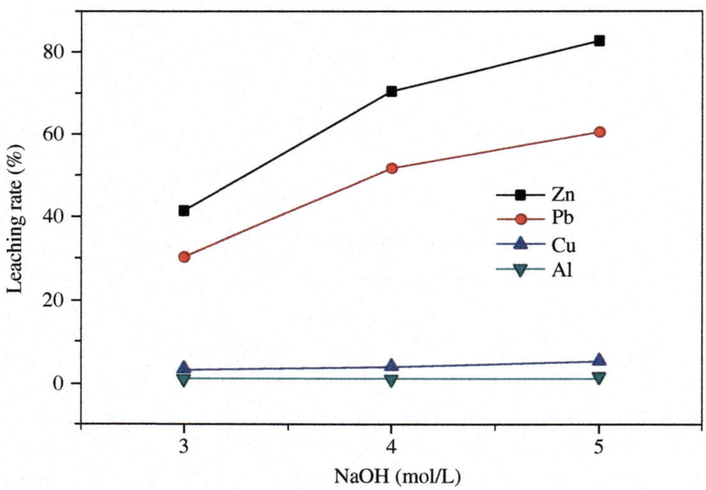

Fig. 4.29 Effects of NaOH concentration on the metal extraction (cycle number = 4, L/S = 10:1) from wastes using microwave

If time was further increased, the leaching rate did not change notably. From Figs. 4.29 and 4.30, it can be seen that the optimal NaOH solution concentration is 5 mol/L and liquid-solid ratio is 10:1 when using microwave as an enhancement of alkaline leaching process. An 8-min microwave irradiation can obtain a high zinc

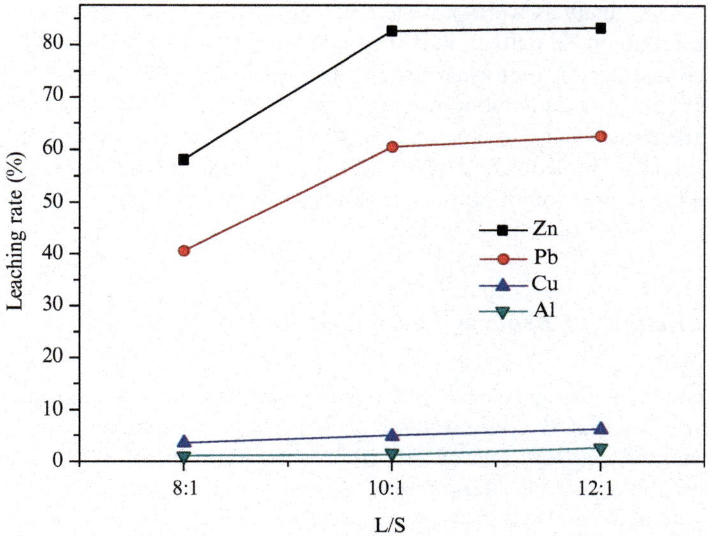

Fig. 4.30 Effects of L/S ratio on the metal extraction (cycle number = 4, NaOH =5 M) from wastes using microwave

Table 4.36 Yates analysis of microwave-assisted leaching of zinc

No.	Control variables A	B	C	Zn % (0)	Yates analysis (1)	(2)	(3)	Response value
1	–	–	–	37.48	91.94	191.61	456	–
A	+	–	–	54.46	99.67	264.39	49.6	12.4
B	–	+	–	42.11	111.34	32.43	49.44	12.36
AB	+	+	–	57.56	153.05	17.17	5.44	1.36
C	–	–	+	53.12	16.98	7.73	72.78	18.19
AC	+	–	+	58.22	15.45	41.71	-15.26	-3.81
BC	–	+	+	70.49	5.1	-1.53	33.98	8.495
ABC	+	+	+	82.56	12.07	6.97	8.5	2.12

recovery rate of 82%, while the leaching rates of Cu and Al were below 5%. With low-concentration alkaline solution and low liquid-solid ratio, microwave radiation can increase the leaching rate notably.

Optimal operation conditions of microwave-enhanced leaching process were determined at the number of microwave irradiation for 4, NaOH concentration of 5 mol/L, and liquid-solid ratio of 10:1. On this basis, main control variables and interactions between variables can be studied by single-factor analysis method, Yates analysis (set total leaching rate of zinc as response value, shown in Table 4.36). The three impact factors are concentration of NaOH solution (A), number of microwave irradiation (B), and liquid-solid ratio (C), the ranges of which are A [4 (–), 5 (+)] mol/L, B [2 (–), 4 (+)], and C [8:1 (–), 10:1 (+)] g/L. The results of Yates analysis were presented at Table 4.36; it shows that the effect of liquid-solid ratio on Zn extraction is the most important among the NaOH concentration, the number of microwave irradiation, and the combined action of microwave irradiation and liquid-solid ratio (C > A,B > BC). And there are little effects on Zn extraction of the combined action of NaOH concentration and liquid-solid ratio (AC), the combined action of NaOH concentration and microwave irradiation (AB), and the interaction of all three impacts (ABC).

4.11.2 *Leaching Process Enhanced by Pressure*

High pressure reactor can also be used as a way to enhance leaching process besides microwave. The NaOH concentration of 3–5 mol/L, reactor pressure of 2.0 or 5.8 atm, and the recovery rate of zinc from wastes are shown in Fig. 4.31. It can be seen that the optimal leaching rate of zinc occurred at 20 min, 2.0 atm with the leaching rate of 70%. With pressure increase to 5.8 atm, the leaching rate cannot improve notably even when prolonging the reacting time to 40 min.

ANOVA method is adopted to NaOH solution concentration and liquid-solid ratio as two main variables to study effects of them and their possible interaction on leaching rate of zinc. The results are shown in Table 4.37.

F-value of different factors can be obtained using ANOVA. In pressure-enhanced alkaline leaching process, the descending sequence of effects on leaching is as follows: NaOH concentration ($F = 213.41$) > liquid-solid ratio ($F = 120.57$) > interaction of concentration of NaOH solution and liquid-solid ratio ($F = 27.75$). Comparing pressure enhancement with mechanical stirring, similar metal leaching rates can be obtained, while pressure enhancement can shorten leaching time (NaOH concentration 4–5 mol/L, L/S 8:1–10:1). When

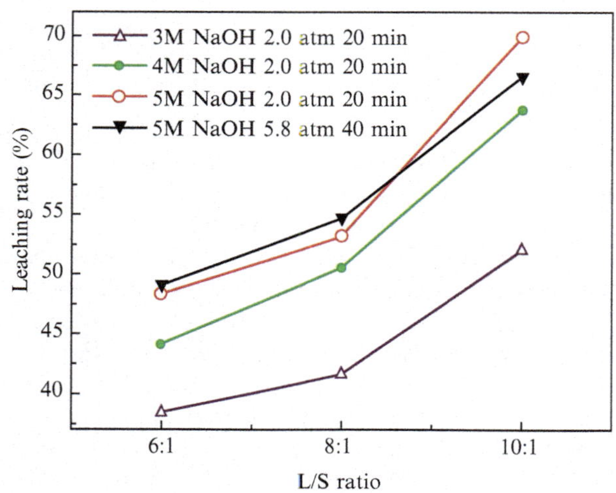

Fig. 4.31 Pressure leaching of zinc at different conditions (20– 40 min, 3–5 M NaOH, 6:1–8:1 L/S) from wastes in alkaline solution

Table 4.37 ANOVA of pressure leaching of zinc from wastes in alkaline solution

Sources of variations	Sum of squares	Degrees of freedom	Mean square	F	P > F
Concentration of NaOH solution	1239.55	2	619.775	213.41	2.61e-008
Liquid-solid ratio	700.3	2	350.152	120.57	3.18e-007
Interaction	322.31	4	80.579	27.75	4.48e-005
Errors	26.14	9	2.904		
Total sum	2288.3	17			

NaOH concentration is 3 mol/L and L/S is 8:1, leaching rate of zinc is 40% after leaching for 20 min. With the same leaching conditions, mechanical stirring can only get 25% zinc from wastes.

4.11.3 Ultrasound-Enhanced Leaching Process

The ultrasonic wave is a mechanical wave with over 20 kHz. Ultrasonic wave can produce mechanical and chemical effects on the reaction media by acoustic cavitation. It can accelerate or cause chemical reactions. The experiments were carried out, and the data were analyzed by a two-degree polynomial regression analysis. According to the principle of central composite design, determination coefficient (R^2) was applied to estimate fitting degree of the model to predict the optimal parameter. The leaching rate of zinc was pointed as response value, and the design of three factors in five levels and five zero points was presented at Table 4.38. The leaching rate of zinc is around 35–81%.

According to the above steps, the multiple regression model equation is obtained:

$$Y = 71.39 + 8.28X_1 + 7.05X_2 + 9.06X_3 - 3.84X_1^2 - 4.43X_2^2 - 4.67X_3^2. \quad (4.15)$$

X_1, X_2, and X_3 are sodium hydroxide concentration, ultrasound-assisted leaching time, and liquid-solid ratio, respectively.

Table 4.39 is the variance analysis of the regression model. The determination coefficient (R^2) of the regression equation is 0.91, and the signal-noise ratio is 10.584. It shows that the model is enough fit to the experimental datas and can be used to analyze the actual experiment results.

As the $P > F$ value of F-test is less than 0.05, it proved that the regression model is reliable significantly, and it can be judged that the regression model is similar to the real data surface. X_1, X_2, and X_3 and X_1^2, X_2^2, and X_3^2 are significant factors, and the interaction terms X_1X_3, X_2X_3, and X_1X_2 are not significant factors.

As shown in Figs. 4.32 and 4.33, three-dimensional response surface methodology of different factors can be plotted by using software (Expert Design). The three-

Table 4.38 Experimental design matrix and results

No.	Variable			Zn leaching rate(%)
	NaOH concentration (M)	Ultrasonic time (h)	L/S	
1	4	1	8	70.68
2	4	1	8	72.31
3	3	1.5	6	54.66
4	5	1.5	6	64.74
5	5	1.5	10	80.76
6	4	1	4.64	35.46
7	2.32	1	8	38.80
8	4	1	8	71.36
9	3	0.5	10	60.88
10	3	1.5	10	62.02
11	3	0.5	6	40.82
12	4	0.16	8	36.53
13	5	0.5	6	54.48
14	4	1	8	71.19
15	4	1.84	8	73.31
16	4	1	8	72.25
17	5	0.5	10	71.55
18	5.68	1	8	74.42
19	4	1	8	71.93
20	4	1	11.36	73.02

Table 4.39 ANOVA for response surface quadratic model for ultrasound-assisted leaching

Source	Sum of square	df	Mean square	F-value	p-value prob > F
Model	3442.35	9	3442.35	9.22	0.0009
X_1	935.91	1	935.91	22.57	0.0008
X_2	679.14	1	679.14	16.38	0.0023
X_3	1120.04	1	1120.04	27.01	0.0004
X_1X_2	2.52	1	2.52	0.061	0.8103
X_1X_3	4.02	1	4.02	0.097	0.7620
X_2X_3	23.63	1	23.63	0.57	0.4677
X_{12}	211.95	1	211.95	5.11	0.0473
X_{22}	283.14	1	283.14	6.83	0.0259
X_{32}	314.69	1	314.69	7.59	0.0203

dimensional response surface methodology of the relationship among the liquid-solid ratio, the ultrasound time, and the leaching rate is shown in Fig. 4.32. The liquid-solid ratio should be set to above 8:1. Because of the low liquid-solid ratio, the internal diffusion mass transfer process can be hindered, and the enhancement to leaching process was obviously slight at high liquid-solid ratio.

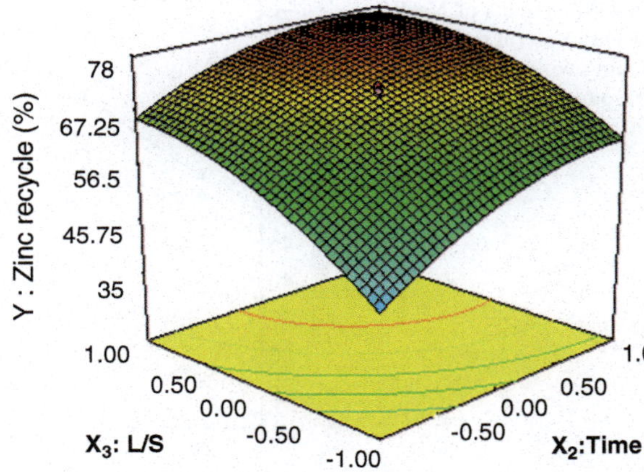

Fig. 4.32 Effects of L/S and ultrasound time on the zinc leaching (4 M NaOH) from wastes

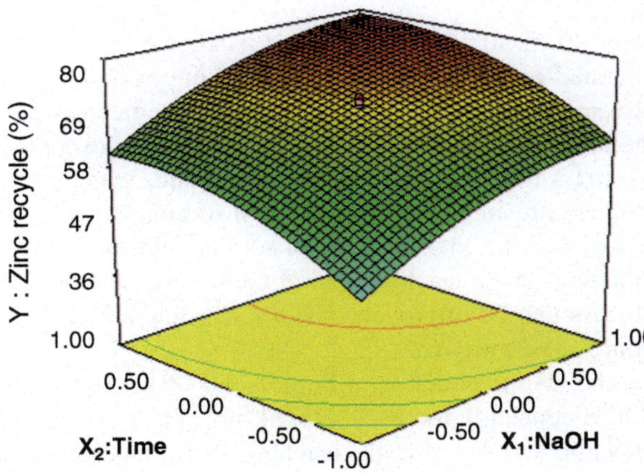

Fig. 4.33 Effects of NaOH concentration and ultrasound time on the zinc leaching (L/S ratio = 8:1) from wastes

Figure 4.33 shows the effects of NaOH concentration and ultrasound time on the alkali leaching process. The recovery rate of zinc could be stabilized by 1 h ultrasound-assisted leaching, while the longer ultrasonic time could not result in higher leaching rate significantly. However, when the concentration of NaOH was 3–4.5 mol/L, the leaching rate of zinc increased with the increasing NaOH concentration. But the trend was not significant at higher NaOH concentration. Accroding to the analysis above, it can be found that there exist the extreme points of the interaction of factors. By using the prediction function of Expert Design, the optimal parameters were obtained: liquid-solid ratio of 8.7:1, NaOH concentration of 5 M, and ultrasonic time of 82 min. Under the conditions, the leaching rate of zinc is 81.3%.

Compared with ultrasound-assisted leaching and mechanical-assisted leaching, on the one hand, the ultrasonic greatly reduced the required temperature of the leaching process, and the temperature required to mechanical stirring is reduced

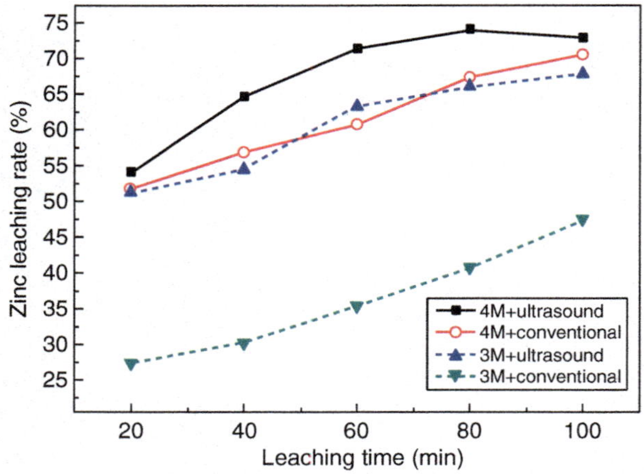

Fig. 4.34 Effects of NaOH concentration on the zinc leaching with or without ultrasound at L/S ratio = 8:1 and T = 50 °C

from 80 to 50 °C. On the other hand, the zinc leaching rate of 81–83% can be obtained by ultrasound-assisted leaching at liquid-solid ratio of 8.7:1, which was also significantly smaller than that of using mechanical-assisted leaching to recover the same amount of zinc. These phenomena were contributed to the improvement of the mass transfer condition by ultrasound, which reduced the requirement of the temperature and the consumption of sodium hydroxide. When the liquid-solid ratio is 10:1 and the concentration of sodium hydroxide is 5 M, the effects of ultrasound radiation on the leaching rate of metal were not significant. However, the leaching process can be improved obviously at low alkali concentration (3–4 M) and low liquid-solid ratio (6:1–8:1).

The results obtained at liquid-solid of 8:1, reaction temperature of 50 °C, and NaOH concentration of 3–4 mol/L are shown in Fig. 4.34. When the concentration of NaOH was 4 mol/L, the leaching rate of zinc could be increased by ~10%. When the concentration of NaOH was as low as 3 mol/L, the enhancement of ultrasound was more obvious, which could increase the leaching rate of traditional mechanical stirring by ~25%. From Fig. 4.34, it can be found that the variation of leaching rate tend to stabilize the ultrasound-assisted leaching after 80 min leaching. However, this trend has not appeared in the conventional leaching process without ultrasound.

Figure 4.35 shows the effects of different liquid-solid ratio on the leaching rate of zinc when the NaOH concentration is 4 mol/L and the temperature is 50 °C. The differences between ultrasonic-assisted leaching and mechanical stirring were analyzed in order to further study the potential advantages of ultrasonic enhancement. In order to facilitate the comparison, the zinc leaching process of the four kinds of conditions is all listed as a function of time variation. The recovery rate of zinc in all kinds of cases increased with time. When the reaction time of the ultrasonic-assisted reaches to 80 min, the leaching rate could be improved by ~15% and ~10% at the liquid-solid ratio of 6:1 and 8:1, respectively.

These phenomena are mainly caused by ultrasonic radiation, enhancing the mass transfer between solid and liquid and speeding up the reaction rate. The effects of

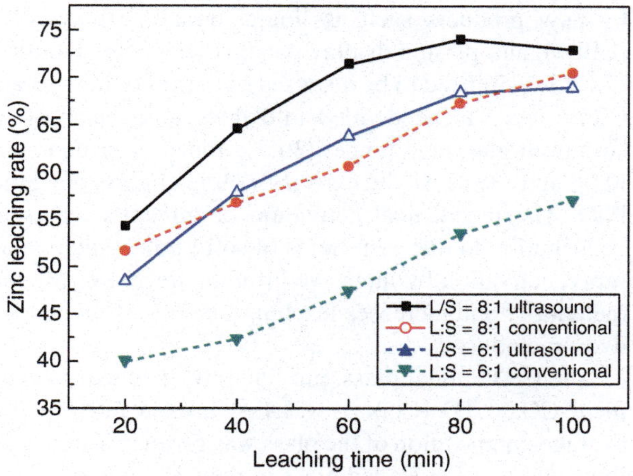

Fig. 4.35 Effects of liquid-solid ratio on the zinc leaching with or without ultrasound using 4 M NaOH solution at 50 °C

washing and crushing on the particle surface, the erosion of micro jet caused by cavitation on solid surface, the formed active group, and the local high temperature are beneficial to increase the contact area between the leaching agent and the particles, so as to speed up the reaction.

The enhancement effect of ultrasonic radiation on leaching kinetics was more significant under the condition of low liquid-solid ratio and low NaOH concentration. Taking the low sodium hydroxide concentration as an example, the following mechanism can be explained. When the concentration of sodium hydroxide is low, the viscosity of the leaching agent is small, and it is easy to produce cavitation bubble. In the process of mass transfer, the energy loss is small, and the ultrasonic wave can promote the cavitation. The asymmetric collapse of the hole near the solid particles accelerates the diffusion within the particles, thus shortening the reaction time.

4.12 Recovery of Lead from CRT Funnel Glass by Mechanochemical Extraction in Alkaline Solution

Lead is an important nonferrous metal required for various applications in batteries, solder, dielectric materials, piezoelectric materials, glass, etc. Lead glass is a variety of glass in which lead replaces the calcium content of a typical potash glass and desirable owing to its decorative properties. Lead glass is one of the very popular glass materials used as a protection material against radioactive, such as in a cathode ray tube (CRT) rays, and as decorative glass owing to its high X-ray absorbing and decorative properties. Due to the potential health risks of lead contained in materials, research and development of lead-free materials, which can replace lead-containing materials, are being vigorously promoted in all industrial fields.

With the rapid development of image display technology, more and more conventional cathode ray tube (CRT) monitors in TVs and PCs have been replaced by new products such as liquid crystal displays (LCDs), light-emitting diode (LEDs), and plasma display panel (PDPs). Predictions indicate that huge numbers of CRTs will need to be disposed of in the coming decades. The main component of CRT glass can be divided into three parts, namely, (1) panel (65%), a barium strontium glass; (2) funnel (30%), a lead glass containing approximately 20 wt % PbO; and (3) neck (5%), a very rich lead glass containing approximately 40 wt % PbO. The mechanical activation of minerals represents nowadays an important contribution to different fields of solid processing technology. In extractive metallurgy, activation by intensive grinding decreases the reaction temperature in pyrometallurgy and increases leaching kinetics of several sulfide and oxide minerals in hydrometallurgy.

The CRT funnel glass, provided by Shanghai Senlan Industrial Waste Management CO., LTD (China), was first broken into small pieces ($\leq$10 mm), and the average composition of the glass was given in Table 4.40. X-ray diffraction analysis indicated that the leaded glass in the CRT funnel glass was nanocrystal formation (Fig. 4.36). Two modes of leaching were studied. The separate mechanical activation and subsequent chemical leaching were applied in the first mode and compared with the combined mechanical activation and chemical leaching (mechanochemical leaching) in the second mode.

Table 4.40 Chemical composition of the CRT funnel glasses in this experiment

Elements	PbO	SiO_2	CaO	Al_2O_3	Na_2O	BaO	K_2O	MgO
Wt. %	21.69	50.31	3.18	2.87	6.05	0.34	7.72	2.08

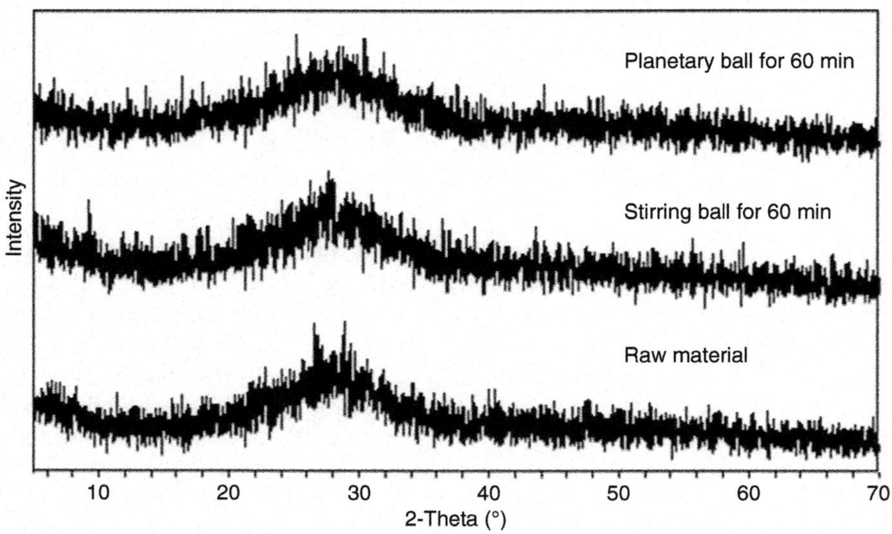

Fig. 4.36 XRD patterns of raw material and activated samples with different mills

Mechanical activation tests of CRT funnel glass were carried out using two mills: planetary ball mill and stirring ball mill. Stainless steel ball was used as grinding bodies with dimensions between 1 mm and 5 mm. Leaching experiments of CRT funnel glass mechanically activated were carried out in a flask placed on a thermostatically controlled magnetic stirrer. To a 200 mL sodium hydroxide solution, CRT funnel glass was added and then leached at constant temperature. The volume was kept constant by adding water.

Mechanochemical leaching was performed in a stirring ball mill under the following conditions: Φ5 mm stainless steel ball as activation medium, mass ratio of ball to raw materials of 25:1, NaOH concentration of 5 mol/L, and room temperature of 70 °C. Since planetary ball mill was difficult to be heated, mechanochemical leaching was performed in a planetary ball mill at room temperature.

4.12.1 *Effects of Activation Modes on Recovery of Lead from CRT Funnel Glass by Mechanochemical Extraction in Alkaline Solution*

The results of leaching of CRT funnel glasses mechanically activated by different mills are shown in Fig. 4.37. Compared with the nonactivated sample, the leaching rate of sample mechanically activated had obvious increase. The effect of activation in planetary ball mill was better than in stirring ball mill. For the nanocrystal leaded glass, nonbridging oxygen hole center (NBOHC, ≡Si–O·) and peroxy radical (≡Si–O–O·) are the most important structural point defects in silicate glass, which are generated during the process of mechanical activation.

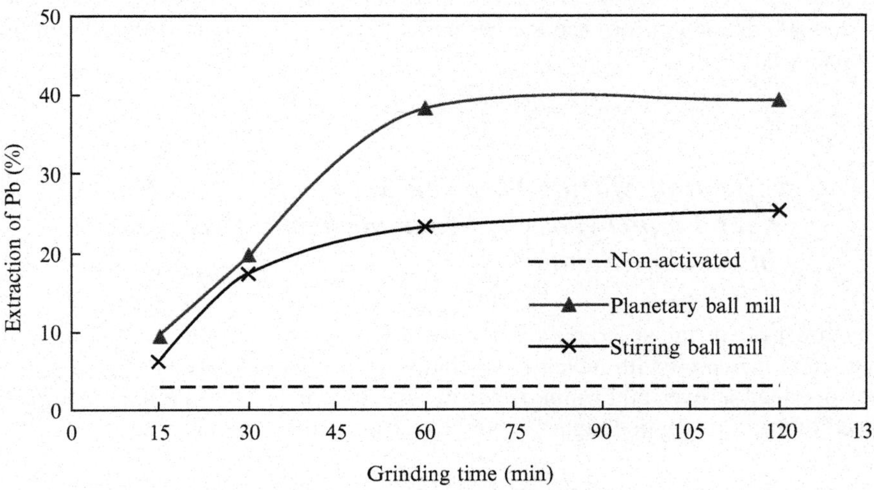

Fig. 4.37 Effects of types of mills for mechanical activation on leaching rate of CRT funnel glasses (NaOH concentration 5 M, temperature 70 °C, leaching time 2 h)

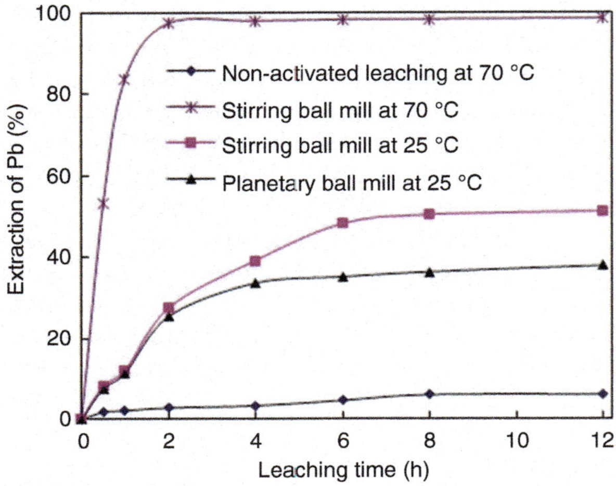

Fig. 4.38 Effects of methods of mechanical-chemical leaching on leaching rate of CRT funnel glasses (NaOH concentration 5 M, mass ratio of ball to raw materials 25:1)

4.12.2 Effects of Mechanochemical Leaching Modes for Recovery of Lead from CRT Funnel Glass by Mechanochemical Extraction in Alkaline Solution

Comparative studies have been performed with mechanochemical leaching in different mills and chemical leaching of nonactivated samples (Fig. 4.38). The results confirmed the favorable effects of mechanochemical leaching on the recovery of lead from CRT funnel glasses. The mechanochemical leaching efficiencies increased with increase of leaching temperatures. 97.39% Pb extraction can be achieved by mechanochemical leaching in stirring ball mill at 70 °C, compared with 2.75% Pb extraction for chemical leaching of nonactivated samples and 39.28% Pb extraction for chemical leaching of activated samples for 120 min by planetary ball mill.

4.12.3 Effects of Stirring Speed on Recovery of Lead from CRT Funnel Glass by Mechanochemical Extraction in Alkaline Solution

The effect of stirring speed on the lead extraction is shown in Fig. 4.39. The lead extraction increased with increase of rotation speed, and 97.39% of lead leaching was obtained at over 500 r/min of stirring speed. It indicated that higher stirring speed of milling supplied higher energy to break chemical bonds of lead glass.

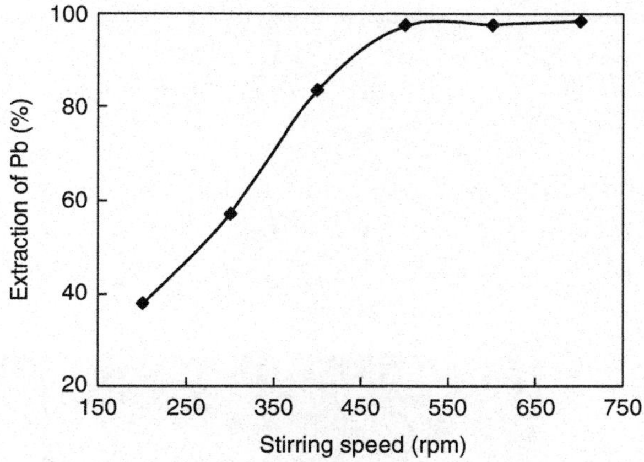

Fig. 4.39 Effects of stirring speed on lead extraction (NaOH concentration 5 M, temperature 70 °C, mass ratio of ball to raw materials 25:1, leaching time 2 h)

4.12.4 Eletrowinning of Lead Powder for Recovery of Lead from CRT Funnel Glass by Mechanochemical Extraction in Alkaline Solution

The electrolysis process was tested for the leaching solution of CRT funnel glasses. The typical composition of leaching solution of CRT funnel glasses is shown in Table 4.41. The electrodeposition was conducted under a galvanostatic condition for 2 h at ambient temperature. Experimental results are shown in Table 4.42. The deposited lead (Pb>97.2%) was in spongy form and gray with a slightly bright metallic luster. Morphological structure of the deposited lead was given in Fig. 4.40.

The leaching solution can be used to produce metallic lead powder by electrowinning, and the NaOH solution was recycled to the next leaching operation after most of the lead was electrowon. The high-purity lead powder with metallic lead over 97% can be obtained. NaOH losses in the whole process were estimated to be around 50–60 g per kg of Pb from titration analysis. The loss may be beyond this figure in practical application as the loss of NaOH was dependent on the type and composition of the leaded glass.

The main cost item of the process may be the energy consumed by mechanochemical leaching and electrowinning. Around 4.0–4.5 kW·h of electricity was needed for mechanochemical leaching to recover 1 kg metallic lead from the CRT funnel glasses. And the energy needed for electrowinning was estimated as 0.6–0.7 kW·h/kg Pb on average. Table 4.43 shows the typical electrowinning conditions for production of lead from the alkaline leaching solution of CRT funnel glasses.

After lead in the waste CRT funnel glasses was removed through alkaline leaching, it is both economically and environmentally friendly to use the leaching residues with fine granularity to prepare the foam glass. To compare with other current methodologies, mechanochemical alkaline leaching-electrowinning used to

Table 4.41 Chemical composition of the CRT funnel glasses in this experiment

Elements	Al	Ba	Ca	Bi	Fe	Na	Pb	Zn	Mg
Wt. %	1.44	0.68	3.10	0.05	0.59	5.86	25.12	0.16	1.09

Table 4.42 Typical composition of leaching solution of the glasses

Elements	Pb	Al	Ba	Ca	Bi	Fe	Zn	Mg
g/L	17.21	0.37	0.048	0.017	0.012	0.007	0.025	0.011

Fig. 4.40 SEM photomicrographs of cathode lead electrowon at 500 A/m^2

Table 4.43 Typical electrowinning conditions for production of lead from the alkaline leaching solution of CRT funnel glasses

Pb (g/L)	Voltage (V)	Current density (A/m^2)	Specific energy consumption (kWh/kg Pb)	Electrical current efficiency (%)	Copper metal purity (%)
17–20	1.7–1.9	400–500	0.6–0.7	97.5%	>97.2

recover lead from CRT funnel glasses was of high efficiency and simple process, which could produce Pb metal powder with high purity from waste CRT funnel glass directly. A new cleaner hydrometallurgy route to deal with the troublesome issue of lead recovery from waste CRT funnel glasses and other leaded glasses has been proposed. Figure 4.41 gives the schematic flow sheet of hydrometallurgical process for leaded glasses by alkaline mechanical-chemical leaching and electrowinning.

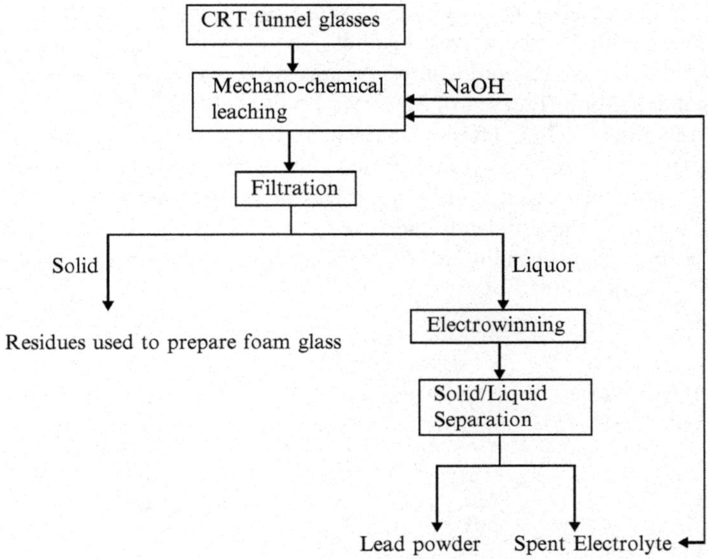

Fig. 4.41 Schematic flow sheet of hydrometallurgical process for leaded glasses by alkaline mechanical-chemical leaching and electrowinning

4.13 Extraction of Lead from Spent Leaded Glass in Alkaline Solution by Mechanochemical Reduction with Metallic Iron

Mechanical activation tests of CRT funnel glass were carried out in a planetary ball mill (QM-QX Planetary Ball Mill, China). Stainless steel ball was used as grinding bodies with dimensions between 1 mm and 5 mm. Leaching experiments of CRT funnel glass mechanically activated were carried out in a flask placed on a thermostatically controlled magnetic stirrer. To a 250 mL sodium hydroxide solution, CRT funnel glass was added and then leached at constant temperature. After leaching, the pulp was centrifuged, and the clear liquid was diluted and subjected to analysis lead by ICP-AES. Leaching behavior of the alkaline leaching residues was determined according to the toxicity characteristic leaching procedure.

4.13.1 Effects of Fe/Leaded Glass Ratios for Extraction of Lead from Spent Leaded Glass in Alkaline Solution by Mechanochemical Reduction with Metallic Iron

Leaded glass was crushed and sieved to <0.15 mm and then leached directly with a 5 M NaOH solution at 80 °C. Less than 5% of the Pb can be released from the glass

(Fig. 4.42). Direct alkaline leaching was not an effective means of extracting Pb. Compared with the nonactivated sample, the Pb extraction of sample mechanically activated increased significantly. Around 67% of Pb was released after 2 h of mechanical activation (Fig. 4.43). Since the Pb in leaded glass cannot be extracted mostly by alkaline leaching after mechanical activation, a mechanochemical reduction with metallic iron was tested. After 2 h mechanochemical reduction with different Fe/glass mass ratios, the sample was leached with a 5 M NaOH solution at 80 °C. The leaching efficiency increased considerably as the metallic iron was added. The maximum leaching efficiency was obtained when the Fe/glass (mass ratio) ratio was over 20%.

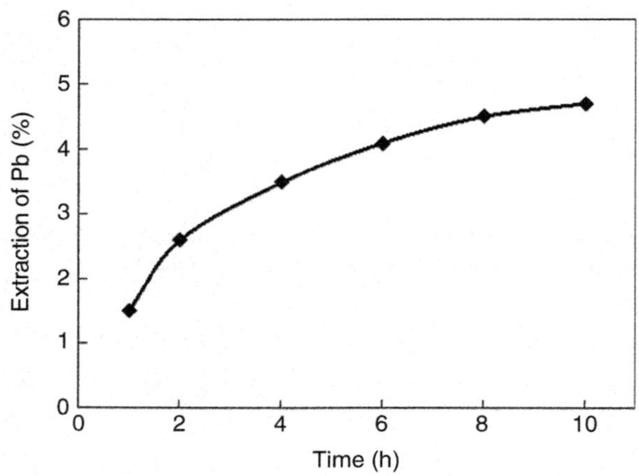

Fig. 4.42 Direct alkaline leaching of spent leaded glass

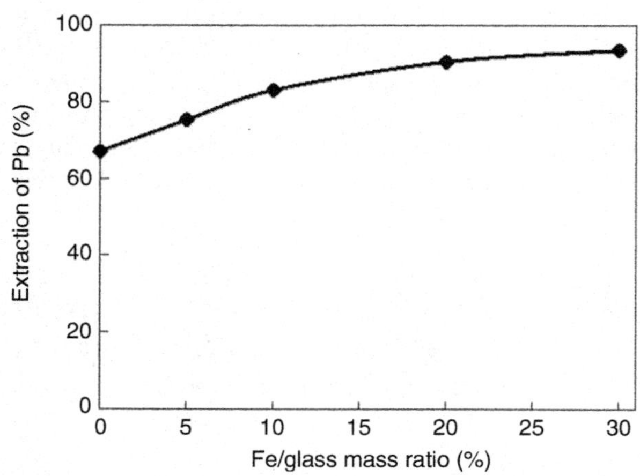

Fig. 4.43 Effects of the Fe/glass mass ratios on the Pb extraction from spent leaded glass using a mechanochemical reduction with metallic iron

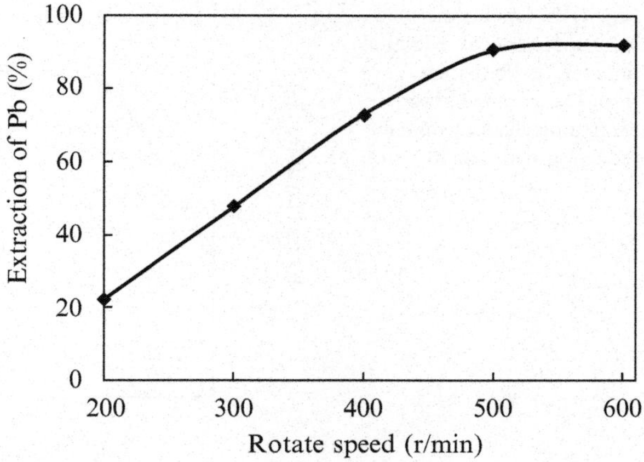

Fig. 4.44 Effects of rotate speed on the Pb extraction from spent leaded glass using a mechanochemical reduction with metallic iron

4.13.2 Effects of Rotate Speed for Extraction of Lead from Spent Leaded Glass in Alkaline Solution by Mechanochemical Reduction with Metallic Iron

In the planetary ball milling processing, different rotate speed supplied inequable energy, which significantly affected the reaction rate of the mechanochemical reduction. The effects of rotate speed on the Pb extraction are shown in Fig. 4.44. The Pb extraction increased with the increase of rotate speed, and 90.55% of Pb can be extracted at rotate speed of over 500 rpm.

4.13.3 Effects of Mechanochemical Reduction Time on Extraction of Lead from Spent Leaded Glass in Alkaline Solution by Mechanochemical Reduction with Metallic Iron

The effects of mechanochemical reduction time on the Pb extraction are shown in Fig. 4.45. The rate of Pb leaching increased with increase of mechanochemical reduction time. No remarkable increase in lead dissolution was observed for the time above 3 h, when 92.75% of lead can be extracted.

4.13.4 Analysis on Physicochemical Changes of Prepared Samples After Mechanochemical Reduction with Iron

The morphologies of mechanically activated and mechanochemical reduction glass samples observed by SEM are shown in Fig. 4.46. Compared with the activated

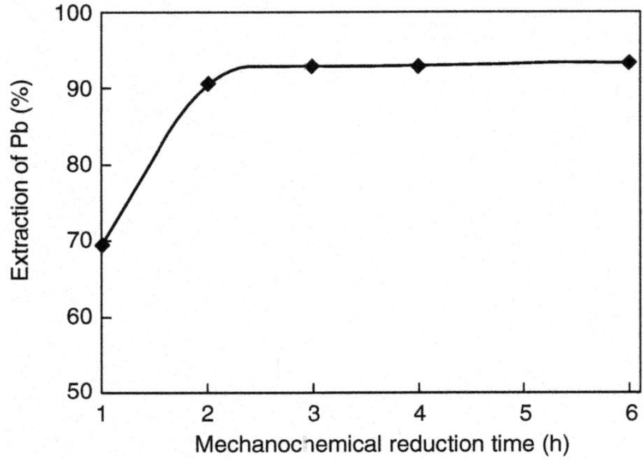

Fig. 4.45 Effects of mechanochemical reduction time on the Pb extraction from spent leaded glass using a mechanochemical reduction with metallic iron

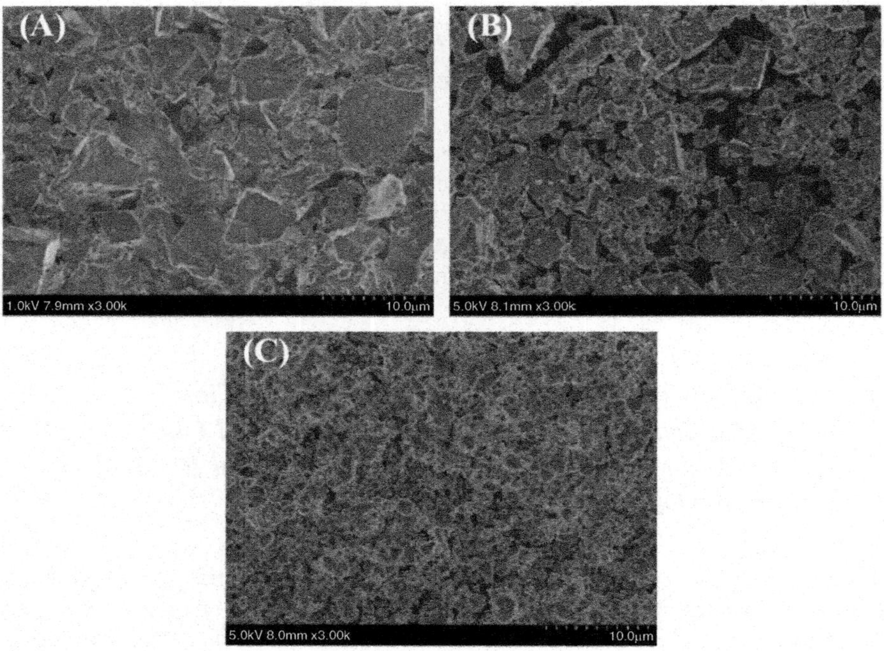

Fig. 4.46 SEM images of CRT funnel from spent leaded glass. (**a**) Raw material, (**b**) activated sample at the speed of 600 rpm, (**c**) mechanochemical reduction sample with metallic iron

sample, the mechanochemical reduction sample was very fine, which indicated that mechanochemical reduction with metallic iron was effective in prompting the destruction of leaded glass.

XRD patterns of the form spent leaded glass after ground with Fe at different Fe/glass mass ratios, time, and rotational speeds are given in Figs. 4.47, 4.48, 4.49,

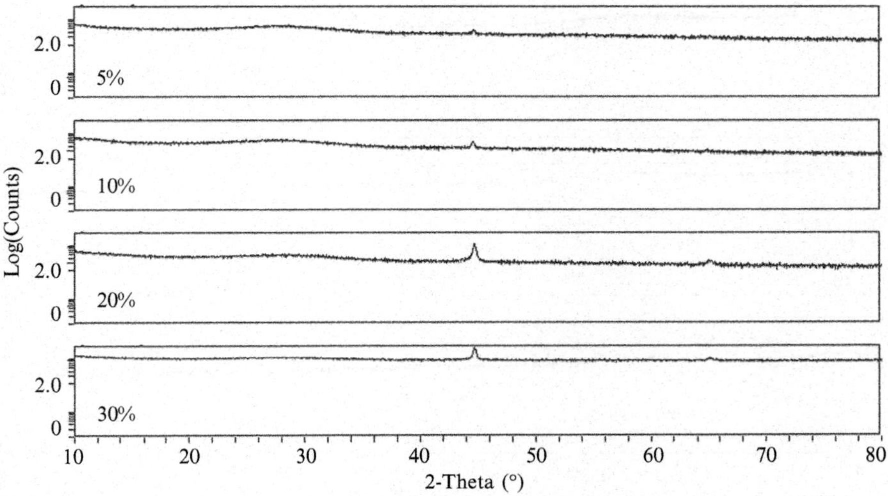

Fig. 4.47 XRD pattern of mechanochemical reduction sample with Fe at different Fe/glass mass ratios (time 2 h, rotational speed 500 rpm)

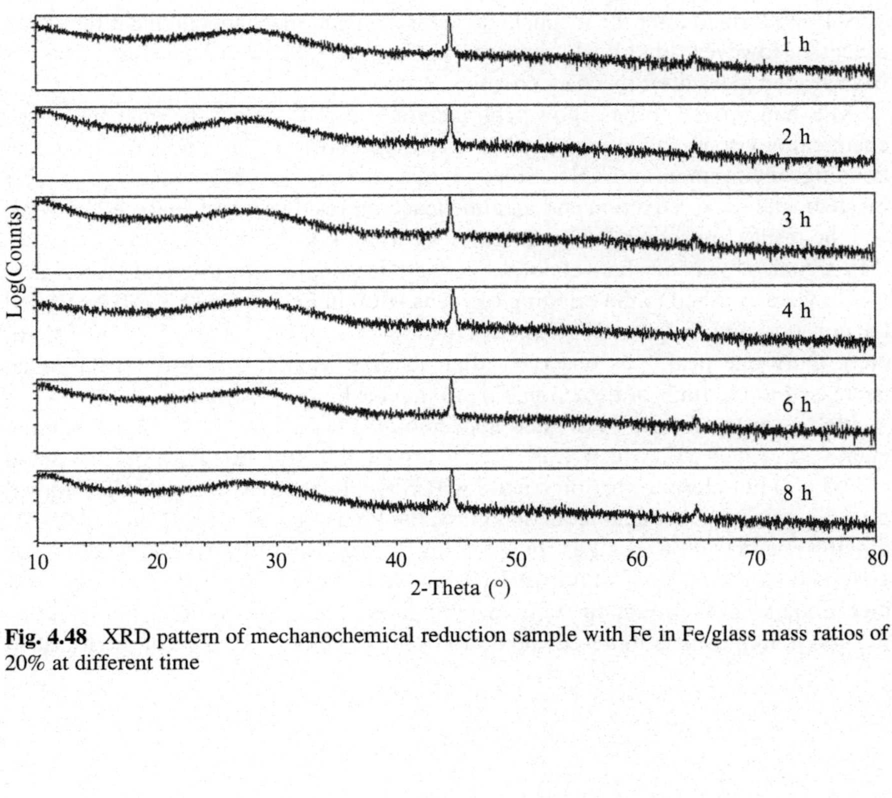

Fig. 4.48 XRD pattern of mechanochemical reduction sample with Fe in Fe/glass mass ratios of 20% at different time

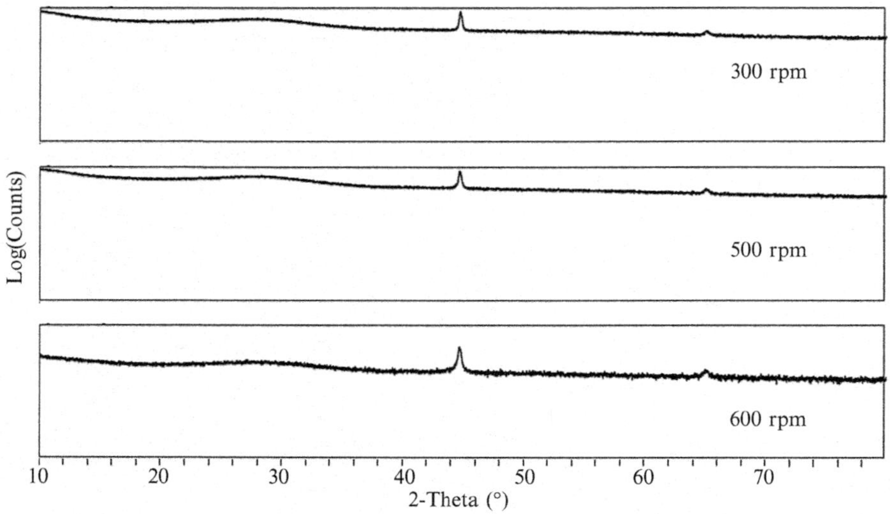

Fig. 4.49 XRD pattern of mechanochemical reduction sample with Fe in Fe/glass mass ratios of 20% at different rotational speeds

and 4.50, which revealed that the initial leaded glass in the form spent leaded glass was nanocrystal formation, and the leaded glass was still amorphous, without crystalline formed after the mechanical activation and mechanochemical reduction process. However, PbO peak appeared in the XRD pattern of leaching residues from leaded glass after mechanochemical reduction with iron.

XPS had proved to be a powerful analytical tool to detect the changes in the chemical environments of elements in glass structures. In this study, the chemical bonding environments of O1s, Si2p, Pb4f, and Fe2p in CRT funnel glass after mechanochemical reduction and alkaline leaching residues were analyzed by XPS, and the results are shown in Fig. 4.51. As shown in Fig. 4.51a, it was found that the O1s signal of two mechanochemical reduction samples consisted of two peaks, which were ascribed to the bridging oxygens (BO) in Si-O-Si (532.3 ± 0.2 eV) and the non-bridging oxygens (NBO) in Si-O-X (530.4 ± 0.2 eV, where X = Na, Pb, Al, etc.). Only one peak was observed after alkaline leaching, and the NBO signal decreased to the limit of detection, due to the leaching of the Pb.

$Pb4f_{7/2}$ of the funnel CRT glass appeared as a symmetric peak with a binding energy of 138.4 eV, which was clearly higher than for the PbO crystal phase (137.5 eV) but close to that obtained for $PbSiO_3$. In comparison with CRT funnel glass, in mechanochemical reduction glass, the $Pb4f_{7/2}$ peaks were shifted to higher binding energies (138.9 eV) as shown in Fig. 4.51b. The shift in the Pb4f peak was governed by the increase in the relative content of Pb modifier with the glass after mechanochemical reduction with metallic iron. Since the Pb-O bond in lead-oxygen pyramids was more covalent, a $Pb4f_{7/2}$ peak at 138.2 eV can be ascribed

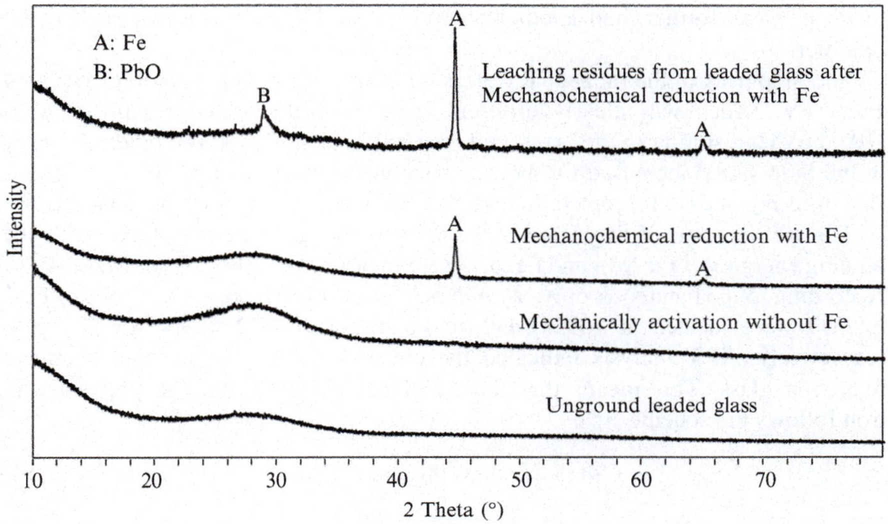

Fig. 4.50 XRD pattern of raw material and activated samples for spent leaded glass

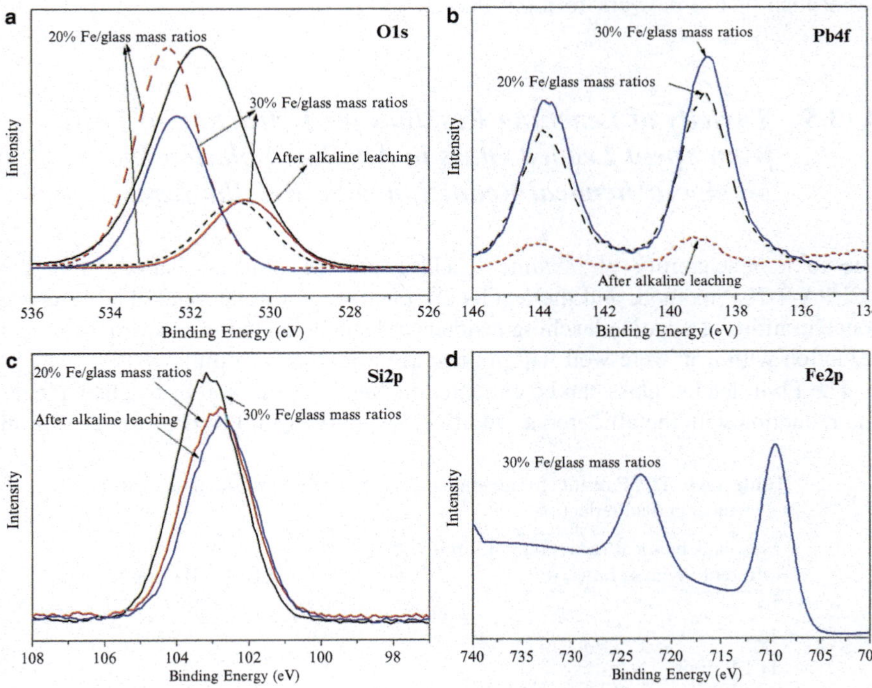

Fig. 4.51 O1s, Si2p, Pb4f, and Fe2p spectra of mechanochemical reduction and leached samples

to Pb network former, and another component at 139.0 eV was ascribed to Pb modifier.

The Si2p peak of the funnel CRT glass was characterized by a binding energy of 102.4 eV, which was clearly different from the value characteristic of SiO_2 (103.5 eV) but similar to the one found for $PbSiO_3$ (102.7 eV). The binding energy of the Si2p signal increased for mechanochemical reduction with metallic iron, as shown in Fig. 4.51c. It seemed that some lead in the glass formed as network.

The iron spectra of mechanochemical reduction glass in Fig. 4.51d with the binding energies of Fe $2p_{3/2}$ and Fe$2p_{1/2}$ were measured as 709.4 eV and 722.94 eV. According to the handbook of X-ray photoelectron spectroscopy, the peaks of Fe^{2+} in FeO were located at 709.30 eV for Fe $2p_{3/2}$ and 722.30 eV for Fe $2p_{1/2}$, respectively. XPS analysis indicated the presence of FeO in mechanochemical reduction glass. This means that the mechanochemical reduction process with iron follows the scheme

$$[PbO_2]+Fe \rightarrow Pb_{mod}\text{-}O + FeO \tag{4.16}$$

where $[PbO_2]$ is the network-forming lead (glass former) and Pb_{mod} is the modifier lead.

The leaching rate of leaded glass increased after mechanochemical reduction with iron indicated that the dissolution of network-modifying Pb by leaching was easier than that of network-forming Pb.

4.13.5 *Toxicity of Leaching Residues for Extraction of Lead from Spent Leaded Glass in Alkaline Solution by Mechanochemical Reduction with Metallic Iron*

The toxic leachability of alkaline leaching residues was measured using the USEPA-TCLP method, and the results are presented in Table 4.44. The leaching concentrations of alkaline leaching residues of leaded glass after mechanochemical reduction with iron were well within the current regulatory limits.

The Pb in leaded glass can be extracted in NaOH solution after mechanochemical reduction with metallic iron as additive. Over 90% of Pb can be extracted from

Table 4.44 TCLP results for alkaline leaching residues of leaded glass after mechanochemical reduction

Mechanochemical reduction with iron under different Fe/glass ratio (wt.%)	Pb (mg/L)	Ba (mg/L)
20	1.21	8.26
30	0.97	7.83
TCLP limit	5	100

the mechanochemically reduced leaded glass, compared with less than 5% of Pb extraction for alkaline leaching of nonactivated samples and 67% Zn extraction for alkaline leaching of activated samples for 2 h by planetary ball mill. After leaching, the solution can be used to produce metallic lead powder by electrowinning, and the leaching residues can be used to prepare the foam glass.

4.14 Alkaline Leaching Process of Fume Dust and Lead Oxide Ore

A huge quantity of lead-bearing wastes, which is classified as hazardous waste in most countries, has been generated in the industries such as hydrometallurgical process of zinc, batteries manufacture, and chemical process. In general, the contents of lead in the wastes may be quite low and usually are disposed of at dumping sites. In addition, thousands of millions of tailings containing low-grade lead are also being stored at the dumping sites in the mining plants. The concentration of lead from these secondary resources may be quite difficult, and their recycling as raw materials has been limited due to the unavailability of cost-effective technologies. Pyrometallurgical process may be one option for the lead extraction from the wastes, but its lead dust pollution has caused environmental constraints. A cleaner hydrometallurgical process using alkaline leaching and electrowinning had been developed for the extraction of lead from the wastes and low-grade lead ores in this work.

4.14.1 Alkaline Leaching Process for Pb and Zn Recovery from Fume Dust

The fume dust comprised of Zn, Pb, and Fe (Table 4.45) was taken from zinc smelters with a size smaller than 0.074 mm, and the main chemical contents were ZnO and $PbCO_3$ (Fig. 4.52).

The leaching rate rapidly increased with the increase of temperature and NaOH concentration. 70 °C and NaOH of 5 mol/L were chosen as the optimum temperature and NaOH concentration (Fig. 4.53a, c). The effect of liquid-solid ratio (L/S) on the leaching efficiency of Zn and Pb was similar. A bigger liquid-solid ratio would lead to a higher leaching efficiency and maintained to a stable value at a ratio

Table 4.45 Chemical composition of fume dust used

Composition	Zn	Pb	Fe	W	Cu	Al	Sn	Ca	Si	As	Sb
Content (%)	45.76	15.66	10.29	1.71	0.57	0.14	1.39	0.52	0.14	0.014	0.019

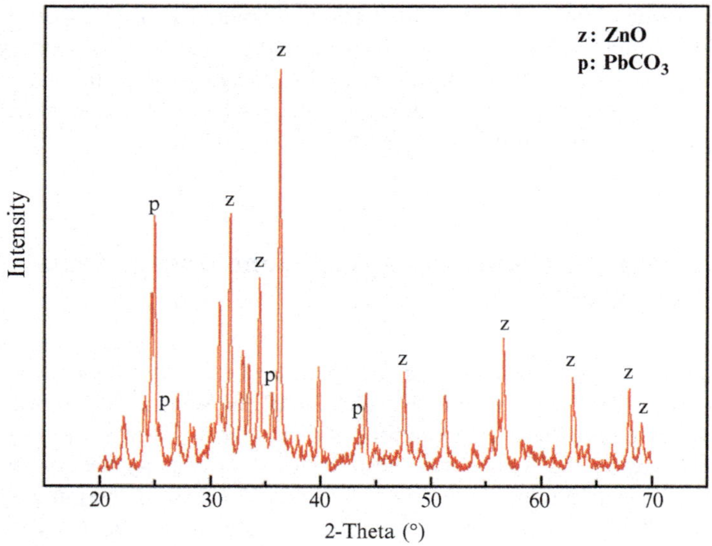

Fig. 4.52 X-ray diffraction pattern of fume dust used

over 11:1 (Fig. 4.53d). The leaching rate was over 90% after 30 min or longer (Fig. 4.53b). Meanwhile, agitation speed exerted a great effect on the leaching as settling of dust in the bottom of reactor may occur at lower agitation strength (Fig. 4.53e). Meanwhile, Al, As, and Sb in the fume dust would be also leached out along with Zn and Pb as given in Fig. 4.54.

4.14.2 Alkaline Leaching Process for Lead Oxide Ore

The elements of lead oxide ore used are shown in Table 4.46, and the XRD analysis showed that lead oxide ore contained complicated components and the major phases were $PbSO_4$, $PbSiO_3$, and PbO (Fig. 4.55).

Pb leaching rate increased with the increase of temperature and NaOH concentration (Fig. 4.56). The optimum leaching can be obtained at over 70 °C and 30 min leaching time, 5 mol/L NaOH concentration, and a suitable liquid-solid ratio. The concentration of Pb in leaching solution at various liquid-solid ratios is presented in Table 4.47. A liquid-solid ratio of 25:1 was used.

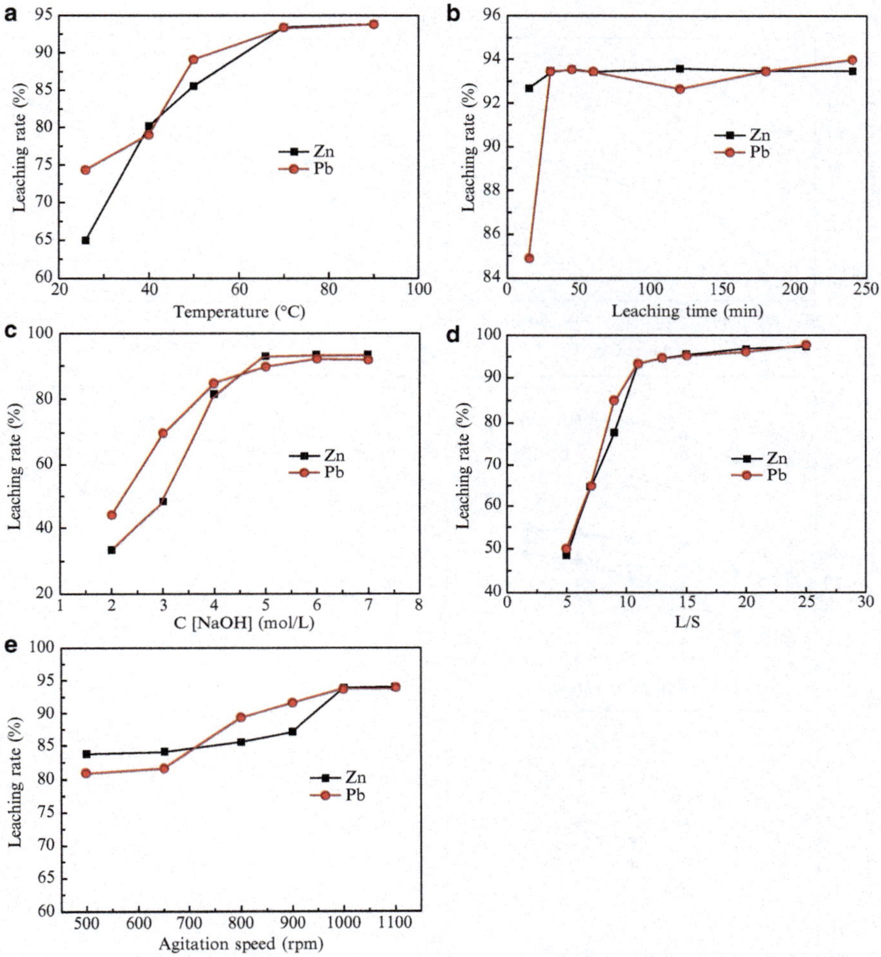

Fig. 4.53 Leaching rate of Pb and Zn from fume dust under different alkaline leaching conditions: (**a**) effects of temperature, (**b**) effects of leaching time, (**c**) effects of NaOH concentration, (**d**) effects of liquid-solid ratio (L/S), (**e**) effects of agitation speed

As shown in Fig. 4.56, the stirring speed and particle sizes affected the leaching process. Over 90% leaching of Pb can be obtained from the dust with the particle size of 0.15 mm. The major chemical reactions of alkaline leaching for lead oxide ore are as follows:

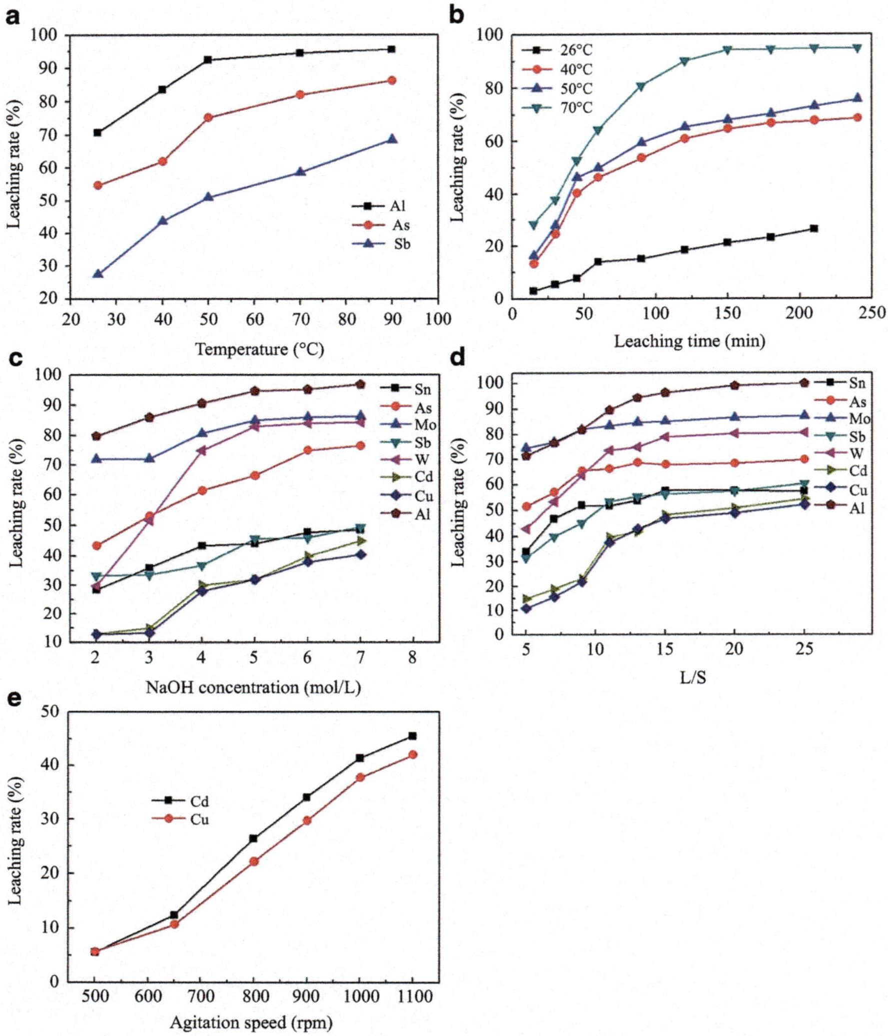

Fig. 4.54 Leaching rate of nontarget metals from fume dust under different alkaline leaching conditions: (**a**) effects of temperature, (**b**) effects of leaching time and temperature on the leaching rate of Cu, (**c**) effects of NaOH concentration, (**d**) effects of liquid-solid ratio (L/S), (**e**) effects of agitation speed on the leaching rate of Cd and Cu

Table 4.46 Chemical composition of lead oxide ore

Composition	Pb	Zn	Sb	Al	Cu	Sn	Ca	Si	As	W
Content (%)	51.61	14.65	2.12	0.47	0.88	0.33	2.27	0.27	0.34	0.26

Fig. 4.55 X-ray diffraction pattern of lead oxide ore

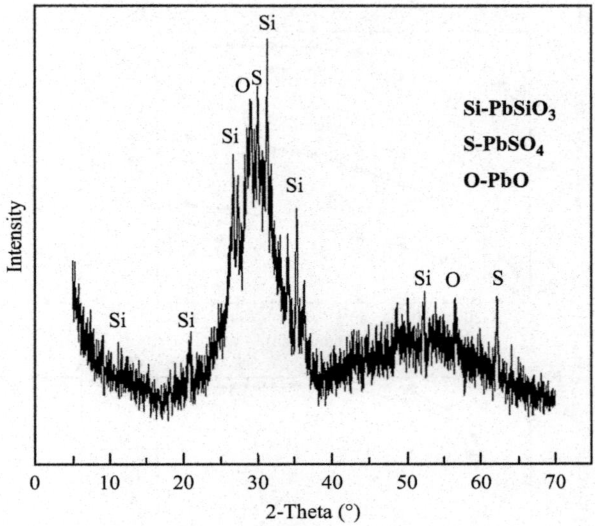

$$PbSiO_3(s) + 3NaOH(aq) = NaPb(OH)_3(aq) + Na_2SiO_3(aq) \tag{4.17}$$

$$PbSO_4(s) + 3NaOH(aq) = NaPb(OH)_3(aq) + Na_2SO_4(aq) \tag{4.18}$$

$$PbO(s) + NaOH(aq) + H_2O(aq) = NaPb(OH)_3(aq) \tag{4.19}$$

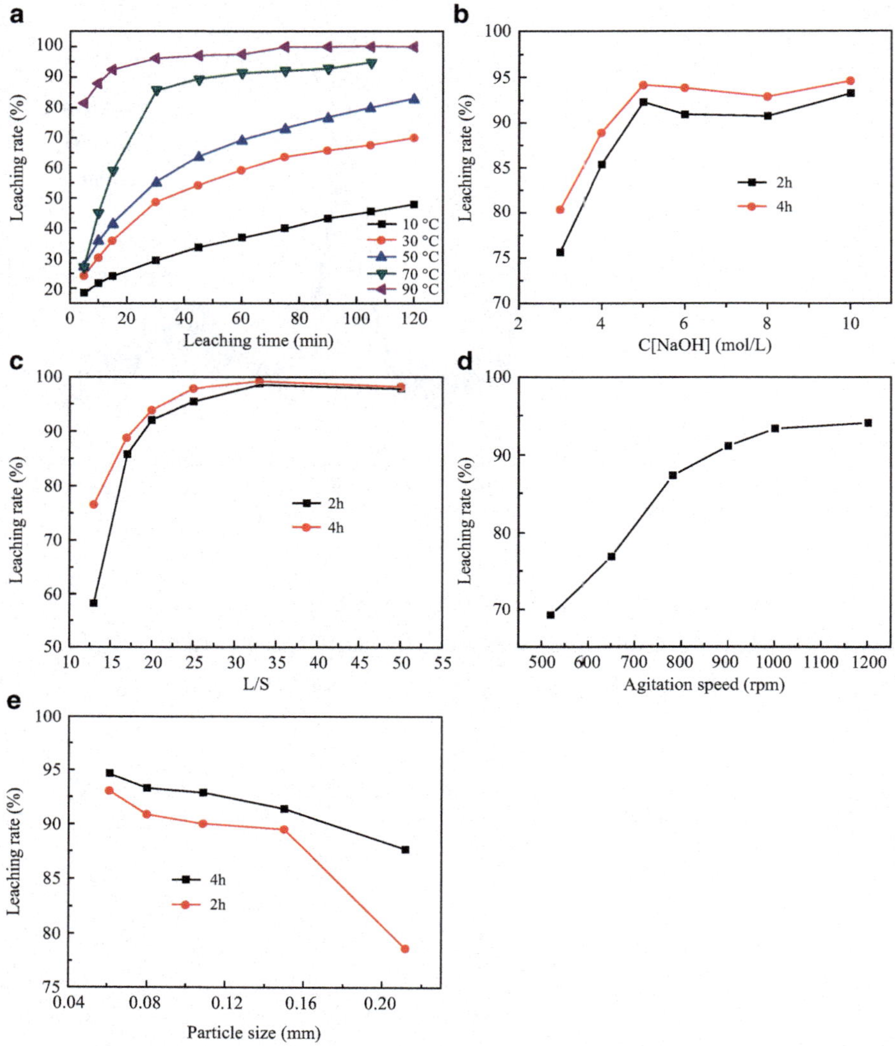

Fig. 4.56 Leaching rate of Pb from lead oxide ore under different alkaline leaching conditions: (**a**) effects of temperature, (**b**) effects of NaOH concentration, (**c**) effects of liquid-solid ratio (L/S), (**d**) effects of agitation speed, (**e**) effects of particle size

Table 4.47 Effects of liquid-solid ratio on the concentration of Pb in the alkaline leaching solution

Liquid-solid ratio	50:1	33:1	25:1	20:1	17:1	13:1
Pb (g/L)	10.11	15.38	20.23	24.20	27.48	31.67

Chapter 5
Purification of Leach Solution of Zinc and Lead in Alkaline Solutions

Abstract Zinc and lead are usually concomitantly present in zinc and lead hazardous wastes such as electric arc furnace (EAF) dust. The selective and quantitative separation of lead from zinc is an important step in these alkaline hydrometallurgy processes. The separation of lead from zinc in alkaline zinc solution, in leaching solutions of zinc and lead hazardous wastes, can be achieved by sodium sulfide precipitation. When the weight ratio of sodium sulfide (average molecular weight 222) added to the lead present in alkaline solution is over 1.8–2.0 (molar ratio approximately 1.5–1.7), lead could be separated selectively and quantitatively, while the zinc remained in the solution without concomitant loss. The residues from the precipitation step are identified as mixtures of PbS and $Na_2Pb(OH)_2S$, with PbS being the predominant compound. Sn can be cemented quantitatively from alkaline zinc solution by the addition of zinc powder at 95 °C, 500 rpm with Zn/Sn molar ratio 6. Al in the leaching solution can be removed by Na_2SiO_3, As can be removed by $Fe_2(SO_4)_3$ at low NaOH concentration, Cl^- can be removed by increasing NaOH content, and Cu can be removed from alkaline lead leaching solution by Pb power replacement.

Keywords Purification • Sn • Al • Cu • Chloride • Impurity separation

5.1 Selective Precipitation and Separation of Lead from Alkaline Zinc Hydroxide Solution

When dust was contacted with NaOH solution, the leaching solutions may generally contain Zn, Pb, Cr, Cu, and Al. In most cases, Zn and Pb are the major elements to be considered for the separation, as the corresponding contents of Cr, Cu, and Al may be usually quite low in comparison with those of Zn and Pb. A unique technique was developed for the selective recovery and separation of lead from strong alkaline zinc hydroxide solution using sodium sulfide as precipitant. The leaching solutions obtained were used directly as the testing solutions for the selective precipitation of lead and zinc. Sodium sulfide was not stable, and the number of hydrates in the crystal was also changed gradually. The sodium sulfide

Y. Zhao, C. Zhang, *Pollution Control and Resource Reuse for Alkaline Hydrometallurgy of Amphoteric Metal Hazardous Wastes*, Handbook of Environmental Engineering 18, DOI 10.1007/978-3-319-55158-6_5

used was produced in Germany (Merck, Pro Analysi). The chemical formula shown in the container was $Na_2S{\cdot}xH_2O$ ($x = 7$–9). The accurate concentration of sodium sulfide was determined by EPA direct colorimetric analysis. It was found that the molar mass of the chemical was about 222 ($x = 8$).

The alkaline mixture solution of zinc and lead may also be prepared by the addition of sodium hydroxide solution to the mixed solution of lead and zinc nitrates. It can be observed that colloidal and bulky zinc and lead hydroxides appear and then dissolved slowly as the alkaline solution was added continuously. Nevertheless, in many cases, there was always a small proportion of the precipitates that could not be redissolved, even if a large excessive sodium hydroxide solution was added. The alkaline zinc and lead hydroxide solutions were also prepared in a different way, i.e., by the addition of zinc and lead nitrate solutions to sodium hydroxide solution drop by drop while stirring with magnetic stirrers. The final pH (or concentration of sodium hydroxide) and contents of zinc and lead in the resulting alkaline solution can be controlled quite easily. It was found that the precipitate of lead and zinc hydroxides appeared when the final pH was decreased to around 12.8. In the testing solutions used, the final pH value was controlled at over 13. The solubility of zinc and lead hydroxide prepared in such a way may strongly relate to the final pH value and the addition of other ions. It was easily to prepare an alkaline solution containing up to 80 g/L Zn and 10 g/L Pb. Nevertheless, it was observed that the powdery precipitates may appear when the solution was kept standing by for 2 h. However, it was found that the leaching solution was still very stable when the zinc concentration was at as high as 60 g/L.

5.1.1 Selection of the Precipitants for Selective Precipitation and Separation of Lead from Alkaline Zinc Hydroxide Solution

The potential and available precipitants for the selective precipitation of lead from alkaline zinc hydroxide solution may include sodium phosphate, sodium carbonate, sodium sulfate, and sodium sulfide. The experimental results using these precipitants are shown in Table 5.1. It can be seen that the effective separation of lead from zinc was obtained when sodium sulfide was used as precipitant. Thus, sodium sulfide was selected and used as the precipitant in the following experiments.

Table 5.1 Selection of the precipitants for the separation of lead from alkaline zinc hydroxide solution

Precipitants	Phosphates	Carbonates	Sulfates	Sulfide
Concentrations (mol/L)	0.01–0.05	0.01–0.05	0.01–0.05	0.0025–0.015
Removal of zinc (%)	0.0	0.0	0.0	0.0
Removal of lead (%)	0.0	0.0	0.0	40.1–100.0

Note: Leaching solution—Zn 14.40 g/L and Pb 2.98 g/L; total volume of 12 mL

It was observed that the precipitates could not form when the other possible precipitants were added into the alkaline solution containing Zn and Pb, indicating that lead sulfate, lead and zinc phosphates, and lead and zinc carbonates, which were insoluble in neutral medium, were soluble in strong alkaline solution.

5.1.2 Purification Mechanism of Na_2S

5.1.2.1 The Existing Form of Sulfur in Alkaline Solution

When Na_2S is dissolved in the aqueous solution, $H_2S(aq)$, HS^- and S^{2-} may exist in the solution (Fig. 5.1), and there is an equilibrium as follows:

$$Na_2S(s) + 2H_2O = 2Na^+ + H_2S(aq) + 2OH^- \quad (5.1)$$

$$\Delta_r G^\theta_{298.15} = -41.4008\ \text{kJ/mol}, \quad K^\theta_{298.15} = 1.7927 \times 10^7$$

$$Na_2S(s) + H_2O = 2Na^+ + HS^- + OH^- \quad (5.2)$$

$$\Delta_r G^\theta_{298.15} = -81.7266\ \text{kJ/mol}, \quad K^\theta_{298.15} = 2.0796 \times 10^{14}$$

$$Na_2S(s) = 2Na^+ + S^{2-} \quad (5.3)$$

$$\Delta_r G^\theta_{298.15} = -88.27\ \text{kJ/mol}, \quad K^\theta_{298.15} = 2.9181 \times 10^{15}$$

According to the standard Gibbs free energy in Eqs. 5.1, 5.2, and 5.3, Na_2S is freely soluble in aqueous medium. S^{2-} as well as HS^- mainly happens in the

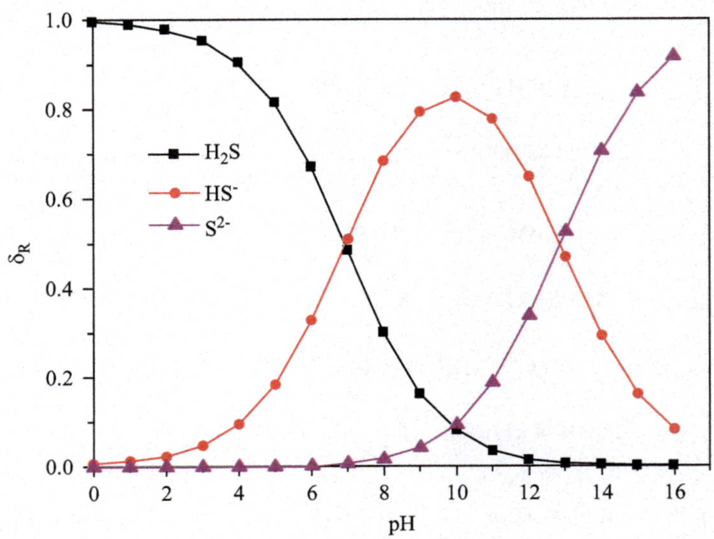

Fig. 5.1 Relationship between pH value and δ_R of different S(II)-containing species

solution with pH >14 (account for more than 99.87%), and hardly any H_2S (aq) exists (account for less than 0.13%). As the further increase of pH, the ratio of S^{2-} rises quickly and accounts for 99.97% when pH = 15. When implying a Na_2S purifying process in practice, the concentration of NaOH always comes to 180 ~ 200 g/L, which can provide an alkali system for the dissolution of Na_2S and generation of abundant S^{2-}. However, Na_2S is generally dissolved first in the water and subsequently added to the purification tank so as to reduce the generation of unwanted and poisonous H_2S.

5.1.2.2 Mechanism of Quantitative Separation of Lead and Zinc by Na_2S

Lead and zinc can dissolve in the alkaline solution, so the next reaction will happen when adding the Na_2S to the leaching solution:

$$\mathrm{Pb(OH)_3^-} + \mathrm{S^{2-}} = \mathrm{PbS(s)} + \mathrm{3OH^-} \tag{5.4}$$

$$\Delta_r G^\theta_{298.15} = -80.85\ \mathrm{kJ/mol},\quad K^\theta_{298.15} = \frac{[\mathrm{OH^-}]^3}{[\mathrm{Pb(OH)_3^-}][\mathrm{S^{2-}}]} = 1.4625 \times 10^{14}$$

$$\mathrm{Pb(OH)_3^-} + \mathrm{HS^-} = \mathrm{PbS(s)} + \mathrm{2OH^-} + \mathrm{H_2O} \tag{5.5}$$

$$\Delta_r G^\theta_{298.15} = -87.3974\ \mathrm{kJ/mol},\quad K^\theta_{298.15} = \frac{[\mathrm{OH^-}]^2}{[\mathrm{Pb(OH)_3^-}][\mathrm{HS^-}]} = 2.05217 \times 10^{15}$$

$$\mathrm{Zn(OH)_4^{2-}} + \mathrm{S^{2-}} = \mathrm{ZnS(s)} + \mathrm{4OH^-} \tag{5.6}$$

$$\Delta_r G^\theta_{298.15} = -50.545\ \mathrm{kJ/mol},\quad K^\theta_{298.15} = \frac{[\mathrm{OH^-}]^4}{[\mathrm{Zn(OH)_4^{2-}}][\mathrm{S^{2-}}]} = 7.17123 \times 10^{8}$$

$$\mathrm{Zn(OH)_4^{2-}} + \mathrm{HS^-} = \mathrm{ZnS(s)} + \mathrm{3OH^-} + \mathrm{H_2O} \tag{5.7}$$

$$\Delta_r G^\theta_{298.15} = -57.0924\ \mathrm{kJ/mol},\quad K^\theta_{298.15} = \frac{[\mathrm{OH^-}]^3}{[\mathrm{Zn(OH)_4^{2-}}][\mathrm{HS^-}]} = 1.00626 \times 10^{10}$$

$$\mathrm{ZnO_2^{2-}} + \mathrm{S^{2-}} + \mathrm{2H_2O} = \mathrm{ZnS(s)} + \mathrm{4OH^-} \tag{5.8}$$

$$\Delta_r G^\theta_{298.15} = -51.651\ \mathrm{kJ/mol},\quad K^\theta_{298.15} = \frac{[\mathrm{OH^-}]^4}{[\mathrm{ZnO_2^{2-}}][\mathrm{S^{2-}}]} = 1.12038 \times 10^{9}$$

$$\mathrm{ZnO_2^{2-}} + \mathrm{HS^-} + \mathrm{H_2O} = \mathrm{ZnS(s)} + \mathrm{3OH^-} \tag{5.9}$$

$$\Delta_r G^\theta_{298.15} = -58.1984\ \mathrm{kJ/mol},\quad K^\theta_{298.15} = \frac{[\mathrm{OH^-}]^3}{[\mathrm{ZnO_2^{2-}}][\mathrm{HS^-}]} = 1.57212 \times 10^{10}$$

Thermodynamic data (Eqs. 5.4, 5.5, 5.6, 5.7, 5.8, and 5.9) exhibits that the Pb (II) and Zn(II) can react with S(II) spontaneously and produces the metal sulfide precipitation. On the basis of thermodynamic equilibrium constants and NaOH concentration (180–200 g/L, $[OH^-] = 4.5 \sim 5$ mol/L), the equilibrium concentration of three kinds of metal complex can be calculated (Eqs. 5.10, 5.11, and 5.12). The concentration of Pb(II) in alkaline solution is much lower than the Zn(II), resulting in the easily generation of PbS compared to ZnS. Therefore, if only a proper amount of Na_2S is added, lead and zinc can be separated theoretically:

$$[Pb(OH)_3^-][S(II)] = 4.87288 \times 10^{-16}[OH^-]^2 + 6.83762 \times 10^{-15}[OH^-]^3$$

$$\left[Zn(OH)_4^{2-}\right][S(II)] = 9.93774 \times 10^{-11}[OH^-]^3 + 1.39446 \times 10^{-11}[OH^-]^4$$

$$[ZnO_2^{2-}][S(II)] = 6.36083 \times 10^{-11}[OH^-]^3 + 8.9255 \times 10^{-11}[OH^-]^4$$

$$[Pb(OH)_3^-][S(II)] = (6.32945 \sim 8.66884) \times 10^{-13} \tag{5.10}$$

$$\left[Zn(OH)_4^{2-}\right][S(II)] = (5.80872 \sim 8.83960) \times 10^{-7} \tag{5.11}$$

$$[ZnO_2^{2-}][S(II)] = (3.71798 \sim 5.65795) \times 10^{-7} \tag{5.12}$$

In practice, the concentration of zinc is adjusted to 40 g/L (0.61 mol/L). For the purpose of minimizing the amount of ZnS during the purifying process, based on the discussion above, the controlled concentration of S(II) should be less than

$$[S(II)] = (1.56175 \sim 2.37665) \times 10^{-6}\text{mol/L}$$

Then, in order to eliminate the Pb(II) thoroughly where retain the Zn(II), the maximum Pb(II) concentration in leaching solution can be

$$[Pb]_T = (4.05279 \sim 3.64751) \times 10^{-7}\text{mol/L} = (8.39737 \sim 7.55763) \times 10^{-5}\text{g/L}$$

5.1.3 *Effects of Mass Ratio of Sodium Sulfide Added to Lead or Zinc in Leaching Solutions*

The selective separation of lead from the alkaline zinc hydroxide solution was found to be always quantitatively without any loss of zinc, when sulfide was added enough to make the mass ratio of sodium sulfide added to the lead present in the solution was at 1.7–1.9 (molar ratio around 1.5–1.7) as given in Fig. 5.2. In practical application, a mass ratio of 1.8 should be used in order to make sure that the lead can be removed quantitatively, while the zinc will be remained in the solutions without loss. After the lead was separated, the zinc in the lead-free solution can be precipitated quantitatively when the weight ratios of sodium sulfide added to the

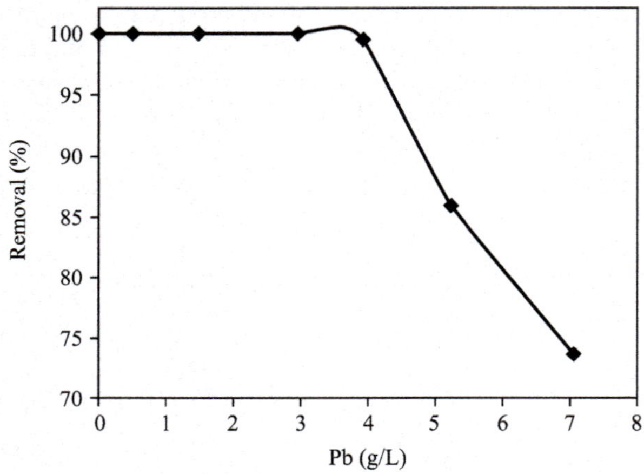

Fig. 5.2 Effects of Pb concentration in Zn-Pb alkaline solution by the addition of sodium sulfide. Sodium sulfide 7.8 g/L (M.W. = 204–240), Zn 14.3 g/L

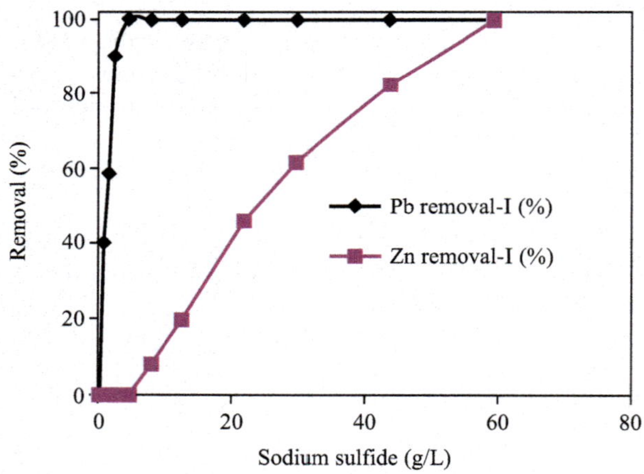

Fig. 5.3 Effects of sodium sulfide concentration on the removal of lead and separation of lead from zinc from alkaline solution. Pb 2.89 g/L, Zn 14.30 g/L, M.W. of sodium sulfide c.a. 222

zinc present were at or higher than 2.6–2.7 (molar ratio around 0.7–0.8) as shown in Fig. 5.3.

The concentration of sulfide ion was determined after lead or zinc was precipitated quantitatively at the abovementioned weight ratios. It was found that the sulfide concentrations present in the solutions were around 1 mg/L for lead case and 32 mg/L for zinc case after precipitation, respectively. Obviously, the sulfide added was consumed completely for the formation of sulfide precipitates of lead or zinc.

Table 5.2 The coremoval of copper from alkaline leach solutions

Cu added (mg/L)	Pb removal (%)	Zn removal (%)	Cu removal(%)
0	100.0	0.0	100.0
48.0	100.0	0.0	100.0
236.0	100.0	0.0	100.0
354.0	100.0	0.0	100.0
473.0	100.0	0.0	100.0
591.0	100.0	0.0	100.0
>600	Precipitates of copper	Hydroxide appear	

Note: Leaching solution—Zn 14.4 g/L, Pb 2.89 g/L, and Cu 20 mg/L; total volume of 12 mL

5.1.4 Co-removal of the Other Possible Soluble Coexistent Elements from Alkaline Zinc Hydroxide Solution Using Sodium Sulfide as Precipitant

In the strong alkaline leaching solution of pyrometallurgical dust and common industrial wastes, the other possible elements present may include Cu, Al, and Cr (III). Among them, Cu can be easily removed from the alkaline solutions together with lead (Table 5.2). In this case, the purity of lead sulfide precipitate may be contaminated by Cu sulfide. Fortunately, the concentration of Cu in the most leaching solution was lower than 5–50 mg/L, while that of lead may be 2–6 g/L. The copper content in the lead precipitate can be lower than 1–2%, which may be decreased further when the lead sulfide was treated in the subsequent process used for the purification of lead compound. However, it was not possible to remove Al and Cr(III) using sulfide, as sulfides of these two elements are instable in the solutions. Hence, the separation of Al and Cr(III) from Zn in alkaline solutions should be carried out by other processes, after the lead has been removed.

5.1.5 Scale-Up Experiments on Selective Precipitation and Separation of Lead from Alkaline Zinc Hydroxide Solution Using Sodium Sulfide as Precipitant

In order to justify the experimental results obtained from a relatively small volume of testing solution, the scale-up tests were done. The results are shown in Table 5.3. It can be seen that the lead was indeed separated selectively and zinc precipitated quantitatively from leaching solution in a relatively large scale.

Table 5.3 Removal and recovery of the lead from the direct leaching solution of the hydrolyzed dust melts by the addition of sodium sulfide

Composition of leaching solution (g/L)	Zn 26.95, Pb 6.53
Leaching solution volume used (mL)	2700
Sodium sulfide added (g)	25.90
Sodium sulfide/Pb (w/w)	1.47
Pb removal (%)	94.50
Approximate molar ratio of the sulfide added to the lead removed	1.58 ~ 1.34
Lead sulfide obtained after washing and drying at 110 °C (g)	21
Content of Pb in the lead sulfide sample (%)	79.58
Weight gain of the lead sulfide sample at 550 °C (%)	20.74
Content of Zn in the lead sulfide sample (%)	1.05
Proposed chemical formula of the sample	hydrates of PbS and Na_2PbS_2

5.1.6 Removal of Lead from Leaching Solution by the Addition of Solid Sodium Sulfide

Both the leaching solutions obtained from the direct leaching of dust and from the leaching of melts of hydrolyzed dust were used for the tests for the selective removal of lead by the addition of sodium sulfide. Table 5.3 shows the experimental results of the direct leaching solutions in which the concentration of lead was higher than that of the leaching solution of melts of hydrolyzed dust. It should be pointed out that less sodium sulfide with mass ratio of sodium sulfide to the lead of 1.47 rather than 1.8 was added to the leaching solution, for it was considered that it may be not necessary to precipitate out all the lead from leaching solution in practical application in order to save the chemical and reduce partly the operation cost. It can be seen that though the weight ratio of 1.47 was used, about 95% of lead was removed.

The precipitate was separated and hydrolyzed in water by heating for 2 h. Analyzed the composition of the dried lead sulfide sample obtained. It was found that the components of the precipitates of lead sulfide were the mixture of lead sulfide and sodium lead hydroxide sulfide, as there was certain content of sodium which was detected in the precipitate. After the lead was removed, more solid sodium sulfide was added to such an extent that the mass ratio of sodium sulfide to the zinc present was from 2.46 to 2.62. It can be seen that the higher the ratio were used, the higher the precipitation rate would be obtained (Table 5.4). The remaining zinc in the filtrate can be recycled to the next leaching step.

The corresponding chemical formula of zinc precipitate was proposed and also given in Table 5.4. It seemed to be the mixtures of ZnS, $Zn(OH)_2$, $Na_2Zn(OH)_2S$, etc. The particles of freshly precipitated lead and zinc sulfides were very fine and very difficult to be separated from the liquid. The precipitates should be washed or hydrolyzed in water for hours while heating to release the entrained NaOH. It was

Table 5.4 Recovery of zinc from the Pb-free alkaline leaching solution by the addition of sodium sulfide

Sample No.	1	2	3
Pb-free leaching solution (mL)	1250 (from the direct leaching of dust)	300 (leaching solution of the melts)	600 (leaching solution of the melts)
Zn content in the leaching solutions (g/L)	26.95	45.00	50.85
Sodium sulfide added (g)	84.00	33.21	80.00
Sodium sulfide/Zn (w/w)	2.49	2.46	2.62
Zinc sulfide obtained after washing and drying at 110 °C (g)	43.93	18.11	38.02
Zinc content in the zinc sulfide sample (%)	72.18	71.45	74.52
Zinc recovery based on the zinc sulfide sample collected (%)	94.12	95.85	99.86
Zinc content in the aqueous solution after recovery of zinc as zinc sulfide (g/L)	1.06	2.32	0.12
Approximate molar ratios of the sulfide added and the zinc precipitated	0.85–0.72	0.82–0.70	0.84–0.71
Weight gain (%)	-4.22	-6.09	-3.73
Proposed chemical formula	Mixtures of ZnS, ZnO, Zn $(OH)_2$, etc.	Mixtures of ZnS, ZnO, Zn $(OH)_2$, etc.	Mixtures of ZnS, ZnO, Zn $(OH)_2$, etc.

found that after the sulfides of lead and zinc were hydrolyzed thoroughly, the composition of the resultant sulfides roughly approaches to the simplest forms of PbS and ZnS.

5.1.7 *Recovery of Zinc from Lead-Free Alkaline Leaching Solutions by Crystallization*

The lead-free leaching solution was evaporated and concentrated to certain extent, and crystal would appear. Filtered, the crystal was dried at 120 °C and then analyzed. The results are given in Table 5.5. Obviously, the recovery of zinc was very low. If the mother liquid was further evaporated, a large amount of NaOH would be precipitated out, in which the major component of the crystal should be NaOH.

When the crystal was hydrolyzed in water, the content of zinc in the resultant solid was determined to be around 73%, which was much higher than that of the crystal. The corresponding hydrolysis parameters are shown in Table 5.6. The zinc contents in the supernatant were found to be quite higher, which indicated that the

Table 5.5 Crystallization of sodium zinc hydroxide from the Pb-free alkaline leaching solutions

Sample No.	1	2
Zn content in the leaching solution (g/L)	26.95	50.85
Original volume of the leaching solution (mL)	1350	600
Mother liquid after filtering out the crystal (mL)	400	400
Content of zinc in the mother liquid (g/L)	65.53	65.78
Crystal obtained (g)	77.48	42.85
Zinc content in the crystal (%)	13.50	9.75
Zinc recovery (%)	28.75	13.69
Weight of the hydrolyzed product of the crystal drying at 110 C (g)	14.76	
Zinc content in the hydrolyzed product of the crystal (%)	72.47	
Possible chemical formula of the dried hydrolyzed product of crystal	Mixtures of ZnO, Zn $(OH)_2$	

Table 5.6 Effects of water/crystal ratios (w/w) on the hydrolysis of the crystal obtained from Pb-free alkaline leaching solution as shown in Table 5.5

No.	Water/ crystal ratios	Zn content in the supernatant (g/L)	Total zinc amount in the supernatant (mg) A	Zinc loss (%) (A/4000 mg × 13.5%)
1	1	22.25	89.03	16.48
2	2	11.13	89.02	16.48
3	2.5	10.04	91.09	16.87
4	3	7.43	89.23	16.52
5	4	5.77	92.35	17.10
6	5.5	4.16	91.52	16.95
7	6.25	3.68	92.04	17.04

Note: Crystal 4 g, Zinc content 13.50%

NaOH concentration in the solution was also quite high, for zinc can only be dissolved in concentrated NaOH solution. It may be considered that crystallization was not suitable for the recovery of zinc from lead-free leaching solution.

Lead can be removed selectively and quantitatively as the mixtures of lead sulfide and sodium lead sulfide from the leaching solution when the mass ratios of sodium sulfide added to the lead present were kept at around 1.8. The content of lead in the sulfides was as high as about 80% and may be possible to be used directly in the lead smelters for the production of metallic lead. The zinc in the lead-free solution can be further precipitated out and recovered quantitatively as mixtures of zinc sulfides and zinc hydroxide. In addition, the recovery of zinc from lead-free leach solution via evaporation and concentration was very low and seemed unsuitable for the practical uses.

5.1.8 Chemical Reactions Taking Place in the Sulfide Precipitation Processes in Alkaline Solution

Around 237.5 kg Zn (3634 mol) in 1000 kg dust will be extracted into alkaline solution when the dust is treated by melting and leaching. The NaOH/(Zn + Pb) molar ratios in the sulfides are ranged from 1.24 (Sample B) to 1.52 (Sample A). The molar ratios of Zn/S in the sulfides are ranged to be around 0.7–1.0 and Pb/S to be around 1.5–2.0. Therefore, the components of the sulfides may be very complex. The following reactions may be proposed.

Precipitation by sulfide:

For Zn,

$$x\mathrm{Na_2Zn(OH)_4(aq)} + y\mathrm{Na_2S(s)} = n\mathrm{ZnS} + b\mathrm{Na_2Zn(OH)_mS_{(4\text{-}m)/2}(s)} + k\mathrm{Zn(OH)_2(s)} + p\mathrm{NaOH\ (aq)} \tag{5.13}$$

For Pb,

$$x\mathrm{Na_2Pb(OH)_4(aq)} + y\mathrm{Na_2S\ (s)} = n\mathrm{PbS_y(s)} + b\mathrm{Na_2Pb(OH)_mS_{(4\text{-}m)/2}(s)} + k\mathrm{Pb(OH)_2(s)} + p\mathrm{NaOH(aq)} \tag{5.14}$$

5.1.9 The Optimized Condition of Na_2S Purification Process

5.1.9.1 Purification Time

Na_2S was added to the leaching solution with a mass ratio (Na_2S to Pb) of 1.8 from the tank and purified for 30, 60, 90, and 120 min, respectively. After the accomplishment of purification, the solution was filtered, and impurity content was determined. The initial time was 30 min, and the removal efficiency was calculated in different purification time (Fig. 5.4). The impurities in the leaching solution all decreased partly (except Cu) with the purification time extension, and the maximum removal efficiency could be acquired at 90 min.

5.1.9.2 Purification Temperature

Four different temperatures of 25, 50, 70, and 90 °C were selected to study the purification temperature. After purifying for an hour, the content of impurities decreased, and 70 °C was the optimized temperature (Fig. 5.5). At the 90 °C, impurities like Cu, Ni, Mn, Mg, Fe, and Cd could not be removed simply because the evaporation of the water concentrated the solution.

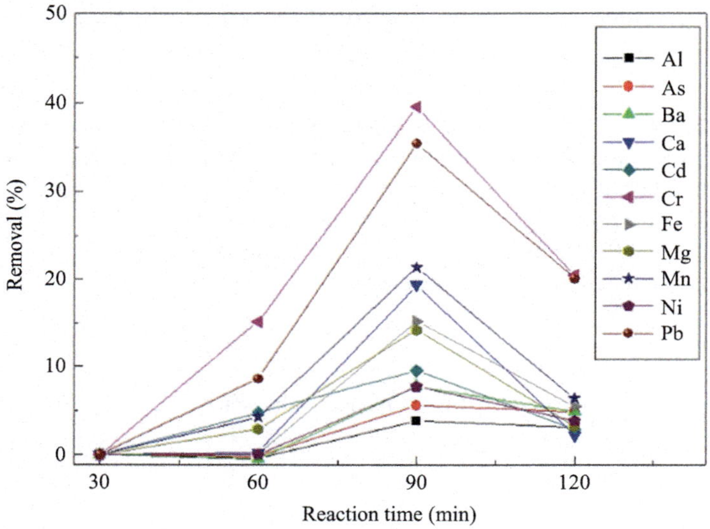

Fig. 5.4 Relationship between purification time and impurities removal efficiency in alkaline solution

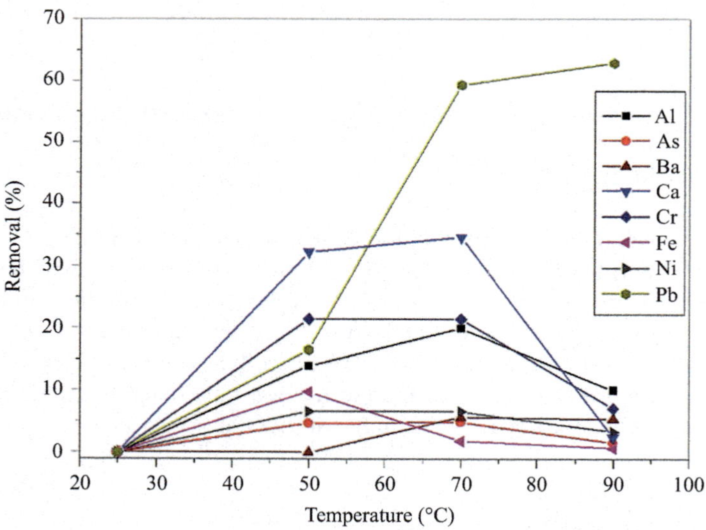

Fig. 5.5 Relationship between purification temperature and impurities removal efficiency in alkaline solution

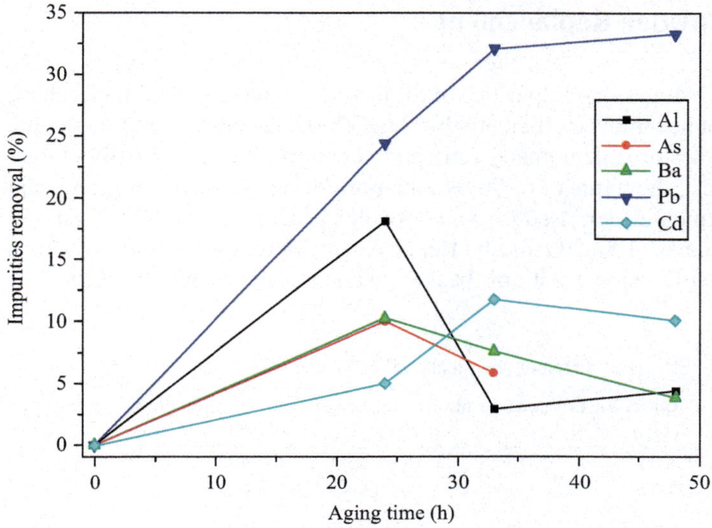

Fig. 5.6 Relationship between aging time and impurities removal efficiency in alkaline solution

5.1.9.3 Aging Time

Leaching solution taken from the tank was aged for 48 h after adding the Na_2S. The removal efficiency of impurities at the aging time of 0, 24, 33, and 48 h, which was calculated based on the initial concentration, is shown in Fig. 5.6. Aging is a thermodynamically stable process, where the particle size and structure of precipitates reform automatically in the solution. Generally, though varying size of precipitates coexisted at the beginning, the fine particles grow up gradually because of the dissolution and recrystallization process. Results showed that the black particles appeared after aging for 24 h, which manifested the growth and precipitation of particles in the solution during the aging process. In addition, the transformation from metastable state to steady state kinetically was another reason. Precipitation process always shows up the metastable state substances, which facilitates the formation of core and turns to lager particles subsequently.

The gradual increase of removal efficiency of Pb meant the decreased Pb in the residue (Fig. 5.6). Therefore, aging process was proved to be feasible to promote the quality of zinc powder, mainly because the lead precipitation turned to a stable form and reduced the effects of S^{2-} and Pb on zinc powder electrolysis. The removal efficiency of some impurities, like Al and As, decreased slightly due to the reduction of specific surface area, making the amount of adsorption decreased. Above all, 48 h was selected as the optimized aging time.

5.2 Removal of Sn from Alkaline Zinc Solution by Zinc Powder Replacement

Table 5.7 shows that impurities of Sn in zinc-containing alkaline leaching solution could not be removed efficiently by $NaHCO_3$, CaO, and $Ca(OH)_2$ under the addition of 1.5 times of stoichiometry ratio and new impurities can possibly be introduced. In contrast, the removal of Sn by zinc powder replacement reaction is apparent.

The zinc powder, used for Sn removal, was consisted of 99.5% Zn with particle size range of 100–200 mesh. Replacement were carried out by stirring Zn-Sn alkaline solution in a 250 mL beaker, using a heater magnetic stirrer (Fig. 5.7).

Table 5.7 Effects of different chemicals on Sn(IV) removal

Chemical	Initial Sn(IV) concentration (g/L)	Dosage (g/L)	Temperature (°C)	Sn(IV) removal (%)
$NaHCO_3$	1.02	2.20	90	15.53
	7.24	15.50	90	21.42
CaO	1.02	0.75	90	31.28
	7.24	5.20	90	39.45
$Ca(OH)_2$	1.02	1.00	90	11.28
	7.24	6.80	90	19.45
Zn	1.02	1.70	90	45.28
	7.24	12.00	90	50.32

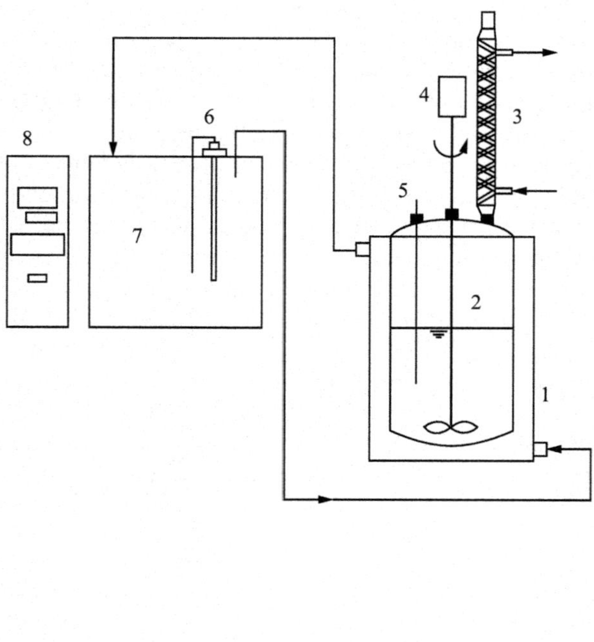

Fig. 5.7 Schematics for Sn removal in alkaline zincate solutions. (*1*) Jacket, (*2*) beaker, (*3*) condenser, (*4*) stirrer, (*5*) thermometer, (*6*) electric contact thermometer, (*7*) thermostatic bath, (*8*) microsystem

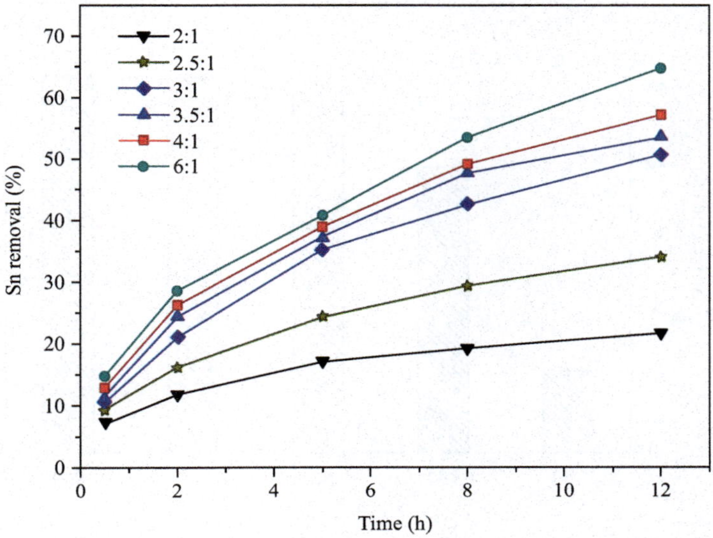

Fig. 5.8 Effects of Zn/Sn ratios on Sn removal from alkaline zincate solutions (Solution A, 70 °C, 500 rpm)

5.2.1 *Effects of Zn/Sn Ratio for Removal of Sn from Alkaline Zinc Solution by Zinc Powder Replacement*

The investigated Zn/Sn molar ratios were 2, 2.5, 3, 3.5, 4, and 6, while keeping the stirring speed at 500 rpm and the solution temperature at 70 °C. As shown in Fig. 5.8, Sn removal rate increased with increase of Zn/Sn ratio up to 6 where 64.9% Sn was removed in 12 h. The Sn removal became increasingly slow at Zn/Sn ratio of 3 with an Sn removal of 50.7%. Hence, a Zn/Sn molar ratio up to 3 was selected in the following tests.

5.2.2 *Effects of Stirring Speed on Removal of Sn from Alkaline Zinc Solution by Zinc Powder Replacement*

A series of stirring speed were investigated, and the Sn removal tests were carried out with Zn/Sn molar ratio of 3 at 70 °C for 12 h. The results are shown in Fig. 5.9. It was found that the Sn removal increased from 11.2% to 52.3% with increasing stirring speed from 100 to 900 rpm. However, the Sn removal did not increase obviously when the stirring speed exceeded 500 rpm. Hence, the stirring speed of 500 rpm was selected in considering energy saving.

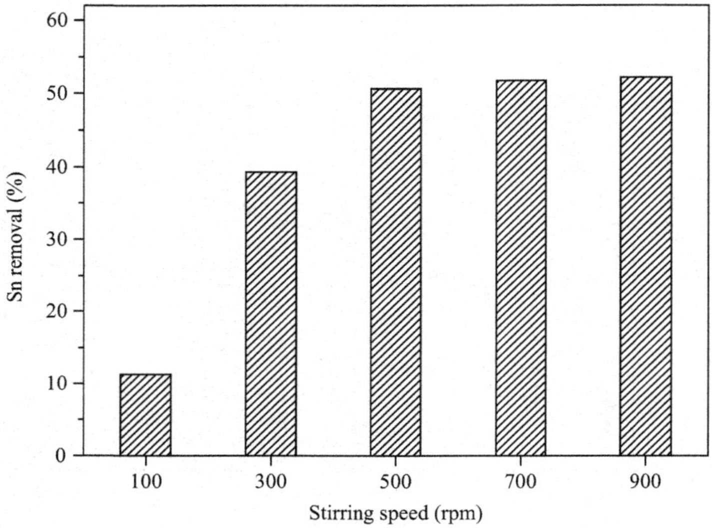

Fig. 5.9 Effects of stirring speed on Sn removal from alkaline zincate solutions (Solution A, Zn/Sn ratio 3, 70 °C)

5.2.3 Effects of Temperature on Removal of Sn from Alkaline Zinc Solution by Zinc Powder Replacement

The effect of temperature on Sn removal is shown in Fig. 5.10. As expected, the Sn removal increased with the increase of solution temperature with 82% Sn removal obtained at 95 °C and 18.9% at 30 °C. At 95 °C, the maximum Sn removal was obtained at 7 h and then decreased to 81.3% and 77.8% at 9 and 12 h, respectively.

5.2.4 Effects of Initial Sn Concentration on Removal of Sn from Alkaline Zinc Solution by Zinc Powder Replacement

The effect of initial Sn concentratios on Sn removal is shown in Fig. 5.11. The Sn removal decreased with decreasing initial Sn concentrations. When the initial Sn concentrations were 0.2, 0.5, and 0.8 g/L, the corresponding maximum Sn removal was 24.1, 30.4, and 40.9%, obtained at 4 h, 4 h, and 8 h, respectively. Then, they went down to zero after reaction at 12 h.

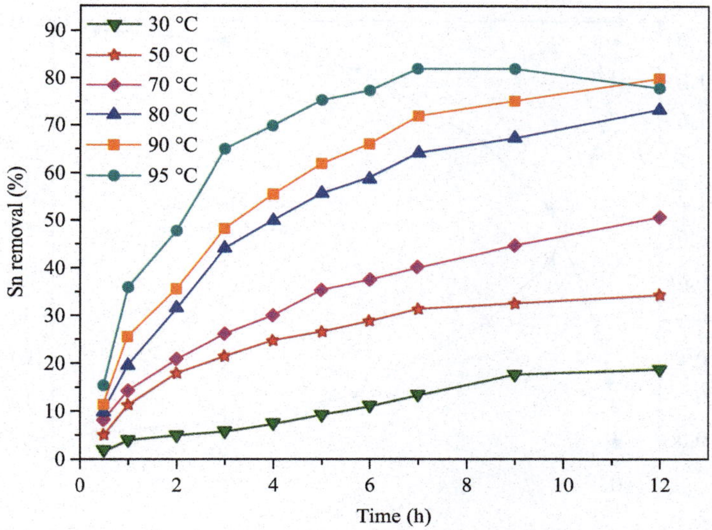

Fig. 5.10 Effects of temperature on Sn removal from alkaline zincate solutions (Solution A, Zn/Sn molar ratio 3, 500 rpm)

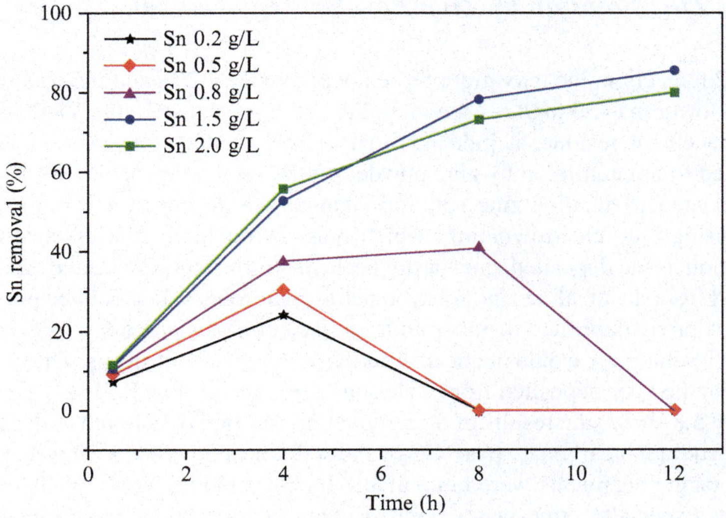

Fig. 5.11 Effects of initial Sn concentration on Sn removal from alkaline zincate solutions (Zn/Sn ratio 3, 500 rpm, 95 °C)

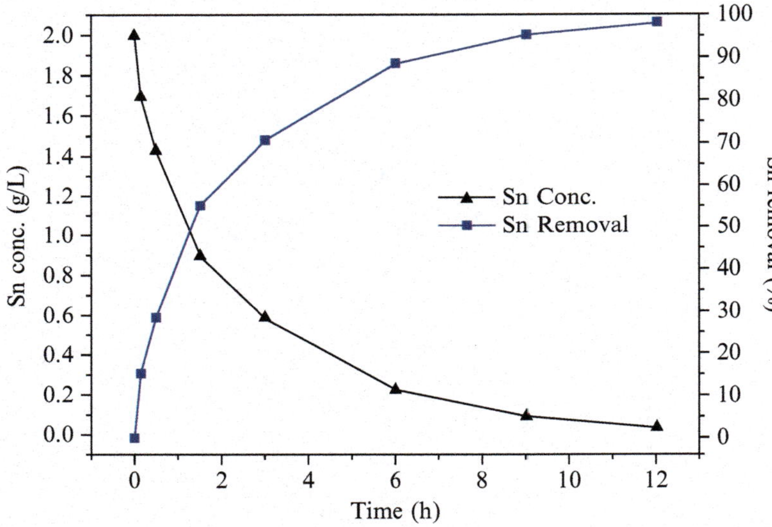

Fig. 5.12 Process optimization for Sn removal in alkaline zincate solutions (Solution A, Zn/Sn molar ratio 6, 500 rpm, 95 °C)

5.2.5 Process Optimization for Removal of Sn from Alkaline Zinc Solution by Zinc Powder Replacement

After 12 h reaction, the maximum Sn removal rate is 98.2%, and the concentration of Sn is lower than 50 mg/L as shown in Fig. 5.12. As shown in the XRD pattern of the replacement residue, it indicates that most of SnO_3^{2-} in solution has been converted into metallic Sn by zinc powder replacement (Fig. 5.13).

Sn existed in alkaline zinc solution changes the deposited zinc into compact form during zinc electrowinning, which makes trouble for the alkaline process application as the deposited zinc hardly peels off from cathodes. Hence, Sn must be removed from the alkaline zinc solution before electrolysis. It should be pointed out that Sn is rarely deposited together with Zn in electrolysis process. The Sn concentration of Solution A would not decrease obviously after electrolysis. The content of Sn in zinc powder deposited from Solution A was lower than 0.71%.

Table 5.8 shows the results of Sn removal by addition of calcium hydroxide (Ca$(OH)_2$), sodium bicarbonate ($NaHCO_3$), and sodium sulfide (Na_2S). It was indicated that all of the chemicals were not suitable for Sn removal from the alkaline zinc solution. Especially, zinc was separated quantitatively by the addition of Na_2S, while tin remained in solution without concomitant loss.

Sn can be separated quantitatively by the addition of zinc powder, and the reaction can be represented as:

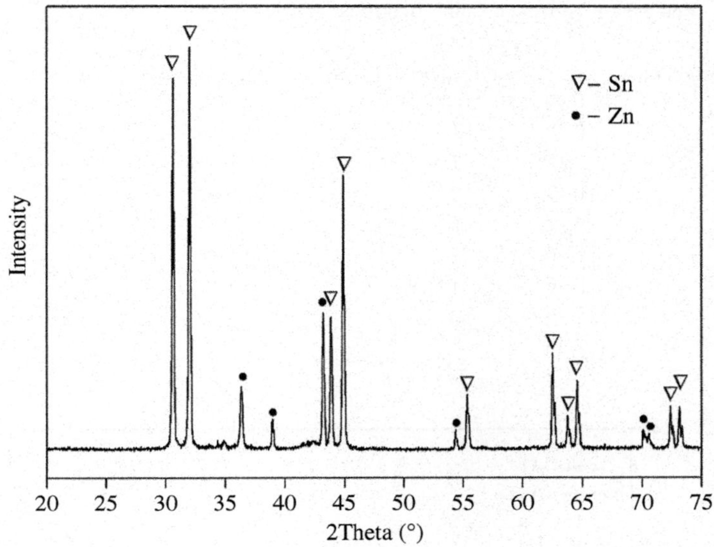

Fig. 5.13 XRD pattern of replacement residue from process optimization for Sn removal in alkaline zincate solutions

Table 5.8 Sn removal from Solution A by addition of chemicals

Chemicals	Reaction conditions	Sn removal (%)
$Ca(OH)_2$	95 °C, 500 rpm, 12 h, $Ca(OH)_2$/Sn ratio 3	8.9
$NaHCO_3$	95 °C, 500 rpm, 12 h, $NaHCO_3$/Sn ratio 6	31.2
Na_2S	95 °C, 500 rpm, 12 h, Na_2S /Sn ratio 3	0

$$Zn(s) + SnO_3^{2-}(aq) + 2OH^-(aq) = Sn(s) + 2ZnO_2^{2-}(aq) + H_2O(l) \quad (5.15)$$

The following adverse reactions occurred when zinc powder is added into Solution A:

$$Zn(s) + 2OH^-(aq) = ZnO_2^{2-}(aq) + H_2(g) \quad (5.16)$$

$$Sn(s) + 2OH^- + H_2O(l) = SnO_3^{2-} + 2H_2(g) \quad (5.17)$$

The zinc powder reacts with NaOH in strongly hot alkaline solution by the reaction of Eq. 5.16. After the zinc powder consumes up, Eq. 5.17 occurs, which means the redissolution of Sn makes Sn removal rate decrease. Hence, zinc powder should be overdosed in order to obtain a high Sn removal.

Table 5.9 Removal of Al from alkaline zinc solution using Na_2SiO_3

Concentration of Al (g/L)	Content of Al (g)	Ratio of Na_2SiO_3 and Al	Content of Al after purification (g/L)	Removal (%)
0.86	0.17	1	0.1844	6.90
0.86	0.17	1.5	0.1474	14.54
0.86	0.17	2	0.1430	17.09
0.86	0.17	3	0.1601	7.15
2.30	0.46	1	0.3569	22.73
2.31	0.46	1.5	0.3595	22.16
2.31	0.46	2	0.3449	25.32
2.31	0.46	3	0.3622	21.58

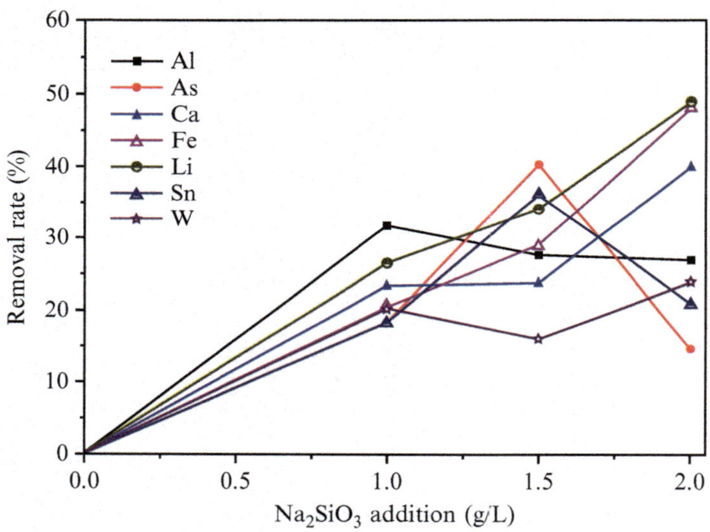

Fig. 5.14 Impurity removals in the real leaching alkaline solution using Na_2SiO_3

5.3 Removal of Al from Alkaline Zinc Solution

Na_2SiO_3 was used to control the content of Al in the electrolyte. The test solution contained NaOH 200 g/L, zinc 20 g/L, and Al. Added Na_2SiO_3 into the solution and stirred at 90 °C for 1 h analyzed the content of Al in the electrolyte by ICP spectrometer. The results are recorded in Table 5.9.

From Table 5.9, it can be seen that Na_2SiO_3 is effective to remove Al from alkaline zinc solution. Took leaching solution from industrial production, added different quality of Na_2SiO_3 into it, and stirred the solution at 90 °C for 1 h analyzed the content of impurities. The results are recorded in Fig. 5.14. It proved that

Na_2SiO_3 can reduce the content of Al in the actual alkaline leaching solution effectively, and the optimum addition amount of Na_2SiO_3 was determined as 1.5 g/L.

5.4 Removal of As from Alkaline Zinc Solution

Arsenic has a great negative effect on electrowining of zinc in the alkaline solution and should be removed as much as possible. Considering that arsenic can precipitate with Na_2S, CaO, and ferric salt, a series of exploratory experiments on removal of arsenic with Na_2S, ferric sulfate, lime milk, and ferric oxide were carried out, finally used in large-scale production line.

Sodium arsenate was added to the pure leach liquor with the alkali concentration of 200 g/L and the zinc concentration of 20 g/L; then the Na_2S, ferric sulfate, and lime milk were added, respectively. The mixture solution was stirred at 90 °C for 1 h, and the content of arsenic in clear supernatant liquid was analyzed by ICP. The results are shown in Table 5.10 and Table 5.11.

The results showed that the removal experiments of arsenic from pure leaching solution by Na_2S, lime milk, and ferric oxide were not working well, and the highest removal rate of arsenic was only 11% by ferric sulfate. The arsenide in the zinc electrolysis can be removed according to the reaction as follows:

$$2AsO_4^{3-} + Fe_2(SO_4)_3 \rightarrow 2FeAsO_4 + 3SO_4^{2-} \tag{5.18}$$

Furthermore, the removal of arsenic with ferric sulfate was carried out. Different amounts of ferric sulfate were added to the leaching solution, respectively, the mixed solution was stirred at 90 °C for 1 h, and then the content of impurities in clear supernatant liquid were analyzed. With the impurities content in leaching

Table 5.10 Removal of As with ferric sulfate, ferric oxide, and lime in alkaline solution

Agents	Concentration of As in leaching solution (mg/L)	Content of As in leaching solution (mg)	Content of As in treated solution (mg/L)	Removal rate (%)
Ferric sulfate, 0.5 g	479.97	95.994	94.64	1.41
Ferric sulfate, 1.0 g	479.97	95.994	85.12	11.33
Lime milk, 20 mL	479.97	95.994	96.57	−0.61
Lime milk, 40 mL	479.97	95.994	91.94	4.22
Ferric oxide, 0.5 g	428.84	85.768	81.257	5.26
Ferric oxide, 1.0 g	428.84	85.768	83.03	3.19

Table 5.11 Removal of As with Na_2S in alkaline solution

Concentration of As in leach liquor (mg/L)	Content of As in leach liquor (mg)	The ratio of Na_2S content to As content	Content of As in treated solution (mg/L)	Removal rate (%)
92.6	18.52	1	17.62	4.85
92.6	18.52	3	18.5	0.11
92.6	18.52	5	17.76	4.11
92.6	18.52	7	18.62	−0.54
333.6	66.72	1	66.8	−0.12
333.6	66.72	3	68.78	−3.08
333.6	66.72	5	68.76	−3.05
333.6	66.72	7	73.44	−10.07
457.6	91.52	1	90.62	0.98
457.6	91.52	3	94.88	−3.67
457.6	91.52	5	97.3	−6.31
457.6	91.52	7	89.18	2.55
600.9	120.18	1	126.6	−5.34
600.9	120.18	3	115.1	4.22
600.9	120.18	5	119.1	0.89
600.9	120.18	7	117.92	1.88

solution as the original point, the removal rate with the increase of the amount of added ferric sulfate was calculated. The results showed that the ferric sulfate cannot only remove arsenic in the leaching solution but also has a good performance on removal of Sb, Ca, Sr, Si, Cu, Li, Mn, Pb, and other elements (Fig. 5.15).

Under acidic conditions ($pH \leq 6$), the arsenic in the solution reacts with ferric hydroxide, which is the hydrolysis product of ferric sulfate, and the reaction is as follows:

$$Fe(OH)_3 + H_3AsO_4 \rightarrow FeAsO_4 \downarrow + 3H_2O \tag{5.19}$$

$$Fe(OH)_3 + H_3AsO_3 \rightarrow FeAsO_3 \downarrow + 3H_2O \tag{5.20}$$

But as the pH increases, part of the iron arsenate can be converted into iron hydroxide or goethite, thus emitting arsenate; the reaction is as follows:

$$FeAsO_4 \cdot 2H_2O + 3OH^- \rightarrow Fe(OH)_3 + AsO_4^{3-} + 2H_2O \tag{5.21}$$

$$FeAsO_4 \cdot 2H_2O + 3OH^- \rightarrow FeOOH + AsO_4^{3-} + 3H_2O \tag{5.22}$$

Figure 5.16 shows the effects of NaOH concentrations on the As removal rate in a composition of 40 g/L Zn^{2+} and 300 mg/L As. The As removal rate increased with NaOH concentration decreasing down to 40 g/L where 91.2% As was removed in 1 h in the form of $FeAsO_4$ which can be redissolved in strong alkaline solution.

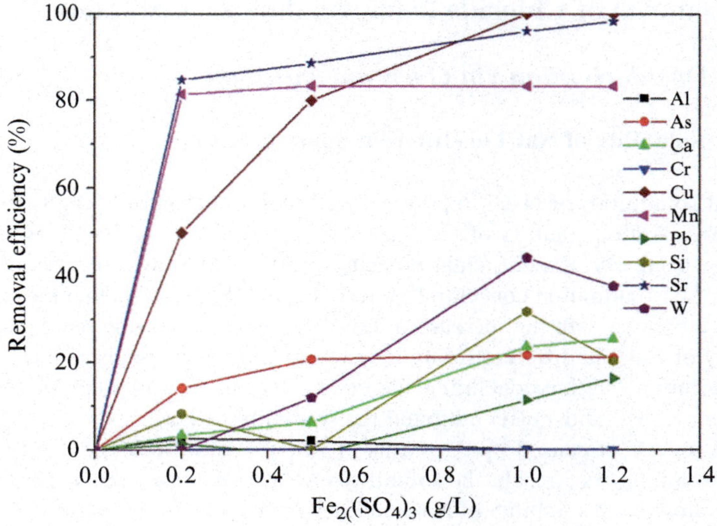

Fig. 5.15 As and other impurities removal efficiency using $Fe_2(SO_4)_3$ in alkaline solution

Fig. 5.16 Effects of NaOH concentration on As removal from alkaline zinc solution

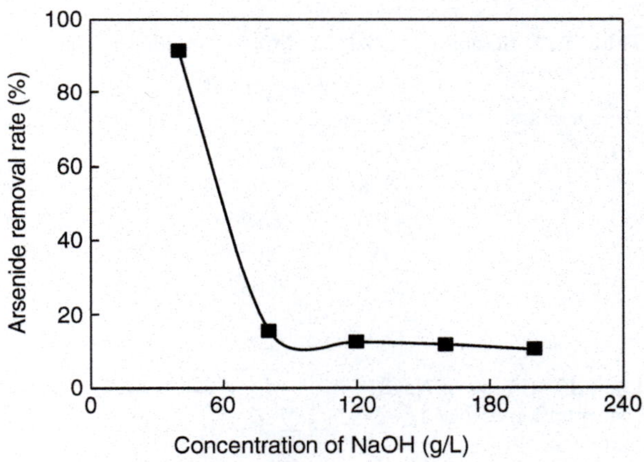

Therefore, the removal of arsenic in alkaline solution is much less effective than in acidic conditions (under acidic conditions, arsenic can be removed to 5 mg/L or less). The electrolytic solution in this process is alkaline, and the solubility of iron is very low. The ferric salt added in solution was converted into ferric hydroxide icon and ferric hydroxide colloid. The main removal action of As in alkaline solution is adsorption, bridging, cross-linking, and coprecipitation, which is exactly the reason why the removal of arsenic from the leaching solution is better than the removal of pure leachate with ferric sulfate. According to the experimental results, the optimum dosage of ferric sulfate was determined to be 1 g/L.

5.5 Removal of Chloride from Alkaline Zinc Solution

5.5.1 Dechlorination via Overconcentration

5.5.1.1 Solubility of NaCl in Different Solution System

Different solubilities of NaCl in pure water, NaOH solution of 200 g/L, and zinc-containing alkaline solution of $C_{NaOH} = 200$ g/L and $C_{Zn} = 40$ g/L were determined, respectively. Experimental procedure and methods are consistent with the method of Cl^- saturation concentration test. Table 5.12 suggests that the solubility of NaCl increases with the increase of temperature. At the same temperature, the solubility of NaCl in different solution systems follows the trends of pure water > NaOH solution > zinc-containing alkaline solution. It was suggested that the existence of OH^- and Zn(II) inhibited the solubility of NaCl, possibly due to the common-ion effect caused by the forms of ZnO_2^{2-} and $Zn(OH)_4^{2-}$ in the strong alkaline solution. As a result, the solubility of NaCl decreased remarkably.

According to the solubility of NaCl in Zn(II)-NaOH-H_2O system (Fig. 5.17), the concentration of NaOH and Zn(II) can increase by heating Zn(II)-NaOH-H_2O system so as to decrease the NaCl solubility and separate NaCl crystals out. For

Table 5.12 Solubility of NaCl in different solution systems

	NaCl solubility (mol/L)		
Temperature (°C)	Pure water	$C_{NaOH} = 5$ mol/L	$C_{NaOH} = 5$ mol/L, $C_{Zn(II)} = 0.6$ mol/L
10	5.4192	2.9523	2.5906
20	5.4278	2.9783	2.6448
30	5.4346	2.9960	2.7184
40	5.4483	3.0082	2.8109
50	5.4620	3.0403	2.9368

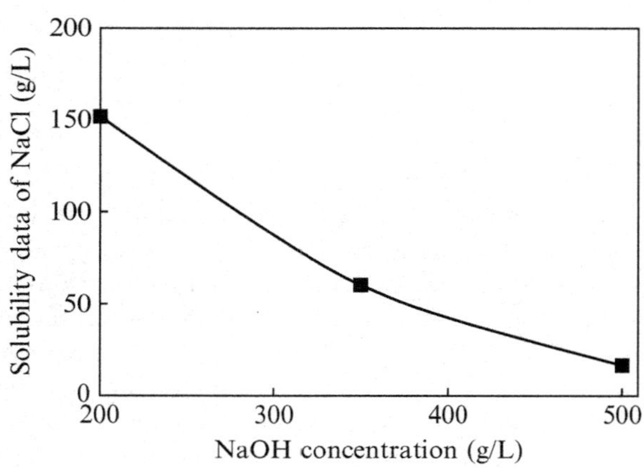

Fig. 5.17 Effects of NaOH concentration on NaCl solubility

Table 5.13 Dechlorination in zinc-containing alkaline solution

Sample	V (mL)	C_{NaOH} (g/L)	$C_{Zn(II)}$ (g/L)	C_{Cl^-} (g/L)
S_1	200	200.23	39.95	64.58
S_2	78	475.11	92.63	11.96
S_3	200	185.29	36.13	4.66

the electrolyte in a composition of 40 g/L Zn^{2+}, 200 g/L NaOH and 65 g/L Cl^-, 92.78% Cl^- can be removed by heating the electrolyte to increase the NaOH concentration to 480 g/L.

5.5.1.2 Dechlorination by Concentration Process

Cl-containing electrolyte ($w_{Cl} > 60$ g/L) was obtained (S_1), as 200 mL electrolyte was concentrated to about 78 mL in water bath at 90 °C with the stirring speed of 500 r/min. The concentrated electrolyte was filtered after standing for 8 h at 20 °C, while the concentration of NaOH, Zn(II), and Cl^- in the filtrate (S_2) were measured. The filter residue was tested by XRD after drying at 105 °C.

Table 5.13 presents the results of chlorine removal from zinc-containing alkaline solution. In order to give a better comparison of the variation of electrolyte concentration of NaOH, Zn(II), and Cl^- before and after chlorine removal, S_3 was used to indicate the recovered solution of 200 mL by deionized water from the concentrated S_2 of 78 mL.

Table 5.13 shows that the concentrations of NaOH and Zn(II) are 475.11 g/L and 92.63 g/L, respectively, after concentration. When cooling, settling, filtration, and attenuation to the original volume, Cl^- concentration in Zn(II)-NaOH-H_2O system can decrease to 4.66 g/L from initial 64.58 g/L, which is far lower than the limiting value of 60 g/L. That is to say, the purified solution can meet the requirement of NaOH, Zn, and Cl^- concentrations in the zinc powder production process by alkali leaching-electrowinning method.

As shown in Fig. 5.18, the main phase in residue is NaCl, and the characteristic peaks occur at 27.3°, 31.7°, 45.5°, 53.9°, 56.5°, and 66.2°. Moreover, characteristic peaks of $Zn_2OCl_2{\cdot}2H_2O$ and ZnO occur at 16.0°, 31.6°, 34.8°, 45.3° and 31.8°, 34.4°, 56.6°, 66.4°, and 68°, respectively, which suggests that these two kinds of substances can be generated at high temperature of 363 K. Moreover, $Na_4ZnO_3{\cdot}4H_2O$ probably also existed, for the peaks occurred at 20.0°, 34.9°, and 54.2°. The crystallization and precipitation of NaCl led to the effective removal of Cl^-, while little loss of NaOH and Zn(II) was observed.

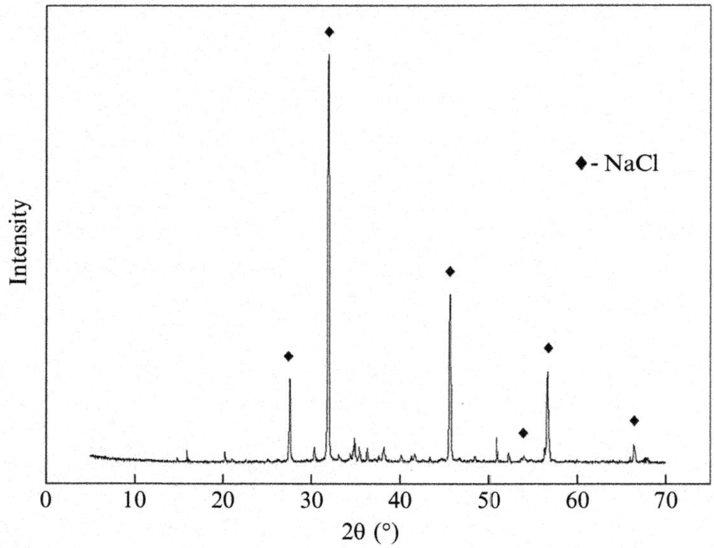

Fig. 5.18 XRD pattern of the filter residue from concentration of zincate solution

5.5.2 *Dechlorination by Washing via Na_2CO_3 Solution*

NaCl is of big solubility in water, while Zn_2CO_3 is insoluble in water. Due to the discrepancy of solubility between NaCl and $ZnCO_3$, both the chlorine removal and zinc conservation can be attained.

Quantified dust was measured, and zincate solution prepared according to certain liquid-solid ratio before quantified Na_2CO_3 was added in a beaker of 1 L. Then the mixture was stirred by a blender under a certain temperature and time. The chlorine content was measured after the reaction.

5.5.2.1 Determination of Na_2CO_3 Concentration

With the liquid-solid ratio (mL:g) of 5:1, 90 °C, and reaction time of 90 min, both removal rate of chlorine and loss rate of zinc changed with the $ZnCO_3$ concentration. As shown in Fig. 5.19, the removal rate of chlorine increases, and the loss rate of zinc decreases, with the increasing of Na_2CO_3 concentrations. It is the reaction between $ZnCl_2$ or other chloride and Na_2CO_3 that resulted in the generation of dissolving NaCl and refractory $ZnCO_3$, which achieved the goal of chlorine removal and zinc recycling. A higher chlorine removal rate can be achieved with a lower Na_2CO_3 concentration up to 4 g/L. Although there is a little increase of chlorine removal rate for higher Na_2CO_3 concentration, a proper concentration of 4 g/L for Na_2CO_3 may be used.

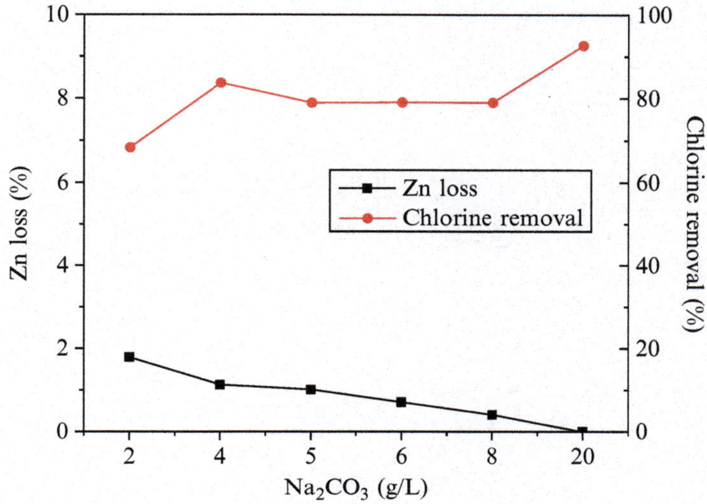

Fig. 5.19 Effects of Na_2CO_3 concentration on removal rate of chlorine and loss rate of zinc in alkaline solution

5.5.2.2 Determination of Liquid-Solid Ratio

The liquid-solid ratio influences both removal rate of chlorine and loss rate of zinc. As shown in Fig. 5.20, the removal rate of chlorine increases, while the loss rate of zinc decreases, with the increasing of the liquid-solid ratios. With the Na_2CO_3 concentration of 4 g/L, 90 °C, and reaction time of 90 min, the removal rate of chlorine could increase to 99% from 79.80%, while the loss rate of zinc decreased to 0.27% from 1.31%.

5.5.2.3 Determination of Reaction Temperature

Figure 5.21 shows the removal of chlorine under the condition of Na_2CO_3 concentration 4 g/L, liquid-solid ratio 8:1, and reaction time 90 min. It can be seen that the impact of temperature on chlorine removal by Na_2CO_3 was quite significant. The removal efficiency of chlorine increased obviously with the increase of temperature. The loss ratio of zinc decreased slightly with the increase of temperature. When the temperature was below 90 °C, the chlorine removal efficiency increased swiftly with increase of temperature. The chlorine removal efficiency increased slightly with the increasing temperature when the temperature kept at 90 °C. The chlorine removal efficiency was only 75.86% when the temperature was 50 °C, while it increased to 92.78% when the temperature increased to 90 °C. It can be concluded that temperature is one of the important factors in this reaction.

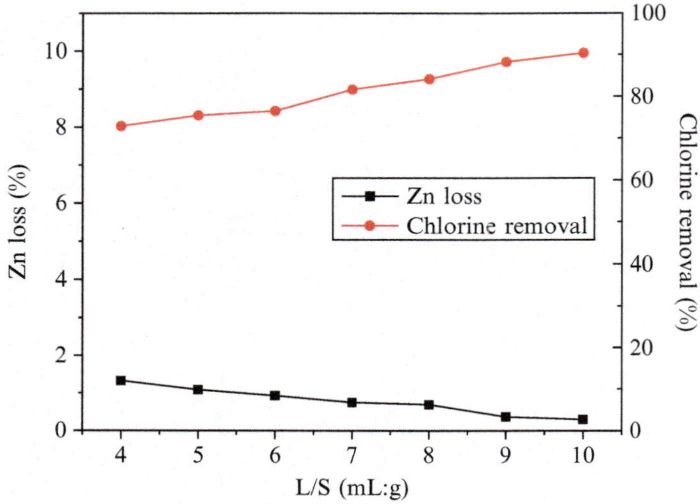

Fig. 5.20 Effects of liquid-solid ratio on removal rate of chlorine and loss rate of zinc in alkaline solution

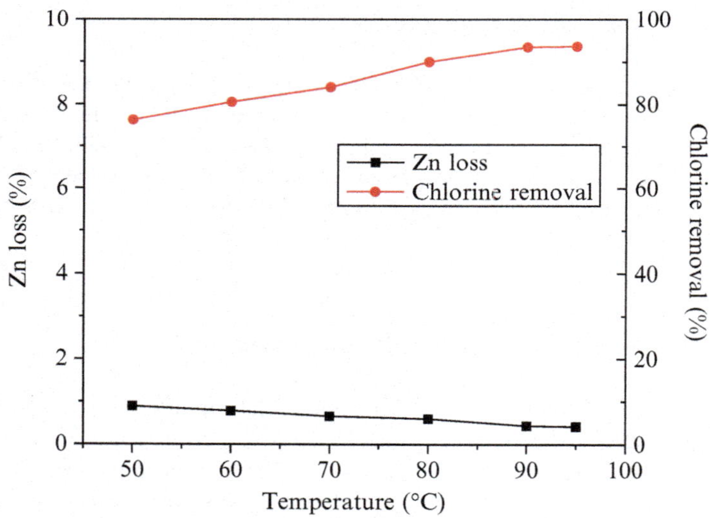

Fig. 5.21 Effects of temperature on removal rate of chlorine and loss rate of zinc in alkaline solution

5.5.2.4 Determination of Reaction Time

The effects of reaction time on chlorine removal were mainly appeared in the initial 90 min. During this period, the concentration of chlorine in the solution decreased greatly, and the removal efficiency could reach to 92.78%. The reaction reached to

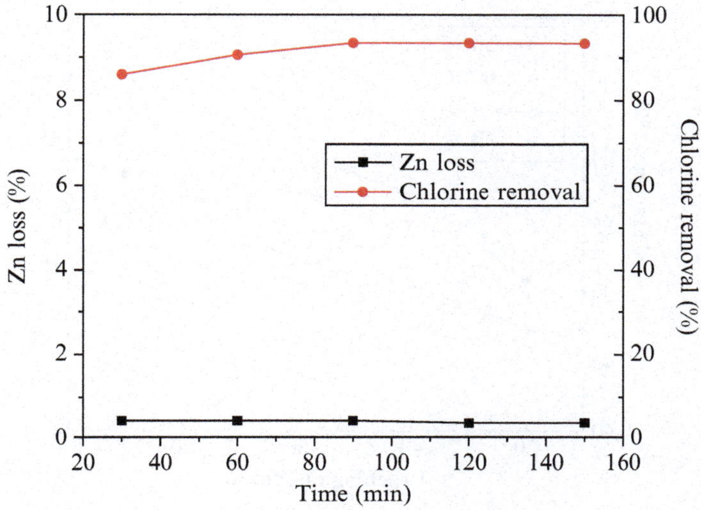

Fig. 5.22 Effects of reaction time on removal rate of chlorine and loss rate of zinc in alkaline solution

the balance after 90 min, and a prolong time would not enhance the extent of reaction. In Fig. 5.22, the chlorine removal efficiency is still maintained approximately 93% when the reaction time is 150 min. The change of loss efficiency of zinc with the reaction time was tiny.

5.5.3 *Dechlorination by Water Washing from the Wastes*

Figure 5.23 shows the effect of temperature on chlorine removal efficiency from wastes that were used for leaching under different washing time at the liquid-solid ratio 10:1.

The chlorine removal efficiency were all low during the 40-min washing period when the temperature was between 50 and 80 °C. With the extension of time, the chlorine removal efficiency increased gradually and achieved to nearly 53% and 62% after 60 min. At 80 min, the chlorine removal efficiency achieved the highest value. The optimum operation condition of dechlorination washing was a liquid-solid ratio of 10:1, temperature of 70 °C, and reaction time of 80 min, with the highest dechlorination efficiency of 63–64%.

A 100 $\sim$ 200 W ultrasonic device is adopted to assist washing chlorine from wastes as given in Fig. 5.24. Ultrasonic enhances the removal rate and achieved the highest value in 40 min. The increase of ultrasonic power also had a positive effect on dechlorination process, and the chlorine removal efficiency was 60% when the power was 100 W and attained the highest value 64% when the power was 150 W.

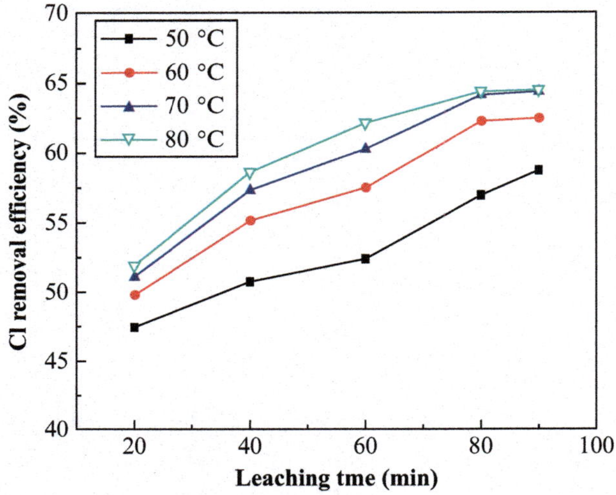

Fig. 5.23 Effects of temperature and leaching time on Cl^- washing from wastes

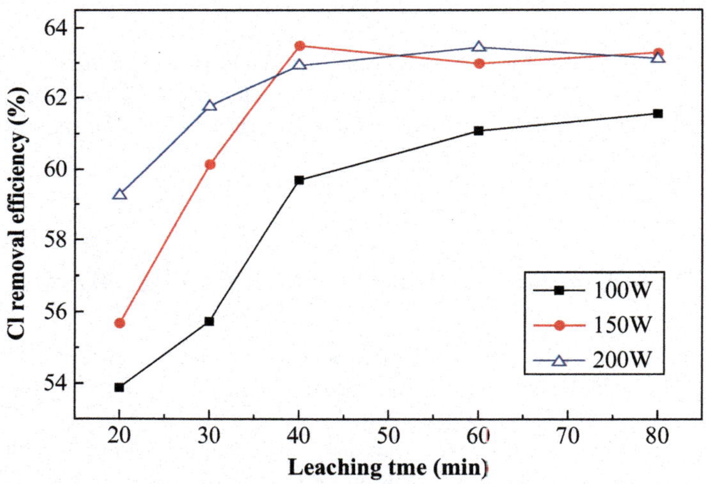

Fig. 5.24 Effects of ultrasound energy on the Cl^- washing from wastes

However, further increase of power cannot change the effect of dechlorination significantly. Hence, the optimal operation condition of ultrasonic was L/S of 10:1, temperature of 50 °C, the ultrasonic power of 150 W, and the ultrasonic assistance time of 40 min.

The effects of ultrasonic decreased the reaction temperature by nearly 20 °C and reduced the operating time by 40 min. Nevertheless, the ultrasonic energy provided by the equipment was not able to increase the total dechlorination efficiency value (Fig. 5.24).

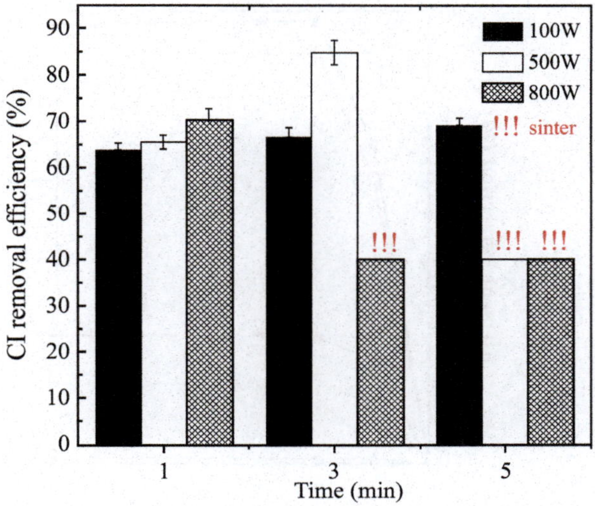

Fig. 5.25 Effects of microwave pretreatment on the Cl^- washing from wastes

In order to enhance the dechlorination efficiency, microwaves (1 kW, 2.45 GHz) were adopted to pretreat the sample. The heating power of microwave was 100 ~ 800 W and the time 1 ~ 5 min. The pretreated sample was put into the ultrasonic assistance washing process. Figure 5.25 shows the combined effect of microwave power and heating time on the dechlorination of sample. Firstly, 1 min microwave heating showed no significant effect; the dechlorination quantity was only increased by 5% at the 800 W power. Then, the power from 100 to 500 W at 3 min was used, and the dechlorination quantity increased greatly and achieved the best value (85%) at 500 W. However, further increase of power would decrease the dechlorination efficiency. If the microwave pretreatment was extended to 5 min, the dechlorination washing could meet the requirements by 70% at 100 W. Melting and other changes may occur at higher power which would decrease removal rate of chlorine.

The dissolutions of metals during the dechlorination washing process are shown in Fig. 5.26. The sampling was washed on the optimum dechlorination condition, at 500 W microwave preheating for 3 min and 150 W ultrasonic assistance washing for 40 min. 80 ~ 90% of Na and K had been washed away, while Na and K in residue reduced to 0.59% and 0.25%, respectively. The removal efficiency of Ca and Cd was 15–25%, and at the same time, traces of Cu, Pb, and Al were washed away. Lower than 1% of Zn entered into the washed liquid with chlorine, and the content of zinc in residue raised by nearly 8%.

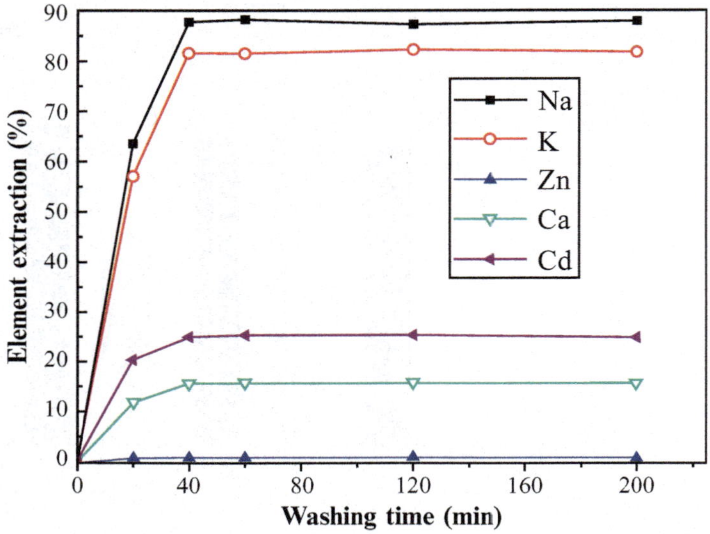

Fig. 5.26 Metals dissolution in the solution during the water washing for wastes

5.6 Deep Purification of Zinc Alkali Leaching Solution

5.6.1 Deep Purification Process of Lead, Aluminum, and Arsenic in Zinc Leaching Solution

In the purified leaching solution using Na_2S for lead removal, ferric sulfate for arsenic removal, and sodium silicate for aluminum removal, CaO was added and stirred for 1 h, the contents of impurities in the supernatant was analyzed, and the results are shown in Fig. 5.27.

As shown in Fig. 5.27, the addition of calcium oxide has a certain impurity removal effect on Pb, Mn, Fe, As, and other impurities. Lead precipitation, ferric hydroxide, and sodium silicate colloid were generated in the purification solution after the addition of Na_2S, iron sulfate, and sodium silicate, respectively. Calcium hydroxide, was generated in the hydrolysis process of calcium oxide and then precipitated from the purification solution with a certain adsorption of lead and colloidal particles. The fine lead precipitation particles were difficult to filter, and the addition of large lime milk particles could effectively improve the filtration performance of purification slag. The optimum dosage of calcium oxide was 0.8 times (mass ratio) of the dosage of sodium sulfide that added in purification solution.

The deep purification process of lead, aluminum, and arsenic in leaching solution is shown in Fig. 5.28. After the leaching solution was heated to 70 °C, sodium sulfide was added, and the addition dosage of sodium sulfide was 1.8 times (mass

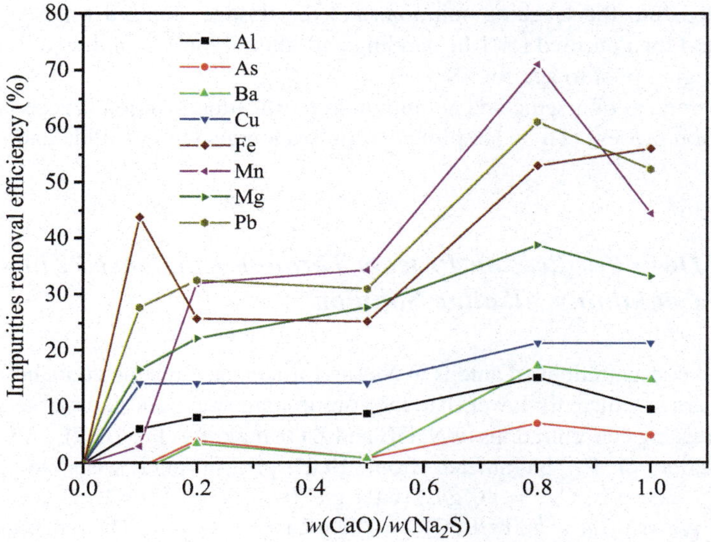

Fig. 5.27 The impurities removal efficiency from purified leaching solution in the presence of CaO

Fig. 5.28 Deep purification process on removing Pb, Al, and As in leaching solution

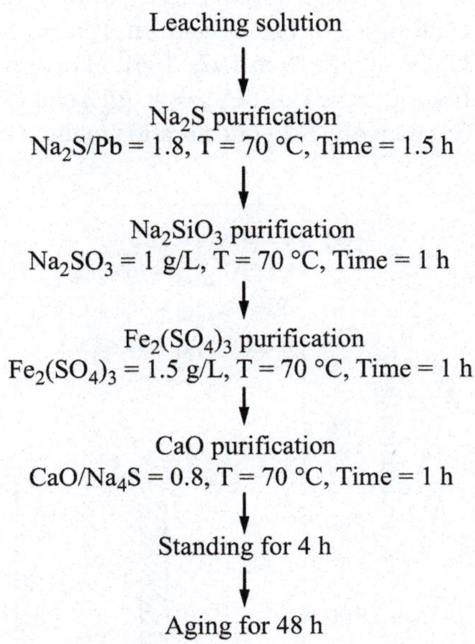

ratio) of the lead content in the leach solution and then purified for 1.5 h. Sodium silicate added into the solution was 1.5 g per liter and then purified for 1 h. Ferric sulfate added into leaching solution was 1 g per liter and then purified for 1 h. Lime was added into the leaching solution, and the dosage was 0.8 times of sodium sulfate and then purified for 1 h, standing and filtered for 4 h, and then transported into the aging pool to age for 48 h.

If the content of arsenic and aluminum in purification solution is acceptable, the purification process can be simplified by only adding sodium sulfide and lime.

5.6.2 Deep Purification Process Through Condensed Zinc-Containing Alkaline Solution

Saturated concentrations of anions in Na salt form existed in zinc-contained alkaline solution are significantly lower than they are in pure water and fall off very rapidly with increasing concentrations of NaOH and Zn in majority (Fig. 5.29). The saturated concentration of F^- plummeted from 19.02 g/L in pure water to 1.12 g/L (C_{NaOH} = 200 g/L, C_{Zn} = 40 g/L), 0.04 g/L (C_{NaOH} = 500 g/L, C_{Zn} = 40 g/L), and 0.01 g/L (C_{NaOH} = 500 g/L, C_{Zn} = 100 g/L), respectively. The other anions had the similar change rule, which showed that if leaching agent had been condensed, the anion crystallization would happen due to a drop in saturated concentration, leading to a high disposal efficiency for anion.

However, saturated concentration of SiO_3^{2-} increased with the growth of concentrations of NaOH and Zn. It reached to 30.46 g/L under C_{NaOH} = 200 g/L and C_{Zn} = 40 g/L from 107.11 g/L in pure water at first and then increased to 141.35 g/L (C_{NaOH} = 500 g/L, C_{Zn} = 40 g/L) and 190.56 g/L (C_{NaOH} = 500 g/L, C_{Zn} = 100 g/L), respectively. So, concentration method can't be used to dispose SiO_3^{2-}. Its maximum

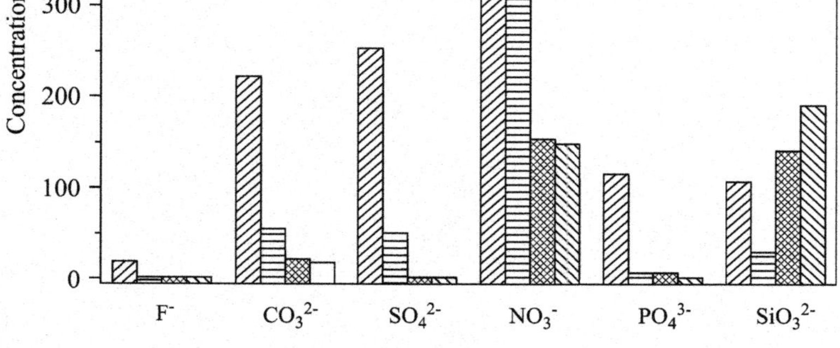

Fig. 5.29 Saturation concentration of impurity ions in different solution systems (25 °C)

concentration might be about 30 g/L based on the concentration of NaOH and zinc in electrolyte. But there was no sign of SiO_3^{2-} accumulation in strong alkali solution in real production, thanks to CaO added during purification section.

$$CaO(s)+H_2O = Ca(OH)_2(aq) \tag{5.23}$$

$$Ca(OH)_2(aq) + SiO_3^{2-} = CaSiO_3(s)+2OH^- \tag{5.24}$$

The CaO quantity added in production was needed to be adjusted based on SiO_3^{2-} concentration for zinc silicate which was used as raw material. While the concentration of SiO_3^{2-} was low (<5 g/L), the addition of CaO can be 0.5–2.0 g/L. The increase of concentration of CO_3^{2-} and F^- can be inhibited, and the reaction equations are shown as follows:

$$Ca(OH)_2(aq) + CO_3^{2-} = CaCO_3(s)+2OH^- \tag{5.25}$$

$$Ca(OH)_2(aq)+2F^- = CaF_2(s)+2OH^- \tag{5.26}$$

Besides, the addition of CaO can also remove Pb, Mn, Fe, As, and other impurities with a flocculation and adsorption of $Ca(OH)_2$.

5.7 Removal of Cu from Alkaline Lead Solution by Lead Powder Replacement

The equilibrium potential of copper is more positive than that of lead. Therefore, in the process of lead electrowinning, Cu will be deposited preferably on the cathode. Therefore, the Cu in electrolytic solution should be removed before lead electrowinning. The method of Pb powder replacement is used to remove Cu in electrolyte. The reaction of removing Cu in the electrolyte by the method of Pb powder replacement is:

$$Cu(OH)_4^{2-} + Pb = Pb(OH)_3^- + Cu + OH^- \tag{5.27}$$

The replacement reaction is an electrochemical reaction; actually, Eq. 5.27 is composed of Eqs. 5.28 and 5.29:

$$Cu(OH)_4^{2-} + 2e^- = Cu + 4OH^- \tag{5.28}$$

$$Pb + 3OH^- = Pb(OH)_3^- + 2e^- \tag{5.29}$$

When the Pb powder is added to the solution containing Cu, the replacement reaction can react automatically because the potential of $\varphi_{Pb(OH)_3^-/Pb}$ is more negative than $\varphi_{Cu(OH)_4^{2-}/Cu}$. The Pb is covered by Cu, which become the cathode of electrochemical reaction, and Pb is the anode. The Cu is deposited on the cathode.

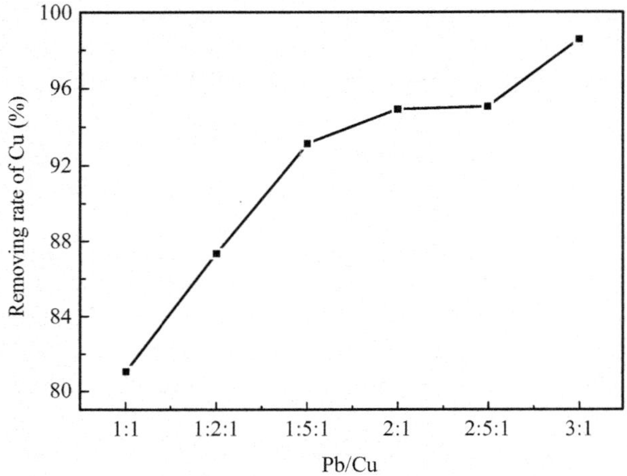

Fig. 5.30 Effects of the dosage of Pb on the removal of Cu from alkaline lead solution

Table 5.14 Copper concentrations in the solution after replacement reaction at different Pb/Cu ratios

Pb/Cu (molar ratio)	1:1	1.2:1	1.5:1	2:1	2.5:1	3:1
Cu concentrations (mg/L)	63.33	42.38	22.98	16.98	16.53	4.83

5.7.1 Effects of Pb/Cu Ratio on Removing Rate of Cu from Alkaline Lead Solution

The conditions were as follows: molar ratios of Pb powder to Cu in solution 1:1, 1.2:1, 1.5:1, 2:1, 2.5:1, and 3:1, respectively, replacement time of 15 min, temperature of 10 °C, NaOH concentration of 5 mol/L, and the initial Cu concentration of 335 mg/L. The results are shown in Fig. 5.30 and Table 5.14. The method of lead powder replacement can effectively remove copper in solution. The copper removal rate increased with the increase of the amount of lead; when the Pb/Cu ratio reached 3:1, the Cu concentration was lower than 5 mg/L. Considering the removal efficiency and dosage of Pb powder, the best ratio of Pb/Cu was set at 2:1.

5.7.2 Effects of Replacement Reaction Time on Removal of Cu from Alkaline Lead Solution

The conditions were as follows: replacement reaction time 15, 30, 45, 60, and 90 min, ratio of Pb/Cu 2:1, temperature 10 °C, NaOH concentration 5 mol/L, and copper. The results are shown in Fig. 5.31. The rate of Cu replacement was fast during the first 45 min, and the effect of replacement reaction time on the removal rate of copper was small. With the increase of reaction time, the removal rate of

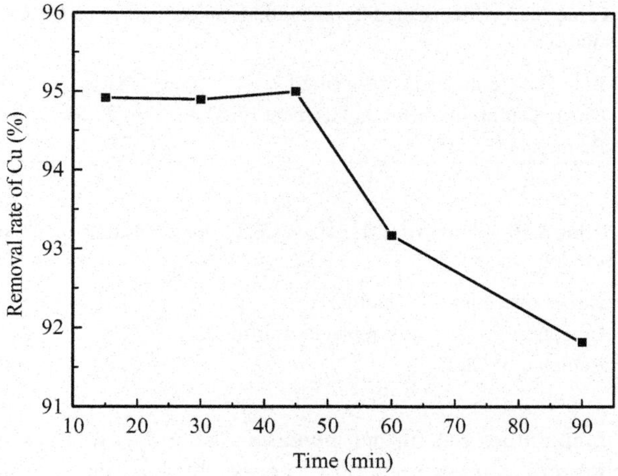

Fig. 5.31 Effects of reaction time on the removal rate of copper from alkaline lead solution

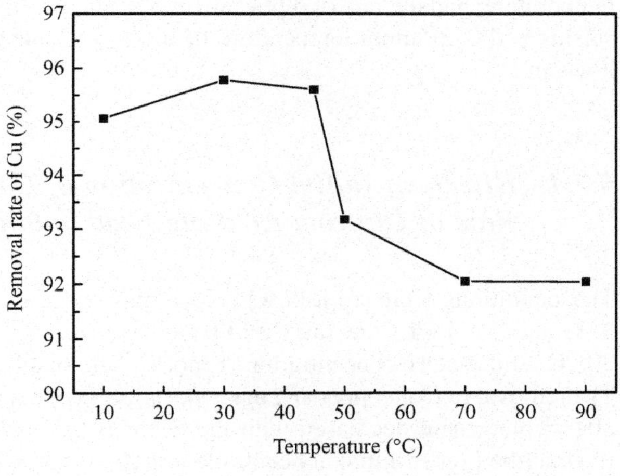

Fig. 5.32 Effects of temperature on the removal of Cu from alkaline lead solution

copper decreased. Considered that the Cu replaced by Pb may dissolve in the alkaline solution, 15 min is the appropriate replacement reaction time.

5.7.3 *Effects of Temperature on Removal of Cu from Alkaline Lead Solution*

The conditions were as follows: temperatures 10, 30, 45, 50, 70, and 90 °C, the ratio of Pb/Cu 2:1, reaction time 30 min, NaOH concentration 5 mol/L, and copper. The results are shown in Fig. 5.32. When the temperature was less than 45 °C, the copper removal rate increased slightly with the increase of temperature. When the

Table 5.15 Effects of different initial concentration of Cu on Cu removal from alkaline lead solution

Initial concentrations of Cu (mg/L)	815	420	335	35
Cu concentrations after replacement (mg/L)	27.58	20.66	14.10	10.74
Removal of Cu (%)	96.62	95.08	95.79	69.49

Table 5.16 Effects of different concentrations of NaOH on the removal of Cu from alkaline lead solution

NaOH concentration (mol/L)	2	3	4	5	6
Cu concentrations after replacement (mg/L)	1.80	2.60	8.47	10.74	13.24
Removal of Cu (%)	94.89	92.61	75.94	69.49	62.39

temperature was further increased, the copper removal rate decreased, meaning that the replacement reaction of copper by lead powder was more favorable at room temperature as the replaced copper will dissolve again in alkaline solution at higher temperature, and the rate of replacement reaction was thus decreased. Thirty degree celsius is the optimum temperature of the replacement process of copper by lead powder.

5.7.4 Effects of Initial Concentration of Copper on Removal Rate of Cu from Alkaline Lead Solution

The conditions were as follows: the initial concentrations of copper 815, 420, 335, and 35 mg/L, the ratio of Pb/Cu 2:1, reaction time 30 min, temperature 30 °C, and NaOH concentration 5 mol/L. The results are shown in Table 5.15. The removal rate of copper and the concentration of copper remained in the solution after replacement decreased with the decrease of initial concentration of copper. Even at the higher initial concentration of copper of 815 mg/L, most of the copper can be removed by lead replacement.

5.7.5 Effects of NaOH Concentrations on Removal of Cu from Alkaline Lead Solution

The conditions were as follows: NaOH concentrations 2, 3, 4, and 6 mol/L, initial concentration of Cu 35 mg/L, the ratio of Pb/Cu 2:1, reaction time 30 min, and temperature 30 °C. The results are shown in Table 5.16. NaOH concentration in the solution had a greater impact on the removal of Cu, and the removal rate of copper increased with the decrease of NaOH concentration.

Chapter 6
Electrowinning of Zinc and Lead from Alkaline Solutions

Abstract Electrowinning of zinc and lead from alkaline solutions depends on zinc or lead concentrations, temperature, current density, NaOH concentration, electrode distance and electrode material on current efficiency, and energy consumption of zinc or lead electrowinning. The optimum conditions for zinc electrowinning are zinc concentration 30–40 g/L, current density 800–1000 A/m^2, NaOH concentration 180–200 g/L, and temperature 30–50 °C, and those for lead electrowinning are lead concentration 20 g/L, 5 g/L lead concentration left in the spent electrolyte, below 50 °C, 400 A/m^2 current density, 5 mol/L NaOH concentration, and stainless steel of electrode material. Impurities ions have great effect on the quality of the metallic zinc or lead powder in alkaline leaching electrowinning process. The electrolysis deteriorates when the concentration of arsenide is over 500 mg/L, but CO_3^{2-}, SO_4^{2-}, SiO_3^{2-}, and F^- have no effects on zinc electrowinning. Copper will be electrodeposited prior to lead in the cathode. Tin, $C_4H_4O_6^{2-}$, CO_4^{2-}, and gluten in the solution decrease the current efficiency of the lead electrowinning process, while As, Sb, and W, as well as SO_4^{2-}, PO_4^{3-}, SO_3^{2-}, SiO_3^{2-}, SCN^-, CO_3^{2-}, Cl^-, and AC^-, have no effects.

Keywords Electrowinning • Zinc • Lead • Alkaline • Impurities ions

6.1 General Electrowinning Production Process of Zinc Powder by Alkaline Hydrometallurgy

Zinc and lead powders can be produced by electrolysis, which is a complex process and depends on zinc or lead concentrations, temperature, current density, NaOH concentration, electrode distance and electrode material on current efficiency, and energy consumption of zinc or lead electrowinning, as well as the equilibrium potential of Zn, As, Sb, W, Cu, and Sn in NaOH solution and the presence of impurities in the solution.

The general operating parameters of alkaline hydrometallurgical process is given in Table 6.1. The zinc existed in raw materials can be leached effectively by caustic alkaline solution, while the existed ZnS and $ZnFe_2O_4$ cannot. After purified by Na_2S precipitation, an alkaline zinc solution is used for electrowinning;

Y. Zhao, C. Zhang, *Pollution Control and Resource Reuse for Alkaline Hydrometallurgy of Amphoteric Metal Hazardous Wastes*, Handbook of Environmental Engineering 18, DOI 10.1007/978-3-319-55158-6_6

Table 6.1 General operating parameters of alkaline hydrometallurgical process

Stages	Parameters	Values	Stages	Parameters	Values
Leaching	Initial NaOH concentration (g/L)	220–260	Electrolysis	Temperature (°C)	30–50
	Temperature (°C)	80–95		Current density (A/m^2)	800–1000
	Time (min)	60–150		Electrolyte updating rate (min)	10–30
	Liquid-solid ratio (L/kg)	7–15		Ultimate zinc concentration (g/L)	<6–10
	Stirring speed (rpm)	300–600	Zinc powder treatment	Washing, ultimate washing water	pH<8
	Ultimate zinc concentration (g/L)	30–40		Drying temperature (°C)	150–180
Purifying	Temperature (°C)	85		Drying vacuum (MPa)	<−0.85
	Time (min)	180		Drying, discharge temperature (°C)	<50
	Stirring speed (rpm)	400		Grinding and screening	As needed

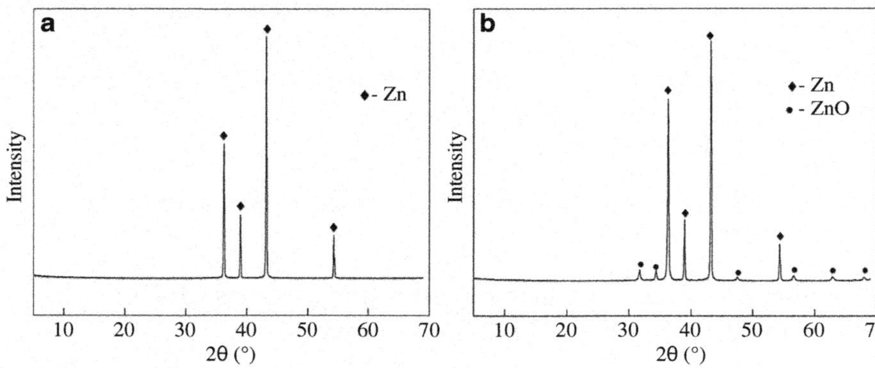

Fig. 6.1 X-ray diffraction patterns of the samples, (**a**) the zinc powders produced by the acidic-air pulverization process; (**b**) alkaline deposited zinc powder products

the zinc in solution can deposit on the cathode rapidly and adhere very lightly, which makes the deposits be removed easily from the cathode by scraping with simple tools, such as plastic stick, or even be detached automatically from the cathode by self-gravity. Considerably high current efficiencies (85–95%) are obtained, with a total energy consumptions of 3.2–3.5 kWh/kg zinc powder, much lower than 4.0–4.4 kWh/kg zinc powder of the conventional acidic-air pulverization process, which results in 10–20% energy savings.

As shown in Fig. 6.1, some metallic Zn was oxidized into ZnO during zinc powder washing and drying stages, which resulted in a slightly decline of metallic

Table 6.2 Chemical composition of alkaline deposited zinc powder in comparison with GB/T 6890–2000

	Chemical composition (wt.%)					
	No less than		No more than			
Grade	Zn	Metallic Zn	Pb	Fe	Cd	Acid leaching residue
I	98	96	0.1	0.05	0.1	0.2
II	98	94	0.2	0.2	0.2	0.2
III	96	92	0.3	–	–	0.2
IV	92	88	–	–	–	0.2
The products	96–99	92–97	<0.08	<0.03	<0.01	<0.1

Fig. 6.2 The surface morphology and shape of zinc powders from the acidic-air pulverization process and alkaline deposited zinc powder products (1000×). (**a**) Zinc powders produced by the acidic-air pulverization process; (**b**) alkaline deposited zinc powder products

Zn content in the zinc powder products. The impurities contents of the products were always lower than the allowed value of the Chinese National Standard GB/T 6890–2000 (Table 6.2). Therefore, the products can reach the needs of Grade III or even II and I.

The surface morphology and shape of the samples are given in Fig. 6.2. From a macroscopic viewpoint, the products were always obtained in powdery form. However, dendritic deposits were observed from a microscopic viewpoint, while the zinc powders produced by the acidic-air pulverization process were composed of spherical particles. The results of the BET tests proved that the dendritic structures offered a greater specific surface area than spherical ones, which resulted in 7.43 m^2/g of the products and 3.45 m^2/g of the zinc powder (<150 μm) produced by the conventional process, respectively. This was very important in zinc powder applications such as displacer, paint and extraction of gold and silver, catalysis, etc., where high surface areas are needed to increase reaction rates. The apparent density of the products (<150 μm) was 2.01 g/cm^3, which are lower than the conventional ones of 2.98 g/cm^3.

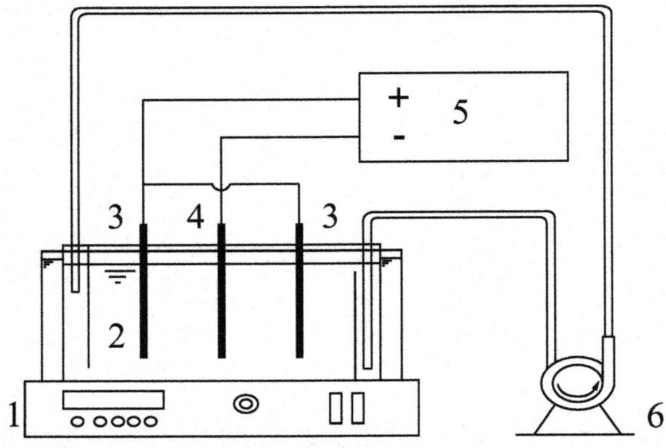

Fig. 6.3 Schematics of the apparatus for the alkaline zinc electrowinning experiments, *1*. Thermostatic bath, *2*. electrolysis cell, *3*. anodes, *4*. cathode, *5*. DC power supply, *6*. pump

6.2 Production of Ultrafine Zinc Powder from Wastes Bearing Zinc by Electrowinning in Alkaline Solution

The electrowinning experiments were carried out at atmospheric pressure as shown in Fig. 6.3. A plexiglas cell having a volume of 1 L and one magnesium cathode flanked by two stainless steel anodes were used. The interelectrode distance was 3 cm. The samples for analysis were taken at the fixed time intervals of 20, 40, and 60 min. The obtained zinc powder could easily be removed from the electrode and was washed thoroughly with water and then with ethanol. Finally, it was dried at 50 °C and weighed. The current efficiency was then determined by dividing the theoretical current required to produce the deposit by the real current supplied. A series of zinc powders with different size distributions were obtained by changing the experimental parameters.

The temperatures of the electrolyte, current density, Zn^{2+} concentration, and NaOH concentration were allowed to vary from 30 to 70 °C, 500 to 1500 A/m^2, 10 to 40 g/L, and 150 to 300 g/L, respectively. Lead oxide was tested as grain refiner in the range of concentrations from 10 to 60 mg/L. Zn deposits were examined by laser particle size analyzer to determine the size distribution and by scanning electron microscopy to identify the surface morphology and microstructure.

The samples of the same deposits were also analyzed from liquid state by ICP-OES, so the solid samples must be broken down in an aqua regia-HF solution. All the samples were diluted in order to be in the calibration range. Electrochemical studies, such as cathodic polarization and cyclic voltammetry, were performed in a three-electrode plexiglas cell. High purity aluminum (>99.95%) was used as the working electrode for the cyclic voltammetry and cathodic polarization experiments. A graphite rod was used as counter and a saturated calomel electrode (SCE) as the reference electrode. All potentials were recorded with respect to the SCE. The potential was scanned from −1.3 to −1.7 V/SCE for cathodic polarization with a scan rate of 10 mV/s. Cyclic voltammetric experiments were carried out by scanning at a scan rate of 2 mV/s in the potential range of −0.7 to −2.0 V/SCE.

In the electrowinning of zinc from alkaline electrolyte, the overall cell reaction is:

$$Zn(OH)_4^{2-} = Zn + H_2O + 2OH^- + 1/2O_2 \uparrow \tag{6.1}$$

According to the following calculation:

$$E = E_{O_2/H_2O} - E_{Zn(OH)_4^{2-}/Zn} \tag{6.2}$$

the value of E in theory for alkaline media electrowinning is lower than that obtained in the typical acid media electrowinning by approximately 0.5 V. The zinc electrowinning process is a one-electron electrode reaction. Since the calculated electron transport number n is about 1, two electrons can be transported by two consecutive reactions. The reaction mechanism may be:

$$Zn(OH)_4^{2-} = Zn(OH)_3^- + OH^- \tag{6.3}$$

$$Zn(OH)_3^- + e = Zn(OH)_2^- + OH^- \tag{6.4}$$

$$Zn(OH)_2^- = Zn(OH)_{aq} + OH^- \tag{6.5}$$

$$Zn(OH)_{aq} + e = Zn + OH^- \tag{6.6}$$

Generally, the zinc electrowinning process was operated under the conditions mentioned in Table 6.3.

Zinc and lead residues contained various metal impurities such as Co, Ni, Cd, Pb, Al, Cr, Ca, Mg, Fe, and Cu. The concentrations of elements (Fe, Ca, Cu, Cd, Mg) in leaching solution were all lower than 0.5 g/L. While Cd exerted a deleterious effect on the electrowinning in the acid leach method, the effect of Cd in the alkaline leaching solution was found to be negligible. Lead can be separated quantitatively from the alkaline leaching solution using sodium sulfide.

In electrowinning of ultrafine zinc powder, nucleation and growth are in a competition to determine grain size. Therefore, a low growth rate and a high nucleation rate for the synthesis of microscale materials are unavoidable. In general, factors that result in an increase of overpotential are expected to promote grain refinement. Increasing the nuclear overpotential will facilitate electrowinning of

Table 6.3 Industrial ranges for current density, voltage, electrolyte temperature, zinc and sodium hydrate concentrations, and current efficiency in zinc electrowinning process

Parameter	Range
Current density (A/m^2)	800–1200
Voltage (V)	2.6–3
Temperature (°C)	30–50
Zinc concentration (g/L)	30–50
Sodium hydrate concentration (g/L)	180–200
Current efficiency (%)	>99

ultrafine zinc powder. The process of zinc powder electrowinning is controlled by mass transport; therefore, the diffusion overpotential, η_d, is focused on:

$$\eta_d = E_r - E_{ir} = \frac{RT}{ZF} \ln \frac{C_0}{C_s} \tag{6.7}$$

$$\frac{C_s}{C_0} = 1 - \frac{i_d}{i_{dl}} \tag{6.8}$$

$$\eta_d = \frac{RT}{ZF} \ln \frac{i_{dl}}{i_{dl} - i_d} = \frac{RT}{ZF} \ln \frac{1}{1 - \frac{i_d}{i_{dl}}} \tag{6.9}$$

$$i_{dl} = \frac{ZFDC_0}{\delta} \tag{6.10}$$

$$D = \frac{KT}{6\pi r\eta} \tag{6.11}$$

where E_r—the equilibrium potential on the cathode, V
E_{ir}—the potential measured in the experiment, V
C_o—the concentration of Zn ion at the surface of the electrode, mol/L
C_s—the bulk concentration of Zn ion, mol/L
i_{dl}—the diffusion-limiting current density, A/m^2
i_d—the apparent cathodic current density, A/m^2
δ—the thickness of the diffusion layer, cm
K—the stoichiometric coefficient of Boltzmann, J/K
R—the radius of solvation of the ion, cm
η—the viscosity of the leach solution, J/(mol·K)

On the basis of the above equations, high current density favors the decrease of the grain size, provided other electrowinning conditions are kept constant. It can also be inferred that the ultrafine zinc powder size can be controlled by adjusting the electrolyte temperature, composition, and viscosity at the given current density.

Scanning electron microscopy (SEM) was used to reveal whether a relationship existed between the Zn ion concentration and zinc electrowinning morphology. Figure 6.4 shows SEM micrographs of zinc electrodeposits produced at 10–40 g/L Zn^{2+}, with a current density of 1000 A/m^2.

It can be seen that decreasing Zn^{2+} concentrations resulted in smaller particle size. This can be explained in terms of nucleation overpotential. Because the cathodic deposition of zinc is controlled by mass transport, any decrease in zinc ion concentrations during the electrowinning can be assumed to account for an increase in overpotential. However, decreasing Zn^{2+} concentrations facilitated hydrogen evolution and exerted a deleterious effect on the current efficiency. A Current efficiency (CE) of 75% was achieved at 10 g/L Zn^{2+}, whereas the CE remained approximately 100% with the Zn^{2+} concentrations of 30 and 40 g/L. The presence of Pb^{2+} (30–50 mg/L) was found to increase the CE to about 95–98% in the range of 25–30 g/L Zn^{2+}.

Surface photographs of the deposited zinc showing the effect of NaOH concentration are given in Fig. 6.5. The electrowon particles were smaller at high NaOH

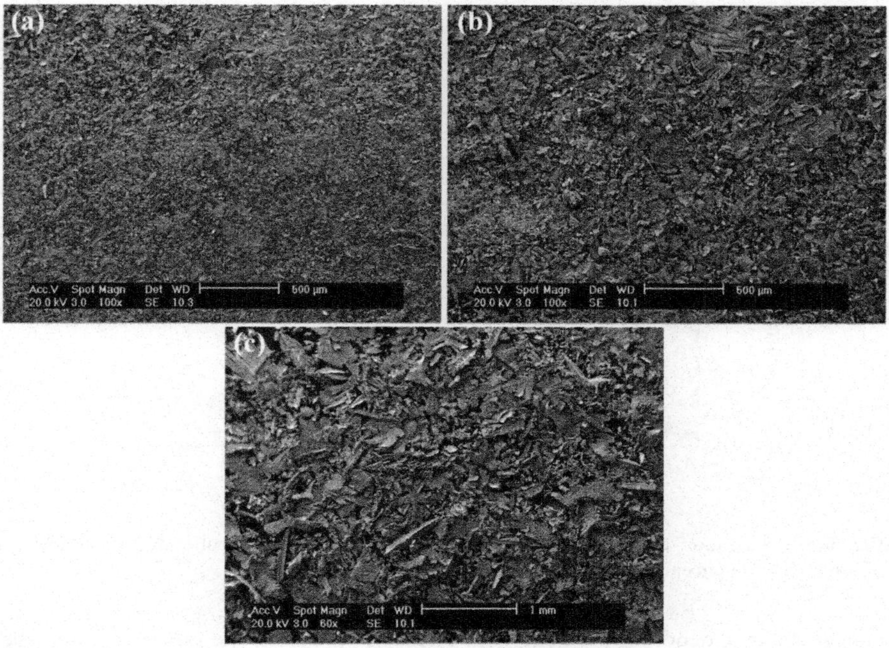

Fig. 6.4 SEM micrographs of the Zn particles synthesized under different alkaline Zn^{2+} concentrations of (**a**) 10, (**b**) 25, and (**c**) 40 g/L

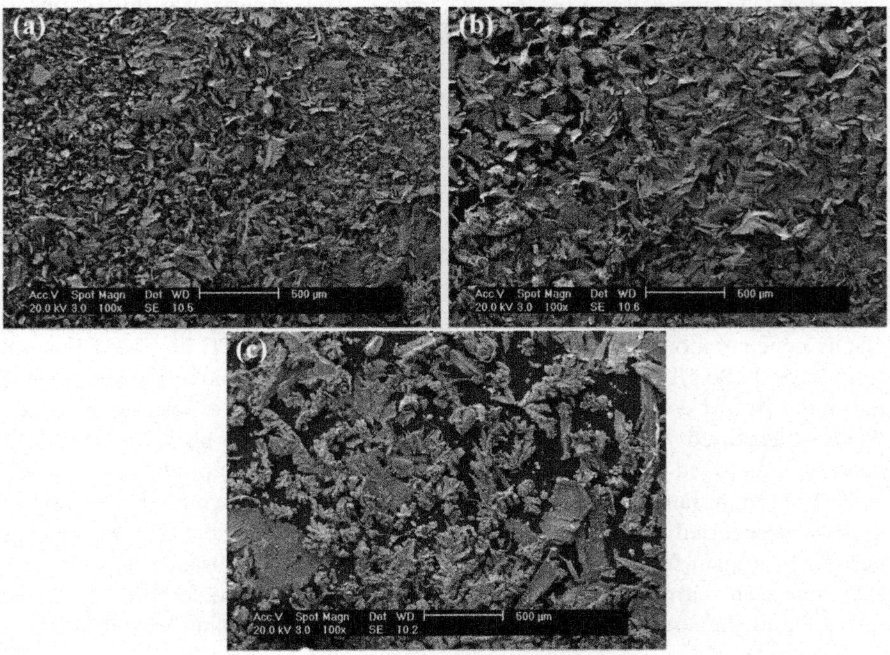

Fig. 6.5 SEM micrographs of zinc particles synthesized at different NaOH concentrations of (**a**) 300, (**b**) 250, and (**c**) 150 g/L

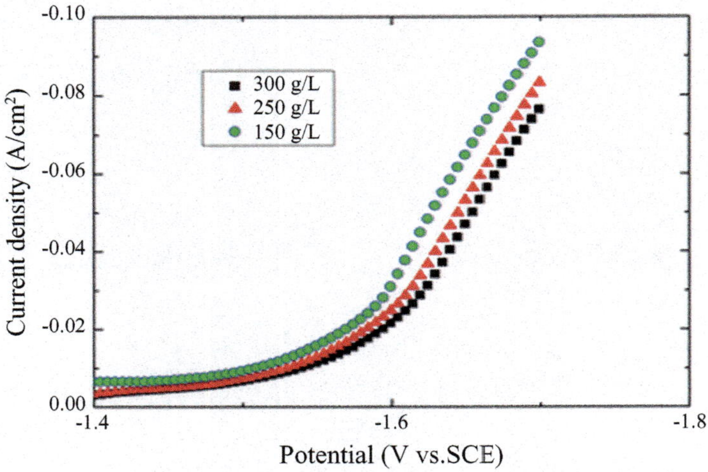

Fig. 6.6 Polarization curves of zinc deposition recorded under potentiostatic conditions at different NaOH concentrations

concentrations, denoting the beneficial effect of NaOH on the grain refining. The decrease in particle size can be correlated to a corresponding increase in the polarization curves. The polarization curves of zinc deposition recorded under potentiostatic conditions at different NaOH concentrations are shown in Fig. 6.6. Comparing the curves, it was obvious that the polarization curve was shifted toward more negative potentials with increasing NaOH, showing an increased polarization.

The change in the overpotential as a result of increasing NaOH concentration can be discussed from two points of view: increase of solution viscosity and increase of $[OH]^-$. (1) High viscosity minimizes the rate of mass transfer, because too much $[OH]^-$ inhibits the discharge rate of zinc, whose electrowinning is diffusion controlled, probably by increasing the viscosity and thus diminishing the diffusion coefficient. (2) The Zn^{2+} ions and the $[OH]^-$ groups could be linked through chelation, thus self-assembling into a chain structure. Moreover, the nucleation and growth of zinc particles depended on the decomposition of n $[Zn\ (OH)_n]^{n-2}$.

CE values below 180 g/L NaOH were found to be poor, i.e., < 90%. As the NaOH concentrations were increased beyond 180 g/L, the CE increased linearly from 90% at 180 g/L NaOH to 100% at 200 g/L NaOH. It was observed that, when the initial NaOH concentrations were over 220 g/L, current efficiency of the Zn deposit decreased dramatically as the NaOH concentrations increased. This decrease was found to be related to the dissolution of the zinc produced.

The optimum range of current density was determined at which the refining of grain size occurred with low energy loss. Particle size distribution analysis was performed to gain further knowledge of the dependence of ultrafine powder electrowinning on current density. The number of smaller particles present in the mixture and the specific surface area increased with increasing current density.

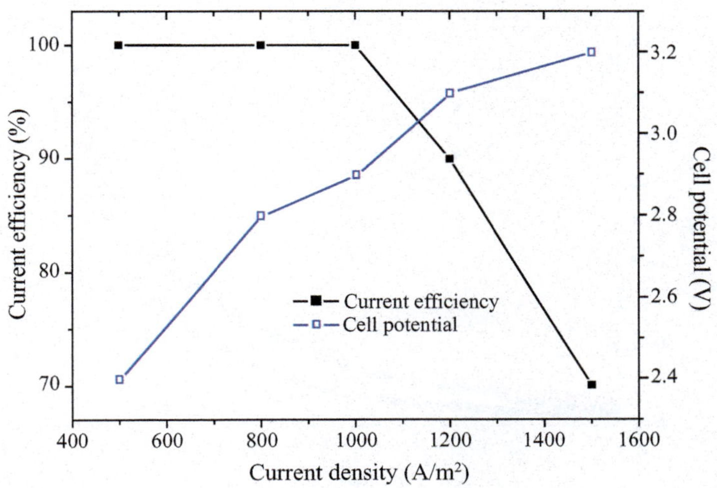

Fig. 6.7 Effects of current density on cell potential and current efficiency in alkaline zincate solutions

This observation may be explained by the easy formation of stable nuclei due to a large number of ions discharged in the cathode. The hydrogen evolution overpotential on the cathode also was high when the current density was high, favoring the discharge of the zinc ions.

Figure 6.7 shows that the CE of the deposition process was approximately 100% in the range of current density from 600 to 1000 A/m^2 but decreased when the current density reached 1000 A/m^2. The linear trend of the cell potential vs. current density from 500 A/m^2 to 1500 A/m^2 is also shown in Fig. 6.7. Clearly, the deleterious effect of increasing current density on the cell potential was proved. A lower temperature was also anticipated to promote grain refinement. A decrease in the temperature was expected to result in a higher overpotential, which would increase the nucleation rate. The cathodic potentiodynamic polarization curves for zinc electrowinning from alkaline solution at different temperatures are presented in Fig. 6.8. The evident increase in the cathode overpotential was dependent on the decreased mass transfer rate at the lower temperature.

There was no evident linear relationship between temperature and current efficiency. It was found that the CE dropped off much more rapidly at temperatures higher than 50 °C. The reason for this decrease was due to the lower hydrogen overvoltage. On decreasing the temperature to lower than 35 °C, a sharp decrease in the CE by 10% was seen, and this phenomenon could be attributed to the growth in the cell potential.

Lead was known to impede hydrogen evolution and prevent the CE from decreasing. In this work, lead was added to the electrolyte to function as a grain refiner. Particle size distribution analysis of electrowinning with 60 mg/L Pb^{2+} showed that the most represented fraction was 4–7 μm; however, the fraction

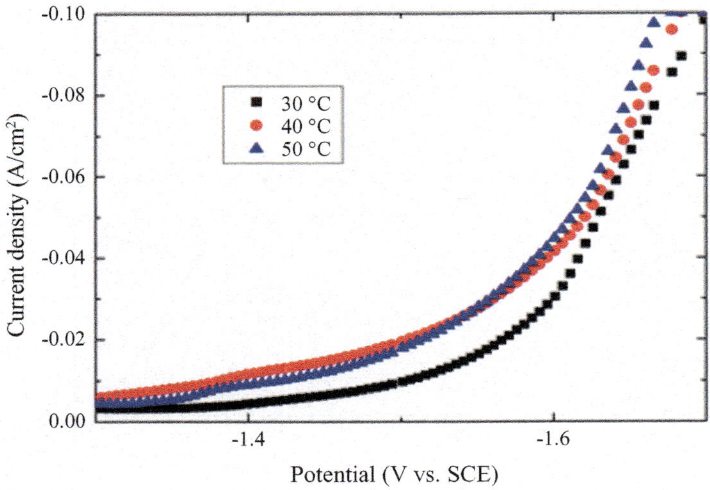

Fig. 6.8 Polarization curves at different temperatures in alkaline zincate solutions

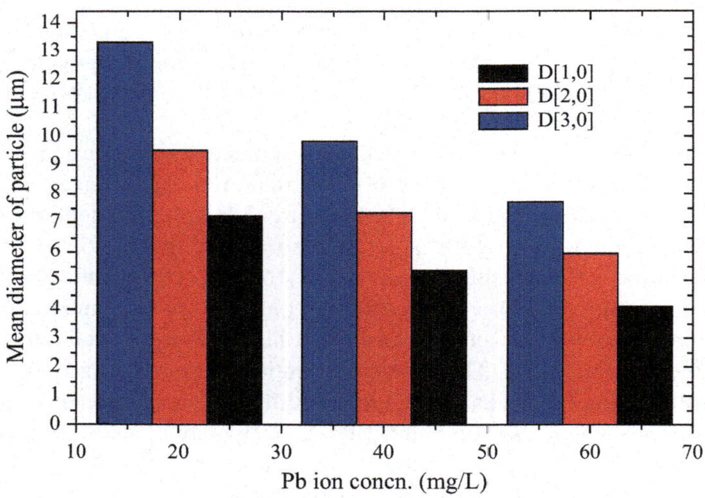

Fig. 6.9 Effects of lead addition on the zinc particle size. D [1, 0] mean diameter by length, D [2, 0] mean diameter by surface, and D [3, 0] mean diameter by volume

0.6–4 μm was also not negligible. In addition, 90% of the zinc particles have a grain size below 9.04 μm.

The mean particle sizes achieved in the tests are shown in Fig. 6.9. The data compiled in Table 6.4 indicated that lead addition decreased the absolute diameter of the nuclei formed on the Mg plate. This result was ascribed to the nucleation overpotential with the addition of lead.

Table 6.4 Effects of Pb^{2+} on current efficiency, cathodic contamination, cell potential, and particle size for zinc electrowinning in the alkaline electrolyte

Pb^{2+} (mg)	Size (μm)	STD (μm)	Current efficiency (%)	Cathodic contamination (%)	Cell potential (V)
20	6.3	7.17	99.10	0.06	2.55–2.65
40	5	5.33	98.88	0.12	2.68–2.78
60	4.3	4.07	98.76	0.18	2.7–2.85

The effects can be attributed to the decrease in the rate of a few reactions involved in the zinc electrowinning in the presence of Pb^{2+}. The major inhibitory effect of Pb^{2+} accounts for the higher electrode polarization necessary for micro-scale zinc particle electrowinning to take place.

As shown in Table 6.4, Pb^{2+} had a negligible effect on the current efficiency of electrowinning in the alkaline electrolyte due to the fact that lead exhibited high hydrogen overvoltages. This result was consistent with that obtained in the typical acid process. The cathodic contamination by lead was counteracted by the diffusion-limiting current density. During the electrowinning process, diffusion control was rapidly achieved, and the Zn^{2+} electrowinning, which was not controlled by the limiting current, tends to be predominant. This mechanism can be explained by the following equation:

$$\omega(\mathrm{Pb}) = \frac{i_{dl}(\mathrm{Pb}) \cdot M(\mathrm{Pb})}{i(\mathrm{Zn}) \cdot M(\mathrm{Zn}) \cdot CE} \tag{6.12}$$

where M (Pb)—the molar mass of lead

M (Zn)—the molar mass of zinc

CE—the current efficiency of electrowinning, %

The cyclic voltammetric curves for zinc electrowinning from alkaline solution containing 25 g/L Zn are depicted in Fig. 6.10. The BCD region corresponded to the electrowinning process, whereas the BEA region concerned the anodic dissolution of previously obtained zinc. With the addition of Pb, the BCD region reflected a larger nucleation overpotential than the case without Pb, confirming the size distribution analysis. In the reverse scan, the anodic peak was shifted toward more positive potential, showing the more noble character of the zinc obtained due to the presence of Pb. Hence, Pb was found to inhibit the process of zinc dissolution under the conditions studied.

Optimum conditions for zinc electrolysis in alkaline solution were determined as:

Current density of 1000–1200 A/m^2
Electrolyte temperature between 30 and 40 °C
Zn concentration of 25–35 g/L
NaOH concentration of 200–220 g/L
Lead level between 20 and 40 mg/L

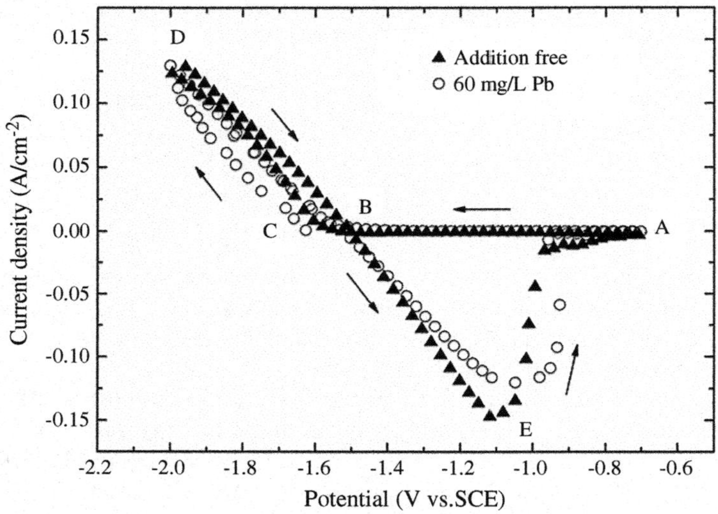

Fig. 6.10 Cyclic voltammetry curves recorded for zinc in 200 g/L NaOH in the presence of lead

A zinc electrolysis current efficiency higher than 97% and a size distribution less than 10 μm were achieved under the above conditions. The negative effect of Pb on the purity of zinc powder was counteracted by the high current density, and low amount of additive and high purity zinc powders (99.5–99.8% Zn) were obtained.

6.3 Effects of Organic Additives on the Electrolytic Zinc Powder Refinement in Alkaline Electrolyte

6.3.1 Effects of Cetyltrimethylammonium Bromide and Sodium Lauryl Sulfate on Zinc Electrowinning in Alkaline Electrolyte

Organic additives, such as surfactants used in zinc electrowinning, serve several purposes. On one hand, they are commonly added to affect the metallic crystal shape and size in order to result in beneficial quality changes on the deposits. On the other hand, surfactants are extensively employed to counteract the negative effects that impurities such as antimony have on CE. Glue and gum arabic are the most common additives used industrially in acid electrolyte. Nevertheless, other additives like ammonium ion, various polyols, and alkyl sulfates have been also employed for zinc electrowinning.

The concentrations of ZnO and NaOH were chosen as shown in Table 6.5. Separately, hexadecyl trimethyl ammonium bromide (CTABr) and sodium lauryl

Table 6.5 Industrial ranges for current density, voltage, electrolyte temperature, zinc and sodium hydrate concentrations, and current efficiency in alkaline zinc electrowinning process

Parameter	Range
Current density (A/m^2)	800–1200
Voltage (V)	2.5–2.8
Temperature (°C)	30–50
Zinc concentration (g/L)	30–50
Sodium hydrate concentration (g/L)	180–200
Current efficiency (%)	>91

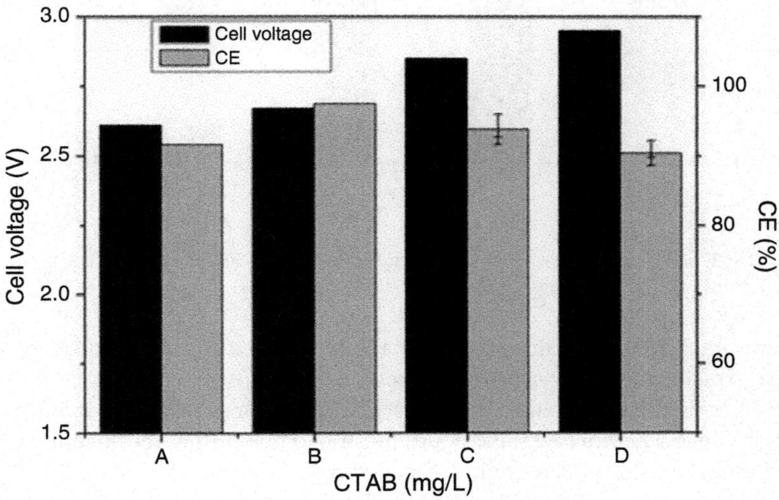

Fig. 6.11 CE and cell voltage of zinc deposit from alkaline electrolyte containing different concentrations of CTABr: (*A*) 0 mg/L, (*B*) 1 mg/L, (*C*) 10 mg/L, and (*D*) 100 mg/L

sulfonate (SLS) were added to this base electrolyte and alkaline solution containing Sb^{3+}. The electrowinning experiments were performed at conditions to give high CE and produce ultrafine zinc powders. The device consists of a Plexiglas cell of 1 L, a magnesium cathode measured 8 cm × 10 cm, and two stainless steel anodes. Moreover, the anode to cathode distance was kept at 3 cm, and all experiments were completed in 1 h period. After electrolysis, the cathode was removed from the cell and rushed thoroughly with water and finally dried in an oven. Thus, the CE value was calculated by weight.

The effects of 0, 1, 10, and 100 mg/L CTABr on CE and on cell voltage during alkaline zinc electrowinning are shown in Fig. 6.11. With addition-free electrolyte, the CE obtained was 91.5%. At 1 mg/L CTABr, the CE rose to about 97.5%. Thus, the CTABr increased the current efficiency since it inhibited H_2 evolution. However, further addition of CTABr did not show any advantage for CE with a drop of ~3.5% at 10 mg/L, ~ 7% at 100 mg/L. These declines in CE could be ascribed to the blocking of the active nucleation sites of the cathode surface by adsorption of the additive, which inhibited the reduction of zinc. In addition, the cell voltage went from 2.6 V at 0 mg/L to 2.95 V at 100 mg/L CTABr, as seen in Fig. 6.11.

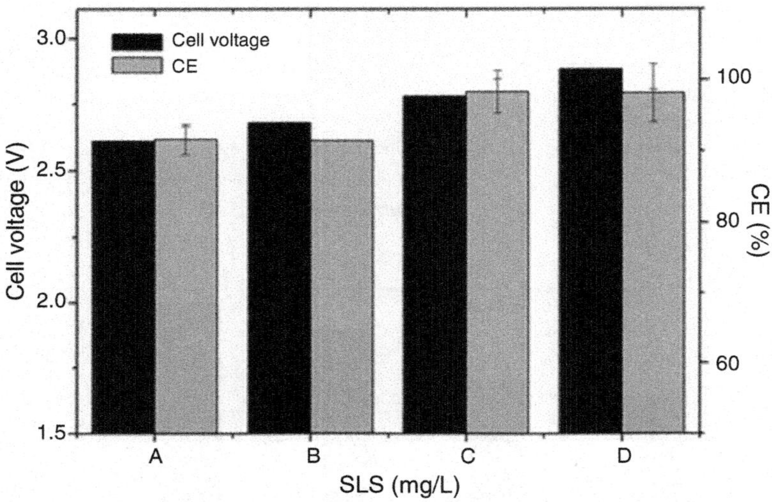

Fig. 6.12 CE and cell voltage of zinc deposit from alkaline electrolyte containing different concentrations of SLS: (*A*) 0 mg/L, (*B*) 1 mg/L, (*C*) 10 mg/L, and (*D*) 100 mg/L

Furthermore, this higher value of cell voltage determined in the presence of CTABr may be explained by the viscosity change in the solution.

For SLS, on the other hand, the highest CE (98.3%) was achieved at 10 mg/L as shown in Fig. 6.12. Moreover, the figure illustrated that further introduction of SLS caused a less dramatic decrease in CE compared to that corresponding to CTABr, reducing it to about 95–97.5% at 100 mg/L. Also, the cell voltage rose progressively with increasing SLS concentrations. Nevertheless, the rate of increase in the cell voltage for SLS was lower than that for CTABr. For example, at 100 mg/L SLS, the cell voltage was ~2.88 V which was considerably less than ~2.95 V achieved with the 100 mg/L CTABr.

Figure 6.13 shows the effects of CTABr and SLS (1 mg/L) on the CE and cell voltage of zinc electrowinning from the solution containing Sb^{3+}. In the alkaline electrolyte studied, the presence of 0.1–1 mg/L Sb^{3+} decreased CE to 85–91%. This was probably due to promotion of the hydrogen evolution reaction, because antimony forms hydrides in the double layer, and consequently the decomposition of hydrides would provide an alternate path for hydrogen evolution.

Figure 6.13 also presents that surfactants employed exert an extremely positive effect on CE, causing increases of ~9 and 13.5% at 1 mg/L, respectively. This may be due to the inhibition effect of surfactants dominating the catalyzing behavior of Sb^{3+} for hydrogen evolution. It was also found that, at 1 mg/L, SLS provided a higher CE than CTABr when Sb^{3+} was also present in the solution. Coinciding with the trend in the above conditions, the cell voltage rose slower for SLS than that for CTABr in the alkaline solution containing Sb^{3+}.

In contrast to what concerned in Fig. 6.13, the addition of 1 mg/L Sb^{3+} increased the CE from ~91.4% to ~97.8% for the alkaline electrolyte containing 1 mg/L SLS.

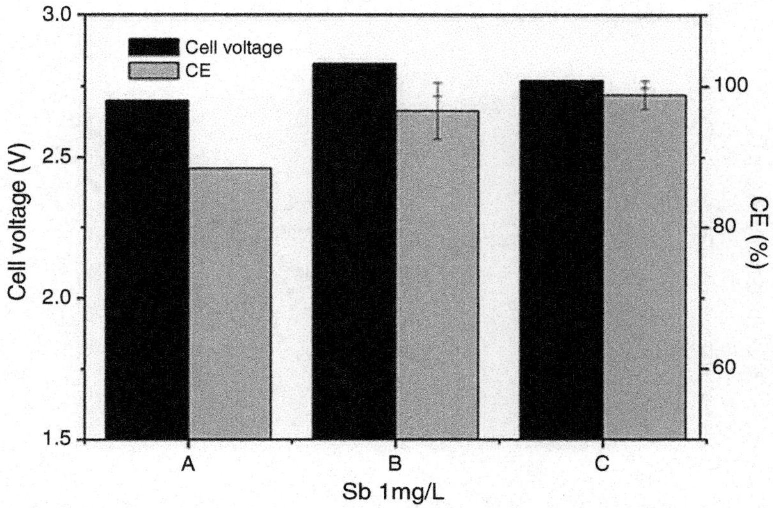

Fig. 6.13 CE and cell voltage of zinc deposit from alkaline electrolyte containing 1 mg/L Sb: (*A*) no addition, (*B*) 1 mg/L CTABr, and (*C*) 1 mg/L SLS

This increase in CE may be explained as follows: when Sb^{3+} was present at the interface, it interfered with the hydrogen adsorption by forming volatile hydride in the double layer, thus preventing hydrogen adsorption from acting as an inhibitor for zinc electrowinning. In the meantime, the SLS counteracted the hydrogen evolution catalyzed by the decomposition of hydride. Hence, under the optimum combination conditions, the presence of low concentration Sb^{3+} increased the CE in the alkaline solution containing SLS.

Figure 6.14 shows voltammograms obtained from alkaline electrolytes containing 0–1 mg/L CTABr and SLS. In the voltammograms, cathodic waves extended from −1.45 V/SCE to −1.60 V/SCE, whereas in the positive sweep, anodic peaks of zinc dissolution can be seen.

As an indicator of the extent of polarization, the nucleation overpotential (NOP) can be determined from the cyclic voltammograms, and it was the difference between the nucleation potential (E_{nu}) and the crossover potential (E_{cru}) as shown in Fig. 6.14. The NOP values increased with the addition of surfactants. For addition-free solution, the NOP value was 30 mV, but in the presence of 1 mg/L SLS, it became 55 mV, which was less than that of CTABr (71 mV at 1 mg/L).

E_{cru}, which, at equilibrium condition, can be related to the Nernst equation, was similar for the three voltammograms ($E_{cru} = -1.42$ V vs. SCE), indicating that the additives did not form complexes with Zn(II), possibly due to their low concentrations. In addition, polarization curves associated with the oxidation process were shifted toward more positive values in the presence of 1 mg/L SLS and CTABr. Thus, CTABr and SLS appeared to hinder zinc dissolution probably by being absorbed on the zinc produced.

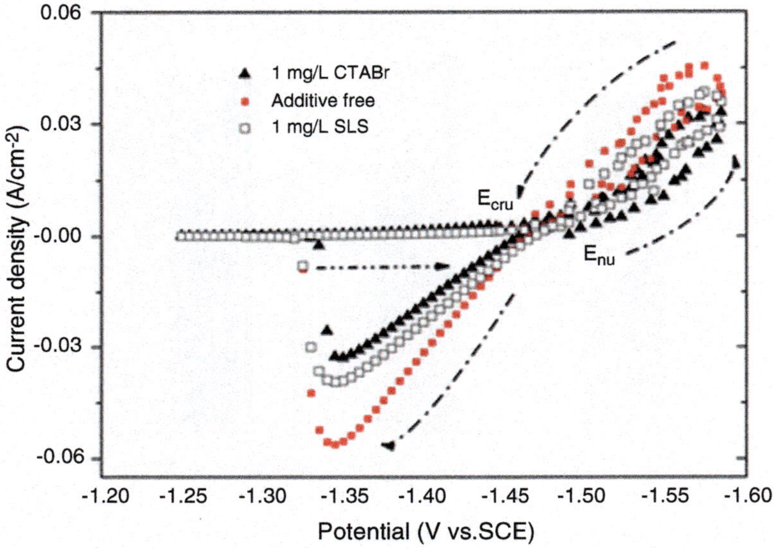

Fig. 6.14 Cyclic voltammograms of alkaline zinc electrowinning in the presence of 0–1 mg/L CTABr of SLS

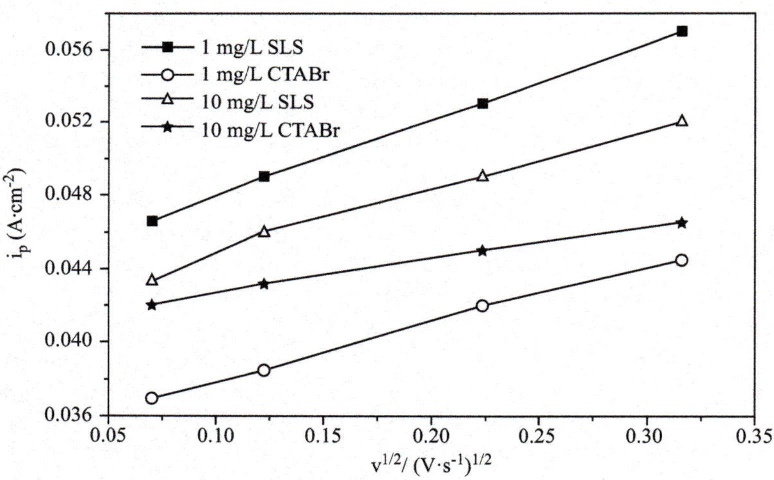

Fig. 6.15 Variation of peak current density with sweep rate in the alkaline electrolytes containing 1–10 mg/L CTABr or SLS

The relationship between the cathodic peak current (i_p) and the square root of scan rate ($v^{1/2}$) is presented in Fig. 6.15. This figure presents an increasing trend of i_p as the rise of sweep rate under four conditions concerned. Furthermore, the variation was linear for 1–10 mg/L additives, and this linearity was expected for

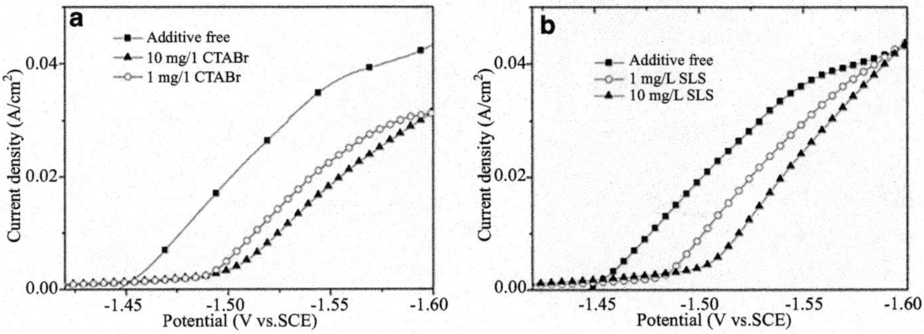

Fig. 6.16 Effects of surfactants on cathodic polarization during zinc electrowinning from alkaline solution: (**a**) 0–10 mg/L CTABr and (**b**) 0–10 mg/L SLS

Table 6.6 Degree of surface coverage obtained at $E = -1.55$ V vs. SCE for the surfactants studied at different concentrations

Surfactant	C (mg/L)	θ
CTABr	1	0.374
	10	0.489
SLS	1	0.181
	10	0.324

a reduction process under mass transfer control. However, the intercepts higher than zero indicated that additional processes other than diffusion occurred.

To complement the cyclic voltammetry measurements and understand the cathodic process more clearly, linear sweep experiments were performed at 5 mV/s with the alkaline electrolytes containing 1–10 mg/L CTABr or SLS. As shown in Fig. 6.16 and Table 6.6, addition of CTABr or SLS (1–10 mg/L) shifted the polarization curves for zinc electrowinning to more negative potentials in the alkaline solutions and thereby inhibited the zinc electrowinning. This result was consistent with that of cyclic voltammetry studies. Moreover, the inhibition effect on the zinc electrowinning, caused by the CTABr and SLS, was due to a blockage of the cathode surface. Furthermore, the degree of surface coverage for the surfactants was evaluated from Eq. 6.13:

$$q = \frac{i - i_s}{i} \tag{6.13}$$

where i and i_s are the current densities (A/m^2) without and with surfactants, respectively, measured at a constant potential value (~1.55 V vs. SCE).

From the cathodic polarization curves in Fig. 6.16, Tafel slopes (b), transfer coefficients (α), and exchange current densities (I_0) were determined (Table 6.7) using the following equations:

$$\eta = a + b\log I \tag{6.14}$$

Table 6.7 Effects of CTABr and SLS on Tafel slopes, transfer coefficients, and exchange current densities in alkaline solution

Additive / $mg{\cdot}L^{-1}$	Tafel slope, b /mV	Transfer coefficient, α	Exchange current density (I_0), mA/cm^2 ($\times 10^{-2}$)
0	−83.1	0.71	5.8
1 mg/L CTABr	−66.4	0.91	4.1
10 mg/L CTABr	−83.0	0.71	3.3
1 mg/L SLS	−71.5	0.83	5.3
10 mg/L SLS	−60.3	0.93	4.2

$$b = RT/\alpha nF \quad (6.15)$$

where η is the overpotential (V) and I the current density (A/m^2).

Small changes in Tafel slopes with the addition of CTABr or SLS indicated that the transfer reactions were not controlled by the concentrations of surfactants in the alkaline electrolyte. Additionally, the transfer coefficient remained essentially unaffected by the presence of additives at almost all concentrations. Moreover, the exchange current densities (j_0) values decreased upon addition of surfactants, and this decrease in j_0 also showed the blocking of the active sites of the cathode. Thus, changes in kinetic parameters and polarization behavior were also reflected in CE, deposit morphology, and crystallographic images.

Although it was not shown in Fig. 6.16, addition of 0.1–1 mg/L Sb(III) was found to significantly depolarize the cathodic process of zinc electrowinning. It was also clear that SLS had a more positive effect on counteracting the depolarization arising from Sb(III) compared to CTABr. Therefore, the CE value associated with the addition of SLS was higher than the one corresponding to the presence of CTABr in alkaline electrolyte containing 0.1–1 mg/L Sb(III). In addition, the difference of surfactants in polarization behavior for alkaline solution containing Sb(III) will result in different deposit morphologies.

SEM was used to reveal whether a relationship existed between the surfactants addition and zinc powders morphology. Figure 6.17 shows SEM micrographs of zinc produced in the presence of 0–1 mg/L CTABr or SLS, with and without Sb(III) from alkaline electrolyte. With additive free, the large grain size and the dendritic growth were quite evident. The addition of 1 mg/L CTABr or SLS did not change the dendrite morphology significantly but decreased the sizes of the platelets considerably. By contrast, grain size of Zn-1 mg/L CTABr was smaller than Zn-1 mg/L SLS. These morphology changes (Fig. 6.17a–c) may be due to their differences in the polarization behavior and thus can be associated to a strong blocking effect of surfactants, which accelerated the nuclei renewal.

The addition of Sb(III) to the additive-free alkaline electrolyte affected the deposit morphology dramatically, as shown in Fig. 6.17d. It can be seen that the aggregation of dendritic deposits diminished, while well-defined hexagonal platelets were facilitated. This basal morphology seems to facilitate hydrogen evolution and may be the reason why Sb(III) caused a significant decrease in CE. The

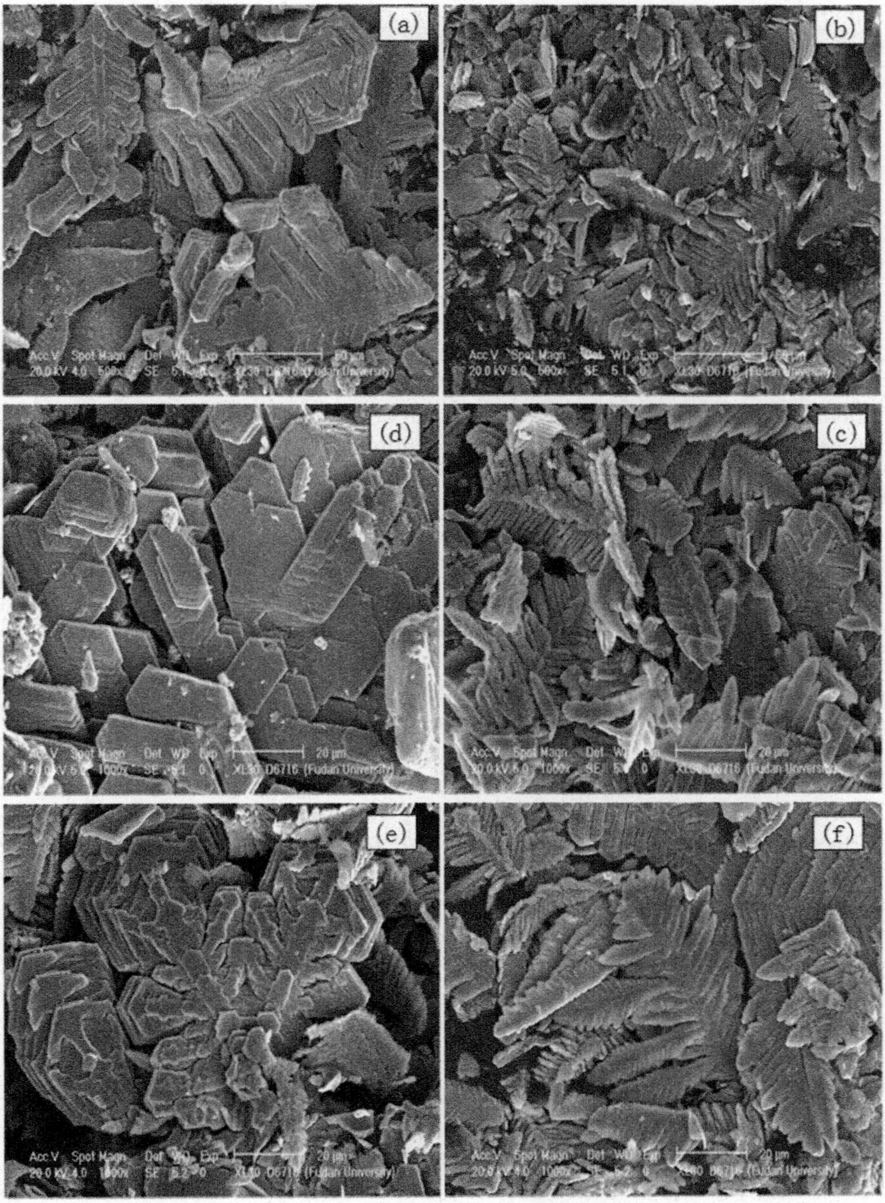

Fig. 6.17 SEM images of zinc powders produced from alkaline electrolyte: (**a**) Blank, (**b**) 1 mg/L CTABr, (**c**) 1 mg/L SLS, (**d**) 1 mg/L Sb(III), (**e**) 1 mg/L Sb(III) + 1 mg/L CTABr, and (**f**) 1 mg/L Sb(III) + 1 mg/L SLS

diminishing of dendritic growth can be associated to the depolarization effect of Sb (III), which reduced the nuclei number. In Fig. 6.17f, the addition of 1 mg/L Sb to solution containing SLS did not affect the platelet morphology markedly, whereas the deposit of Fig. 6.17e became somewhat irregular by comparison with Fig. 6.17b. This result confirmed the fact that SLS had a more positive effect on counteracting the effects of Sb(III) in the alkaline electrolyte.

The effects of the surfactants employed in this work on the crystallographic orientation of zinc obtained from alkaline electrolyte with and without Sb(III) were examined using X-ray diffraction as given in Fig. 6.18. The addition of 1 mg/L CTABr or SLS had minor effects on the order of preferred crystal plane (i.e., an orientation dominated by (101), (002), and (100)). However, the presence of surfactants caused a relatively large increase in the intensity of the highest diffraction line; the intensity value reached ~17,000 and ~14,000 at 1 mg/L CTABr and 1 mg/L SLS, respectively. This variation can be due to an increase of crystallinity, which was associated with zinc ion concentrations at the substrate interface. In addition, this increase in the peak intensity was also reflected in the surface morphologies (Fig. 6.18a–c).

On the other hand, Fig. 6.19 presents that the addition of 1 mg/L Sb(III) promotes the growth of (002) planes substantially by comparison with Fig. 6.19a. This plane corresponded to a basal morphology which is shown in Fig. 6.19d. Moreover, the development of the basal (002) plane can be attributed to a decrease of the surface energy of this plane. On the basis of comparison between Fig. 6.19b, c, the growth of (101) plane was found to be more favored in the case of 1 mg/L SLS for the alkaline solution with Sb(III). For example, the peak intensity ration (I/I_{max}) of (101) rose from 49.2% (Fig. 6.19a) to 83.5% (Fig. 6.19c).

The CE for zinc deposition remained constant (at 0–1 mg/L SLS) or increased (at 0–1 mg/L CTABr) with addition of additives, but it decreased as the surfactants concentration went past 10 mg/L for both two additives. Surfactants additions were found to polarize the cathode and decreased the anionic peak charge. These findings were explained by the surfactant adsorption on the electrode surface. Moreover, the polarization effect of 0–10 mg/L CTABr was greater than that of 0–10 mg/L SLS.

SEM results showed that the surfactants worked as grain refiners of zinc powder when presented in baths at only 1 mg/L. This fact may be justified by the polarization effect as the electrochemical analysis confirms. At 1 mg/L Sb(III), the morphology of zinc shifted from dendritic plane to basal plane. In alkaline electrolytes containing 1 mg/L Sb(III), it was recommended that 1 mg/L SLS should be present since this combination increased CE to ~97.8% and kept cell potential around 2.6 V. This positive effect of SLS was also reflected in the polarization data and morphology results. The presence of CTABr and SLS did not change the preferred crystallographic orientations of the crystal planes, whereas the Sb(III) promoted substantially the growth of (002) plane which corresponded to a basal morphology.

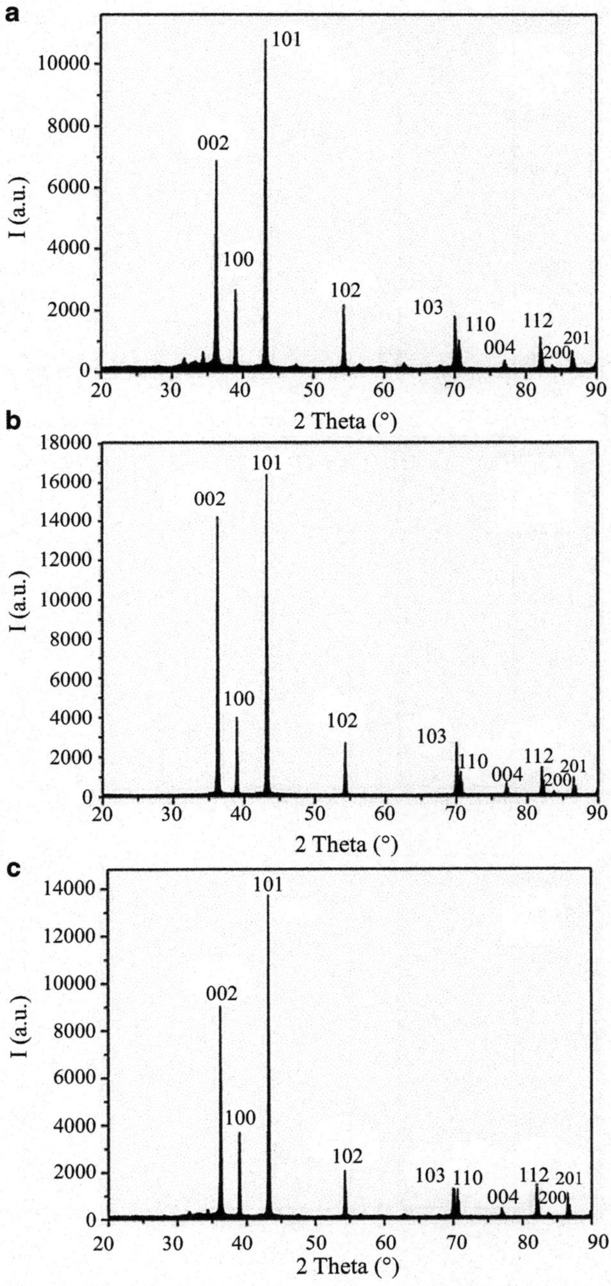

Fig. 6.18 XRD patterns of electrodeposited samples from alkaline solution in absence (**a**) and presence of 1 mg/L CTABr (**b**) or 1 mg/L SLS (**c**)

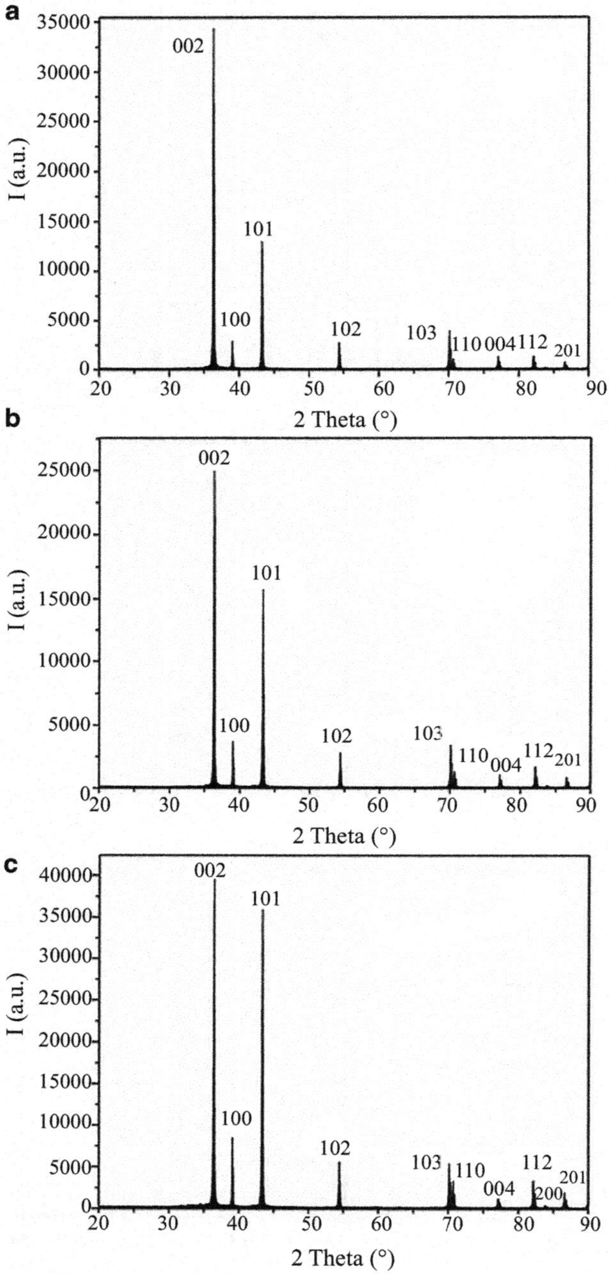

Fig. 6.19 XRD patterns of electrodeposited samples from alkaline solution containing Sb(III), in absence (**a**) and presence of 1 mg/L CTABr (**b**) or 1 mg/L SLS (**c**)

6.3.2 *Effects of Single Organic Additive Among β-CD, SDS, Gelatin, Casein, and Thiocarbamide in the Alkaline Electrolyte*

The effects of the presence of organic additives, including β-cyclodextrin (β-CD), sodium dodecyl sulfate (SDS), polyethylene glycol (PEG), gelatin, casein, thiocarbamide, and tween 80 (T-80), in alkaline zincate solutions on the electrolytic zinc powder refinement were examined. Moreover, three kinds of PEG were employed with different molecular weight, i.e., PEG 400, PEG 1000, and PEG1500.

The electrolyte of 0.55 M Zn^{2+} and 5 M NaOH was selected to be in accordance with the industrial practice. The electrolyte was prepared by dissolving 0.55 mol ZnO and 6.1 mol NaOH (0.55 M ZnO requires 1.1 M NaOH for dissolving reaction) in distilled water and then adding additives with different concentrations as designed in each experiment. Finally, the volume of the electrolyte was fixed to 1 l with distilled water. The electrowinning experiments were carried out in a 1 L cell. Zn electrodeposit was carried out for 60 min at a constant current density of 1000 A/m^2 and an electrolyte temperature of 50 °C. After electrolysis, the Zn electrodeposits with powdery structure forms were removed from the cathode and washed thoroughly with distilled water and then with ethanol. Finally, it was dried in a vacuum drying oven (85 °C, 80 Pa). A laser particle size analyzer (Mastersizer 2000, Malvern, UK) was employed to detect the size distribution of the electrolytic zinc powder, while a standard vibrating screen was used for the mass particle size distribution analysis.

Figure 6.20 shows the effect of additive with different concentrations on the grain refinement of the Zn electrodeposits. The weight percentage of undersize would be increased with reducing the grain size of the zinc deposits. Hence, the grain refinement effect could be described with the weight percentage of undersize. Zn deposits with 18.58% of undersize of 150 μm (100 mesh) was obtained in the absence

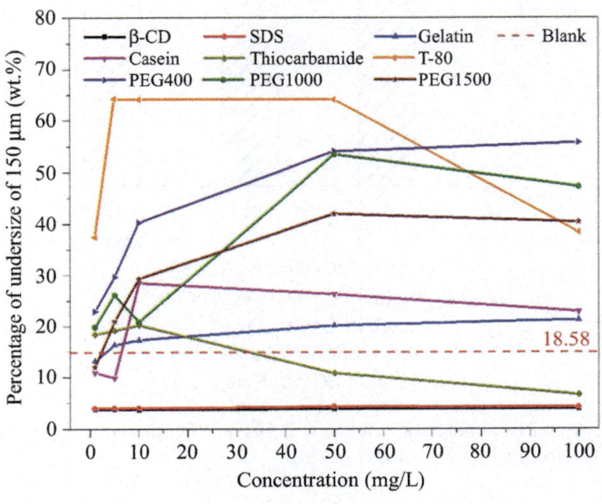

Fig. 6.20 Effects of single organic additive among β-CD, SDS, gelatin, casein, and thiocarbamide to the alkaline electrolyte with different concentrations on Zn electrodeposits refinement

of organic additives. Comparatively, the addition of β-CD, SDS, gelatin, casein, and thiocarbamide to the alkaline electrolyte had no apparent effect on grain refining of the Zn deposits or even had a counter influence. On the other hand, the presence of small amounts of T-80 (5 mg/L) and PEG 400 (50 mg/L) in the alkaline electrolyte markedly reduced the grain size of Zn deposits. An electrolytic zinc powder with 64.25% and 54.06% of undersize of 150 μm can thus be obtained, respectively.

6.3.3 *Synergistic Effects of T-80 and PEG Organic Additives to the Alkaline Electrolyte on Zn Electrodeposits Refinement*

Since both T-80 and PEG had notable effect on grain refinement, a further experiment was carried out to examine the co-effect of T-80 and PEG on the Zn electrodeposits refinement, and the results are shown in Fig. 6.21. Zinc powder with a 71.55% of undersize of 150 μm could be obtained when 5 mg/L T-80 and

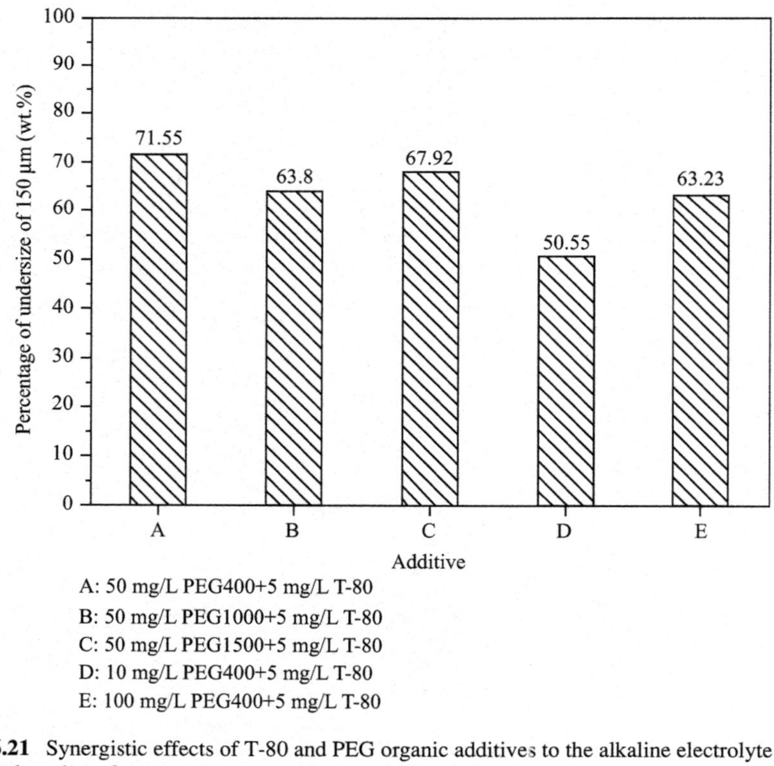

Fig. 6.21 Synergistic effects of T-80 and PEG organic additives to the alkaline electrolyte on Zn electrodeposits refinement

50 mg/L PEG 400 were added together. However, other co-additions of 5 mg/L T-80 with PEG at different concentrations seemed to have less function on the Zn electrodeposits refinement.

6.3.4 Zinc Powder Particle Size Distribution in the Presence of Organic Additives in Alkaline Electrolyte

Figure 6.22 shows the particle size distribution of Zn electrodeposits obtained with different organic additives. The particle size with the largest frequency density shifted to be smaller when 50 mg/L PEG 400, 5 mg/L T-80, and 50 mg/L PEG 400 + 5 mg/L T-80 were added successively. The mean diameter of volume, number, and surface area became smaller accordingly as shown in Table 6.8.

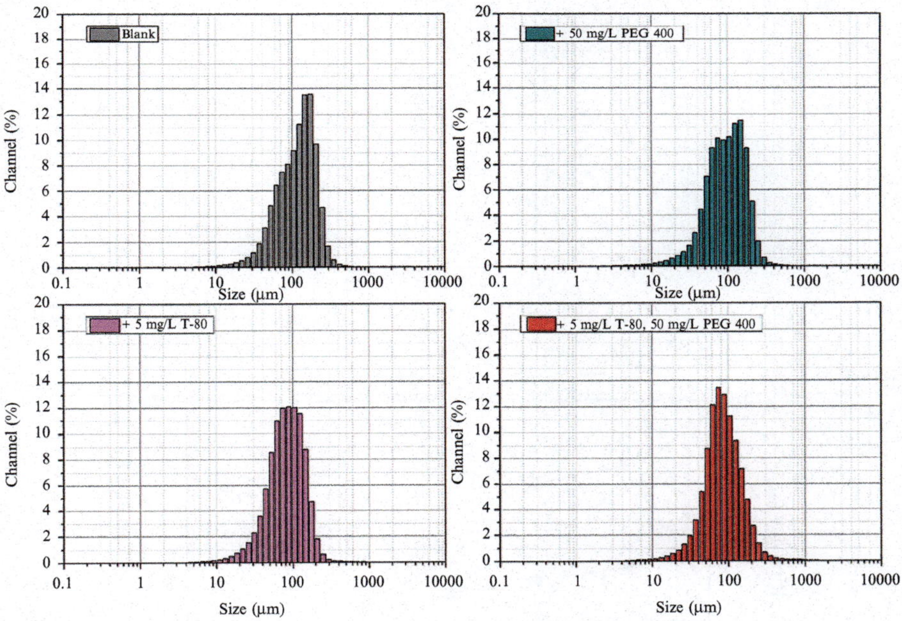

Fig. 6.22 Particle size distribution of Zn electrodeposits obtained with different additives

Table 6.8 Summary data of the particle size distribution, diameter by volume (MV), diameter by number (MN), and diameter by surface area (MA)

Additives	MV, μm	MN, μm	MA, μm
Blank	119.9	15.65	81.56
50 mg/L PEG 400	99.31	15.14	63.84
5 mg/L T-80	89.73	14.99	61.91
50 mg/L PEG 400 + 5 mg/L T-80	84.83	6.70	60.08

6.4 Effects of Ion Impurities on Zinc Electrowinning Process in Alkaline Leaching Solution

6.4.1 Effects of Sn on Zinc Electrolysis in Alkaline Leaching Solution

The presence of tin ion makes power consumption increase dramatically, and cell voltage is greater than 2.90 V, as shown in Figs. 6.23 and 6.24. At the same time, tin has a great effect on zinc electrowinning process, causing declined quality of zinc

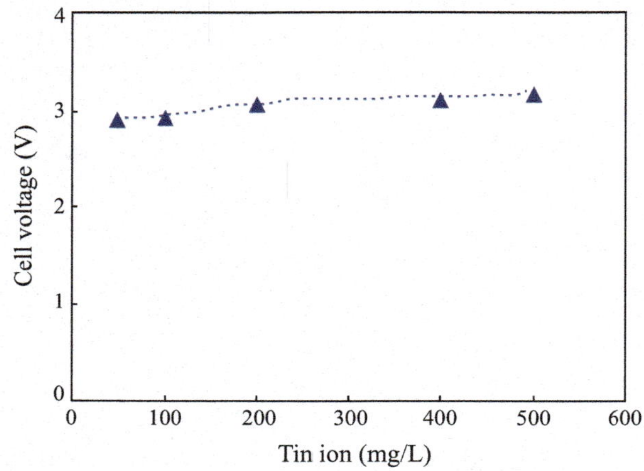

Fig. 6.23 Effects of tin ion concentration on cell voltage in zinc electrowinning process

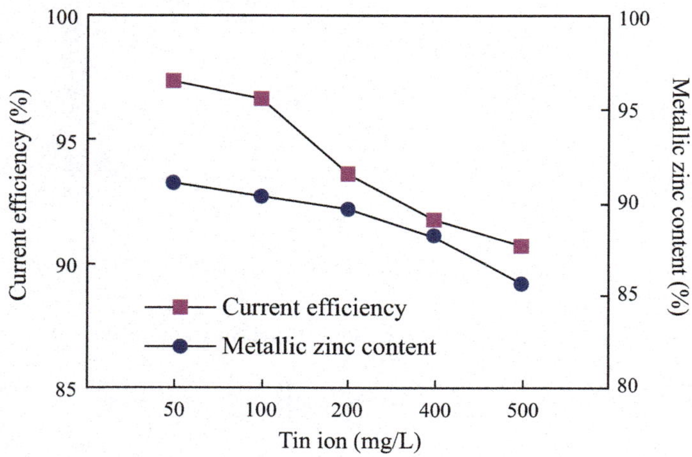

Fig. 6.24 Effects of tin ion concentration on current efficiency and zinc content in electrowinning process

Fig. 6.25 Harm of tin ion on zinc electrowinning in alkaline solution. (**a**) Sn free, (**b**) 2 g/L Sn

as well as declined electrical efficiency obviously. In addition, the presence of tin ion also makes power consumption increase dramatically.

When tin ion concentration was 400 mg/L, zinc content was 88.13%; the Sn contents in the metallic zinc powder would reach to Class 4 quality standard. While tin ion concentration rose up to 500 mg/L, zinc content in the zinc powder decreased to 85.54%. When tin ion concentration was 400 mg/L, the current efficiency was 91.76%, reduced by about 8% compared with no tin. Therefore, it is suggested that the tin ion concentration in the electrolyte should be below 400 mg/L under strict control.

Tin is one of the impurity elements which do great harm to zinc electrowinning process as shown in Figs. 6.25 and 6.26. It can be seen that when tin ion concentration was relatively high, sediments on plate became loose, and holes formed. There were even some falling off to the bottom of the beaker, and a large number of bubbles in the solution appeared at the same time, hindering the zinc from separating out on the cathode board.

The following reactions will happen in alkaline zincate solution through tin ions:

$$Sn^{2+} + 2e \rightarrow Sn \tag{6.16}$$

$$HSnO_2{}^{-} + H_2O + 2e \rightarrow Sn + 3OH^{-} \tag{6.17}$$

$$Sn + 4H^{+} + 4e \rightarrow SnH_4 \tag{6.18}$$

Alkaline zinc solution (40 g/L Zn, 200 g/L NaOH) with 2 g/L Sn concentration (marked as Solution A) was chosen as research sample according to previous leaching results of a Zn-Sn-containing waste. Electrolyte composition of 40 g/L Zn and 200 g/L NaOH (marked as Solution B) was selected in electrolysis tests as a comparison. The zinc electrolysis tests were operated in a 0.5 L volume Plexiglas

Fig. 6.26 Effects of tin ion on the morphology of deposition in alkaline solution. (**a**) Sn free, (**b**) 2 g/L Sn

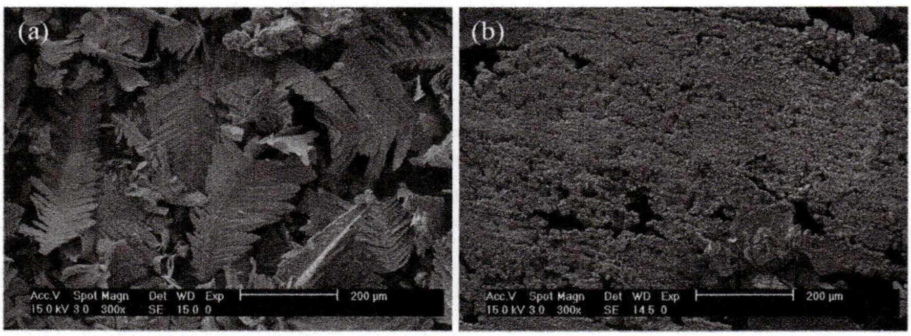

Fig. 6.27 Stereomicroscope and ESEM micrographs showing the surfaces of the zinc deposited at 1000 A/m^2 and 40 °C with impurity content of (**a**) 0 g/L Sn and (**b**) 2 g/L Sn (300×) from alkaline zincate solutions

cell where anode and cathode were placed 30 mm away from each other. Stainless electrode was used as anode and magnesium alloy as cathode with dimensions of 70 × 70 mm.

In the zinc electrolysis tests, Solution A and B were treated with 1000 A/m^2 current density at 40 °C for 1 h. Considerably high current efficiencies (over 98%) were both obtained with A and B. At the end of 1 h electrolysis, the average cell potential values of 2.63 and 2.65 V were obtained for Solution A and B, respectively. Surface photographs of the zinc deposited, shown in Fig. 6.27, displayed the effects of Sn on zinc electrolysis.

Deposited zinc is in powdery form and peeled off easily from the substrate in pure electrolyte (Solution B). By contrast, dendritic growth is restricted, and the

Table 6.9 Effects of Sn(IV) on alkaline zinc electrowinning

Sn(IV) (g/L)	Current efficiency (%)	Energy consumption (kWh/kg Zn)	Content of deposited zinc powder (%)		
			Metal zinc	Zn	Sn
0	100	2.27	98.28	99.56	0
0.98	98.15	2.49	97.91	98.97	0.52
8.12	97.70	2.61	98.06	98.54	0.71

deposited zinc is in compact form and hardly stripping from cathode in the presence of Sn. The effects of Sn on current efficiency, energy consumption, and metallic Zn of deposits from alkaline zinc solution are shown in Table 6.9. The current efficiency and metallic Zn of deposits decreased, while energy consumption increased with increasing Sn(IV) concentrations in the electrolytes.

6.4.2 *Effects of Al on Zinc Electrolysis in Alkaline Leach Solution*

The solubility of some amphoteric elements in strong alkaline solution decreases in the following sequences Zn > Pb > Al > Cr(III) > Cu. The solubility of Cr(III) and Cu are found to be negligible in the presence of zinc and lead. Furthermore, Pb can be precipitated quantitatively without concomitant loss of zinc by sulfide precipitation using sodium sulfide as precipitant. Nevertheless, Al cannot be removed together with lead when sodium sulfide is added to leach solution. So, the effects of Al on current efficiency, polarization behavior of the cathode, deposit morphology, and crystallographic orientation in the production of zinc powder by alkaline electrowinning process are explored.

1. Current efficiency and power consumption

   Figure 6.28 shows the current efficiency as a function of the percentage of recovered zinc in the absence and presence of 50–1000 mg/L Al(III). It should be explained that the lines that connect points only make data reading easier without indicating a linear fit or area integration. It is observed that at lower Al concentration in the range of 0–50 mg/L, the CE does not change significantly, but it decreased progressively with the further addition of Al.

   Additionally, Fig. 6.28 also shows a low current efficiency in the presence and absence of additives at the onset of electrowinning. For example, the CE was ~75% in the Al(III)-free solution, and it was practically unaffected by the presence of Al(III) at <50 mg/L. With 200 mg/L Al, the current efficiency was low (~45%) but higher than in the presence of 1000 mg/L Al (~40%). In the region of 30% zinc recovery, the current efficiencies increased drastically,

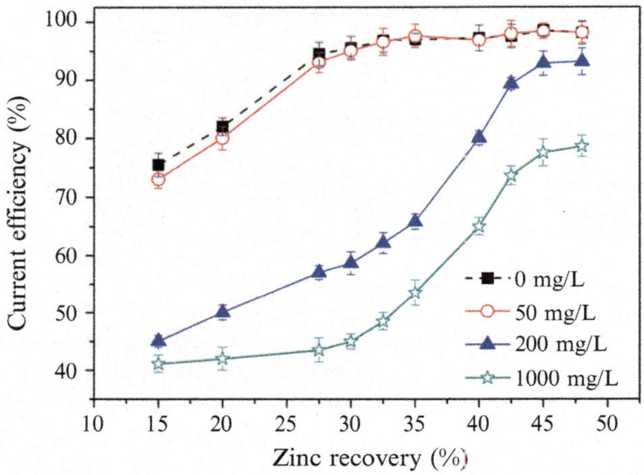

Fig. 6.28 Current efficiency as a function of zinc recovered with the addition of Al for Zn(II)/Zn process in alkaline media

Table 6.10 Effects of Al on CV, PC, and viscosity during electrowinning of zinc from alkaline solution

Additive (mg/L)	PC (kWh/t)	Viscosity (cP)	CV (V)
0	2033	1.92	2.43
50	2049	1.93	2.45
200	2159	1.93	2.55
1000	2366	1.96	2.67

reaching high values at 0–50 mg/L Al (92%), 200 mg/L Al (61%), although it was smaller at 1 g/L Al. In the electrowinning region of over 45% zinc recovery, the greatest efficiencies was achieved at 0–50 mg/L Al (97.8%), 200 mg/L Al (93.5%), and 1 g/L Al (78.3%). In alkaline electrolyte, the adverse effects of high concentration Al(III) (>1 g/L) may be related to the changes in viscosity of electrolyte and in surface morphology of zinc powder obtained.

The effects of Al on cell voltage (CV) and power consumption (PC) during zinc electrowinning are presented in Table 6.10. PC is determined by the following equation:

$$PC = \frac{CV}{q \times CE} \times 1000 \tag{6.19}$$

where q is electrochemical equivalent of zinc, 1.220 g/ (Ah).

With 0–50 mg/L Al, the PC obtained was 2030–2050 kWh/t; addition of 200 mg/L Al increased PC by about 100 kWh/t. In the case of 1 g/L Al, the highest PC (2366 kWh/t) was achieved. Similarly, CV increased as the Al(III) concentration rose from 0 to 1 g/L. Additionally, the presence of 0–200 mg/L Al (III) had little effect on the viscosity of electrolyte. At very high additive concentrations, the electrolyte conductivity was decreased, and consequently a

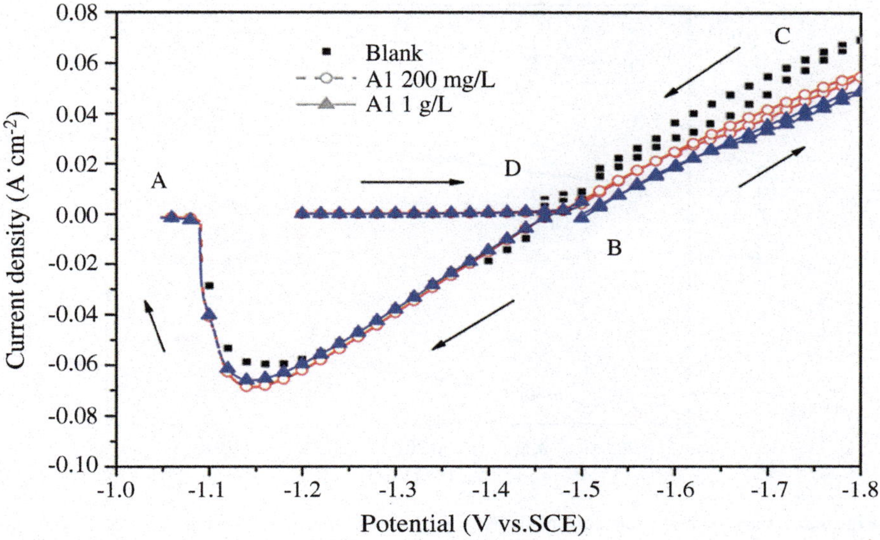

Fig. 6.29 Cyclic voltammograms of zinc electrowinning in alkaline solution with the addition of different concentration Al

higher driving for zinc ion reduction was required, usually resulting in a substantial increase in CV and hence PC.

2. Polarization behavior

The effects of Al on the cathodic polarization of zinc electrowinning were investigated by linear sweep and cyclic voltammetry.

Cyclic voltammograms for zinc reduction process in the presence of Al are shown in Fig. 6.29. The voltammograms were initiated at point "A" (−1.10 V vs. SCE (Fig. 6.29)), scanned in the negative direction and reversed at −1.80 V vs. SCE in the positive direction. Then the current increased sharply to point "C" where it was reversed. Thus, this gave rise to a decrease in current which subsequently reached zero at point "D" and the current then became anodic corresponding to the dissolution of the deposited zinc powder previously formed. The values of nucleation overpotential (NOP) can be determined from the potential difference between the points "B" and "D". Furthermore, the points "B" and "D" are referred to as the NOP and the crossover potential, respectively.

An increase in the NOP along with a progressive shift of deposit potential to more negative value was observed with increasing additive concentration. For addition-free solution, the NOP value is 15 mV. But in presence of 200 mg/L Al, it became 28 mV, whereas at 1 g/L Al, the NOP value became 80 mV. This polarization effect could be attributed to the low rate of zinc deposition in the presence of Al, and it was also confirmed by the deposit morphology where the decrease in the grain size was evident.

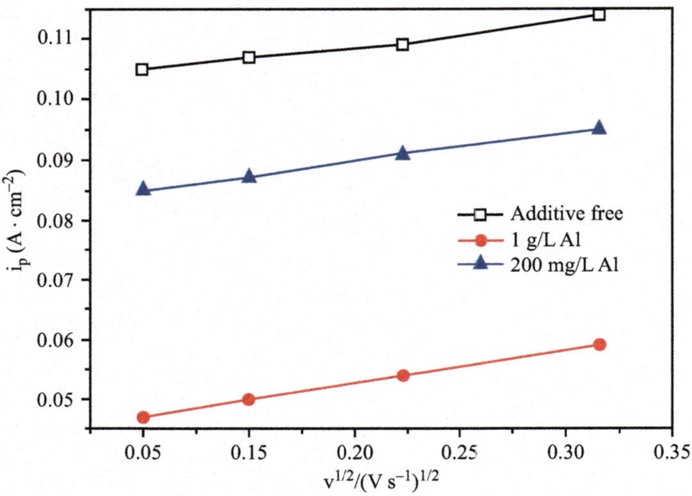

Fig. 6.30 Variation of peak current density with sweep speed in alkaline zinc electrowinning process

Table 6.11 Zinc deposition polarization current as a function of Al(III) concentration in alkaline zinc electrowinning process

	E (V vs. SCE) at current			
Additive concentration, mg/L	−250 mA	−500 mA	−750 mA	−1000 mA
0	1.58	1.64	1.70	1.76
50	1.59	1.65	1.71	1.78
200	1.61	1.67	1.73	1.82
1000	1.63	1.70	1.79	1.96

Additionally, the results showed that the crossover potential remained constant in the above conditions concerned. Since in these circumstances, the crossover potential was equal to the metal/metal-ion reversible potential; it can be assumed that the equilibrium potential for Zn/Zn^{2+} was around −1.40 V vs. SCE. This value was more negative than that calculated from Nernst equation, due to crystallization overpotential related to the Mg substrate. Moreover, the relationship between the cathodic peak current and the square root of sweep speed ($\nu^{1/2}$) is presented in Fig. 6.30. The linearity variation is expected for a reduction process under mass transfer control.

To complement the cyclic voltammetry measurements and understand the cathodic process more clearly, linear sweep experiments were performed at 5 mV/s with the alkaline electrolytes containing 0–1000 mg/L Al(III). The data are summarized in Table 6.11 for four values of the cathodic potential. A comparison of the data in Table 6.11 indicates that 1000 mg/L Al causes a significant polarization since it gives the smallest currents at the selected cathode potentials. In addition, the Al of 0–50 mg/L had almost a similar polarization characteristic.

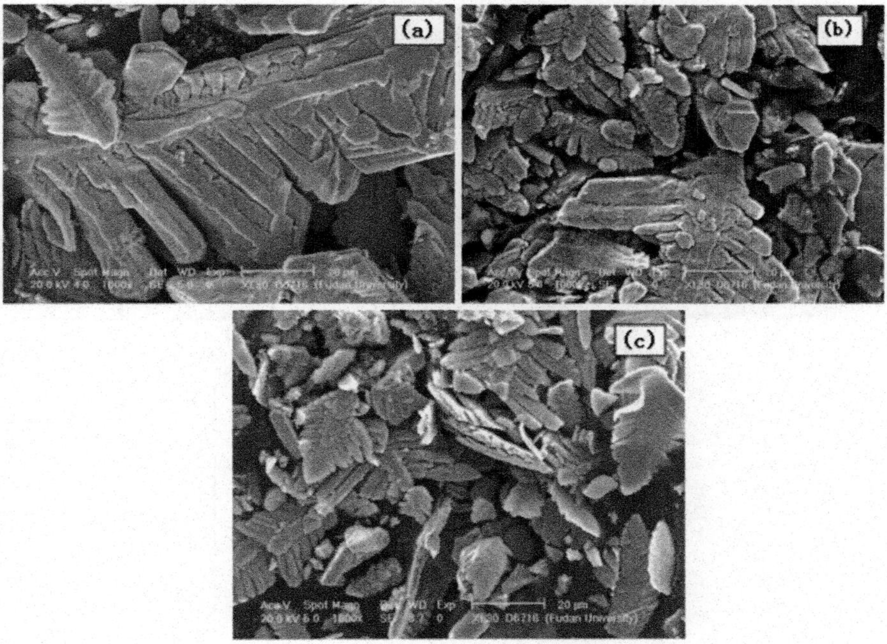

Fig. 6.31 Micrographs of zinc powders electrowon from alkaline solution, with the addition of Al: (**a**) addition free, (**b**) 200 mg/L Al, and (**c**) 1 g/L Al

3. Deposit morphology and crystal orientation

The presence of Al in the alkaline electrolyte resulted in 1.5 h zinc deposits consisting of distinct morphology types (Fig. 6.31). These morphology types had certain preferred zinc deposit orientations.

The morphology type shown in Fig. 6.31a is commonly associated with solution without impurity addition. It was found that the zinc deposit was brittle and consisted of dendrite particles. Moreover, this morphology, which corresponded to a high CE value, may be due to a comparatively high rate of zinc deposition and the adhesion of particles. In addition, the deposits obtained exhibited a ((101), (002), (100)) preferred orientation.

At higher Al(III) concentration (>200 mgL^{-1}), growth of (100) plane was favored resulting in corroded zinc powders consisting of round-edged crystals over which tiny crystallites were spread, as shown in Fig. 6.31b. In this case, the zinc platelets were reduced in size. The addition of 1000 mg/L Al to the electrolyte resulted in the deposit morphology shown in Fig. 6.31c. As indicated the particle had more rounded edges, and its size became smaller further. According to the results shown in Fig. 6.32, the preferred surface orientation observed for this case was still (101); however, some planes were found to shift from (101) to (002). This morphology type seemed to facilitate hydrogen evolution and may be associated with the decrease in CE.

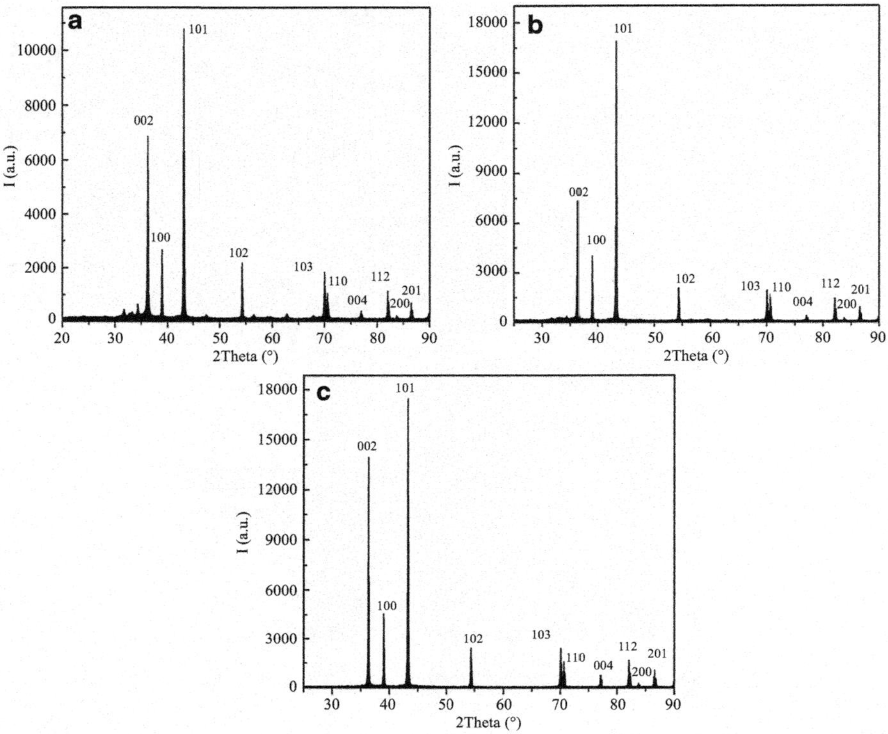

Fig. 6.32 XRD patterns of electrodeposited samples during 2 h, in absence (**a**) and presence of Al: 200 mg/L (**b**) and 1 g/L (**c**) in alkaline zinc electrowinning process

As expected, for all the samples, the zinc is the main phase. However, the Al also has a marked effect on the other orientations. At 10 mg/L Al, no measurable growth was observed in this case. Increasing Al concentration to 200 mg/L enhanced the growth prismatic (100) planes, as the I % value rose to 79.8. Further increase in Al concentration (> 1 g/L) did not change the preferred deposit orientation (101), but the intensities of (002) were increased substantially. The change in the crystallographic orientations is a consequence of surface energy differences, which are responsible for the selective growth of the grains that have the lowest surface free energy.

6.4.3 Effects of As on Zinc Electrolysis in Alkaline Leaching Solution

Electrolyte composition of 40 g/L Zn and 200 g/L NaOH solution was selected in electrolysis tests. The impurities of concern included arsenide (50, 100, 200, 300, 500 mg/L arsenide as Na_3AsO_4). The zinc electrolysis tests were operated in a 0.5 L

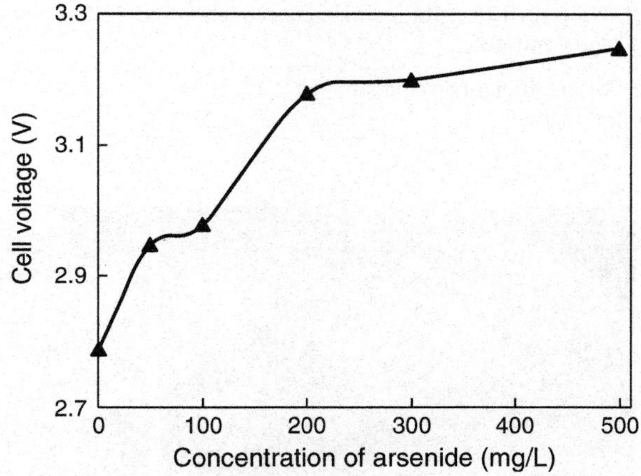

Fig. 6.33 Effects of concentration of arsenide on cell voltage in alkaline zinc electrowinning process

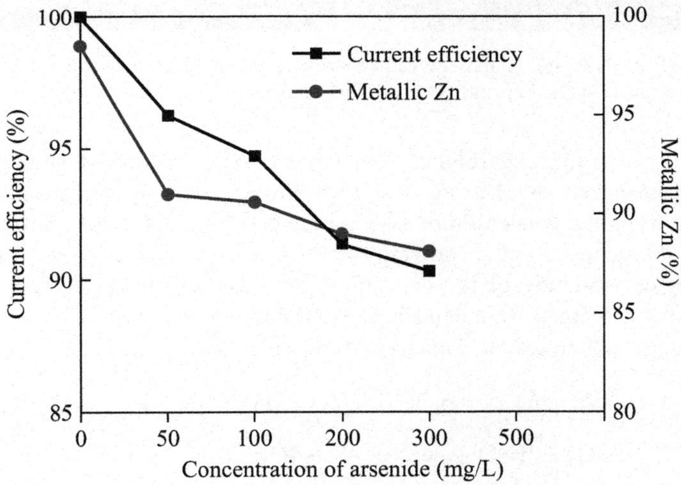

Fig. 6.34 Effects of concentration of arsenide on current efficiency and purity of zinc deposits in alkaline zinc electrowinning process

volume plexiglas cell where anode and cathode were placed 30 mm away from each other. Stainless electrode was used as anode and magnesium alloy as cathode with dimensions of 70 × 70 mm. The cell was placed in a thermostatic bath to keep electrolyte temperature at 40 °C. Electrolysis was carried out for 1.5 h using a DC power supply and current densities at 1000 A/m^2. Electrolyte circulation (1.0 L/min) was controlled by a pump.

The effects of arsenide on the cell voltage (CV), current efficiency (CE), purity specific power consumption (SPC), and metallic Zn of deposits for 1 h zinc deposits electrowon at 1000 A/m^2 from alkaline zinc solution are shown in Figs. 6.33 and 6.34 and Table 6.12. The CV and SPC increased with increasing arsenide

Table 6.12 Effects of arsenide concentrations on SPC of zinc deposits in alkaline zinc electrowinning process

Concentration of arsenide (mg/L)	50	100	200	300	500
SPC (kWh/kg)	2.97	3.07	3.09	3.12	3.15

Fig. 6.35 SEM photomicrographs of cathode zinc at 1000 A/m^2 ((**a**) without arsenide; (**b**) 300 mg/L arsenide) in alkaline zinc electrowinning process

concentrations in the electrolytes. The CE and SPC decreased with increasing arsenide concentrations. The electrolyte billowed seriously to restrain the zinc depositing when the concentration of arsenide was over 500 mg/L. Surface photographs of the deposited zinc showing the effect of arsenide is given in Fig. 6.35. Morphological structure of the deposited zinc proved that the characteristics of crystal growth changed from dendritic to nodular and unstable.

The reactions that arsenic can cause during electrowinning include

$$AsO_4^{3-} + 2H_2O + 2e^- \rightarrow AsO_2^- + 4OH^- \quad E^\theta = -0.67V \tag{6.20}$$

$$AsO_2^- + 2H_2O + 3e^- \rightarrow As + 4OH^- \quad E^\theta = -0.68V \tag{6.21}$$

Electrode potential of arsenic reactions in alkali solution is near to that of zinc electrolysis as well as reaction form. After the arsenic reaches a certain concentration, $AsO_4{}^{3-}$ and AsO_2^- get the electrons (Eqs. 6.20 and 6.21) for redox reaction, thus destroying the zinc electrolytic balance and seriously affecting the precipitation of zinc.

In Fig. 6.36, the (b) is zinc powder from the pure electrolyte, and the (a) is zinc powder from impure electrolyte containing arsenic. It can be seen clearly that arsenic has a great effect on zinc electrowinning process, resulting to the poor quality of zinc powder, finer particle, and no metallic luster.

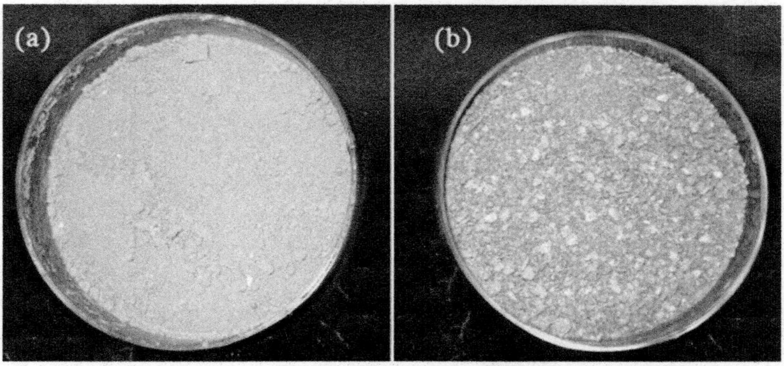

Fig. 6.36 The macrographs of zinc powders prepared from As-contained electrolyte (**a**) and pure electrolyte (**b**)

Table 6.13 Effects of CO_3^{2-} on zinc electrowinning in alkaline process

No.	Concentration of CO_3^{2-} (g/L)	Quality of zinc powder (g)	Current efficiency (%)	Zinc (%)	Metal zinc (%)
1	0	13.3855	93.58	99.91	95.76
2	5.66	14.6725	95.45	99.45	96.50
3	11.32	14.2188	92.50	99.23	96.48
4	16.98	14.3417	93.30	99.73	96.28
5	22.64	15.1181	98.35	99.34	95.28

Note: Electrolytic current 4.2 A, electrolytic voltage 14.2 V

6.4.4 *Effects of* CO_3^{2-}, SO_4^{2-}, *and* SiO_3^{2-} *Concentrations on Zinc Electrolysis in Alkaline Leaching Solution*

Put 348 g NaOH and 60 g ZnO into the beakers, dissolved in distilled water while stirring, and then kept the constant volume of solution for 1200 mL. As a result, a leach solution with NaOH 240 g/L and Zn 40 g/L was got. Added Na_2CO_3, Na_2SO_4, $Na_2SiO_3{\cdot}9H_2O$, NaF, NaCl, and Na_2S into the leaching solutions and electrowon to explore the effect of these impurities on the electrowinning for the production of zinc powder. The results are given in Tables 6.13, 6.14, and 6.15, respectively. It can be found that the presence of CO_3^{2-}, SO_4^{2-}, and SO_3^{2-} exerted negligible effects on the zinc powder quality and current efficiencies.

Table 6.14 Effects of SO_4^{2-} on zinc electrowinning in alkaline process

No.	Concentration of SO_4^{2-} (g/L)	Quality of zinc powder (g)	Current efficiency (%)	Zinc (%)	Metal zinc (%)
1	0	21.6150	92.28	99.06	92.07
2	1	22.0331	94.06	98.45	92.63
3	5	21.7283	92.76	98.05	91.94
4	10	21.1878	90.45	98.14	92.99
5	20	21.8718	93.37	98.45	92.9

Note: Electrolytic current 4.8 A, electrolytic voltage 15.3 V

Table 6.15 Effects of SiO_3^{2-} on zinc electrowinning in alkaline process

No.	Concentration of SiO_3^{2-} (g/L)	Quality of zinc powder (g)	Current efficiency (%)	Zinc (%)	Metal zinc (%)
1	0	16.4426	93.59	97.96	92.93
2	1.65	16.4777	93.80	98.24	92.33
3	3.3	16.6376	94.70	97.92	91.44
4	6.6	16.6302	94.66	97.7	92.22

Note: Electrolytic current 4.8 A, electrolytic voltage 12.4 V

Table 6.16 Effects of F^- on zinc electrowinning in alkaline process

No.	Concentration of F^- (g/l)	Quality of zinc powder (g)	Current efficiency (%)	Zinc (%)	Metal zinc (%)
1	0	16.0695	91.47	98.73	92.96
2	0.05	16.0447	91.85	96.24	92.04
3	0.1	16.1600	91.99	98.34	92.03
4	5	16.1683	92.03	97.67	92.07

Note: Electrolytic current 4.8 A, electrolytic voltage 14.3 V

Table 6.17 Effects of Cl^- on zinc electrowinning in alkaline process

No.	Concentration of Cl^- (g/l)	Quality of zinc powder (g)	Current efficiency (%)	Zinc (%)	Metal zinc (%)
1	0	16.7009	95.06	98.62	92.02
2	0.5	16.4216	93.47	98.84	92.56
3	0.1	16.6330	94.68	98.83	91.17
4	1	16.5149	94.01	98.89	91.84
5	5	16.6994	95.63	98.46	91.60

Note: Electrolytic current 4.8 A, electrolytic voltage 15.2 V

6.4.5 *Effects of F^- and Cl^- Concentrations on Zinc Electrolysis in Alkaline Leach Solution*

Fluorine and chlorine are harmful impurities for the acidic hydrometallurgy due to their strong corrosion on electrode. The effect of fluoride and chlorine on zinc deposit in alkaline medium is given in Tables 6.16 and 6.17. It can be seen that fluorine and chlorine had a great impact on the quality of zinc powder even at 5 g/L.

6.4.6 *Effect of Hypochlorite on Zinc Electrolysis in Alkaline Leaching Solution*

The effect of hypochlorite addition on zinc electrowinning process is shown in Fig. 6.37. It indicated that the cell voltage increased with the increase of volume fraction of NaClO. From Fig. 6.38, it can be seen that the quality of zinc power and current efficiency are greatly influenced by the content of hypochlorite. When the volume of hypochlorite accounted for 2‰, the content of zinc in zinc powder decreased to 87.96%, which could not satisfy the IV grade standard of zinc powder. Meanwhile, the current efficiency decreased by 7.00–91.80% compared with

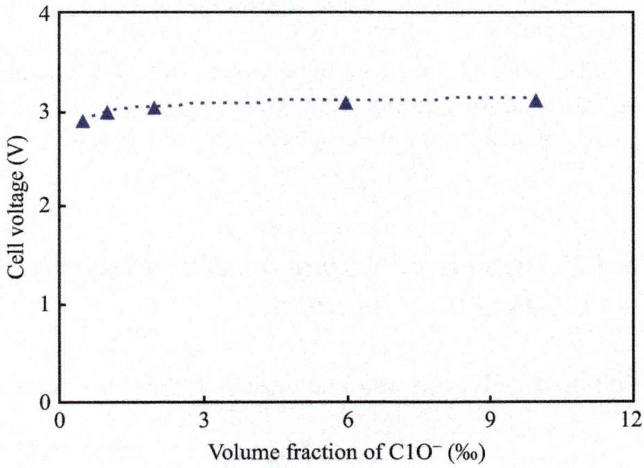

Fig. 6.37 Effects of hypochlorite addition on cell voltage in zinc electrowinning process

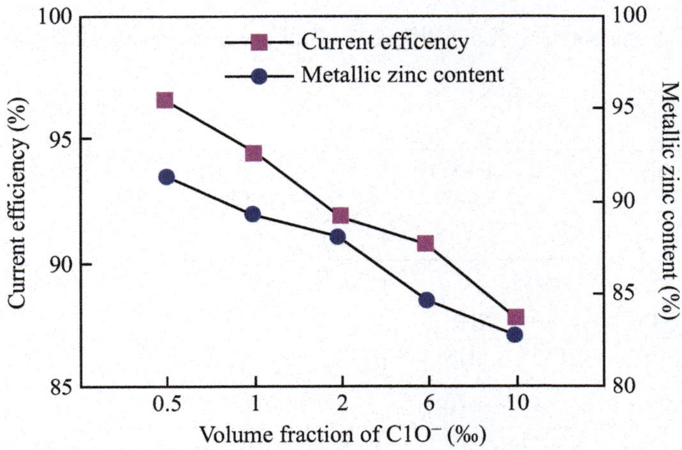

Fig. 6.38 Effects of hypochlorite addition on current efficiency and zinc content in zinc electrowinning process

Table 6.18 Effects of ClO^- volume fraction on energy consumption in zinc electrowinning process

Volume fraction of hypochlorite (‰)	0.5	1	2	6	10
Energy consumption kWh/kg	2.66	2.79	2.86	2.93	2.99

control experiment. Hence, the volume fraction of hypochlorite should be below 1‰. Meanwhile, the energy consumption increased obviously with the addition of hypochlorite (Table 6.18).

The behavior of electrical efficiency decline and energy consumption increase with hypochlorite addition could be explained as the following reaction:

$$2ClO^- + 2H_2O + 2e \rightarrow Cl_2(g) + 4OH^- \tag{6.22}$$

Cl^- and ClO^- had a great negative effects on zinc electroplating process because of the reaction among Cl^-, ClO^-, and Cl_2. The Cl^- and ClO^- had a strong corrosion effects on anode plate, and the latter was more greatly.

6.4.7 Effects of Sulfide on Zinc Electrolysis in Alkaline Leaching Solution

Sulfide dissolves in water solution and the reaction are as follows:

$$S^{2-} + H_2O \rightarrow HS^- + OH^- \tag{6.23}$$

$$HS^- + H_2O \rightarrow H_2S + OH^- \tag{6.24}$$

The relations between K_b of different sulfur speciation and $[OH^-]$ or pH are as follows:

$$\delta_{[S^{2-}]} = \frac{[S^{2-}]}{[S]_T} = \frac{[S^{2-}]}{[S^{2-}] + [HS^-] + [H_2S]} = \frac{1}{1 + \frac{K_{b_1}}{[OH^-]} + \frac{K_{b_1} \cdot K_{b_2}}{[OH^-]^2}}$$
$$= \frac{1}{1 + K_{b_1} \cdot 10^{14-pH} + K_{b_1} K_{b_2} \cdot 10^{28-2pH}} \tag{6.25}$$

$$\delta_{[HS^-]} = \frac{[HS^-]}{[S]_T} = \frac{[HS^-]}{[S^{2-}] + [HS^-] + [H_2S]} = \frac{1}{\frac{1}{K_{b_1} \cdot 10^{14-pH}} + 1 + K_{b_2} \cdot 10^{14-pH}} \tag{6.26}$$

$$\delta_{[H_2S]} = \frac{[H_2S]}{[S]_T} = \frac{[H_2S]}{[S^{2-}] + [HS^-] + [H_2S]} = \frac{1}{\frac{1}{K_{b_1} \cdot K_{b_2} \cdot 10^{28-2pH}} + \frac{1}{K_{b_2} \cdot 10^{14-pH}} + 1} \tag{6.27}$$

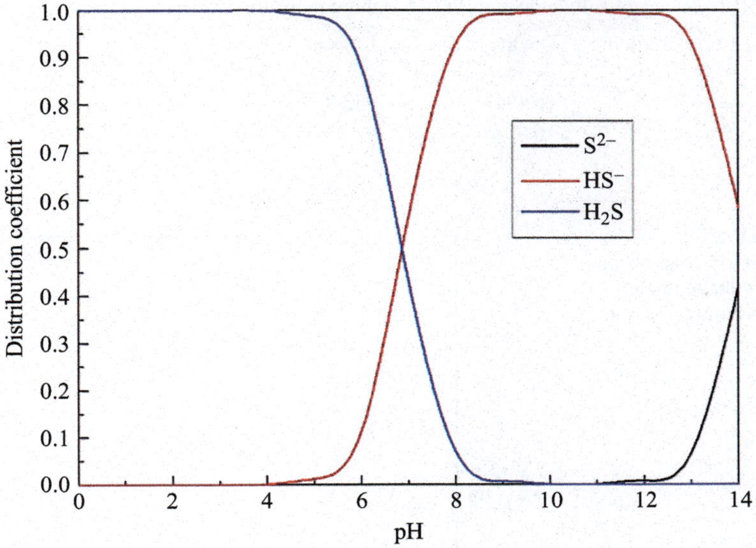

Fig. 6.39 Distribution curve of sulfur speciation with different pH

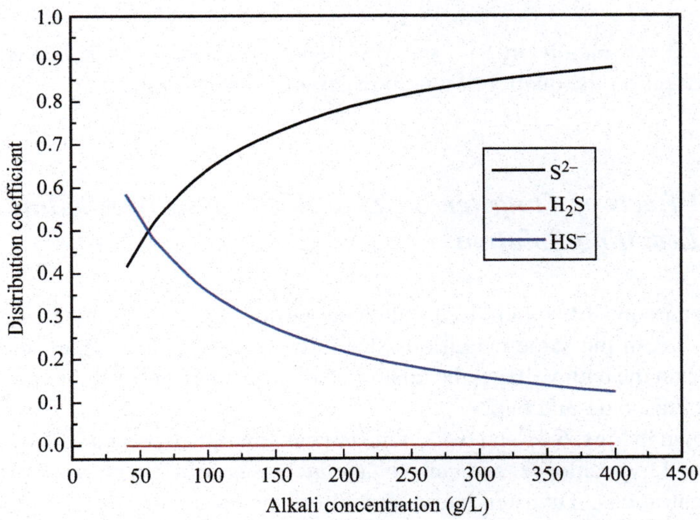

Fig. 6.40 Distribution curve of sulfur speciation with different alkali concentrations

The distribution diagram of sulfur speciation with different pH and alkali concentration is shown in Figs. 6.39 and 6.40.

The electrowinning experiment was conducted at current 4.8 A and voltage 6.1 V. Results are given in Table 6.19. It was found that the sulfide in the alkaline

Table 6.19 Effects of sulfide on zinc electrowinning in alkaline process

No.	Concentration of S^{2-} (g/L)	Quality of zinc powder (g)	Current efficiency (%)	Zinc (%)	Metal zinc (%)
1	0	16.4943	93.89	99.22	92.68
2	Trace	16.0278	91.23	98.96	87.76

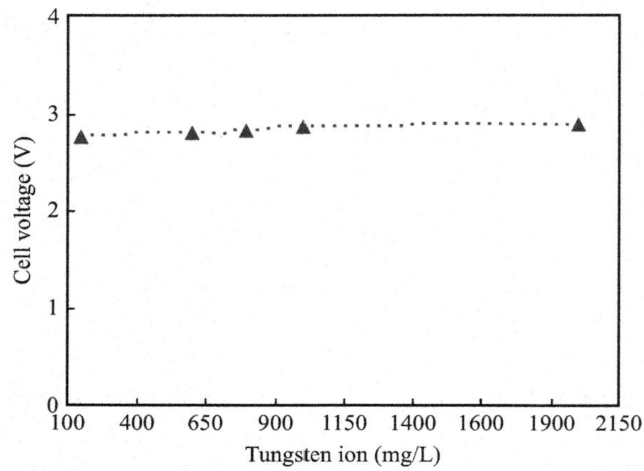

Fig. 6.41 Effects of tungsten ion concentration on cell voltage in zinc electrowinning process

solution existed mainly in S^{2-} and HS^{-}, and the presence of sulfur exerted an adverse effect on the quality of zinc powder and current efficiency.

6.4.8 Effects of Tungsten on Zinc Electrolysis in Alkaline Leaching Solution

Different amounts of sodium tungstate were put into pure electrolyte, and electrolyzed for 2 h in the same condition after fully dissolved. The effects of tungsten impurities on the bath voltage, the quality of zinc powder, and current efficiency are shown in Figs. 6.41 and 6.42.

As shown in Figs. 6.41 and 6.42, tungsten has smaller effects on cell voltage than chlorine and hypochlorite, and cell voltage increases with the increase of tungsten ion concentrations. Tungsten had a certain effects on quality of zinc powder and current efficiency. When tungsten ion concentration was 800 mg/L, zinc content was 89.46%. While tungsten ion concentration rose up to 1 g/L, zinc content decreased to 87.01%. When tungsten ion concentration was 800 mg/L, current efficiency was 94.34%, reduced by about 5% compared with no tungsten. In order to meet standard of the quality zinc powder, tungsten ion concentration in the electrolyte should be no more than 800 mg/L. The presence of tungsten ion also makes power consumption increase slightly, whose details are shown in Table 6.20.

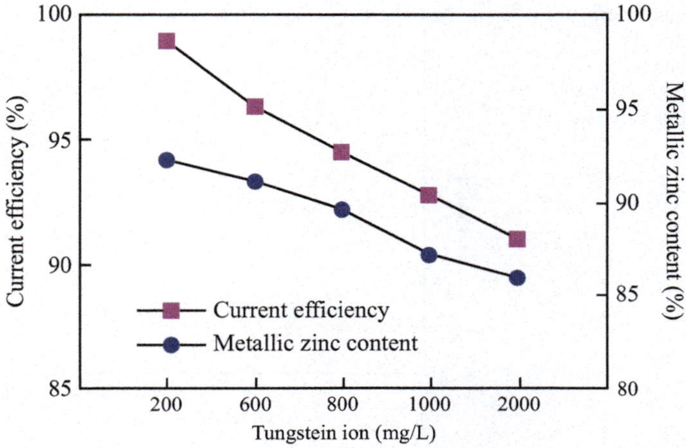

Fig. 6.42 Effects of tungsten ion concentration on current efficiency and zinc content in electrowinning process

Table 6.20 Effects of tungsten ion concentration on power consumption of zinc electrowinning

Tungsten ion concentration (mg/L)	200	600	800	1000	2000
Power consumption (kWh/kg)	2.73	2.77	2.81	2.80	2.83

Tungsten in alkaline solution was in the form of WO_4^{2-}. Due to the electrode potential, the following three reactions were more positive than that of zinc; tungsten ion would have priority for the electronics on the cathode where discharge reaction occurred. Apart from zinc, part W and WO_2 may be separated out on the cathode plate with the presence of tungsten, so that the quality of zinc declined:

$$WO_4^{2-} + 4H_2O + 6e^- \rightarrow W + 8OH^- \quad (6.28)$$

$$WO_4^{2-} + 2H_2O + 2e^- \rightarrow WO_2 + 4OH^- \quad (6.29)$$

$$WO_2 + 2H_2O + 4e^- \rightarrow W + 4OH^- \quad (6.30)$$

6.4.9 *Effects of Molybdenum on Zinc Electrolysis in Alkaline Leaching Solution*

Different amounts of sodium molybdate were put into pure electrolyte, and electrolyzed in the same condition after fully dissolved. The effects of molybdenum ion on the bath voltage, the quality of zinc powder, and current efficiency are shown in Figs. 6.43 and 6.44. The presence of molybdenum ion also makes power consumption increase dramatically.

Fig. 6.43 Effects of molybdenum ion concentration on cell voltage in zinc electrowinning process

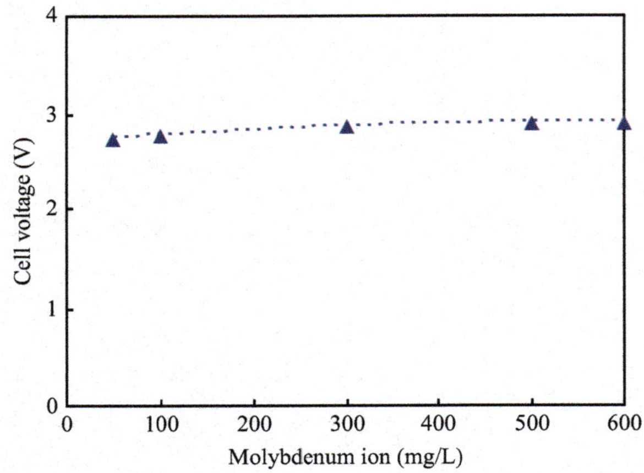

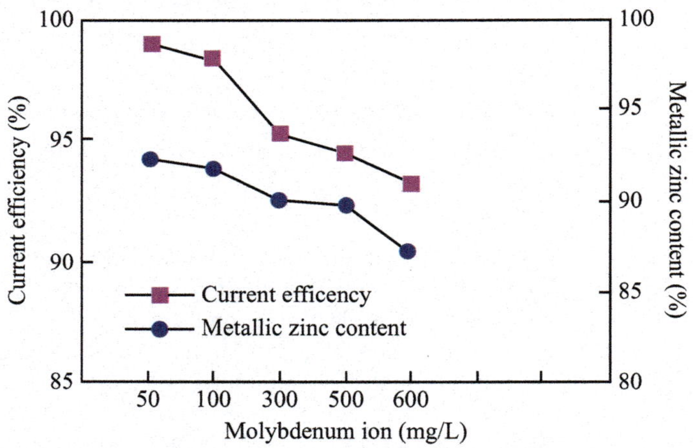

Fig. 6.44 Effects of molybdenum concentration on current efficiency and zinc content in electrowinning process

When molybdenum ion concentration was 500 mg/L, zinc content reduced from 92.27% to 89.64%. While molybdenum ion concentration rose up to 600 mg/L, zinc content decreased to 87.03%, which was lower than the required quality of zinc powder. When molybdenum ion concentration was 500 mg/L, current efficiency reduced from 99.87% to 94.36%, by about 5%. Overall consideration, molybdenum ion concentration in the electrolyte should be no more than 500 mg/L (Table 6.21).

Due to the electrode potential of the molybdenum ion were more positive than that of zinc, part of the molybdenum may be separated out on the cathode plate, so as to decline quality of zinc as well as decline electrical efficiency and increase

Table 6.21 Effects of molybdenum ion concentration on power consumption of zinc electrowinning

Molybdenum ion concentration (mg/L)	50	100	300	500	600
Power consumption (kWh/kg)	2.67	2.70	2.74	2.79	2.81

power consumption. The following reaction will happen in alkaline zincate solution through molybdenum ions:

$$MoO_4^{2-} + 4H_2O + 6e^- \rightarrow Mo + 8OH^- \tag{6.31}$$

6.5 Effects of Zinc Powder Redissolution on Zinc Electrowinning Process

There are two stages for the solid-liquid separation between cathode zinc and spent electrolyte in production. At the beginning, cathode zinc shed from polar plate and accumulated in the bottom of the electrolytic cell. Following that, zinc powder and spent electrolyte entered the transfer pool by flow gravity from discharge outlet of electrolytic cell and then were transported the separation process by alkali-resistance peristaltic pump under stirring conditions. Zinc redissolved in the solution through following reactions:

$$2Zn(s) + O_2(g) = 2ZnO(s) \tag{6.32}$$

$$ZnO(s) + 2OH^- = ZnO_2^{2-} + H_2O \tag{6.33}$$

$$Zn(s) + 2OH^- = ZnO_2^{2-} + H_2(g) \tag{6.34}$$

Zinc redissolution harms yield of zinc powder, increases production cost, and impacts the subsequent operations with high zinc concentration in spent electrolyte. The speed and extent of zinc redissolution are related to many factors, such as stirring speed, NaOH concentration, temperature of spent electrolyte, reaction time, metal speciation and yield of zinc powder, etc. this section tries to find out the most significant factor among the parameters above, in order to reduce the quantity of redissolving zinc through optimizing these parameters.

6.5.1 Effects of Stirring Speed on Zinc Powder Redissolution

The experiments were carried out at the initial NaOH concentration of 240 g/L, Zn of 0 g/L, temperature of 50 °C, and liquid-solid ratio of 25:1. And as the effects of different stirring speed (0, 300, and 500 rpm) on zinc redissolution were

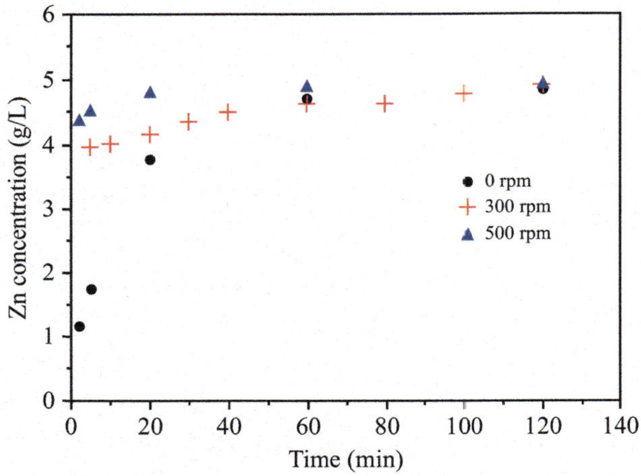

Fig. 6.45 Effects of stirring speed on the redissolution of zinc

investigated, the results are shown in Fig. 6.45. Zinc redissolution rate increases with the growing stirring speed and zinc concentration within 60 min after reaction starting, while zinc concentrations under different conditions tend to perform in line at around 4.9 g/L an hour later. Furthermore, zinc redissolution happens in the initial 5 min at stirring condition, while it emerges in the first 20 min without stirring. Therefore, stirring speed needs to subject to making zinc powder suspended as overhigh speed would speed up the redissolution and increase the cost, and overlow would cause zinc sedimentation, hampering separation process. Overall, 300 rpm was chosen for subsequent experiments and researches.

6.5.2 Effects of Temperature on Zinc Powder Redissolution

The effects of temperature on zinc redissolution were investigated at the initial NaOH concentration of 240 g/L, Zn of 0 g/L, stirring speed of 300 rpm, and liquid-solid ratio of 25:1. The quantity of zinc redissolution increases with growing temperature; however, at the first 5 min, the impact from temperature is not significant as zinc concentration rocketed to about 4 g/L (3.76 g/L at 30 °C and 4.48 g/L at 90 °C) from zero at any temperature (Fig. 6.46). As reaction time was prolonged, the promotion role temperature played was much more important. As temperature climbed from 30 °C to 40 °C, 50 °C, 70 °C, and 95 °C, zinc concentration increases to 4.48, 4.90, 5.28, and 6.65 g/L, respectively, from 4.34 g/L. So the temperature should not remain too high, usually lower than 50 °C while separating zinc from spent electrolyte.

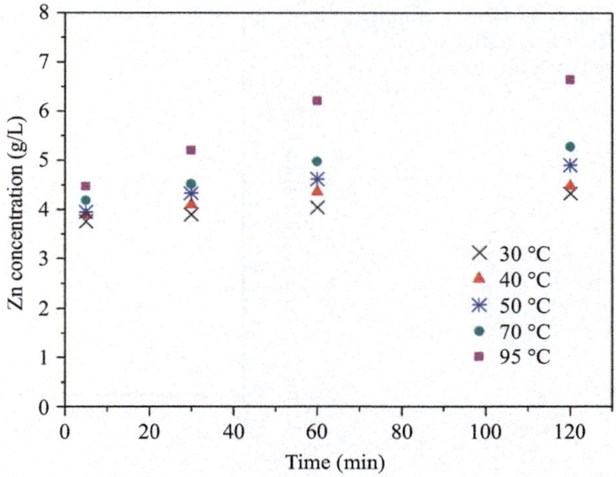

Fig. 6.46 Effects of temperature on the redissolution of zinc

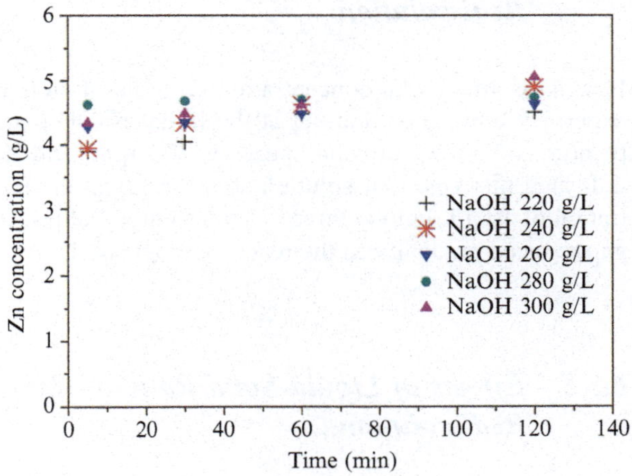

Fig. 6.47 Effects of initial NaOH concentration on the redissolution of zinc

6.5.3 *Effects of Initial Concentration of Sodium Hydroxide on Zinc Powder Redissolution*

The effects of initial NaOH concentration on zinc redissolution were investigated at the initial zinc concentration of 0 g/L, temperature of 50 °C, stirring speed of 300 rpm, and liquid-solid ratio of 25:1. Figure 6.47 presents that the speed of zinc redissolution is quite fast within first 5 min after reaction beginning as zinc concentration surged from 0 to 4 g/L and is in accordance with the change of NaOH concentration. Five minutes later after reaction, growth of zinc concentration and redissolution speed slow down. When NaOH concentration increased from 280 g/L to 300 g/L, however, zinc redissolution speed decreased slightly. High NaOH concentration makes solution viscous which does harm to ion diffusion and slows down the reaction speed.

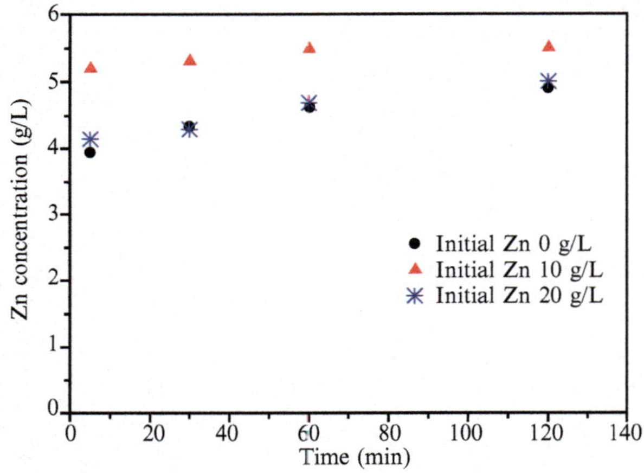

Fig. 6.48 Effects of initial Zn concentration on the redissolution of zinc

6.5.4 Effects of Initial Concentration of Zinc on Zinc Powder Redissolution

Effects of initial zinc concentration on redissolution are shown in Fig. 6.48. The experiments were carried out at the initial NaOH concentration of 240 g/L, temperature of 50 °C, stirring speed of 300 rpm, and liquid-solid ratio of 25:1. It indicated that zinc redissolution ability reached the highest at initial zinc concentration of 10 g/L among three different initial values (0, 10, and 20 g/L), while the gap reduced gradually as the reaction progressed.

6.5.5 Effects of Liquid-Solid Ratio on Zinc Powder Redissolution

The experiments were carried out at the initial NaOH concentration of 240 g/L, zinc of 0 g/L, temperature of 50 °C, stirring speed of 300 rpm, and contact time of 60 min. Based on the fact that the liquid-solid ratio between spent electrolyte and zinc powder was generally lower than 25:1, the effect of four different L/S ratio, 100:1, 100:2, 100:3, and 100:4, on zinc redissolution is given in Fig. 6.49. There existed a linear correlation equation between zinc concentration from redissolution and L/S ratio with $R^2 = 0.98713$. The quantity of zinc redissolution would increase, if the zinc from electrolytic deposition increased. Nevertheless, the amount of zinc was still very low compared with the quantity from electrolytic deposition. For example, when electrolytic deposition zinc increased from 10 g/L to 40 g/L, the redissolution amount would reach 4.62 g/L from 1.30 g/L.

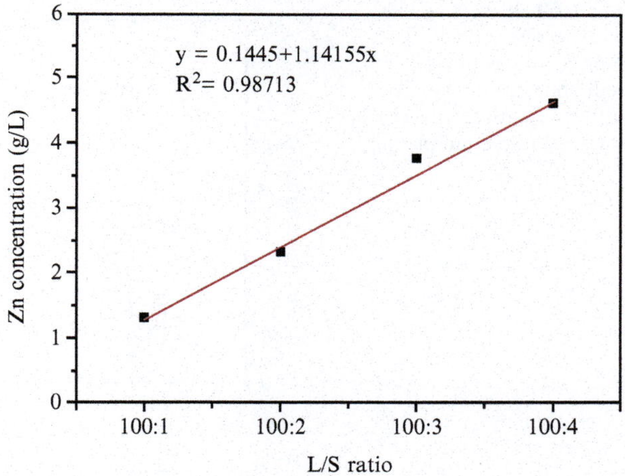

Fig. 6.49 Effects of L/S ratio on the redissolution of zinc

6.5.6 *Effects of Particle Size and Morphology of Zinc on Zinc Powder Redissolution*

The particle size and morphology of zinc powder would change with the determination of electrolysis conditions. The experiments were carried out at the initial NaOH concentration of 240 g/L, zinc of 0 g/L, temperature of 50 °C, stirring speed of 300 rpm, and contact time of 60 min. The effects of zinc powder morphology on the redissolution of zinc is shown in Fig. 6.50. The redissolution condition were remarkably different among three different zinc powder sizes (referred to as small, middle, and large size) obtained from changing electrolysis conditions. Figure 6.50a shows that the finer the zinc powder is, the smaller the redissolution amount will be. Figure 6.50b represents the effects of different sources of zinc with the same particle size on zinc redissolution. The results shows that the redissolution amount of commercial zinc powder is much lower than the electrowinning zinc produced in this work.

6.5.7 *Effects of Contact Time on Zinc Powder Redissolution*

According to the practical production, the zinc powder can be separated from the waste electrolyte in 20–30 min. However, when multiple electrolytic cells stop electrolysis simultaneously, the limited number and capacity of centrifugers can prolong the separation time to 2–3 h and even more. The experiments were carried out at the initial NaOH concentration of 240 g/L, zinc of 0 g/L, temperature of 50 °C, stirring speed of 300 rpm, and liquid-solid ratio of 25:1.The effects of time on zinc redissolution were investigated. As shown in Fig. 6.51, the redissolution

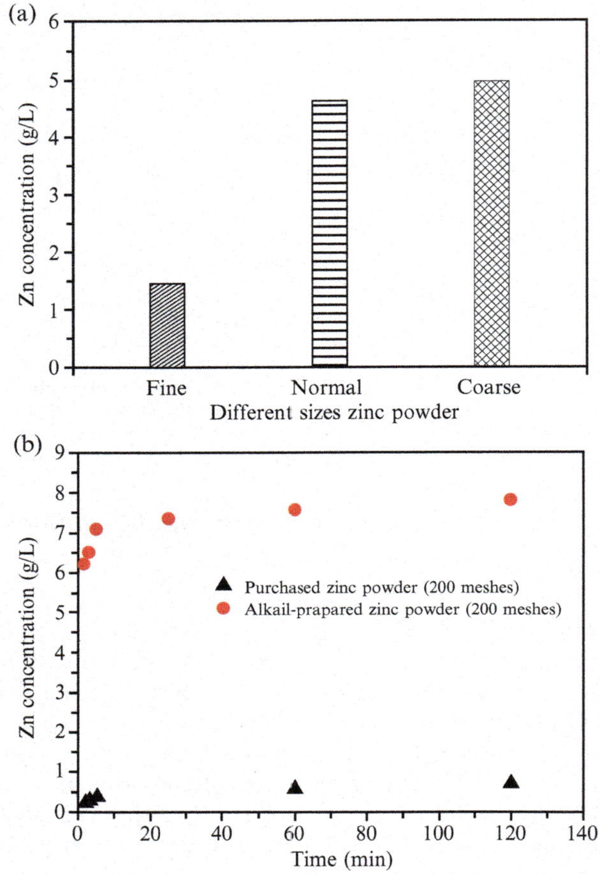

Fig. 6.50 Effects of particle size and morphology on the redissolution of zinc. (**a**) different particle size, (**b**) different micro morphology

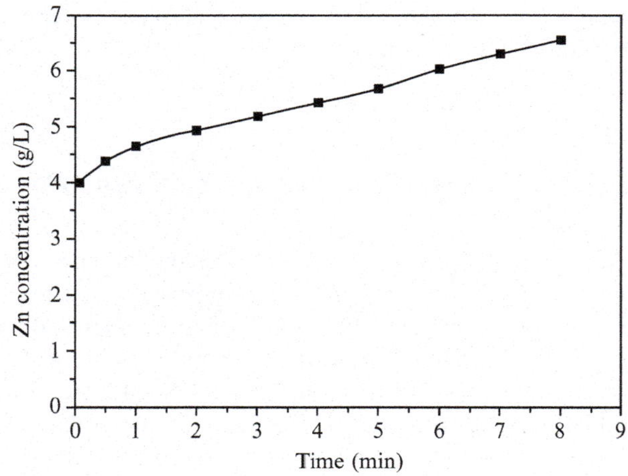

Fig. 6.51 Effects of contact time on zinc redissolution

rate of zinc was much quicker in the initial 5 min, and then the rate went slowly. The amount of redissolution was pretty considerable due to the longer time. When the time prolonged from 5 min to 2, 5, and 8 h, the variation of Zn concentration in electrolyte can increase from 3.95 g/L to 4.90, 5.62, and 6.51 g/L, respectively. Therefore, it was necessary to cut the contact time of zinc powder and waste electrolyte to reduce the amount of zinc powder redissolution.

6.6 Electrowinning of Lead from Alkaline Solutions

6.6.1 Cyclic Voltammetry of Lead Electrowinning Process

The morphological analysis of lead in NaOH solution indicated that the main form of lead is $Pb(OH)_3^-$ at pH > 12. The main cathodic reaction is the deposition of lead in cathode:

$$Pb(OH)_3^- + 2e = Pb + 3OH^- \tag{6.35}$$

and the electrolysis reaction of H_2O:

$$2H_2O + 2e = H_2 + 2OH^- \tag{6.36}$$

Hence, current efficiency will be reduced due to the hydrogen evolution reaction. The main anodic reaction is the bubbling of O_2:

$$4OH^- - 4e = 2H_2O + O_2 \tag{6.37}$$

Meanwhile, a series of side reactions also occur in anode:

$$Pb(OH)_3^- + OH^- - 2e = PbO_2 + 2H_2O \tag{6.38}$$

$$3Pb(OH)_3^- - 2e = Pb_3O_4 + 4H_2O + OH^- \tag{6.39}$$

$$Pb(OH)_3^- - 2e = PbO_3^{2-} + 3H^+ \tag{6.40}$$

The oxidation reaction of $Pb(OH)_3^-$ in anode decreased the concentration of lead in electrolyte and current density while increased the power consumption. The overall reaction of electrowinning process of lead in alkaline solution is:

$$2Pb(OH)_3^- = 2Pb + O_2 + 2OH^- + 2H_2O \tag{6.41}$$

Lead in alkaline solution exists mainly as $Pb(OH)_3{}^-$ complex. As electrode reaction of $Pb(OH)_3^- + 2e^- = Pb + 3OH^-$, the equilibrium electrode potential could be calculated according to the Nernst equation:

$$\varphi_{Pb(OH)_3^-/Pb} = \varphi^{\theta}_{Pb(OH)_3^-/Pb} + \frac{RT}{zF}\ln\frac{\alpha_{Pb(OH)_3^-}}{\alpha^3_{OH^-}} = \varphi^{\theta}_{Pb(OH)_3^-/Pb} + \frac{RT}{zF}\ln\frac{\gamma_{Pb(OH)_3^-}C_{Pb(OH)_3^-}}{(\gamma_{OH^-}C_{OH^-})^3} \quad (6.42)$$

where α—ionic activity, mol/L

γ—activity coefficient

As the concentration of the applied electrolyte solution was high enough, and activity coefficient can be calculated by revised Debye-Hückel equation:

$$\lg\gamma_{\pm} = -\frac{A|z_+z_-|\sqrt{I}}{1 + B\mathring{a}\sqrt{I}} + bI \quad (6.43)$$

where $\gamma_{\pm}$—the average activity coefficient

A, B—constants that are related to the dielectric constant of the solution and temperature

z_+, z_-—valence number of positive ion and negative ion

I—ionic strength, $I = \sum\frac{1}{2}c_iz_i^2$, mol/L

$\mathring{a}$—effective radius of the central ion, Å

b—adjustment coefficient, approximately 0.055 in pure water at 25 °C

While the concentrations of NaOH and PbO are 5 and 0.1 mol/L, respectively, the ionic strength of solution is:

$$I = \sum\frac{1}{2}c_iz_i^2 = \frac{1}{2}\left[5\times1^2 + 4.9\times(-1)^2 + 0.1\times(-1)^2\right] = 5\ \text{mol/L} \quad (6.44)$$

In 25 °C water, A is about 0.5115, B is 0.3291, and the effective radius of the ion is about 4 Å, and $\gamma_{\pm}$= 0.9661 according to Eq. 6.43.

$\varphi^{\theta}_{Pb(OH)_3^-/Pb} = -0.54$ V, and the equilibrium potential of $\varphi_{Pb(OH)_3^-/Pb}$ is calculated as −0.63 V.

$\varphi^{\theta}_{H_2O/H_2} = 0$ V, and the concentration of NaOH is 5 mol·L^{-1}; the equilibrium potential of φ_{H_2O/H_2} is calculated as −0.85 V.

$\varphi^{\theta}_{O_2/H_2O} = 1.229$ V, and the concentration of NaOH is 5 mol·L^{-1}; the equilibrium potential of φ_{O_2/H_2O} is calculated as 0.36 V.

6.6.2 Catholic Reaction of Electrowinning of Lead in Sodium Hydroxide Solution

Three electrodes system, i.e., working electrode and auxiliary electrode are stainless steel electrodes with surface area of 1 and 2 cm^2, respectively, while the reference electrode is saturated calomel electrode. The working electrode should

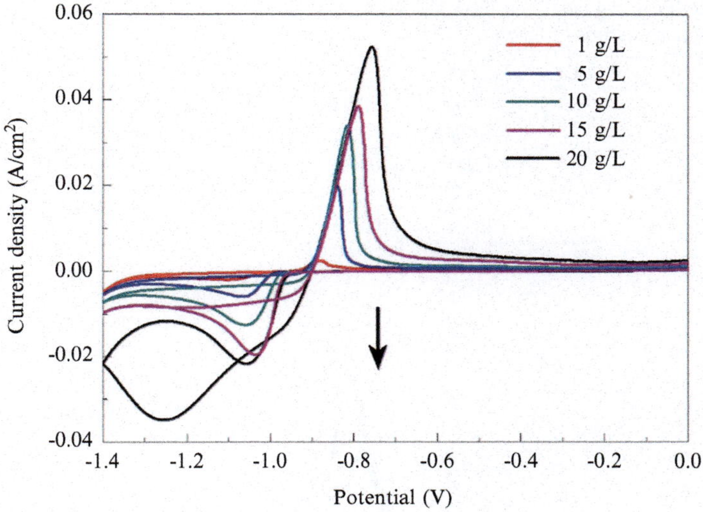

Fig. 6.52 Cyclic voltammetry curves for different lead concentrations in alkaline solution

be polished using metallographic sandpaper and cleaned in ultrasonic washer, then rinsed by distilled water prior to test.

CHI660A electrochemical workstation used for electrochemical testing system in this work had fast digital signal generator, high-speed data acquisition system, voltage current signal filter, multistage signal gain, IR drop compensation circuit, and the potential/constant current meter. The cyclic voltammetry curve, obtained at lead concentrations of 1, 5, 10, and 20 g/L, respectively, and NaOH concentration of 5 mol/L, scanning velocity of 50 mV/s, is presented at Fig. 6.52. The potential was scanned from the initial potential of 0 V to the negative at the same speed, and then scanned to the opposite direction return to the initial potential after reaching to −1.4 V.

The current value was always 0, while the potential ranged from 0 V to −0.95 V. The equilibrium potential of lead was −0.87 V vs. SCE, which indicated overpotential existed in the precipitation process of lead on the stainless steel electrode. When the potential was further scanned to the negative, the current value gradually increased, and the lead was precipitated on the cathode. The precipitation rate of lead reached to the maximum value at the point of the reduction peak. As the potential was further reduced, the concentration of lead in the electrode surface decreased due to the effects of mass transfer, and the current was reduced as well. However, when the potential reached to about −1.3 V, the potential became negative, and the current began to increase. The current increased obviously when the concentration of lead was 20 g/L. The current value was further increased, and the hysteresis were formed when scanned from −1.4 V to the opposite direction, which indicated the secondary crystallization of lead.

Table 6.22 Peak potential and peak current of electrodeposited lead from alkaline solutions at different concentrations of lead

Lead concentration (g/L)	E_{pc} (v)	E_{pa} (v)	ΔE_p(mv)	$j_{pc}(10^{-2}A)$	$j_{pa}(10^{-2}A)$
1	−1.102	−0.891	211	−0.1755	0.2476
5	−1.064	−0.839	225	−0.5884	2.050
10	−1.058	−0.815	243	−1.249	3.629
15	−1.036	−0.790	246	−1.955	4.265
20	−1.056	−0.756	300	−2.168	5.198

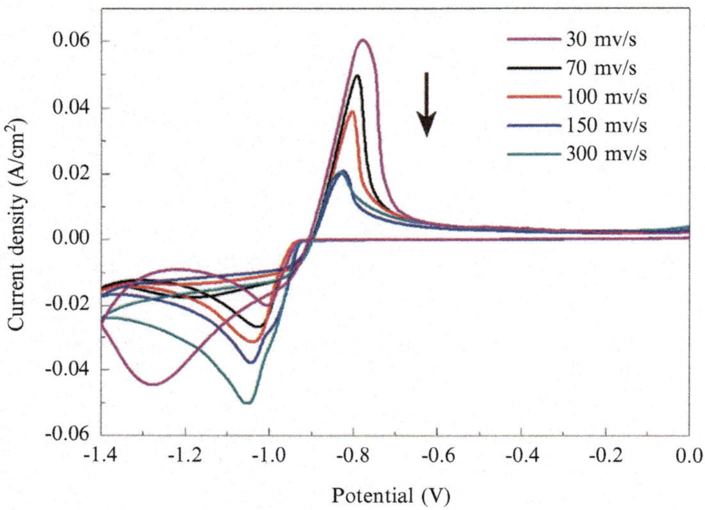

Fig. 6.53 Cyclic voltammetry curves of electrodeposited lead at different scanning velocity in alkaline solutions

The current value of cathodic peak increased with the increase of lead concentration, and the potential of cathodic peak shifted positively. Therefore, in the industrial accumulation process of lead, the selection of relative higher concentration could lead to a higher deposition rate, which might increase the production efficiency and reduce the energy consumption. The peak potential and peak current of cathode at different concentration of lead are listed in Table 6.22. The difference of potential of lead between the reduction peak and the oxidation peak was much larger than that of the cathode peak and anodic peak in reversible electrode process (59/2 = 29.5 mV). Accordingly, the electrode reaction process of lead is a kind of quasireversible electrode process.

The cyclic voltammetry curve, obtained at different scanning velocity (30, 70, 100, 150, 300 mV/s), lead concentration of 20 g/L, and concentration of NaOH 5 mol/L, is presented at Fig. 6.53. With the increase of scanning velocity, the current of cathodic peak increased rapidly, and the potential of cathodic peak shifted negatively; hence, it can be concluded that the reaction of lead in cathode

Table 6.23 Peak potential and peak current of electrodeposited lead at different scanning velocity in alkaline solutions

Scanning velocity (mV/s)	E_{pc} (v)	E_{pa} (v)	$j_{pc}(10^{-2}A)$	$j_{pa}(10^{-2}A)$	$j_{pc}/v^{1/2}$	j_{pa}/j_{pc}
30	−1.005	−0.778	−2.014	6.015	0.368	2.99
70	−1.027	−0.792	−2.659	5.632	0.318	2.12
100	−1.039	−0.804	−3.140	4.644	0.314	1.48
150	−1.044	−0.824	−3.762	2.868	0.307	0.76
300	−1.046	−0.832	−4.927	3.071	0.284	0.62

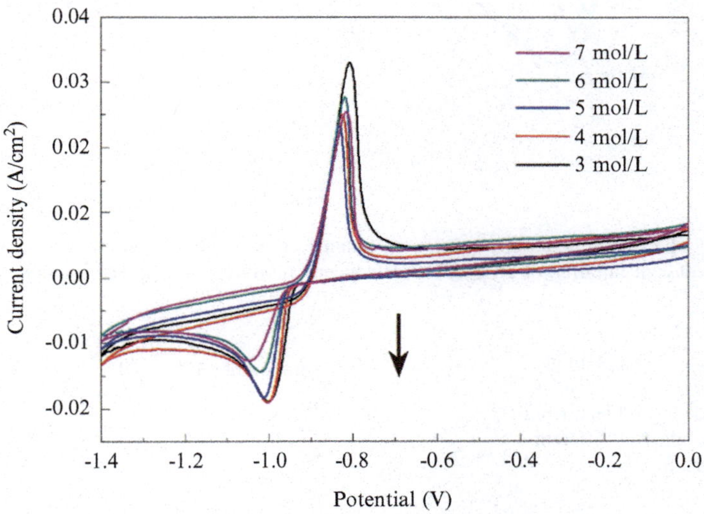

Fig. 6.54 Cyclic voltammetry curves for different NaOH concentrations

is quasireversible. The hysteresis of the cathode occurred at sweep speed of 30 mV/s indicated the existence of two crystallization of lead. The peak potential and peak current of electrodeposited lead at different scanning velocity is listed in Table 6.23.

The cyclic voltammetry curves obtained at different NaOH concentrations (3, 4, 5, 6, 7 mol/L), lead concentration of 10 g/L, and scanning velocity of 50 mV/s are presented at Fig. 6.54. Under the same condition of lead concentration, with the increase of NaOH concentration, the current value of cathodic peak decreased, and the potential shifted to negative. Hence, a relatively low concentration of NaOH had a better performance in industrial electrowinning process of lead. However, it should be taken into account that the solubility of lead decreased in lower concentration NaOH solution, and the electrolysis efficiency would be reduced.

The cyclic voltammetry curve, obtained at different temperatures (30, 40, 50, 60, 70 °C), lead concentration of 20 g/L, NaOH concentration of 5 mol/L, and scanning velocity of 50 mV/s, is presented in Fig. 6.55. The potential continued to get

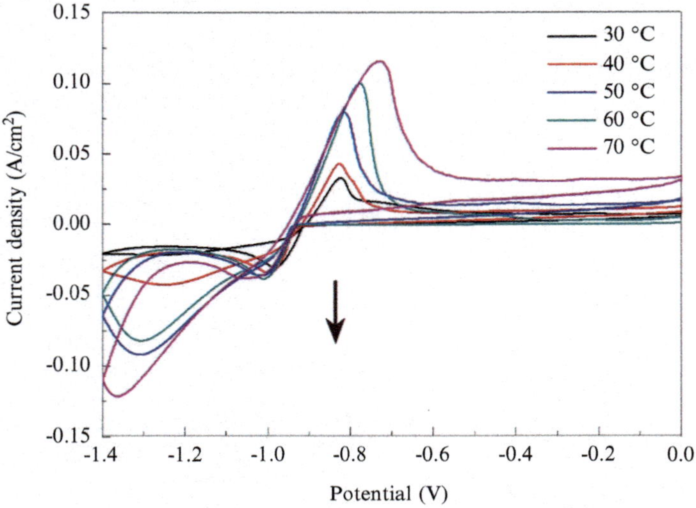

Fig. 6.55 The cyclic voltammetry curves obtained at different temperatures (30, 40, 50, 60, 70 °C), lead concentration of 20 g/L, NaOH concentration of 5 mol/L, and scanning velocity of 50 mV/s

Table 6.24 Peak potential and peak current of electrodeposited lead from alkaline solution at different temperature

Temperature (°C)	E_{pc} (v)	j_{pc} (10^{-2}A)
30	−0.983	−2.967
40	−0.997	−3.292
50	−1.015	−3.547
60	−1.016	−3.828
70	−1.058	−4.435

negative after the appearance of the cathodic peak. Affected by mass transfer, current value was decreased. However, the current began to increase during the further scan process of potential, and with the increase of temperatures, the potential turned to be positive at the point that the current began to increase, which indicated that generation overpotential of H_2 at high temperature reduced. The peak potential and peak current of cathode at different temperatures are listed in Table 6.24. The peak potential of cathode shifted negative, and the peak current value increased with the increase of temperatures.

6.6.3 *Anode Reaction of Electrowinning of Lead in Sodium Hydroxide Solution*

The cyclic voltammetry curve, obtained at lead concentration of 5 g/L, NaOH concentration of 5 mol/L, scanning velocity of 50 mV/s, and potential scanning range of 0–0.8 V, is presented at Fig. 6.56. The potential was scanned forward from

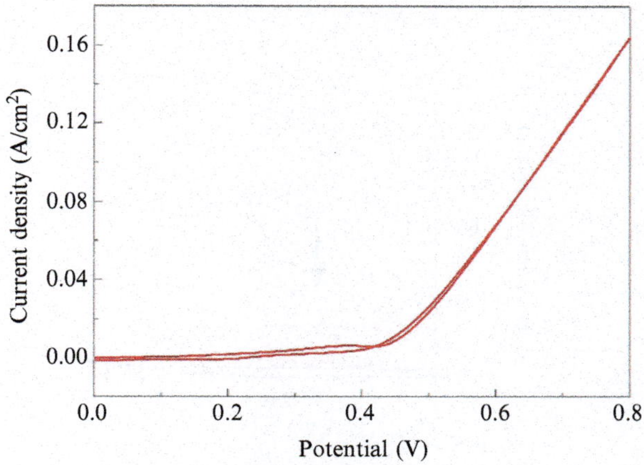

Fig. 6.56 Cyclic voltammetry curve for lead concentration of 5 g/L in alkaline solution

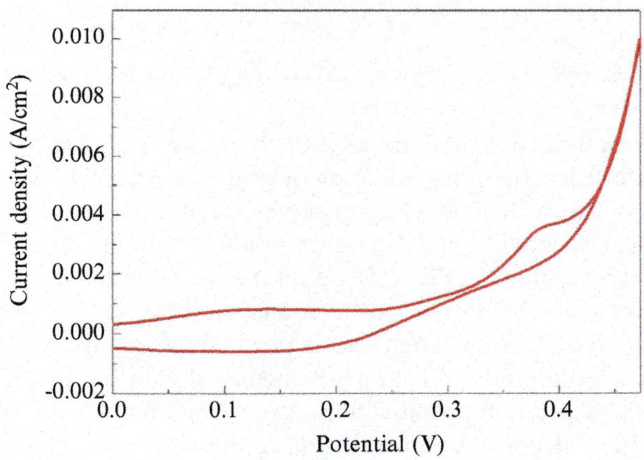

Fig. 6.57 Cyclic voltammetry curve for lead concentration of 5 g/L at high sensitivity in alkaline solution

0 V. Anode current was almost 0 at the range of 0–0.3 V. When keeping on scanning forward, current increased rapidly with increasing potential from 0.442 to 0.8 V. Oxygen evolution, occurred in anode, resulted in the increase of current. For the high concentration of NaOH, the concentration of OH^- did not decrease with increasing current, and there was no current peak of oxygen evolution. The equilibrium potential of φ_{O_2/H_2O} was 0.12 V vs. SCE, and the overpotential for oxygen evolution was 0.322 V on stainless steel electrode plate.

At Fig. 6.57, there was a weak peak in the vicinity of 0.38 V. A little lead oxide was found deposit on the working electrode. As a result of the decreasing concentration of Pb in the solution, the current efficiency for electrowinning of lead on cathode would be reduced. Another cyclic voltammetry curve of high sensitivity was thus scanned and presented at Fig. 6.58. There was a peak at the potential of 0.39 V, which indicated the oxidation process of lead on the anode. However, when

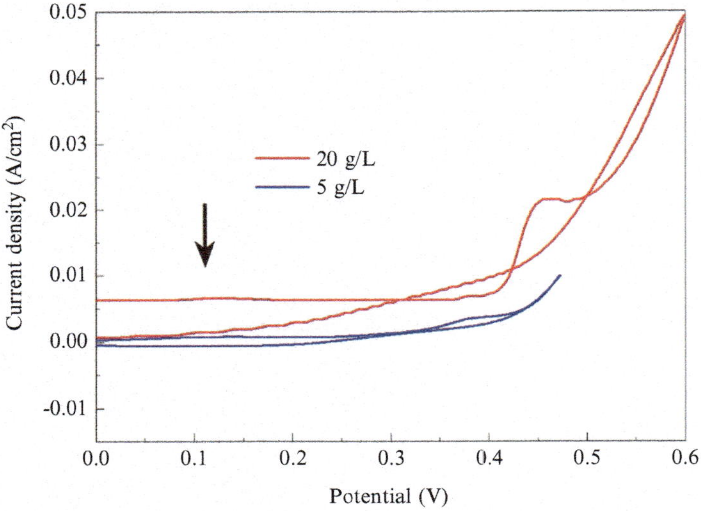

Fig. 6.58 Cyclic voltammetry curves for different lead concentrations in alkaline solution

scanning potential backward, there were almost no appropriate peaks, which illustrated that the oxidation of lead was irreversible.

The cyclic voltammetry curves, obtained at lead concentration of 5 g/L and 20 g/L, respectively, NaOH concentration of 5 mol/L, and scanning velocity of 50 mV/s, are presented at Fig. 6.58. With the increasing concentration of lead, the oxidation peak of lead shifted positive, and the peak current increased sharply. The results showed that although the deposition potential of lead oxide was positive, the deposition rate of lead oxide increased with increasing concentration of lead, and the deposition potential was lower than the oxygen evolution potential. Then, lead oxide deposited on anode during the electrowinning of lead and did harm to the process. However, the results also indicated that the amount of lead deposition on anode was very little.

The cyclic voltammetry curves, obtained at lead concentration of 20 g/L, scanning velocity of 50 mV/s, and NaOH concentration of 4 mol/L, 5 mol/L, 6 mol/L, and 7 mol/L, respectively, are presented at Fig. 6.59, showing an anode peak at different NaOH concentration. With the increasing NaOH concentration, the anode peak shifted positive, and the anode peak current dropped. The phenomenon illustrated that the oxidation rate of lead lowered and deposition potential rose with increasing NaOH concentration.

The cyclic voltammetry curves, obtained at lead concentration of 20 g/L, scanning velocity of 50 mV/s, NaOH concentration of 5 mol/L, and temperature of 30, 40, and 50 °C, respectively, are presented at Fig. 6.60. With the decreasing temperature, the anode peak shifted positive, and the anode peak current dropped. The phenomenon illustrated that low temperature could inhibit the oxidation of lead on anode.

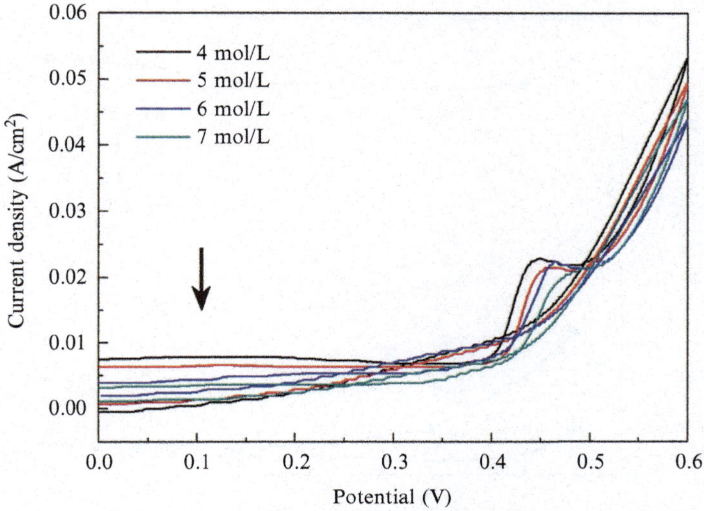

Fig. 6.59 Cyclic voltammetry curves of anode for different NaOH concentrations in alkaline solution

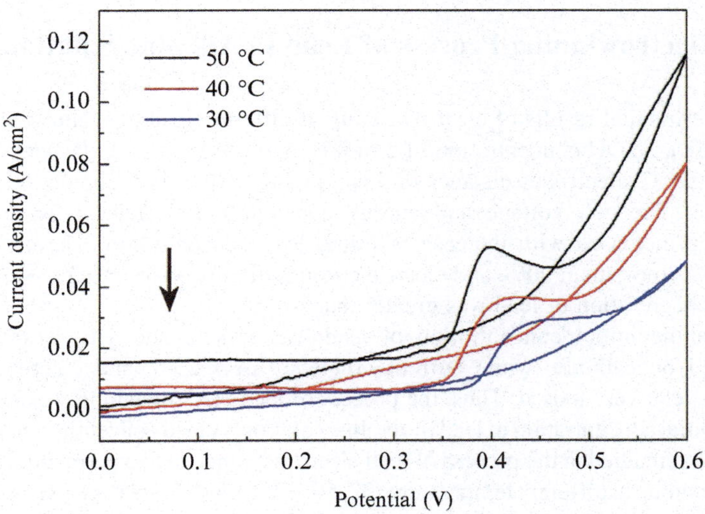

Fig. 6.60 Cyclic voltammetry curves of anode at different temperatures in alkaline solution

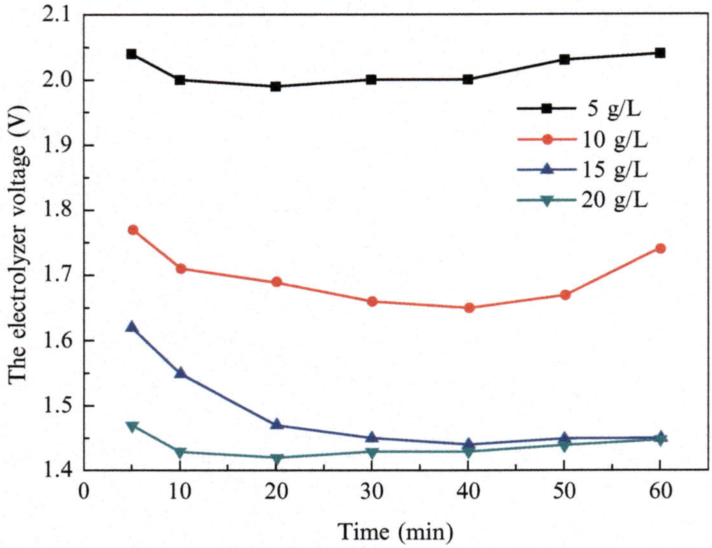

Fig. 6.61 Effects of lead concentrations on cell voltage for electrowinning process of lead in alkaline solution

6.7 Electrowinning Process of Lead in Alkaline Solution

The experimental results of electrowinning at different lead concentrations (5, 10, 15, 20 g/L), NaOH concentration of 5 mol/L, current density of 200A/m^2, temperature of 27 °C, electrodes distance of 3 cm, and time of 1 h are shown in Figs. 6.61 and 6.62. The cell voltage and energy consumption decreased, while current efficiency increased with increase of initial lead concentration. The current efficiency was more than 90%, and electric consumption was lower 0.5 kWh/kg as the initial concentration of lead was greater than 10 g/L.

When the initial concentration of lead was less than 5 g/L, hydrogen was generated on cathode, while current efficiency decreased, and electrodeposited lead was relatively loosed. Thus, the process of lead electrowinning was better with higher initial concentration of lead in alkaline solution. A lead concentration over 5 g/L should be remained in the process of lead electrowinning. The experimental results of electrowinning at different temperature (32, 50, 60, 70, 80 °C), NaOH concentration of 5 mol/L, current density of 400A/m^2, lead concentrations of 20 g/L, electrodes distance of 5 cm, and time of 1.5 h are shown in Figs. 6.63 and 6.64.

The cell voltage and current efficiency decreased with increase of temperatures. The cell voltage was higher at the initial stage of electrolysis and then fell quickly as increase of effective area of electrodeposited lead which was loose on cathode. The energy consumption increased with the increase of temperatures over 50 °C. The lowest energy consumption was 0.485 kWh/kg at 50 °C.

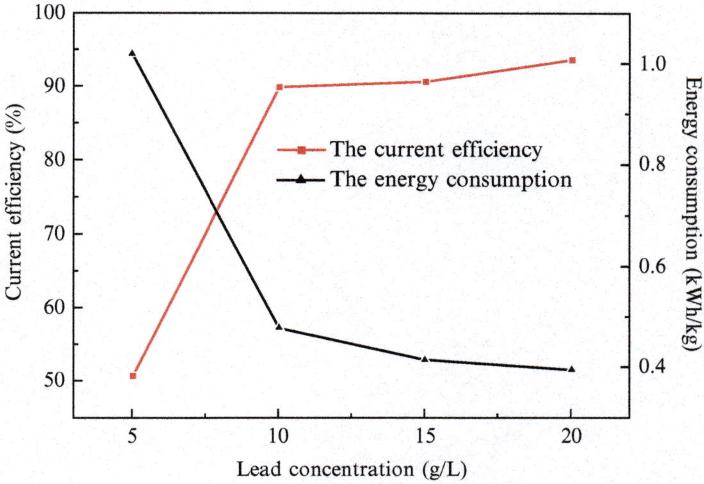

Fig. 6.62 Effects of lead concentrations on current efficiency and energy for electrowinning process of lead in alkaline solution

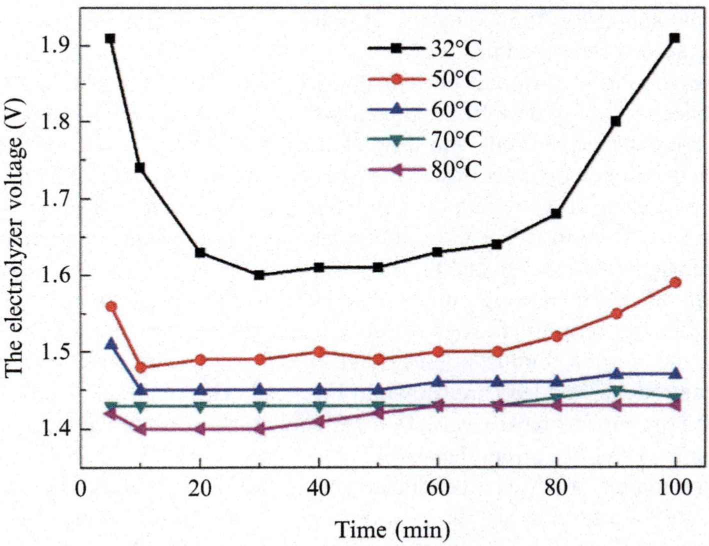

Fig. 6.63 Effects of temperatures on cell voltage for electrowinning process of lead in alkaline solution

The effects of temperature on surface morphology of electrodeposited lead is shown in Fig. 6.65. The electrodeposited lead was dendritic and loose when the temperature was 32 °C, and transformed to sheet with the increase of temperature, and then was lumpish and compact as it reached to 80 °C. The short circuit of

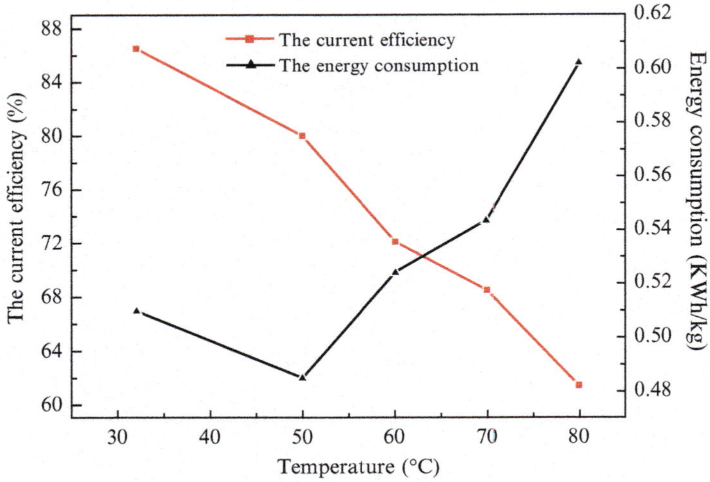

Fig. 6.64 Effects of temperatures on the current efficiency and energy consumption for electrowinning process of lead in alkaline solution

cathode and anode was formed easily when the electrodeposited lead was dendritic and loose at lower temperature.

Electrowinning at different current density (200, 400, 600, 800, 1000 A/m^2), NaOH concentration of 5 mol/L, temperature of 32 °C, lead concentrations of 20 g/L, electrode distance of 5 cm, and time of 1.5 h are shown in Figs. 6.66 and 6.67. The current efficiency was lower on current density of 200 A/m^2, for the electrodeposited lead was sheet closely whose specific surface area and effective current density were low. The current efficiency did not change significantly when current density was higher than 400 A/m^2. The cell voltage and energy consumption increased with the increasing current density; thus, it was useful for energy conservation in the condition of lower current density. The current density of 400 A/m^2 should be the optimal condition. The effects of current density on surface morphology of electrodeposited lead are shown in Fig. 6.68. The electrodeposited lead was sheet when the current density was 200 A/m^2 and transformed to dendritic and loose with the increasing of current density.

Electrowinning of Pb was conducted on the following conditions of distance between anode and cathode of 3, 4, 5, and 6 cm, current density 200 A/m^2, temperature of electrolyte 28 °C, initial lead concentration 20 g/L, NaOH concentration 5 mol/L, electrolysis time 1 h, and no electrolyte recycling, respectively, and the results are shown in Figs. 6.69 and 6.70. The cell voltage increased with the increase of distance of electrodes, due to increase of resistance of electrodes, leading to cell voltage increase. Current efficiency kept stable when the distance of electrodes changed, and the energy consumption on Pb electrowinning rose. The

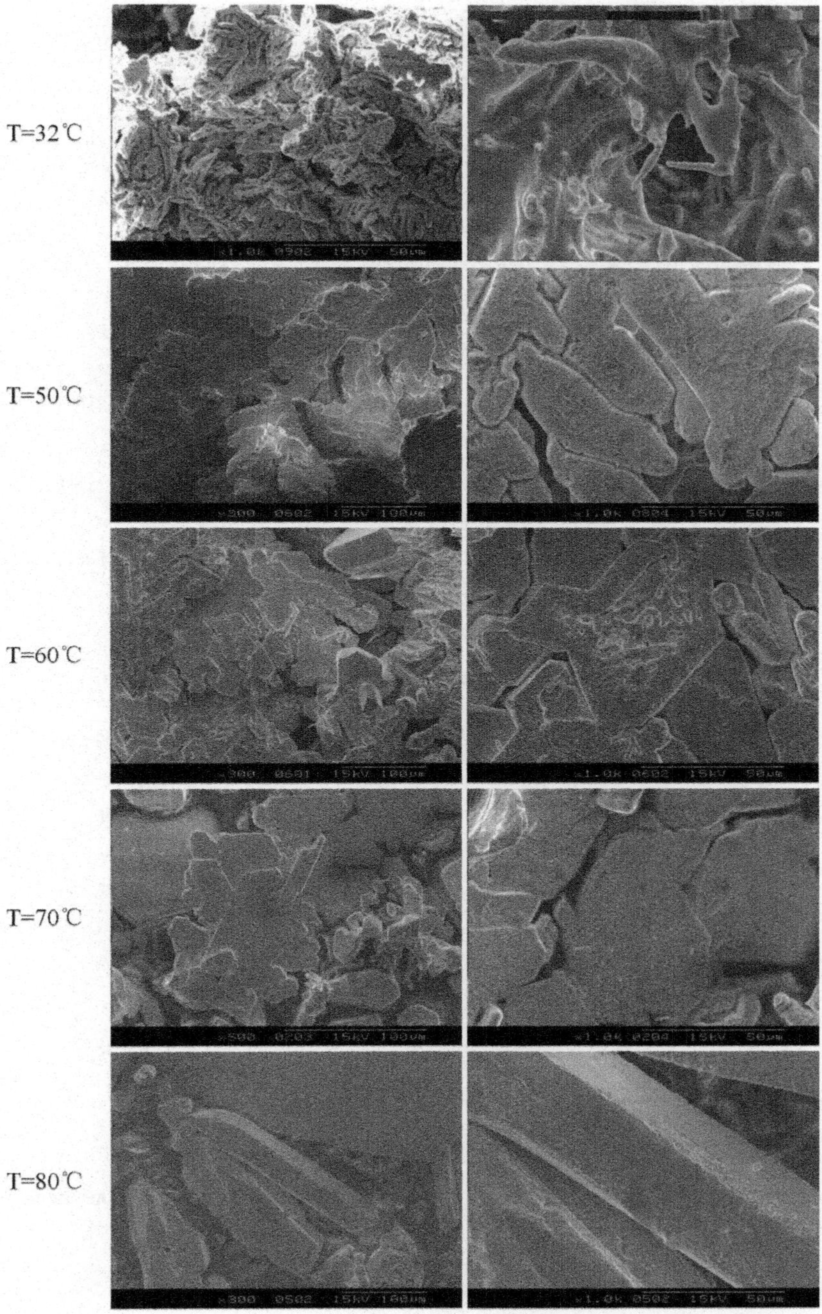

Fig. 6.65 Effects of temperature on surface morphology of electrodeposited lead in alkaline solution

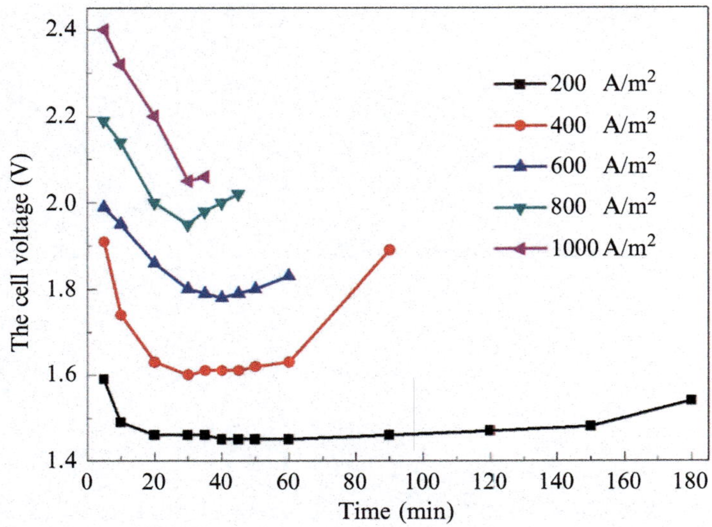

Fig. 6.66 Effects of current density on cell voltage for electrowinning process of lead in alkaline solution

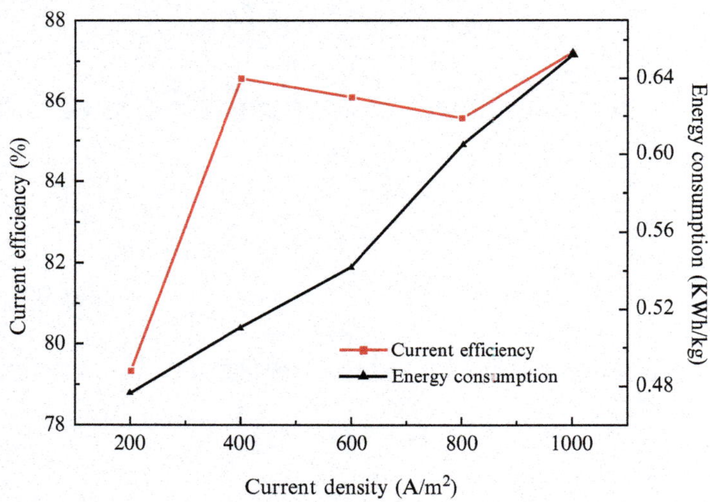

Fig. 6.67 Effects of current density on the current efficiency and energy consumption for electrowinning process of lead in alkaline solution

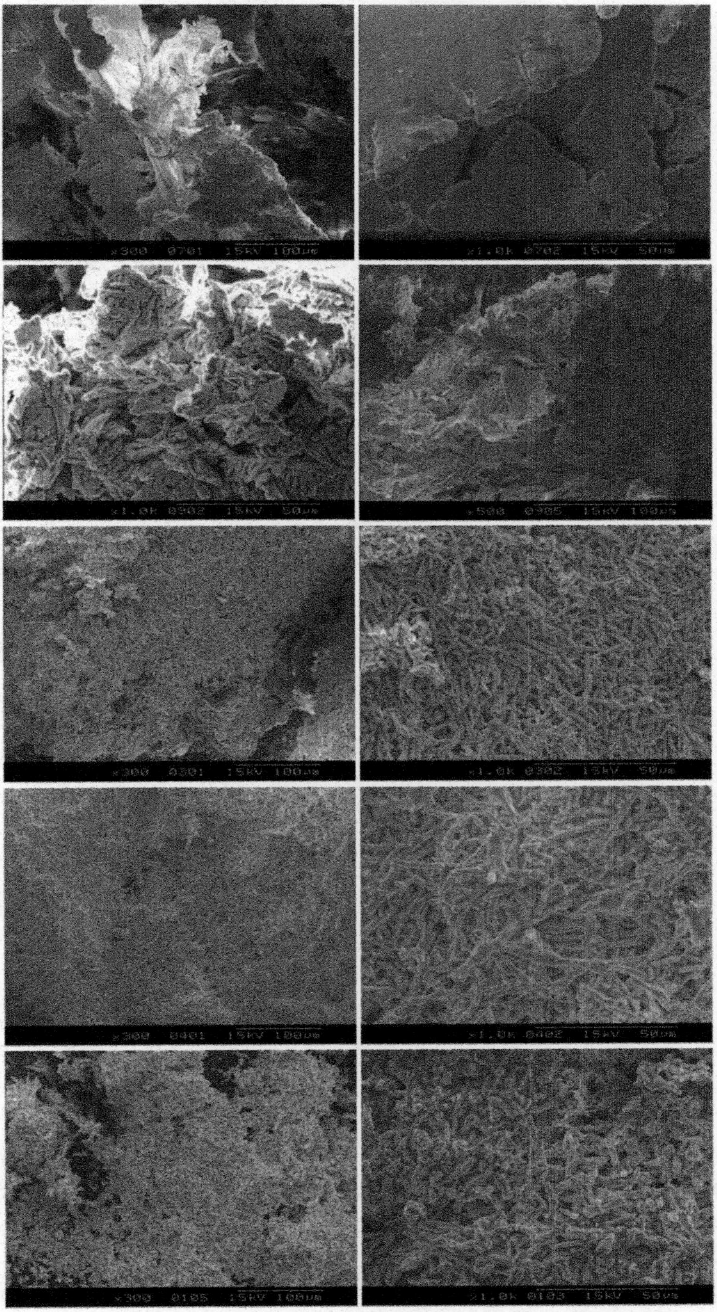

Fig. 6.68 Effects of current density on the surface of electrodeposited lead in alkaline solution

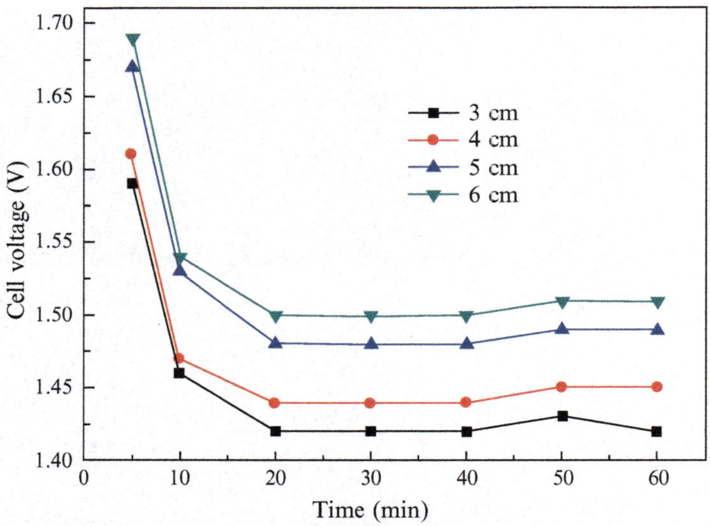

Fig. 6.69 Effects of electrodes distance on cell voltage for electrowinning of Pb in alkaline solution

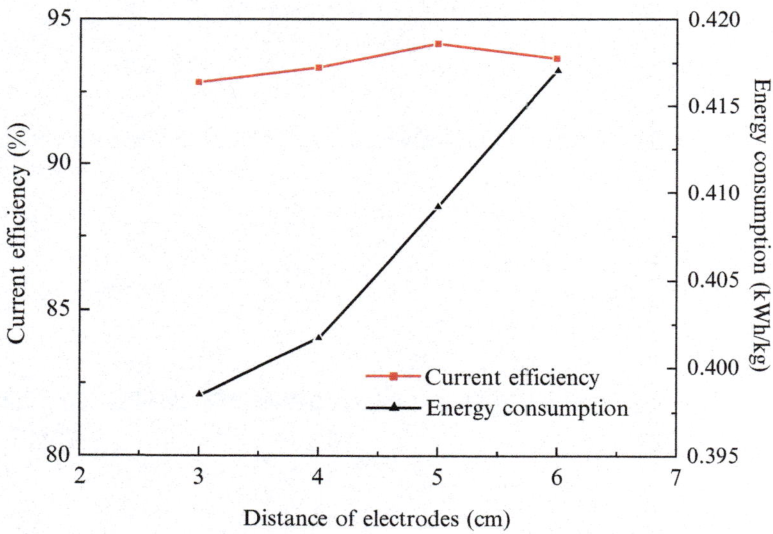

Fig. 6.70 Effects of electrodes distance on current efficiency and energy consumption for electrowinning of Pb in alkaline solution

shorter distance of electrodes should be chosen in Pb electrowinning; however, this would lead the short circuit, so both factors should be considered comprehensively in order to find the appropriate distance.

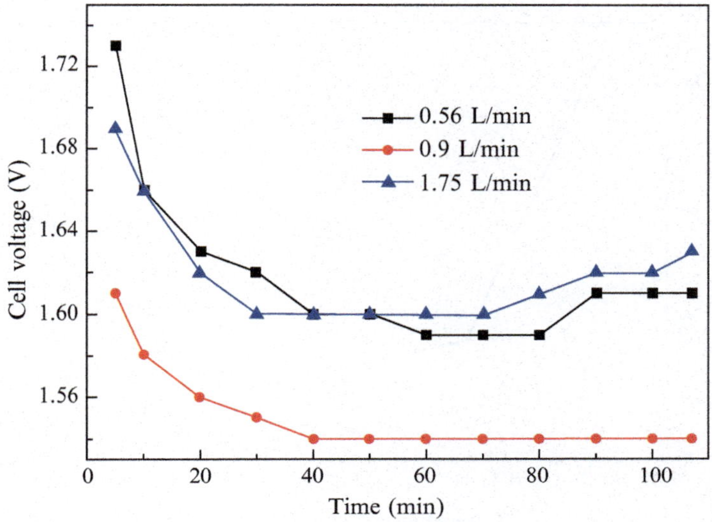

Fig. 6.71 Effects of electrolyte circulating velocity on cell voltage for electrowinning of Pb in alkaline solution

Effects of electrolyte circulating velocity on cell voltage for electrowinning of Pb in alkaline solution were tested with circulating velocity 0.56, 0.9, and 1.75 L/min, distance of electrodes 5 cm, current density 400 A/m^2, electrolyte temperature 30 °C, initial lead concentration 20 g/L, NaOH concentration 5 mol/L, and electrolysis time 1.8 h, and the results are shown in Figs. 6.71 and 6.72.

Figure 6.71 shows that the circulating velocity has great effects on the cell voltage, and increase of flow rate of NaOH solution in the electrolysis cell could reduce the cell voltage. It is mainly caused by hydraulic mixing in the NaOH solution, which makes the concentration of Pb in the electrolysis cell well distributed and increases the concentration of Pb around the electrodes. However, the cell voltage increased at circulating velocity of 1.75 L/min. Thus, with increasing of circulating velocity of NaOH solution, the structure of Pb will change. At high circulating velocity, the structure of Pb become slice shape, more closely thick, and smooth sheet. Figure 6.72 shows that the current efficiency goes up with the increasing circulating velocity, meanwhile the energy consumption goes down. As a result, it can be seen that the electrolytic solution should be circulated to evenly increase the concentration of Pb around the electrodes.

Effects of NaOH concentration on cell voltage for electrowinning of Pb in alkaline solution at NaOH concentration 2, 3, 4, 5, and 6 mol/L, initial lead concentration 10 g/L, current density 200A/m^2, electrolyte temperature 20 °C, electrolysis time 1 h, electrodes distance 5 cm, and without electrolyte circulation are shown in Figs. 6.73 and 6.74. When the concentration of NaOH was less than 4 mol/L, conductivity of electrolyte rose, and cell voltage decreased with the increasing concentrations of NaOH. On the contrary, when NaOH concentration

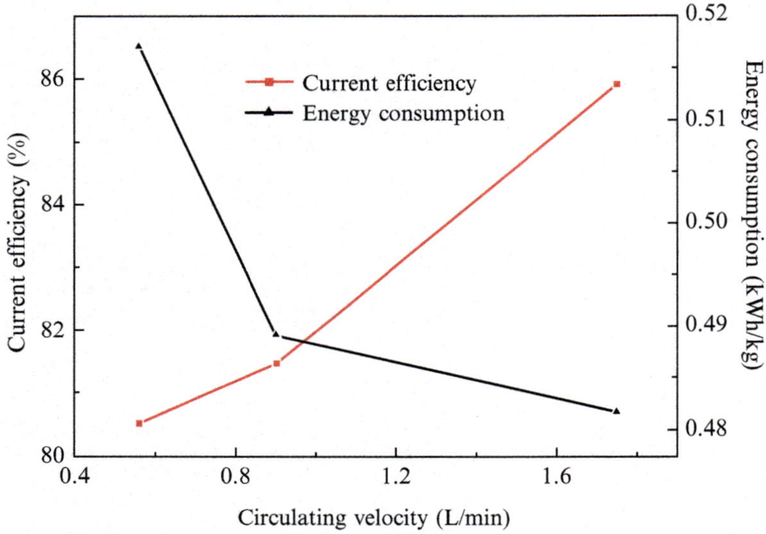

Fig. 6.72 Effects of electrolyte circulating velocity on current efficiency and energy consumption in alkaline solution

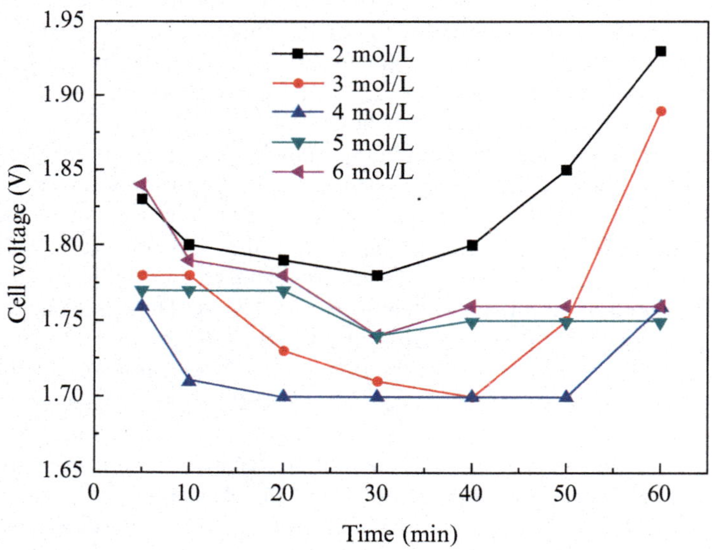

Fig. 6.73 Effects of NaOH concentration on cell voltage for electrowinning of Pb in alkaline solution

was over 4 mol/L, the cell voltage and solution viscosity increased with the increase of NaOH concentration. When NaOH concentration was less than 4 mol/L, the current efficiency increased, and energy consumption decreased with the increasing

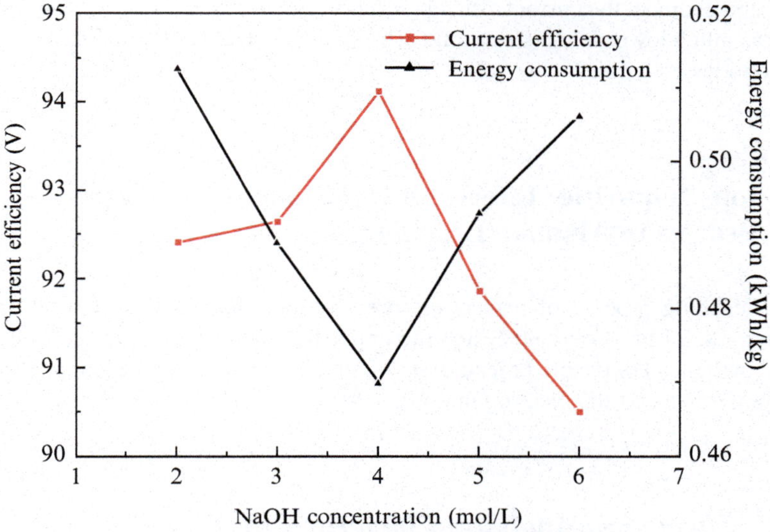

Fig. 6.74 Effects of NaOH concentration on current efficiency and energy consumption for electrowinning of Pb in alkaline solution

Table 6.25 Effects of electrodes materials on current efficiency and energy consumption for electrowinning of Pb in alkaline solution

Cathode	Anode	Average cell voltage (V)	Current efficiency η (%)	Energy consumption (kWh/kg)
Ti	Ti	4.0	90.08	1.1487
Stainless steel	Stainless steel	1.8	92.17	0.5052
Stainless steel	Ti	3.2	90.48	0.9149
Ti	Stainless steel	3.0	90.39	0.8586

of NaOH concentration, and the reverse results were got when NaOH concentration was over 4 mol/L. In Pb electrowinning process, 4–5 mol/L NaOH is appropriate.

The stainless steel and Ti alloy were used as anode and cathode, respectively, to study their effects on current efficiency and energy consumption in the process of Pb electrowinning with Pb concentration of 10 g/L, current density 200 A/m^2, temperature 20 °C, distance of electrodes 5 cm, and without electrolyte circulation, as shown in Table 6.25. The material of electrodes had little effects on current efficiency in the process of electrowinning of Pb, while it had great effects on cell voltage. When Ti alloy was used as both anode and cathode, the cell voltage and energy consumption had a maximum value. When the stainless steel used as both anode and cathode, the current efficiency was pretty high, and the energy consumption was much lower at the same time. The stainless steel is appropriate to be used as electrodes material.

6.8 Ions Impurities Effects on Lead Electrowinning Process in Alkaline Leaching

In the leaching process of lead oxide ore and lead dust in NaOH solution, the leaching rate of Pb is over 90%, and the impurities such as Zn, Cu, As, Sb, W, Sn, and Al in the lead oxide ore or lead dust will be partly dissolved, which may cause a uncertain effects on the electrowinning of Pb.

6.8.1 Theoretical Analysis of Impurity Effect on Electrowinning of Pb

$Zn(OH)_4^{2-}$ is the main existence form of Zn in NaOH solution. The electrode reaction of Zn deposited on cathode is:

$$Zn(OH)_4^{2-} + 2e = Zn + 4OH^- \tag{6.45}$$

$$\varphi_{Zn(OH)_4^{2-}/Zn} = \varphi^{\theta}_{Zn(OH)_4^{2-}/Zn} + \frac{RT}{zF}\ln\frac{\alpha_{Zn(OH)_4^{2-}}}{\alpha^4_{OH^-}} \tag{6.46}$$

where $\varphi^{\theta}_{Zn(OH)_4^{2-}/Zn}$ is -1.285 V.

When the concentration of NaOH is 5 mol/L, the concentration of $Zn(OH)_4^{2-}$ is 0.5 mol/L; the equilibrium electrodes potential $\varphi_{Zn(OH)_4^{2-}/Zn}$ will come to -1.375 V. The deposition potential of Zn in NaOH solution is lower than Pb deposition potential of 0.75 V; the Zn will not deposit on cathode in the process of Pb electrowinning.

$Cu(OH)_4^{2-}$ is the main form of Cu in NaOH solution. The electrode reaction of Cu deposited on cathode is:

$$Cu(OH)_4^{2-} + 2e = Cu + 4OH^- \tag{6.47}$$

$$\varphi_{Cu(OH)_4^{2-}/Cu} = \varphi^{\theta}_{Cu(OH)_4^{2-}/Cu} + \frac{RT}{zF}\ln\frac{\alpha_{Cu(OH)_4^{2-}}}{\alpha^4_{OH^-}} \tag{6.48}$$

where $\varphi^{\theta}_{Cu(OH)_4^{2-}/Cu}$ is -0.137 V.

When the concentration of NaOH is 5 mol/L, and the maximum solubility of Cu dissolved in alkaline solution is 0.8 g/L for the leaching of dust containing Pb. The equilibrium potential of $\varphi_{Cu(OH)_4^{2-}/Cu}$ is -0.275 V. The deposition potential of Cu in

NaOH solution is more than Pb deposition potential of 0.36 V; the Cu in electrolyte will deposit first on cathode in the process of Pb; thus, Cu that existed in electrolyte should be removed before Pb electrowinning.

SnO_3^{2-} and $Sn(OH)_3^-$ are the main forms of Sn in NaOH solution. The electrode reaction of Sn deposited on cathode is:

$$SnO_3^{2-} + 3H_2O + 2e = Sn(OH)_3^- + 3OH^- \quad (6.49)$$

$$Sn(OH)_3^- + 2e = Sn + 3OH^- \quad (6.50)$$

$$\varphi_{SnO_3^{2-}/Sn(OH)_3^-} = \varphi^{\theta}_{SnO_3^{2-}/Sn(OH)_3^-} + \frac{RT}{zF}\ln\frac{\alpha_{SnO_3^{2-}}}{\alpha_{Sn(OH)_3^-}\alpha^3_{OH^-}} \quad (6.51)$$

$$\varphi_{Sn(OH)_3^-/Sn} = \varphi^{\theta}_{Sn(OH)_3^-/Sn} + \frac{RT}{zF}\ln\frac{\alpha_{Sn(OH)_3^-}}{\alpha^3_{OH^-}} \quad (6.52)$$

When the concentration of NaOH is 5 mol/L, the concentration of Sn in alkaline solution is 0.1 mol/L, the equilibrium potential of $\varphi_{SnO_3^{2-}/Sn(OH)_3^-}$ is −0.66 V, and $\varphi_{Sn(OH)_3^-/Sn}$ is −1.00 V. The equilibrium potential of $\varphi_{SnO_3^{2-}/Sn(OH)_3^-}$ in NaOH solution is very near to the equilibrium potential −0.63 V of Pb; the existence of SnO_3^{2-} will affect the current efficiency of Pb deposited in cathode. Meanwhile, the equilibrium potential of $\varphi_{Sn(OH)_3^-/Sn}$ is lower than deposition potential of Pb by 0.37 V, the metal of Sn cannot be deposited on cathode during the Pb electrowinning, and it will not affect the purity of Pb deposited on cathode.

AsO_4^{3-} and AsO_2^- are the main forms of As in NaOH solution. The electrode reactions of arsenate on cathode are:

$$AsO_4^{3-} + 2H_2O + 2e = AsO_2^- + 4OH^- \quad (6.53)$$

$$AsO_2^- + 2H_2O + 3e = As + 4OH^- \quad (6.54)$$

$$As + 3H_2O + 3e = AsH_3 + 3OH^- \quad (6.55)$$

$$\varphi_{AsO_4^{3-}/AsO_2^-} = \varphi^{\theta}_{AsO_4^{3-}/AsO_2^-} + \frac{RT}{zF}\ln\frac{\alpha_{AsO_4^{3-}}}{\alpha_{AsO_2^-}\alpha^4_{OH^-}} \quad (6.56)$$

$$\varphi_{AsO_2^-/As} = \varphi^{\theta}_{AsO_2^-/As} + \frac{RT}{zF}\ln\frac{\alpha_{AsO_2^-}}{\alpha^4_{OH^-}} \quad (6.57)$$

$$\varphi_{As/AsH_3} = \varphi^{\theta}_{As/AsH_3} + \frac{RT}{zF}\ln\frac{1}{p_{AsH_3}\alpha^3_{OH^-}} \quad (6.58)$$

where $\varphi^{\theta}_{AsO_4^{3-}/AsO_2^-}$ is −0.67 V, $\varphi^{\theta}_{AsO_2^-/As}$ is −0.68 V, and $\varphi^{\theta}_{As/AsH_3}$ is −1.37 V.

When the concentration of NaOH is 5 mol/L, the concentration of As in alkaline solution is 400 mg/L, the equilibrium potential of $\varphi_{AsO_4^{3-}/AsO_2^-}$ is −0.75 V, $\varphi_{AsO_2^-/As}$ is −0.73 V, and φ_{As/AsH_3} is −1.41 V. AsH_3 is a poisonous gas, which needs to be controlled strictly to avoid its generation during the Pb electrowinning. The equilibrium potential of φ_{As/AsH_3} in alkaline solution is lower than that of Pb by 0.78 V; AsH_3 cannot be generated on cathode in the process of Pb electrowinning.

SbO_2^- is the main form of Sb in NaOH solution. The electrodes reaction of antimonate in electrodes is:

$$SbO_4^- + 2H_2O + 3e = Sb + 4OH^- \tag{6.59}$$

$$\varphi_{SbO_2^-/Sb} = \varphi^{\theta}_{SbO_2^-/Sb} + \frac{RT}{zF}\ln\frac{\alpha_{SbO_2^-}}{\alpha^4_{OH^-}} \tag{6.60}$$

When the concentration of NaOH is 5 mol/L, the concentration of Sb in alkaline solution is 200 mg/L, and the equilibrium potential of $\varphi_{SbO_2^-/Sb}$ is −0.77 V.

WO_4^{2-} is the main form of W in NaOH solution. The electrodes reaction of W on cathode is:

$$WO_4^{2-} + 4H_2O + 6e = W + 8OH^- \tag{6.61}$$

$$\varphi_{WO_4^{2-}/W} = \varphi^{\theta}_{WO_4^{2-}/W} + \frac{RT}{zF}\ln\frac{\alpha_{WO_4^{2-}}}{\alpha^8_{OH^-}} \tag{6.62}$$

The equilibrium potential of $\varphi^{\theta}_{WO_4^{2-}/W}$ is −1.074 V. When the concentration of NaOH is 5 mol/L, $WO_4{}^{2-}$ concentration is 0.01 mol/L, and the equilibrium potential of $\varphi_{Zn(OH)_4^{2-}/Zn}$ is −1.148 V. The equilibrium potential of W deposited on cathode in alkaline solution is lower than that of Pb by 0.52 V, the metal of W cannot be deposited on cathode in the process of Pb electrowinning, and it will not affect the current efficiency during the Pb electrowinning.

6.8.2 Cyclic Voltammetry (CV) of Pb Electrowinning in the Presence of Zn

Cyclic voltammetry (CV) of Pb electrowinning in alkaline solution at concentration of Zn 0.1 mol/L, concentration of NaOH 5 mol/L, and concentration of Pb 10 g/L is shown in Fig. 6.75. Voltage was scanned from 0 V to negative potential, when the potential reached to −1.037 V, the cathodic current reached to the maximum, meaning the Pb dissolved in electrolyte was deposited in cathode. When the voltage scanned to more negative potential, the cathodic current decreased due to the decrease of Pb^{2+} concentration around the electrodes.

However, when the potential is further negative, the cathode current increased rapidly, and a weak cathodic peak appeared near the −1.52 V corresponding to the precipitation of zinc on the cathode. As the potential continued to scan, the current was further increased due to the hydrogen evolution reaction. In the process of positive scanning, the potential was positive, and the cathode current was reduced slowly to −1.43 V. From Fig. 6.75, it can be concluded that the lead and zinc

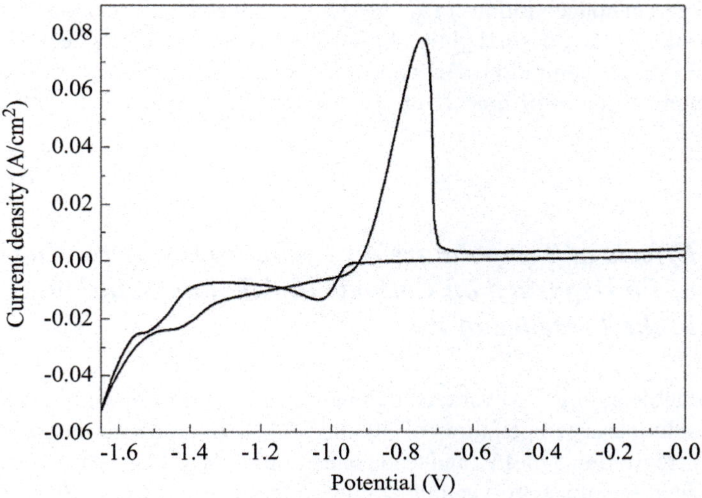

Fig. 6.75 Cyclic voltammetry curve of cathode during the Pb electrowinning at the concentration 0.1 mol/L of Zn in alkaline solution

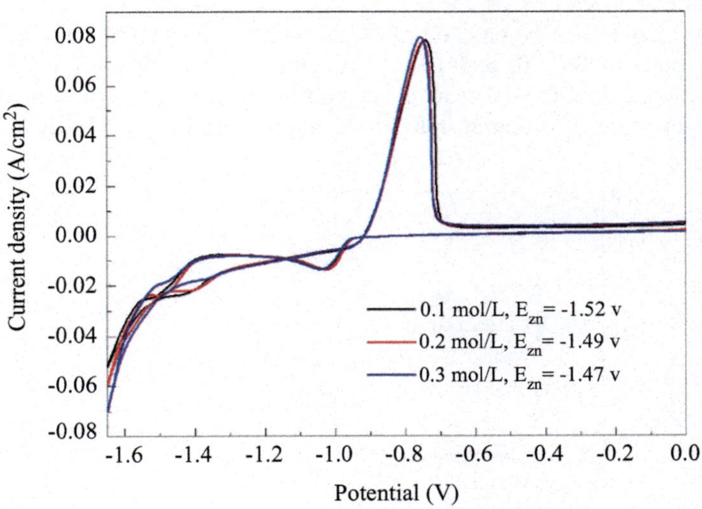

Fig. 6.76 Cyclic voltammetry curves for Zn concentrations effect on Pb electrowinning in alkaline solution

cathode potential have a great difference, Zn has a little effect on the electrowinning of Pb in practical uses.

Cyclic voltammetry curves for Zn concentrations effect on Pb electrowinning was done at 0.1, 0.2, and 0.3 mol/L, NaOH concentration mol/L, and lead concentration 10 mol/L is shown in Fig. 6.76. The cathodic cyclic voltammetry curves of lead at different concentrations of Zn were almost unchanged, and the precipitation potential of zinc could be slightly shifted toward the positive direction with the increase of Zn concentration. The deposition potential of lead and zinc was still a big difference; zinc will not affect the lead in the cathode of the precipitation process.

6.8.3 *Effects of Temperature on Cyclic Voltammetry Curves of Pb Deposited on Cathode in Alkaline Solution in the Presence of Zn*

Effects of temperatures on cyclic voltammetry curves of Pb deposited on cathode in alkaline solution at temperatures of 30 and 50 °C, zinc concentration 0.2 mol/L, NaOH concentration 5 mol/L, and lead concentration 10 g/L are shown in Fig. 6.77. With the increase of electrolyte temperature, cathode peak current of zinc increased and the deposition rate increased at high temperature. But temperature had little effect on the deposition potential of zinc. Therefore, the zinc does not affect the deposition process of lead on the cathode under the condition that the electrolyte temperature is 30–50 °C.

Effects of sweep speed on cyclic voltammetry curves of cathodic Pb at different scanning rates of 30, 70, and 100 mV/s, zinc concentration 0.2 mol/L, NaOH concentration 5 mol/L, and lead concentration 10 g/L are shown in Fig. 6.78. With the increase of scanning rate, the cathode potential of both lead and zinc

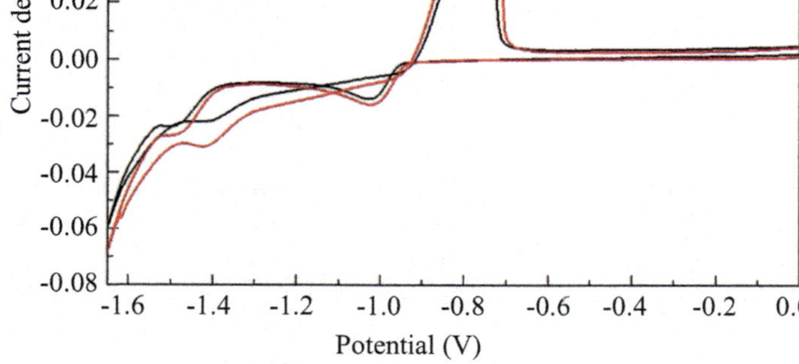

Fig. 6.77 Effects of temperature on cyclic voltammetry curves of Pb deposited on cathode in alkaline solution in the presence of Zn

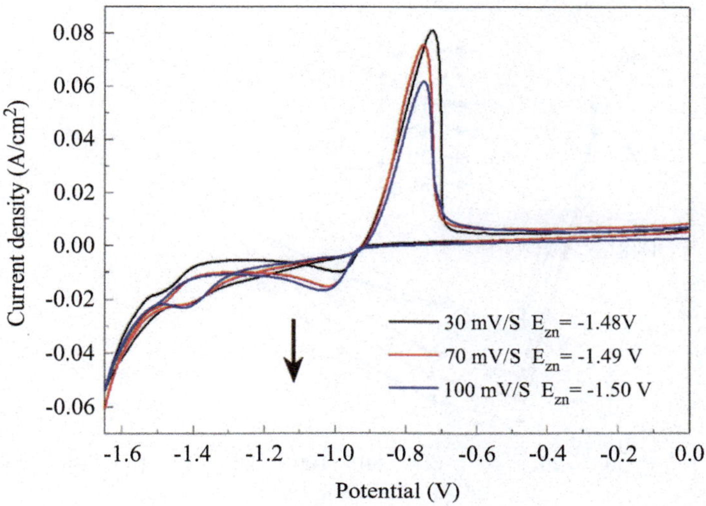

Fig. 6.78 Effects of sweep speed cyclic voltammetry curves of cathodic Pb at different in alkaline solution in the presence of Zn

were negative offset, but the difference was still large. The increase of current peak value means that the high scanning speed increases the deposition rate of lead and zinc. At different scanning rate, the current density was increased in the process of lead deposition in industry, and the zinc will not affect the deposition of lead on the cathode.

6.8.4 Effects of Zn Concentration on Pb Electrowinning in Alkaline Solution

Effects of Zn concentrations on Pb electrowinning in alkaline solution were explored at concentration of zinc 1, 5, 10, 15, 20, and 30 g/L, lead concentration 20 g/L, NaOH concentration 5 mol/L, current density 200 A/m^2, temperature 30 °C, distance of electrodes 4 cm, and electrolytic time 3 h and are shown in Fig. 6.79 and Table 6.26 with respect to cell voltage, current efficiency, energy consumption, and zinc content in lead solutions. Cell voltage increased with the increase of the zinc concentration in the electrolyte, but the zinc in the electrolyte had no effect on the current efficiency, energy consumption, and purity of the lead.

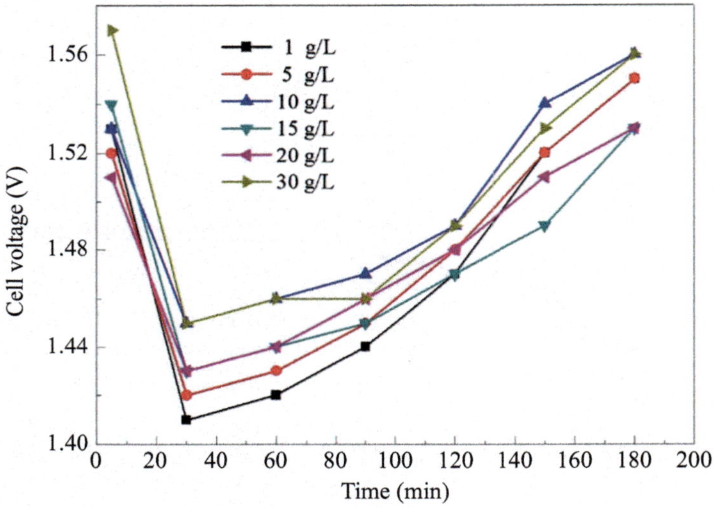

Fig. 6.79 Effects of Zn concentrations on Pb electrowinning in alkaline solution with respect to cell voltage, current efficiency, energy consumption, and zinc content in lead

Table 6.26 Current efficiency, energy consumption, and purity of cathode Pb in lead electrowinning under different concentrations of Zn

Zn content (g/L)	1	5	10	15	20	30
Current efficiency (%)	84.68	84.88	83.68	83.54	83.98	83.57
Energy consumption (kWh/kg)	0.4491	0.4481	0.4575	0.4552	0.4528	0.4551
Zn content in cathodic Pb (%)	–[a]	–[a]	–[a]	0.0033	0.0033	0.0041

[a]Zn is not detected

6.8.5 *Effects of Electrolysis Time on Pb Electrowinning in Alkaline Solution in the Presence of Zn*

The conditions were set as Zn concentration 0.2 mol/L, lead concentration 20 g/L, NaOH concentration 5 mol/L, and current density 200 A/m^2, temperature 35 °C, distance of electrodes 5 cm, electrolysis time 1, 2, 3, 4, and 5 h, respectively, and the results are shown in Fig. 6.80 and Table 6.27. With the increase of electrolysis time, the cell voltage began to decrease during the first 30 min of electrolysis and then remained stable. In the later stage of electrolysis, due to the decrease of the lead concentration in the solution, the cell voltage increased. The current efficiency decreased and energy consumption increased with the increase of the electrolysis time. Meanwhile, with the increase of the electrolysis time, the purity of the cathode lead reduced after 3 h due to the deposition of trace Zn on the cathode as Pb concentration decreased to some extents.

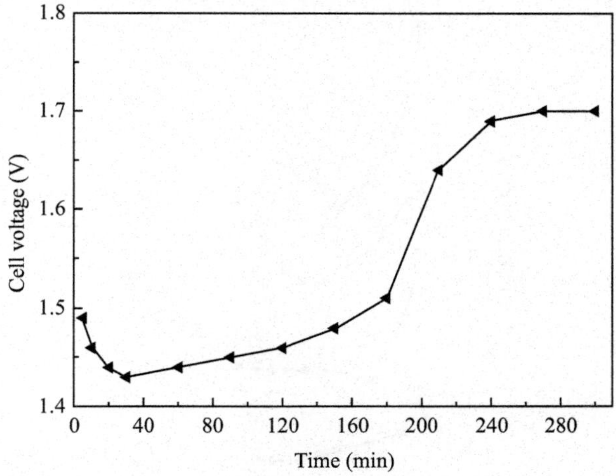

Fig. 6.80 Effects of electrolysis time on Pb electrowinning in alkaline solution in the presence of Zn

Table 6.27 Effects of Zn on current efficiency, energy consumption, and purity of cathode Pb in the process of Pb electrowinning at different time of electrolysis

Electrolysis time (h)	1	2	3	4	5
Current efficiency (%)	87.29	82.93	80.55	79.65	76.22
Energy consumption (kWh/kg)	0.4357	0.4523	0.4689	0.4839	0.5023
Zn content in cathode Pb (%)	0.0008	0.0009	0.0001	0.0088	0.0180

6.8.6 Effects of Current Density on Pb Electrowinning in Alkaline Solution in the Presence of Zn

At the current density 200, 400, 600, 800, and 1000 A/m^2, electrolyte temperature 30 °C, initial Pb concentration 20 g/L, concentration of NaOH 5 mol/L, Zn concentration 8 g/L, distance of electrodes 5 cm, and the current density effect on Pb electrowinning in alkaline solution with respect to cell voltage, current efficiency, energy consumption, and purity of Pb are shown in Fig. 6.81 and Table 6.28. With the increase of the current density, the cell voltage, current efficiency, and energy consumption increased, while purity of the cathode lead was not affected.

6.8.7 Effects of Temperature on Pb Electrowinning in Alkaline Solution in the Presence of Zn

At electrolyte temperature 32 °C, 50 °C, 60 °C, 70 °C, and 80 °C, respectively, current density 400 A/m^2, initial concentration of Pb 20 g/L, concentration of NaOH 5 mol/L, Zn concentration 8 g/L, electrolytic time 1.5 h, distance of

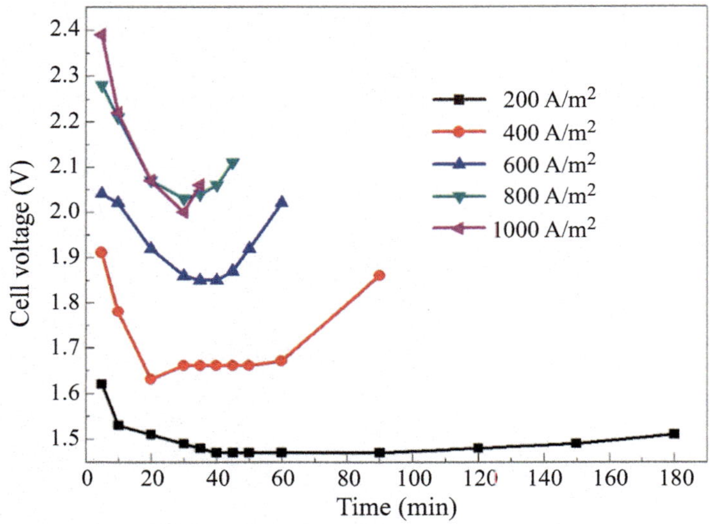

Fig. 6.81 Effects of current density on Pb electrowinning in alkaline solution in the presence of Zn

Table 6.28 Effects of current density on cell voltage, current efficiency, energy consumption, and purity of Pb in Pb electrowinning process in alkaline solution

Current density (A/m^2)	200	400	600	800	1000
Current efficiency (%)	80.7	85.22	86.63	87.67	88.34
Energy consumption (kWh/kg)	0.4712	0.5141	0.5674	0.6079	0.6061
Zn content in cathode Pb (%)	0.0040	0.0193	0.0115	0.0097	0.0100

electrodes 5 cm, the effect of temperature on cell voltage, current efficiency, energy consumption, and purity of Pb in Pb electrowinning process are shown in Fig. 6.82 and Table 6.29. With the increase of temperature, the cell voltage and current efficiency decreased, the energy consumption increased, and presence of Zn had no effect on the cathode lead.

6.8.8 *Effects of Sn on Lead Electrowinning Process in Alkaline Solution*

Cyclic voltammetry curves of lead electrowinning process in the presence of Sn in alkaline solution were drawn at concentration of Sn 100 mg/L, NaOH concentration 5 mol/L, and lead concentration 10 g/L, and the results are shown in Figs. 6.83 and 6.84. There is a cathodic current peak at -1.03 V whether the electrolyte contained Sn or not, and the peak value was almost unchanged. However, the peak current

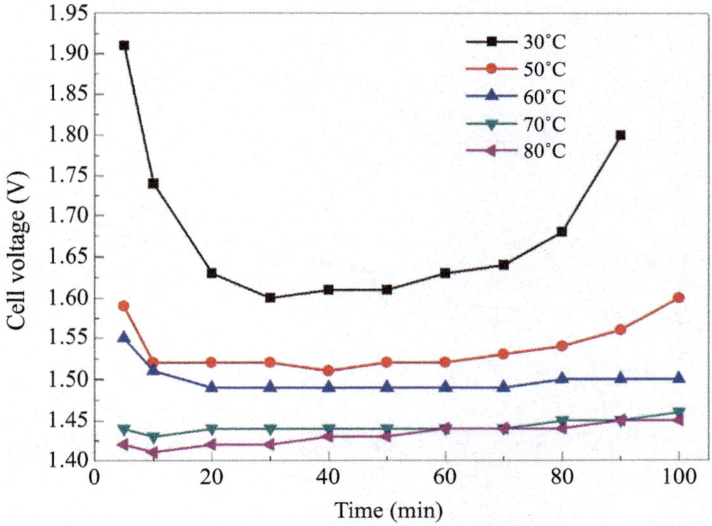

Fig. 6.82 Effects of temperature on Pb electrowinning in alkaline solution in the presence of Zn

Table 6.29 Effects of temperature on cell voltage, current efficiency, energy consumption, and purity of Pb in Pb electrowinning process in the presence of Zn

Temperature (°C)	30	50	60	70	80
Current efficiency (%)	85.22	77.53	75.89	68.90	57.05
Energy consumption (kWh/kg)	0.5141	0.5138	0.5148	0.5407	0.6530
Zn content in cathode Pb (%)	0.0193	[a]	0.0009	0.0017	0.0013

[a]Zn is not detected

increased when the electrolyte contained Sn, which corresponded to the precipitation of lead and reduction of SnO on the cathode. Increasing the scanning velocity, the cathode peak potential shifted to the negative direction, and the peak current value increased.

Effects of Sn on current efficiency and energy consumption in Pb electrowinning process were explored at concentrations of Sn 50, 100, 200, 300, and 400 mg/L, respectively, Pb concentration 14 g/L, NaOH concentration 5 mol/L, current density 500 A/m^2, temperature 40 °C, distance of electrode 5 cm, electrolysis time 70 min, final Pb concentration in the electrolyte 1.4 g/L, and the results are shown in Fig. 6.85. The current efficiency decreased, while energy consumption increased, with the increase of Sn concentrations. Therefore, the presence of Sn in the electrolyte will have an adverse effects on the lead electrowinning process.

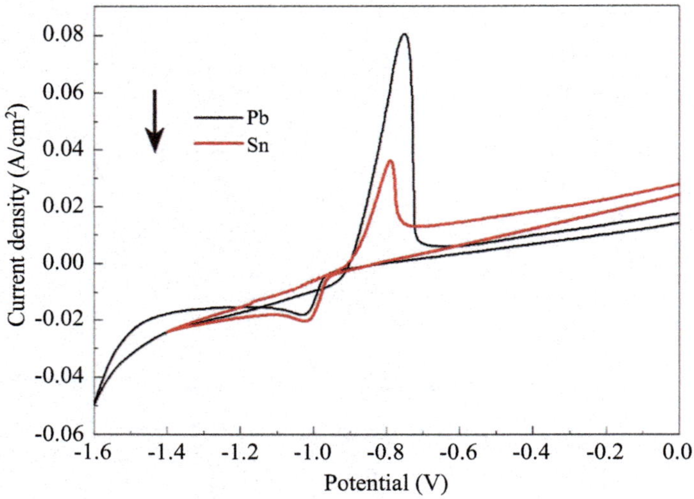

Fig. 6.83 Cyclic voltammetry curves of lead electrowinning process in the presence of Sn in alkaline solution

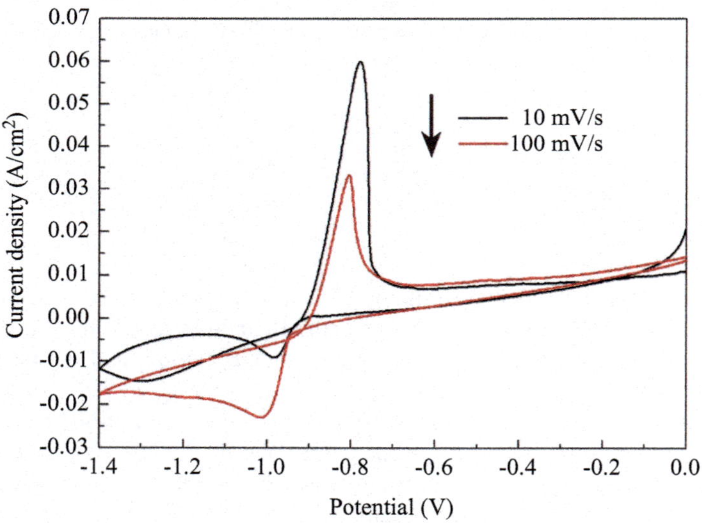

Fig. 6.84 Effects of Sn on lead electrowinning process under different scanning velocity in alkaline solution

When Sn concentration of the electrolyte was 400 mg/L, XRD of the cathode lead product obtained is shown in Fig. 6.86 which shows that Sn is not precipitated on the cathode. The impurity metal tin in the electrolyte has no effect on the purity of the lead product.

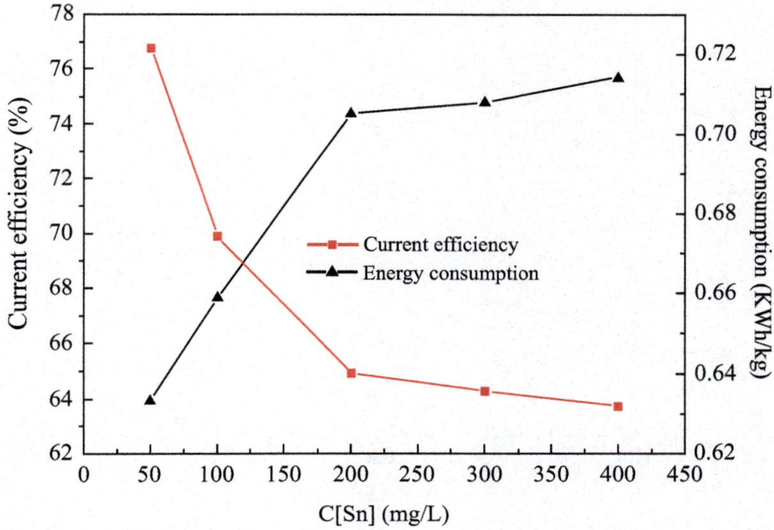

Fig. 6.85 Effects of Sn on current efficiency and energy consumption in Pb electrowinning process were explored at various concentrations of Sn

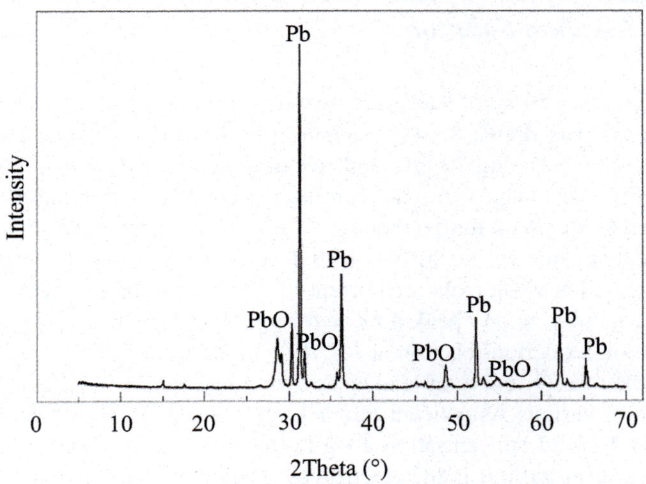

Fig. 6.86 XRD of cathode Pb obtained from lead electrowinning process in the presence of Sn in alkaline solution

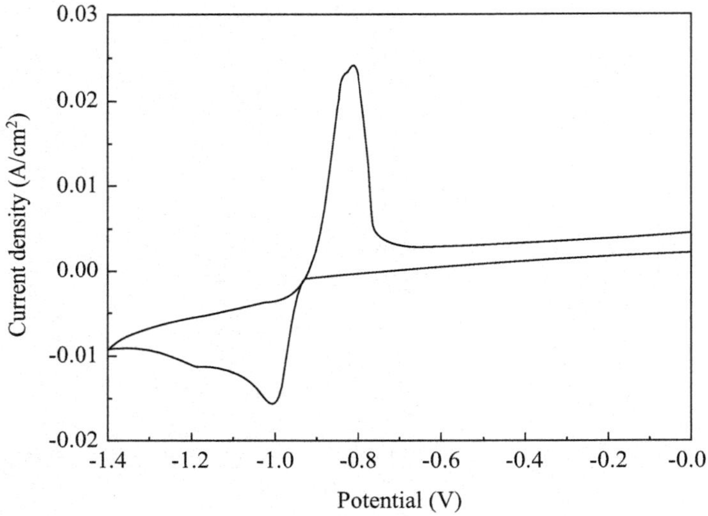

Fig. 6.87 Cyclic voltammetry curve of lead electrowinning process in alkaline solution in the presence of As

6.8.9 *Effects of As on Lead Electrowinning Process in Alkaline Solution*

Cyclic voltammetry curve of lead electrowinning process in alkaline solution in the presence of As was drawn at As concentration 400 mg/L, NaOH concentration 5 mg/L, lead concentration 10 g/L, and scanning velocity 70 mV/s, the results are shown in Fig. 6.87. The cathodic current peak at the potential of −1.003 V corresponded to the precipitation reaction of Pb on the cathode. When the potential become negative, due to the diffusion of lead in the electrolyte limits, cathode current decreased at weak cathodic current of −1.18 V. When the voltage scanned to positive potential, anode peak appeared two adjacent current peaks, and As in electrolyte took reduction reaction at −1.18 V in the cathode.

Cyclic voltammetry curve of lead electrowinning process in alkaline solution in the presence of various As concentrations is shown in Fig. 6.88 at NaOH concentration 5 mol/L, lead concentration 10 g/L, and scanning velocity 70 mV/s. The different As concentrations in the electrolyte solution had no effect on the cathodic lead cyclic voltammetry curve.

Cyclic voltammetry curve of lead electrowinning process in alkaline solution in the presence of As with various scanning velocity is shown in Fig. 6.89 at As concentration 400 mg/L, NaOH concentration 5 mol/L, and lead concentration 10 g/L. With the increase of c, the precipitation potential of Pb shifted to negative, and the peak current value of the cathode increased. When the scanning velocity was 10 mV/s, the recrystallization happened at the voltage scan to positive potential

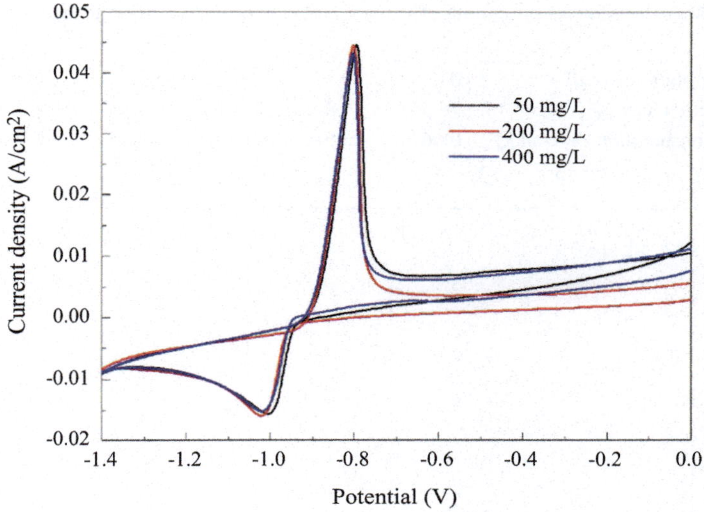

Fig. 6.88 Cyclic voltammetry curve of lead electrowinning process in alkaline solution in the presence of various As concentrations

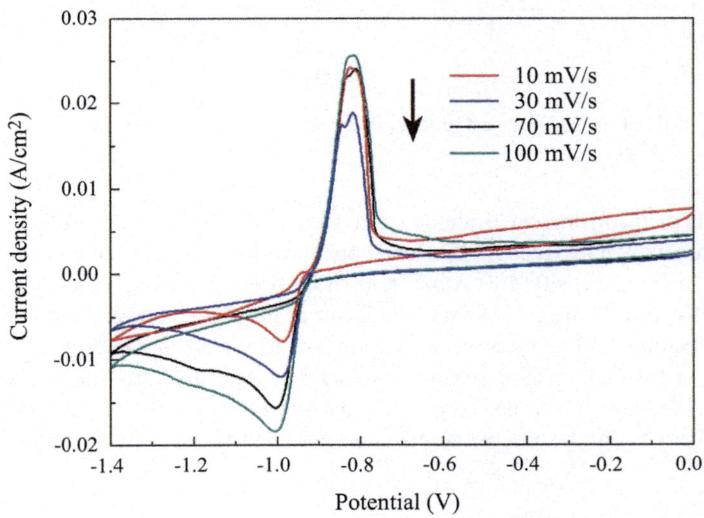

Fig. 6.89 Cyclic voltammetry curve of lead electrowinning process in alkaline solution in the presence of As with various scanning velocity

on the cathode. The cathodic peak was not observed on the cathode at scanning velocity below 70 mV/s. The diffusion rate of lead in electrolyte solution is slower than that of the surface reaction rate on the cathode surface with the increase of scanning velocity, which affect the precipitation reaction on the cathode.

Table 6.30 Effects of As on current efficiency and energy consumption in Pb electrowinning process

As concentrations (mg/L)	0	50	100	200	300	400
Current efficiency (%)	76.56	76.84	76.27	77.08	76.74	76.42
Energy consumption (kW·h/kg)	0.6679	0.6639	0.6701	0.6544	0.6642	0.6684

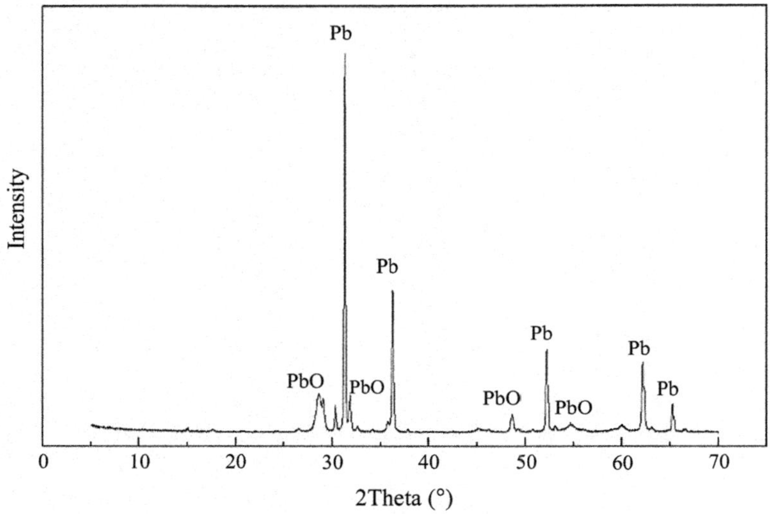

Fig. 6.90 XRD of the cathode lead product in the presence of As in alkaline electrolyte

Effects of As on current efficiency and energy consumption in Pb electrowinning process are shown in Table 6.30 at lead concentrations 14 g/L, NaOH concentration 5 mol/L, current density 500 A/m^2, temperature 40 °C, electrodes distance 5 cm, electrolytic time 70 min, and final Pb concentration 1.4 g/L. The presence of As in the electrolytic process had no effect on the current efficiency and energy consumption.

XRD of cathode of lead products obtained at As concentration of 400 mg/L is shown in Fig. 6.90. No As peak appeared and As would not precipitate in the cathode. The impurity As in the electrolyte had no effect on the purity of the cathode lead.

6.8.10 *Effects of Sb on Lead Electrowinning Process in Alkaline Solution*

Cyclic voltammetry curve of lead electrowinning process in alkaline solution in the presence of Sb at Sb concentration 200 mg/L, NaOH concentration 5 mol/L, Pb concentration 10 g/L, scan rate 50 mV/s is shown in Fig. 6.91. Cathodic current

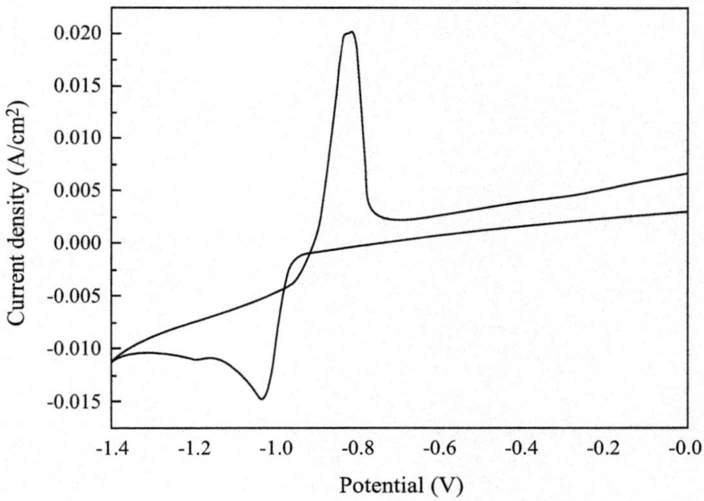

Fig. 6.91 Cyclic voltammetry curve of lead electrowinning process in alkaline solution in the presence of Sb

peak at the potential of −1.035 V corresponded to the precipitation reaction of lead on the cathode. A very weak cathodic current peak appeared near −1.2 V when the potential proceeded further negative. The anodic peak corresponded two adjacent current peaks at voltage scan to the positive, meaning Sb in the electrolyte took the reduction reaction at −1.2 V on the cathode.

Cyclic voltammetry curve of lead electrowinning process in alkaline solution in the presence of various Sb concentrations at NaOH concentration 5 mol/L, lead concentration 10 g/L, and scanning rate 50 mV/s is shown in Fig. 6.92. When the Sb concentration was 200 mg/L, it can be observed that the very weak cathodic reduction peak appeared at −1.2 V, and the peaks become weaker with the decrease of Sb concentration in the electrolyte. It cannot be observed at Sb 50 mg/L. Sb in the electrolyte had great effects on the lead deposition process, with the precipitation potential difference between the Pb and the Sb of about 0.2 V.

Effects of Sb on current efficiency and energy consumption in Pb electrowinning process are shown in Table 6.31 at Sb concentration 10, 50, and 100 mg/L, Pb concentration 14 g/L, NaOH concentration 5 mol/L, current density 500 A/m^2, temperature 40 °C, distance of electrodes 5 cm, electrolysis time 70 min, and final Pb concentration 1.4 g/L. Sb presence in the electrolytic solution had no effect on the current efficiency and energy consumption.

XRD of the cathode lead product obtained at Sb 100 mg/L is shown in Fig. 6.93. There is no intensity peak of Sb in the XRD, and Sb would not precipitate on the cathode. Hence, the impurity Sb in the electrolyte had no effect on the purity of the Pb product.

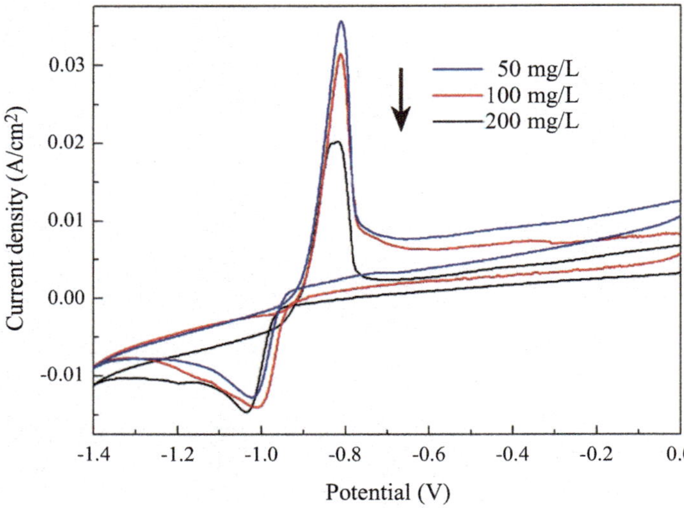

Fig. 6.92 Cyclic voltammetry curve of lead electrowinning process in alkaline solution in the presence of various Sb concentrations

Table 6.31 Current efficiency and energy consumption of lead deposition in different concentrations of Sb in alkaline solution

Sb concentrations (mg/L)	10	50	100
Current efficiency (%)	76.93	75.69	77.44
Energy consumption (kWh/kg)	0.5784	0.5742	0.5779

6.8.11 Effects of W on Lead Electrowinning Process in Alkaline Solution

Cyclic voltammetry curve of lead electrowinning process in alkaline solution in the presence of various W concentrations is shown in Fig. 6.94 at NaOH concentration 5 mol/L, Pb concentration 10 g/L, and scanning rate 50 mv/s. The cathodic peak near -1.03 V corresponded to the precipitation reaction of Pb on the cathode. When the potential was further reduced, the current value was not increased until scanning to the potential of -1.4 due to the occurrence of hydrogen evolution reaction. W in the electrolyte had no effect on the cathodic lead cyclic voltammetry curve.

Effects of W on current efficiency and energy consumption in Pb electrowinning process are shown in Table 6.32 at W concentrations 100, 300, and 500 mg/L, Pb concentration 14 g/L, NaOH concentration 5 mol/L, current density 500 A/m^2, temperature 40 °C, distance of electrodes 5 cm, electrolysis time 70 min, and final Pb concentration 1.4 g/L.

XRD of cathode lead product obtained at the W concentration of 500 mg/L is shown in Fig. 6.95. No peak of W appeared, and W was not separated from the cathode. The impurity W in the electrolyte had no effect on the purity of the lead product.

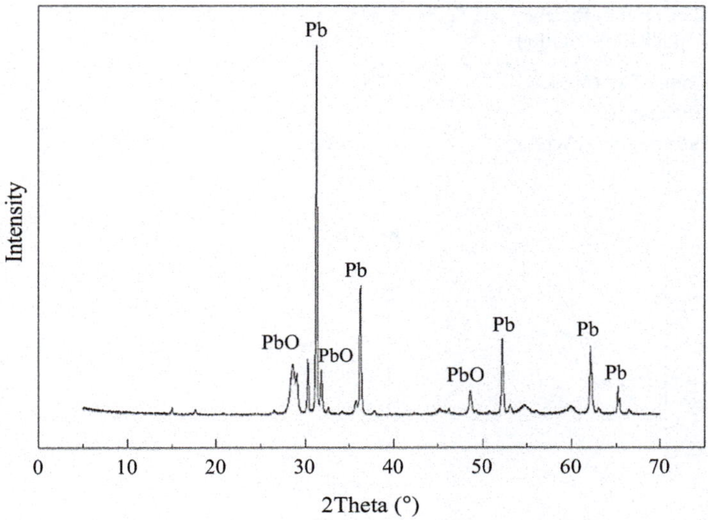

Fig. 6.93 XRD of cathode Pb obtained at Sb 100 mg/L in alkaline solution

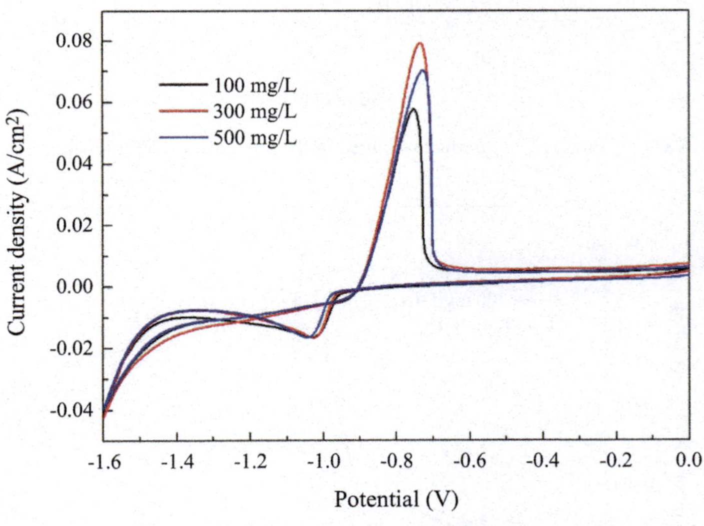

Fig. 6.94 Cyclic voltammetry curve of lead electrowinning process in alkaline solution in the presence of various W concentrations

6.8.12 *Effects of Cu on the Pb Electrowinning Process in Alkaline Solution*

The leaching solutions of Pb dust with Cu concentration of 200 mg/L and without Cu were used as the testing solutions, and their cyclic voltammetry curve on the Pb electrowinning process is shown in Fig. 6.96.

Table 6.32 Current efficiency and energy consumption of Pb deposition in different concentrations of W in alkaline solution

Concentration of W (mg/L)	10	50	100
Current efficiency (%)	76.66	77.23	76.30
Energy consumption (kWh/kg)	0.5782	0.5788	0.5749

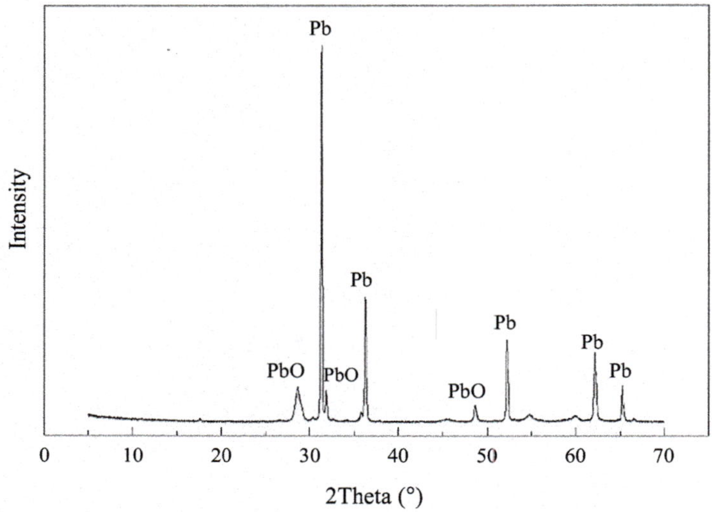

Fig. 6.95 XRD of cathode lead product obtained in the presence of W in alkaline solution

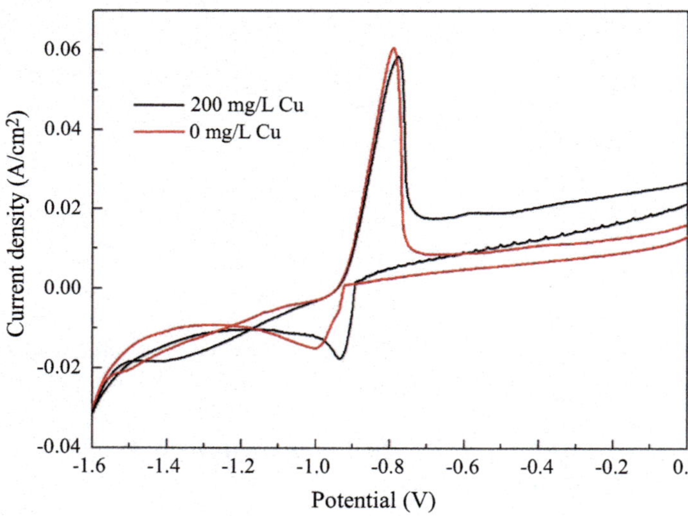

Fig. 6.96 Cyclic voltammetry curve of Pb electrowinning process in the presence of Cu in alkaline solution

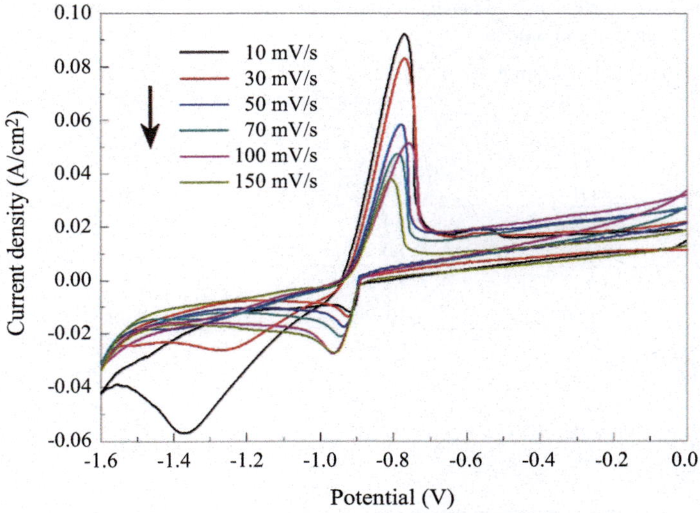

Fig. 6.97 Effects of copper in alkaline solution on cyclic voltammetry curves of lead electrowinning at different scanning velocity

When the concentration of Cu in solution was 200 mg/L, the potential was −0.935 V. Cathode peak potential shifted positive, and Cu precipitates on the cathode. However, the potential was −1.002 V without Cu. The cyclic voltammetry in different scanning velocity is shown in Fig. 6.97 at Cu 200 mg/L. At scanning velocity of 10 mV/s, the potential was −0.769 V, higher than the equilibrium potential of Pb, meaning Cu can precipitate. With the increasing the scanning velocity, the cathodic peak potential shifted negative, and the peak current increased, showing higher scanning velocity led to lower effect caused by Cu in Pb electrowinning.

6.8.13 *Effects of Anions on Pb Electrowinning Process in Alkaline Solution*

Current efficiency and energy consumption of Pb electrowinning process in alkaline solution in the presence of anions are shown in Table 6.33 at Pb concentration 20 g/L, NaOH concentration 5 mol/L, current density 200 A/m^2, temperature 45 °C, distance of electrodes 5 cm, and electrolysis time 4 h. The anions tested included PO_4^{3-}, SO_4^{2-}, SO_3^{2-}, SiO_3^{2-}, SCN^-, CO_3^{2-}, Cl^-, $C_4H_4O_6^{2-}$, $C_2O_4^{2-}$, and AC^-. It was shown that these anions in electrolyte had little effect on the current efficiency and energy consumption in the process of Pb electrowinning. However, the existence of $C_4H_4O_6^{2-}$and $C_2O_4^{2-}$in the electrolyte reduced the current efficiency and increased energy consumption.

Table 6.33 Current efficiency and energy consumption of Pb electrowinning process in alkaline solution in the presence of anions

Impurity	Concentration of impurity	Current efficiency (%)	Energy consumption (kWh/kg)
None	–	81.63	0.5071
Na_3PO_4	10 g/L	81.16	0.5100
Na_2SO_4	10 g/L	81.32	0.5090
Na_2SO_3	10 g/L	84.12	0.4920
$Na_2SiO_3·9H_2O$	10 g/L	80.80	0.5122
KSCN	$10\ g·L^{-1}$	81.21	0.5097
Na_2CO_3	1 mol/L	81.63	0.5075
NaCl	1 mol/L	81.63	0.5075
$KNaC_4H_4O_6$	1 mol/L	62.33	0.7056
$Na_2C_2O_4$	1 mol/L	55.59	0.7911
NaAc	1 mol/L	83.71	0.4955

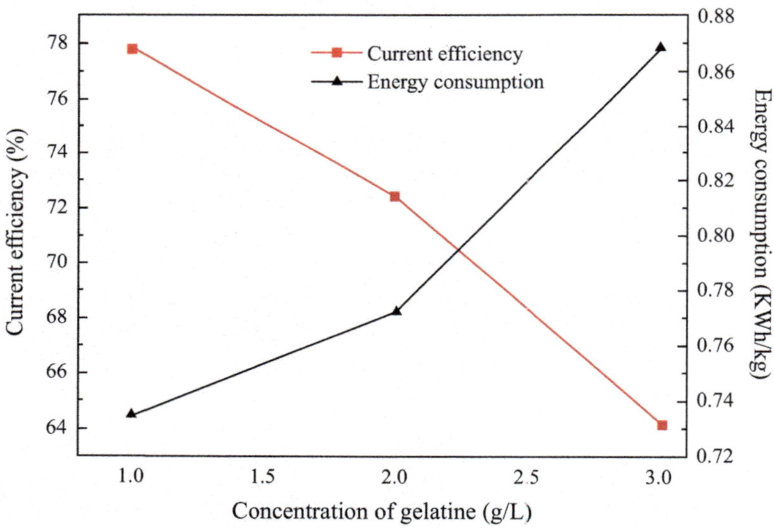

Fig. 6.98 Effects of gelatin on the current efficiency and energy consumption in the process of Pb electrowinning in alkaline solution

Effects of gelatin on the current efficiency and energy consumption in the process of Pb electrowinning are shown in Fig. 6.98 at Pb concentration 20 g/L, NaOH concentration 5 mol/L, current density 300 A/m^2, temperature 45 °C, distance of electrodes 5 cm, and electrolysis time 1 h. Adding gelatin as an additive to the electrolyte can reduce the current efficiency and increase the energy consumption, which will lead to a negative effect on the lead electrowinning process. At the same time, adding additives to the electrolyte can improve the external morphology of the cathode lead in the initial stage of lead deposition with a smooth surface. However, with the increase of the electrolysis time, the structure of the cathode lead was still relatively loose, so it was not recommended to add gelatin in the process of lead electrowinning in industry.

Chapter 7
Alkaline Hydrometallurgy of Low-Grade Smithsonite Ores

Abstract Low-grade smithsonite ores are leached using alkaline leaching solution. The zinc and lead concentrates can be obtained from lean oxidized zinc ores by alkaline leaching followed by two-step sulfide precipitation of lead and zinc sulfides. It is found that over 85% of both Zn and Pb and less than 10% of Al can be leached from the ore when the leaching operation is conducted at over 95 °C using 5 M NaOH solution as leaching agent. The dissolution of impurities such as Fe, Ca, etc. is negligible. Leaching of Pb can be improved remarkably with addition of NaCl to the leaching systems. Optimum conditions of lead precipitation are found to be 1.8 Na_2S/Pb weight ratio, 90 °C, 60 min, and those for zinc precipitation are 2.4 Na_2S/Zn weight ratio, 90 °C, 180 min. According to scale-up experiment, total extractions of both lead and zinc reach above 80% from the ores, and the quality of lead and zinc concentrates can meet industrial requirements for roasting process. The Pb-free solution is then used for the electrolysis of metallic Zn using stainless steel electrodes. Zn metal powder with purity higher than 99.5% is obtained. The specific energy for the Zn electrolysis in alkaline leach solution is around 2.4–2.6 kWh/kg Zn, which is lower than the energy consumption of 3.3 kWh/kg Zn required in the conventional process of Zn electrolysis in acidic sulfate electrolytes.

Keywords Smithsonite ores • Zinc • Lead • Alkaline hydrometallurgy • Concentrates

7.1 Alkaline Hydrometallurgy of Low-Grade Smithsonite Ore Process

Currently, most of zinc is produced from zinc sulfide ores because the sulfides are easily to be separated from gangue and concentrated by conventional flotation techniques. Oxidized zinc ores, such as smithsonite ($ZnCO_3$), willemite (Zn_2SiO_4), hydrozincite ($2ZnCO_3{\cdot}3Zn(OH)_2$), zincite (ZnO), and hemimorphite ($Zn_2SiO_3{\cdot}H_2O$), have also long been an important source of zinc. However, their concentration is difficult, and, until relatively recently, only rich ores are exploited,

Y. Zhao, C. Zhang, *Pollution Control and Resource Reuse for Alkaline Hydrometallurgy of Amphoteric Metal Hazardous Wastes*, Handbook of Environmental Engineering 18, DOI 10.1007/978-3-319-55158-6_7

using limited concentration by washing and gravity methods. For the low-grade oxidized zinc ores, the exploitation and metallurgy are very limited.

The concentration of oxidized zinc ores is usually realized by the flotation processes used in sulfide treatment by the addition of sodium sulfide before flotation is practiced. The recovery of zinc in this process is not high because the transfer of oxidized zinc to sulfide zinc is usually incomplete and ineffective. The oxidized zinc ores may be treated by acidic leaching processes with mineral acidic solution. However, besides the zinc and lead which are to be leached, most of the other soluble elements such as Ca, Fe, Mg, CO_3^{2-}, etc. may also be dissolved together. As a result, a large quantity of extra acids will be consumed and a leach solution with complex components will be obtained. A lot of unit operations may be needed for the purification of such an acidic leach solution, with a loss of zinc, including neutralization, chemical precipitation, and even solvent extraction, before the acidic solution is used for electrowinning. Therefore, this process would not be cost-effective when a low-grade oxidized zinc ore is used as raw materials.

For the case of smithsonite ores, a member of the oxidized ores family, the content of zinc and lead is usually low (<20% Zn and 5% Pb, respectively). The high-grade concentrates cannot be obtained without heavy losses, regardless of the concentration techniques used. Therefore, in many locations using smithsonite ores as raw materials for the production of zinc, especially in developing countries, high-grade concentrates of zinc are produced from low-grade ores by the volatilization techniques at high temperature in a Waelz-type kiln. The zinc content in the collected dust can be as high as 65–70%, in comparison with the original zinc content of around 20% in the feed ore. The dust are dissolved directly in sulfuric acid solution and then electrowon to produce zinc metal powder after necessary purification. This volatilization process is very effective, but heavy pollution may occur. The capital investment is also rather high, and it seems difficult for small and medium enterprises to use this process to produce zinc metal.

The samples (smithsonite) used for study were taken directly in situ from the ore plants in which the volatilization process is being used to concentrate the zinc content as dust (around 70% Zn in the dust). The ore samples were dried and roughly ground to the particles with around 0.2–0.5 mm before chemical analysis by ICP and using for leaching studies. The composition (%) was found as Zn 20.66, Pb 2.83, Cd 0.013, Cu 0.038, Mg 2.46, Al 1.53, and Fe 7.45.

The average composition of the lean oxide ore used in the experiments is given in Table 7.1. The leaching rates of lead and zinc were over 88% and 85%, respectively. While the extractions of the other metal elements Cu, Fe, Cd, Cr, Ca, Mg, and Al were all less than 5% (Table 7.2).

Typically, when the leaching processes were carried out at room temperature, a given weight (usually 3 g) of ore sample (and additives if applicable) was added into a glass bottle containing a given volume of 5 M NaOH solution and then rotating the bottles with an orbital shaker for 42 h (15 rpm). Analyze the supernatant and residue if necessary. For the leaching experiments at high temperature, 10 g of ore sample was mixed with 70 mL of 5 M NaOH and heated at boiling point for 2 h

Table 7.1 Chemical composition of the lead-zinc oxide ore

Elements	Zn	Pb	Cu	Fe	Cd	Ni	Mn
wt. (%)	17.3	4.1	0.06	3.71	0.09	0.02	0.11
Elements	Cr	As	Ba	Ca	Mg	Al	water
wt. (%)	0.01	0.11	0.65	4.05	1.28	0.77	4.96

Table 7.2 Chemical composition of the leach solution and extraction rate

Components	Concentration	Extraction (%)
NaOH	160.57 g/L	–
Pb	7.28 g/L	88.34
Zn	29.63 g/L	85.14
Cu	2.24 mg/L	1.86
Fe	13.51 mg/L	0.18
Cd	2.05 mg/L	1.13
Ni	0.49 mg/L	1.22
Mn	4.45 mg/L	2.06
As	70.17 mg/L	31.73
Ba	11.37 mg/L	0.87
Ca	138.06 mg/L	1.70
Mg	14.29 mg/L	0.56
Al	34.49 mg/L	2.23
Cr	0.12 mg/L	0.60

while stirred with magnetic stirrer. The volume was kept unchanged by the constant addition of water.

After the leaching operation was completed and filtered, the filtrate was used for the separation of lead by sulfide precipitation before electrolysis. Stainless electrodes were used as both anode and cathode. The obtained zinc metal powder can be easily removed from the electrode and was washed firstly thoroughly by water and then by ethanol, dried at 50 °C, weighed and analyzed.

7.2 Effects of NaOH Concentration on the Extraction of Zinc from of Low-Grade Smithsonite Ores

The experiments were carried out at the different temperatures of 25 °C and 100 °C and the liquid-solid ratio (mL/g) of 7 ~ 8. The extraction of zinc from the ores increases sharply with the increase of NaOH concentration in leaching solution when the initial NaOH concentration was less than 5 M and then increased only slightly when the initial NaOH concentration increased further (Fig. 7.1). In fact, zinc oxide, zinc carbonate, zinc phosphate, and the corresponding lead complexes may be dissolved effectively in NaOH solution with relatively high concentration (such as over 4 M). From Fig. 7.1 it can be seen that at room temperature, the maximum leaching and extraction of zinc and lead (65–75%) can be obtained when

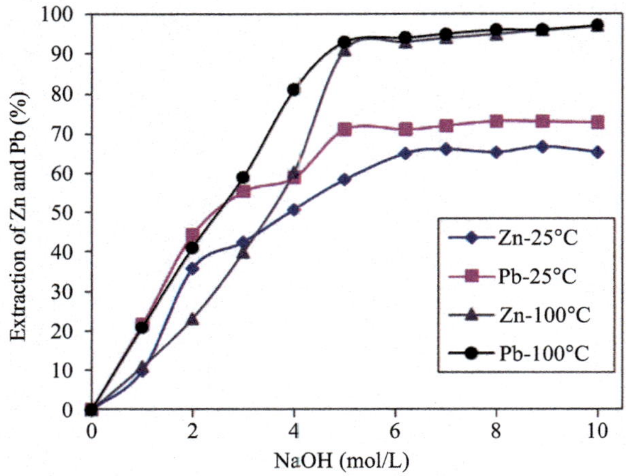

Fig. 7.1 Effects of NaOH concentration on the leaching of ore at different temperatures

the initial concentration of NaOH in the leaching solution was over 5 M which will be used in the following leaching experiments. Higher NaOH concentration in leaching solution may be unfavorable to the subsequent filtration due to the high viscosity of the resultant solutions and a bigger loss of NaOH.

When the leaching operations were carried out at higher temperatures, such as at boiling point of water, the leaching efficiencies will be increased considerably from 60–75% at room temperature to 80–90% at the higher temperature (Fig. 7.1). It can be also seen that the leaching behavior of zinc and lead was very similar. In fact, zinc and lead always coexist in most of the ores containing zinc and/or lead.

7.3 Effects of Liquid-Solid Ratio on the Extraction of Zinc from Low-Grade Smithsonite Ores

Higher liquid-solid ratio of NaOH solution to the ores will facilitate the dissolution of zinc and lead and improve the surface contact between the solution and particles of ores, thus increasing the extraction efficiencies of zinc and lead (Fig. 7.2). But higher liquid-solid ratio will increase the NaOH quantity required and also reduce the zinc and lead concentrations in the resultant leaching solutions. Therefore, a liquid-solid ratio of 6–8 may be suitable. The zinc and lead left in the leaching residues can be leached again by NaOH solution.

The dependence of zinc and lead extraction on liquid-solid ratio is also related closely to the leaching temperature used. At room temperature, the leaching efficiencies always increase with the increase of liquid-solid ratio. Nevertheless, for the extraction at the boiling point of water, the leaching efficiencies increase sharply when the phase ratio is lower than 7 and increase slightly when the phase ratio increases further. Hence, one-stage leaching may be enough for the maximum extraction for both zinc and lead, if the phase ratio of 7 and higher temperature is used.

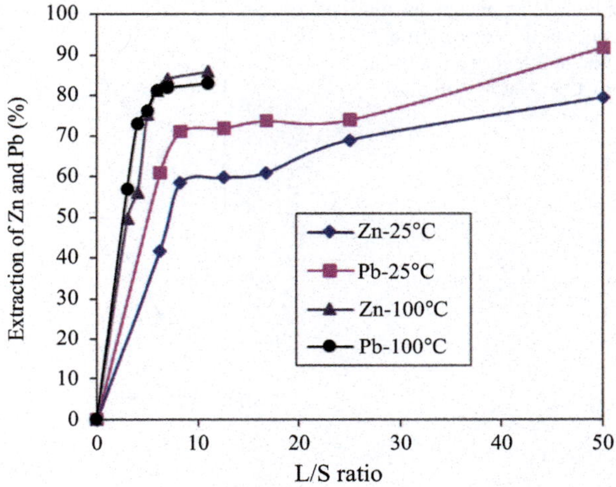

Fig. 7.2 Effects of liquid-solid ratio (mL/g) on the leaching of ore in NaOH solution

7.4 Effects of Leaching Time on the Extraction of Zinc and Lead from Low-Grade Smithsonite Ores

The experiments were carried out at the liquid-solid ratio (mL/g) of 8. The transfer of zinc and lead from ores to leaching solution involves the solid-liquid reaction and physical dispersion. Relatively longer leaching time between the ores particles and liquid is required in order to obtain a maximum extraction of zinc and lead. From Fig. 7.3 it can be seen that over 24 h leaching is enough to reach extraction and leaching equilibrium at room temperature.

However, the leaching equilibrium will be reached very quickly when the leaching operation is carried out at the boiling point of water (Fig. 7.4). Around 2 h heating time was enough for the maximum extraction of zinc and lead from the ore. In this case, over 85% of zinc and 84% of lead can be leached from the ore.

7.5 Effects of Additives on the Extraction of Zinc from Low-Grade Smithsonite Ores

Organic and inorganic salts were added into the leaching systems and the experimental results are shown in Table 7.3. It can be seen that the extraction of lead from ores will be enhanced by the addition of NaCl, which may be contributed to the formation of soluble Pb-Cl complexes, but slightly depressed by the addition of trisodium citrate. The leaching of zinc was remarkably depressed by the presence of trisodium citrate, but no effect was found for the addition of other additives, including NaCl.

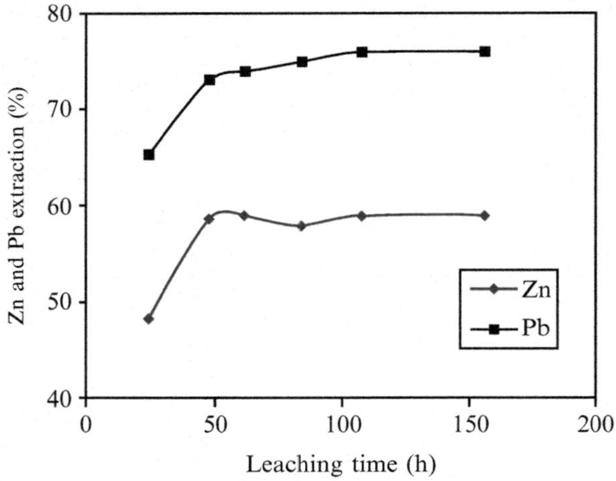

Fig. 7.3 Effects of leaching time on the extraction of zinc and lead from ores at 25°C in NaOH solution

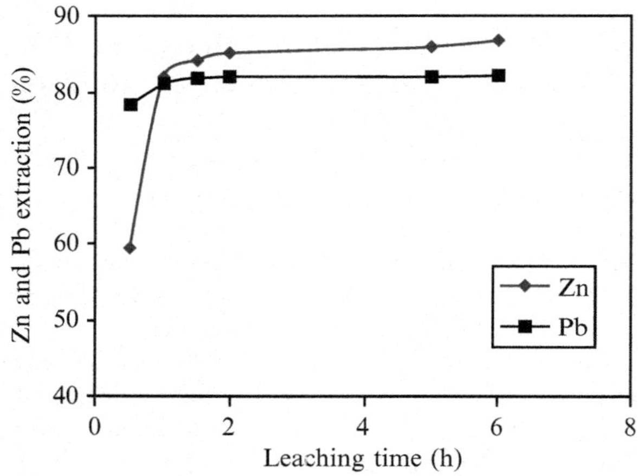

Fig. 7.4 Effects of leaching time on the extraction of zinc and lead from ore at 100°C in NaOH solution

7.6 Typical Contents in the Leach Solution and Leaching Residues for Low-Grade Smithsonite Ores

Under the optimum conditions of 5 M NaOH solution as leaching agent, liquid-solid ratio (mL/g) of 7:1, leaching at over 95 °C for 2 h, the typical contents of leaching solution were determined to be (g/L) 24 Zn, 3 Pb, and 0.5 Al. The corresponding contents of leaching residue were found to be (%) 2.4 Zn, 0.5 Pb, 0.042 Cd, 0.025 Cu, and 12.9 Fe. When NaCl was added to the leaching system with NaCl/ore weight ratio of 1:3, the content of Pb in the leaching residue will be decreased to lower than 0.05–0.2%. In this case, the residue can be disposed of safely in conventional landfill or dumping sites.

Table 7.3 Effects of the addition of salts on the alkaline leaching of zinc and lead from ores

No.	Salts added/weight (g)	Leaching temperature (°C)	Zinc extraction (%)	Lead extraction (%)
1	NaCl/0.0	25	57.98	72.10
2	NaCl/1.0	25	58.62	76.85
3	NaCl/2.0	25	58.89	88.67
4	NaCl/3.0	25	57.96	100.00
5	NaCl/0.0	100	85.89	82.01
6	NaCl/1.0	100	85.89	90.23
7	NaCl/2.0	100	86.09	95.10
8	NaCl/3.0	100	85.98	98.10
9	$CaCO_3$/1.0	25	58.54	76.59
10	$CaCO_3$/2.0	25	58.87	75.99
11	$CaCO_3$/3.0	25	58.98	76.76
12	Tri-Na-Citrate/1.0	25	51.92	82.67
13	Tri-Na-Citrate/2.0	25	50.25	70.94
14	Tri-Na-Citrate/3.0	25	40.20	70.84
15	Na-acetate/1.0	25	58.78	76.85
16	Na-acetate/2.0	25	58.78	75.78
17	Na-acetate/3.0	25	57.89	78.67

7.7 Electrolysis of Zinc from Lead-Free Leaching Solution for Low-Grade Smithsonite Ores and Economic-Technological Analysis

In this work, the electrolysis process was tested, as the leaching solution used was more complex than that prepared by the dissolution of zinc oxide in NaOH solution. The experimental results are shown in Table 7.4. The zinc metal deposited on the cathode was gray with slightly bright metallic luster. It was easy to be removed from the cathode.

The ore should be broken into smaller particles before leaching. Stirring is necessary while heating in order to speed up the transfer of zinc and lead from the ore particles into solution. After 2 h leaching, filtration operation should be carried out immediately, followed by the addition of sodium sulfide to separate the lead in the leaching solutions. The formation of lead sulfide precipitates was very fast and can be filtered soon when all the added sodium sulfides were dissolved. The electrolysis can be carried out when the filtration operation of lead sulfide is finished.

In order to save energy, the whole process can be conducted in the adiabatic environment. All the operations including leaching, filtration, precipitation, and electrolysis will be improved if being carried out at higher temperature. It was found that there was low loss of NaOH during the transfer of zinc compounds from ores to NaOH solutions and the subsequent electrolysis. The loss of NaOH in the precipitation of lead by sulfide was also very low because most of the leads were

Table 7.4 Typical electrolysis conditions for the zinc from lead-free leaching solution

Zn (g/L)	NaOH (M)	Electrode	Voltage (V)	Current density (A/m^2)	Specific energy (kWh/kg Zn)	Zinc metal purity (%)
24–30	5	Stainless steel	2.4–2.7	700–1100	2.5–2.6	99.0–99.9

precipitated as PbS, and the contents of lead in leach solutions are always lower than 3 g/L which was much lower than those of zinc. According to chemical titration, the NaOH loss in the whole process was estimated to be less than 50 g for the production of 1 kg Zn.

The most part of costs for the process developed in this work may be the energy consumed for the electrolysis. Around 2.5–2.6 kWh of electricity will be used for electrolysis of 1 kg of zinc from leach solution. The other energy will be used for heating the leaching systems. However, this proportion of energy can be saved to a great extent if the whole process is conducted in an adiabatic environment. In fact, the electrolysis of zinc may be improved when it is carried out at 75 °C or higher.

Hence, the operational costs would only include several sources such as electrolysis of zinc (2.5 kWh/kg Zn), heating of leaching system (estimated as less than 1 kWh/kg Zn on average), the loss of NaOH (50 g NaOH/kg Zn because of the entrained in the leaching residue and the lead sulfide precipitate), and labor. Also, the equipment required for the process is simple and commercially available. Obviously, the operational and capital costs would be quite low, in comparison with volatilization-dissolution in sulfuric acid-purification-electrolysis process. It should be a cost-effective process for small- and medium-scale companies to produce zinc metal and high purity lead sulfide, if the ore is available.

Table 7.5 presented the comparisons among different processes for the production of metallic zinc from oxidized zinc ores. It can be seen that the alkaline process may be the optimum, especially for the small- and medium-scale companies. The costs shown in Table 7.5 did not exclude the profit of selling the pure lead sulfide, which may offset most of the costs of sulfide consumed for the lead precipitation. Compared with the volatilization acidic leaching processes, the energy for alkaline leaching would be cut down considerably. It is considered that the alkaline leaching-electrowinning processes are worth being recommended as a cost-effective and environmentally friendly technology for the production of zinc from oxidized zinc ores.

The zinc in the ores may exist as zinc oxide, zinc carbonate, and zinc sulfide. Only the former two types of zinc can be dissolved into NaOH solution. Moreover, the dissolution rates for zinc carbonate, the major component in the smithsonite ores, may be slow, and the dissolution effectiveness may be increased at relatively high temperature and higher NaOH concentration. Even at a more critical condition such as boiling the leaching system, there is also around 10% of zinc that cannot be leached (the zinc content in leaching residue is lower than 2.4%). This proportion of zinc may be present in ores as zinc sulfide.

Table 7.5 Comparisons for different process for the production of metallic zinc from oxidized zinc ores (based on the production of 1 kg Zn)

Processes selected	Alkaline process developed in this work	Acidic process	Volatilization-acidic process
Energy for electrolysis	2.5–2.6 kWh	3.3	3.3
Energy for heating leaching system	Estimated as 1 kWh	1 kWh	1 kWh
Energy for volatilization	None	None	1 kWh
Loss of NaOH	Negligible	Loss for neutralization and purification, perhaps quite large	Loss for neutralization and purification, perhaps quite large
Loss of zinc powder for lead removal	None	Necessary	Necessary
Loss of sodium sulfide for lead removal	Necessary	None	None
Loss of Acids	None	Large quantity	Large quantity
Capital costs	Small to large	Small to large	Large
Operational costs	Small	Quite large	Quite large
Feasibility	Small- to large-scale company	Small- to large-scale company	Large-scale company
Complexity of whole process	Simple	Quite complex	Complex
Removal of lead	By sulfide precipitation	Zinc metal powder addition	Zinc metal powder addition
Total costs	Lowest	Low	High

Besides Zn and Pb, the oxides of Al, Cr, and Cu may be also dissolved in NaOH solution. It had found that the solubility of oxides of Cu and Cr in NaOH solution is very low (<1 g/L), when Zn and Pb are present. Part of oxides of Al may be dissolved with Zn and Pb but depressed by the presence of Zn and Pb. The solubility of the five elements in NaOH solution can be given as follows:

$$Zn > Pb > Al > Cr \sim Cu$$

Therefore, some of Al in the ores will be transferred into NaOH solution, but the concentration is always lower than 1 g/L, which cannot be removed together with Pb by sulfide precipitation method and may accumulate in the solution after the Zn is electrowon and the solution is recycled.

The oxides and carbonates of zinc and lead in smithsonite ores can be leached and transferred into NaOH solution effectively, while the sulfides remain in the leaching residues. The addition of NaCl can improve considerably the leaching and extraction of lead from ores. The lead in the leaching solution can be quantitatively separated from zinc as sulfides. The zinc in the lead-free solution can be recovered by electrowinning, and the lead- and zinc-free solution can be recycled into the next leaching step. The alkaline leaching-sulfide separation-electrowinning process is considered to be cost-effective and of industrial significance.

7.8 Treatment and/or Recycling of Wastewaters and Solid Wastes Generated from Low-Grade Ores and Dust Processing in Alkaline Solutions

The quantity and composition of the wastewaters generated depend on the flow sheets used in industry. For the low-grade ores, all the water and wastewaters can be completely recycled and without wastewater will be generate and discharged. For electric arc furnace dust, two processes can be used for the recovery of lead and zinc, i.e., direct leaching-melting of the residues resulted from the direct leaching-melting-leaching of the melts of residues (DML), and hydrolysis-melting of hydrolyzed dust-leaching of the melts (HML). In DML process, the dust is leached firstly; secondly, the resultant leaching residue is melted with NaOH solid followed by the leaching of the melts. The leach solutions are combined together and followed by the recovery of lead and zinc as sulfide precipitates or electrowinning. The resultant lead- and zinc-free NaOH solutions are recycled to the leaching steps for dust and melts. In principle, there is relatively small amount of wastewaters in this process.

In HML process, a relatively large quantity of wastewaters may be generated (Tables 7.6 and 7.7). The wastewaters are generated in the hydrolysis of dust and washing of hydrolyzed dust, which can be reused after neutralizing to pH around 7. Therefore, there is actually little wastewater in the entire processes. The wastewaters that cannot be reused or recycled are easily treated to reach the discharged standards.

The unique solid waste is the leaching residues in both DML and HML processes. In DML, the lead and zinc content in the final residue was lower than 0.1% or 2% and can be considered to be a common non-harmful and nontoxic waste. As the iron content in the residue is as high as 39% (52% iron oxide), it may be used as the feed materials for steelmaking plants. For HML process, the zinc and lead contents in the final residue are about 1–2% and 0.4%. It may be also considered to be non-harmful and nontoxic and is used as the feed materials for steelmaking plants. The residues may be used as the materials for brick making under high pressure. The brick may be used for the construction of road and walls of nonresident houses. The ultimate ways for the residues to go may depend on the market availability. It may be also a problem for the process, if the residues cannot be treated properly.

Table 7.6 The wastewaters generated in direct leaching-melting-leaching process based on 1000 kg of dust

Steps involved	Water required	Wastewaters generated and major components	Treatment methods
Pb- and Zn-free leaching solutions	4500 L in average		Recycled completely to leaching steps
Washing of leaching residues (cakes)	5 M NaOH solution 200 L		Recycled completely to leaching steps
Washing of leaching residues (cakes)	200 L		Recycled to the preparation of NaOH solution or discharged after treating with sodium sulfide and neutralization
Washing of Pb and Zn sulfides precipitates	100 L		Recycled for the preparation of NaOH solution
Melting of hydrolyzed dust			The steam generated is cooled down and used as hydrolysis agent

Table 7.7 The wastewaters generated in hydrolysis-melting-leaching process for the treatment of 1000 kg of dust

Steps	Water required	Wastewaters generated and major components	Treatment methods
Hydrolysis	1000 L	625 L (<3 mg/L of Cu, Zn, Pb, and <1 mg/L of Cd)	Adding sodium sulfide to remove the heavy metals, neutralized to pH around 7, and then recycled to the hydrolysis step
Washing for the hydrolyzed dust	500 L	500 L (<1 mg/L of Cu, Cd, Zn, Pb, etc.)	Used directly in the hydrolysis step, and/or used for preparation of NaOH solution, or discharged directly
Pb- and Zn-free leaching solutions	4500 L in average		Recycled completely
Washing of leaching residues (cakes)	5 M NaOH solution 200 L		Recycled completely
Washing of leaching residues (cakes)	200 L		Recycled to hydrolysis step or discharged after treating with sodium sulfide and neutralization
Washing of Pb and Zn sulfide precipitate	100 L		Recycled to hydrolysis step or for the preparation of NaOH solution
Melting of hydrolyzed dust			The steam generated is cooled down and used as hydrolysis agent

In the practical cases, the leaching residues may contain zinc and lead up to 5%, which can be treated to recover the zinc and lead using vaporization techniques. Hence, these residues can be sold.

All the wastewaters and solid wastes generated in the entire processes can be easily reused, recycled, or treated properly for discharge. There are no environmental problems that may appear when the processes developed are used in industry.

7.9 Lead Precipitation from Leaching Solution for Low-Grade Smithsonite Ores

Sulfide in aqueous solution is present as S^{2-}, HS^-, and H_2S at different pH values. In neutral medium, the main shape of sulfide is H_2S and HS^-, and H_2S is ionized into HS^- gradually with the increasing of pH. In the weekly alkaline solution, HS^- is predominant. While in the strong alkaline solution, sulfide is fully ionized into S^{2-}, and S^{2-} becomes the dominant shape with the increasing of NaOH concentration. The solubility of ZnS and PbS is 2×10^{-23} and 1×10^{-28}, respectively, and PbS is much more insoluble than ZnS. This relative insolubility accounts for one reason why lead can be preferably precipitated prior to zinc. However, little selectivity was observed for zinc and lead separation by sulfide precipitation in neutral medium and weakly alkaline solution. Both zinc and lead precipitated after the addition of sodium sulfide. Pb and Zn cannot combine into HS^- and H_2S to form PbS and ZnS directly and completely. The fractional dissociation of sulfide may be the other cause of the different selectivity for zinc and lead separation by sulfide precipitation in different acidic and basic medium.

Chemical reactions taking place in the sulfide precipitation processes are proposed. The molar ratios of Pb/S and Zn/S are around 1.5 and 0.6, respectively. Therefore, the components of the sulfides may be very complex. The following reactions may be proposed:

$$
\begin{aligned}
x\mathrm{NaZn(OH)_4(aq)} + y\mathrm{Na_2S(s)} = {} & n\mathrm{ZnS(s)} + b\mathrm{NaZn(OH)}_m\mathrm{S}_{(4-m)/2}\mathrm{(s)} \\
& + k\mathrm{Zn(OH)_2(s)} + p\mathrm{NaOH(aq)} \qquad (7.1)
\end{aligned}
$$

$$
\begin{aligned}
x\mathrm{NaPb(OH)_4(aq)} + y\mathrm{Na_2S(s)} = {} & n\mathrm{PbS(s)} + b\mathrm{NaPb(OH)}_m\mathrm{S}_{(4-m)/2}\mathrm{(s)} \\
& + k\mathrm{Pb(OH)_2(s)} + p\mathrm{NaOH(aq)} \qquad (7.2)
\end{aligned}
$$

When the sodium-zinc hydroxide sulfide and its lead analogues are hydrolyzed in water after having been heated for hours, the species may be transformed into PbS or ZnS and hydroxyl ions be released.

The lead precipitation is a function of reaction time, Na_2S/Pb weight ratios, concentration of NaOH, and temperature. The experimental results are shown in Figs. 7.5, 7.6, 7.7, and 7.8. It can be seen that lead can be precipitated quantitatively

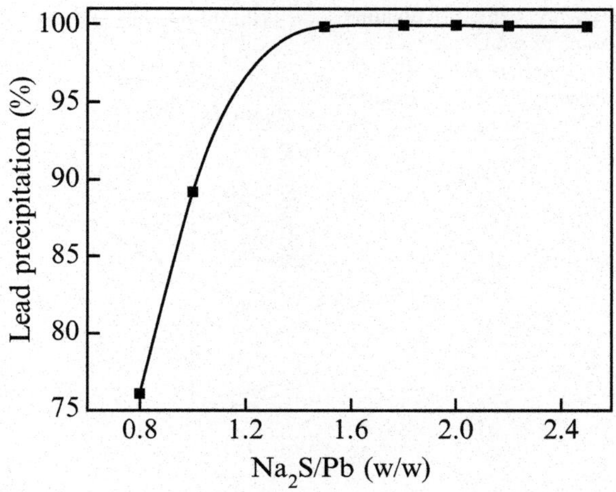

Fig. 7.5 Effects of the Na_2S/Pb mass ratio on the lead precipitation. (NaOH concentration 160.57 g/L, temperature 90 °C, time 60 min)

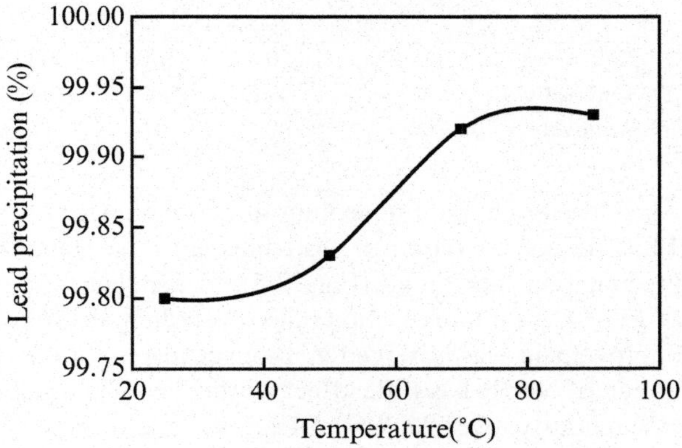

Fig. 7.6 Effects of temperature on the lead precipitation. (NaOH concentration 160.57 g/L, time 60 min, Na_2S/Pb (w/w) = 1.8)

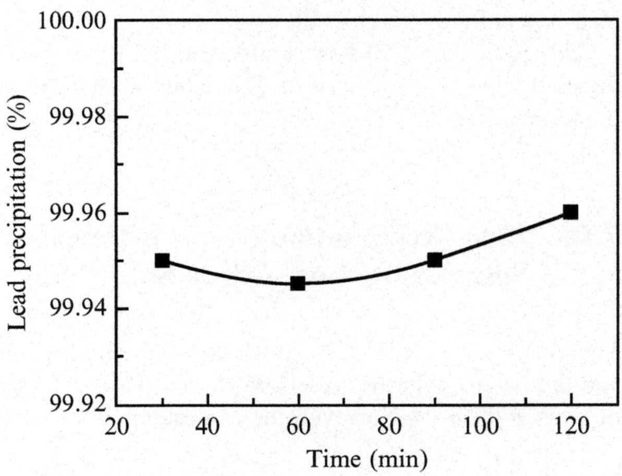

Fig. 7.7 Effects of reaction time on the lead precipitation (NaOH concentration 160.57 g/L, temperature 90 °C, Na_2S/Pb (w/w) = 1.8)

Fig. 7.8 Effects of alkaline concentration on the precipitation of lead (Na_2S/Pb(w/w) = 1.8, temperature 90 °C, time 60 min)

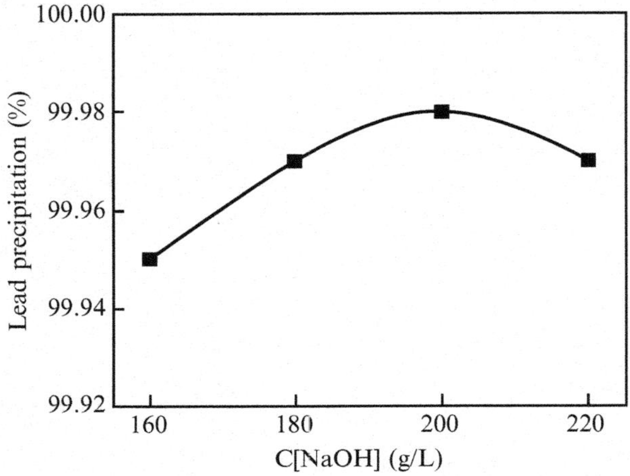

Table 7.8 Separation of lead from zinc in leaching solution by sulfide precipitation

No	1	2	3	4	5	6
$Na_2S\cdot(7\sim9)H_2O$/Pb mass ratio	0.75	1.00	1.55	1.85	2.11	2.33
Pb Removal %	47.68	68.34	95.12	99.91	100.00	100.00
Zn Removal (%)	0.0	0.0	0.0	0.0	2.0	4.0

when the weight ratio of sodium sulfide/Pb in leach solution is slightly higher than 1.8. The concentration of lead falls into the stable value of 6 ~ 8 mg/L after precipitation operation (Fig. 7.5), dependent on temperatures up to 90 °C (Fig. 7.6). A 60-min reaction time was used in the following experiments, and little improvement was observed at a longer reaction time (Fig. 7.7). Also, the concentration of NaOH has little effect on the lead precipitation (Fig. 7.8).

Obviously, over 99.90% of lead in the leaching solution can be precipitated by sodium sulfide at the optimum conditions, and the solid precipitated phase that contains a mixture of lead sulfide and much less sodium-lead disulfide forms the lead concentrates (Table 7.8).

The composition of lead sulfide formed from the leaching solution was determined to be the mixtures of PbS and $Na_2Pb(OH)_2S$, but the former was the predominant.

7.10 Zinc Precipitation from Pb-Depleted Zinc Alkaline Solution for Low-Grade Smithsonite Ores

The zinc can be precipitated from the lead-depleted filtrate using sodium sulfides, and the NaOH solution free from Zn and Pb can be recycled to the next leaching process of the ores. However, the zinc leaching rate will be seriously affected if the circulating solution contains sulfide ion. The concentration of zinc should be held above 4 g/L after sulfide precipitation. When the mass ratio of Na_2S/Zn was 2.4, the residual concentration of zinc was 5.19 g/L, and when the ratio was 2.8, the concentration was below 4 g/L (Fig. 7.9). Therefore, a mass ratio of 2.4 may be used. Furthermore, as shown in Fig. 7.10, the reaction time had little effect on the zinc precipitation, yet it was a benefit to regenerate the alkaline in the liquid of zinc precipitated. The regeneration of NaOH increased rapidly as the reaction time extended for up to 180 min. The longer reaction time reduced the weight of alkaline instead because of the evaporation of the solution.

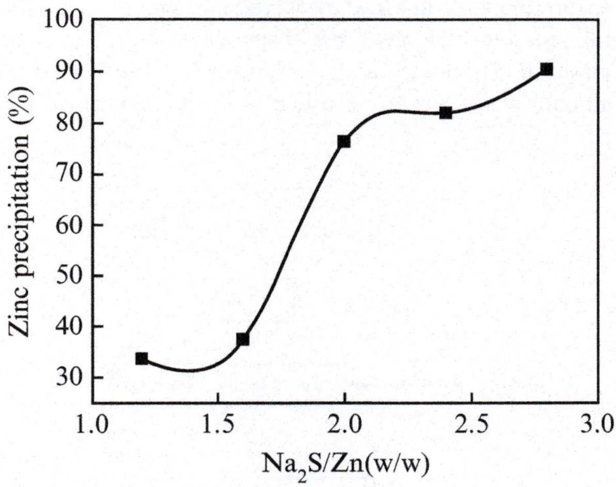

Fig. 7.9 Effects of the Na_2S/Zn mass ratio on the zinc precipitation (NaOH concentration 199.26 g/L, Zn concentration 28.80 g/L, temperature 90 °C, time 60 min)

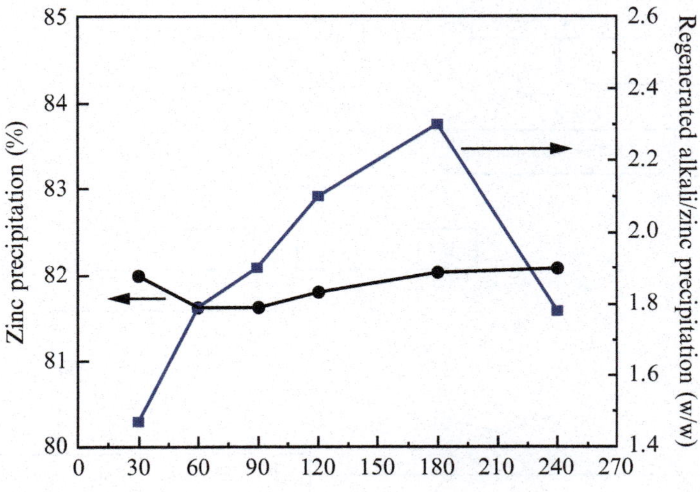

Fig. 7.10 Effects of reaction time on the precipitation of zinc and regeneration of alkaline (NaOH concentration 199.26 g/L, Zn concentration 28.80 g/L, temperature 90 °C, Na_2S/Zn (w/w) = 2.4)

7.11 Lead and Zinc Concentrate Production from Low-Grade Pb-Zn Oxidized Ore

7.11.1 Technological Process

The preparation process of zinc concentrate and lead concentrate is shown in Fig. 7.11. The poor/complex zinc oxide ore and waste-bearing lead and zinc were ground to 0.15 ~ 0.2 mm firstly. Then, the particles were added to the NaOH solution (240 g/L) and stirred for 120 min at the temperature of 80 ~ 100 °C to ensure the zinc and lead dissolved completely. The obtained leaching solution was filtered; the residues were disposed of by landfill or brick making after water washing. The Na_2S as the precipitant was added to the leaching solution, and its amount was 1.8 times as much as the content of lead in the solution (calculated by

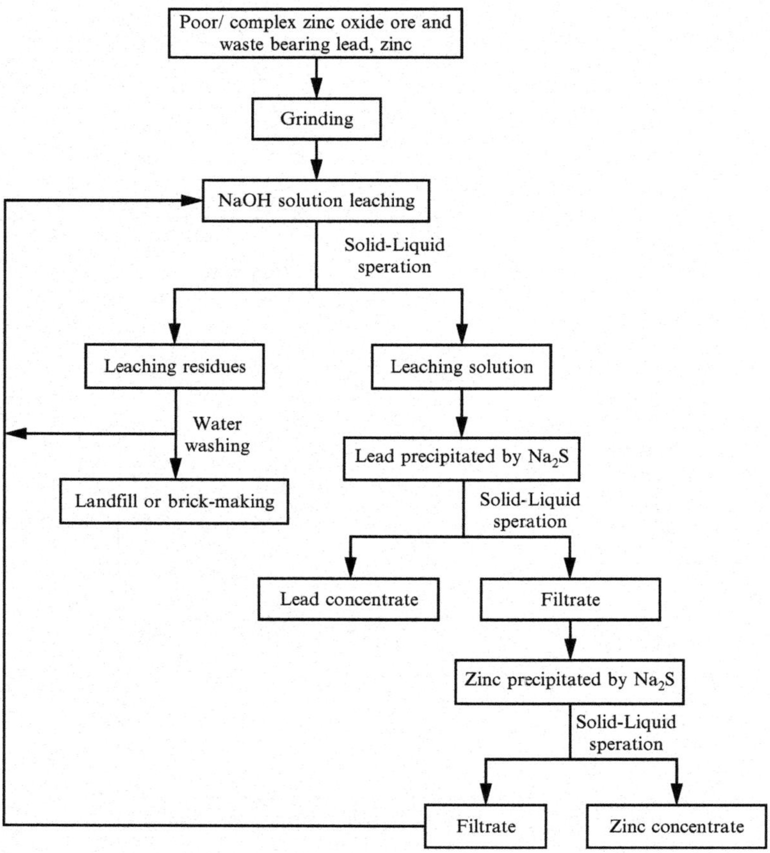

Fig. 7.11 Technological process of preparing zinc concentrate and lead concentrate

mass ratio) and stirred for 1 h at the temperature of 70 ~ 90 °C. After filtration, the lead concentrate was obtained and met the Chinese industry standard (YS/T319–1997 < Lead concentrate>). Next, the Na_2S were added to the zinc-contained filtrate to precipitate zinc; the addition of Na_2S was 2.4 times as much as the difference value between zinc concentration and 4 g/L and stirred for 3 h at the temperature of 70 ~ 90 °C. After filtration, the zinc concentrate was obtained and met the Chinese industry standard (YS/T320–1997 < zinc concentrate>). The final filtrate and washing wastewater reflowed to the leaching process and reused.

7.11.2 Principle of Process

1. Leaching process

In the leaching process, the zinc and lead in feedstock are dissolved by NaOH solution, and possible reactions are as follows:

$$5ZnO + 10NaOH + H_2O \rightarrow 2Na_2ZnO_2 + 3Na_2Zn(OH)_4 \quad (7.3)$$

$$5ZnCO_3 + 20NaOH \rightarrow 2Na_2ZnO_2 + 3Na_2Zn(OH)_4 + 5Na_2CO_3 + 4H_2O \quad (7.4)$$

$$ZnO \cdot SiO_2 + 20NaOH \rightarrow 2Na_2ZnO_2 + 3Na_2Zn(OH)_4 + 5Na_2SiO_3 + 4H_2O \quad (7.5)$$

$$2NaOH + PbO \rightarrow Na_2PbO_2 + H_2O \quad (7.6)$$

$$2NaOH + PbO + H_2O \rightarrow Na_2Pb(OH)_4 \quad (7.7)$$

$$4NaOH + PbCO_3 \rightarrow Na_2PbO_2 + H_2O + Na_2CO_3 \quad (7.8)$$

$$4NaOH + PbCO_3 \rightarrow Na_2Pb(OH)_4 + Na_2CO_3 \quad (7.9)$$

$$4NaOH + PbSiO_3 \rightarrow Na_2PbO_2 + H_2O + Na_2SiO_3 \quad (7.10)$$

$$4NaOH + PbSiO_3 \rightarrow Na_2Pb(OH)_4 + Na_2SiO_3 \quad (7.11)$$

2. Lead precipitation

In the process, the Na_2S is used as precipitant to recover the lead in the leaching solution, and the main reactions are as follows:

$$xNaPb(OH)_4(aq) + yNa_2S(s) \rightarrow nPbS_y(s) + bNaPb(OH)_mS_{(4-m)}(s) + kPb(OH)_2(s) + pNaOH(aq) \quad (7.12)$$

$$xNaPbO_2(aq) + yNa_2S(s) \rightarrow nPbS_y(s) + bNaPb(OH)_mS_{(4-m)}(s) + kPb(OH)_2(s) + pNaOH(aq) \quad (7.13)$$

According to the determined parameters of lead precipitation, the lead concentrate was prepared, and the chemical composition analysis is presented in Table 7.9. The lead content reached to 82.8%, while the zinc content was only

Table 7.9 The chemical composition analysis of the prepared lead concentrate (%)

Chemical composition	Pb	Al	As	Ca	Cd	W	Zn
Lead concentrate	82.8	0.03	0.01	0.19	0.04	0.12	1.9
Chemical composition	Cr	Cu	Fe	Mg	Mn	Sb	Sn
Lead concentrate	0	0.02	0.08	0.08	0	0.04	0

Table 7.10 The chemical composition analysis of the prepared zinc concentrate (%)

Chemical composition	Zn	Al	As	Ca	Cd	W	Pb
Zn concentrate	47.05	0.04	0.01	0.39	0	2.52	0.20
Chemical composition	Cr	Cu	Fe	Mg	Mn	Sb	Sn
Zn concentrate	0	0	0.02	0.10	0	0	0

1.9%, and other impurities were also at low content. The product can meet the first-grade demand in the Chinese industry standard (YS/T319–1997 < Lead concentrate>).

3. Zinc precipitation

In the process, the Na_2S is used as precipitant to recover the zinc in the leaching solution, and the main reactions are as follows:

$$xNaZn(OH)_4(aq) + yNa_2S \rightarrow nZnS + bNaZn(OH)_mS_{(4-m)/2}(s) + kZn(OH)_2(s) + pNaOH(aq) \quad (7.14)$$

$$xNaZn(OH)_4(aq) + yNa_2S \rightarrow nZnS + bNaZn(OH)_mS_{(4-m)/2}(s) + kZn(OH)_2(s) + pNaOH(aq) \quad (7.15)$$

According to the determined parameters of zinc precipitation, the zinc concentrate was prepared, and the chemical composition analysis is presented in Table 7.10. The zinc content reached to 47.05%, the lead content was only 0.20%, and other impurities were also at low content. The product can meet the third-grade demand in the Chinese industry standard (YS/T320–1997 < Zinc concentrate>).

7.12 Scale-Up Experiments for Lead and Zinc Concentrate Production from Low-Grade Pb-Zn Oxidized Ore in Alkaline Solution

When 1200 g lean zinc oxidized ore from Guizhou Province was used, similar extraction and leaching efficiencies were also obtained, as given in Table 7.11. It can be seen that the leaching rate of zinc and lead can reach around 90%. After two-step sulfide precipitation, 81% of zinc and 83% of Pb can be extracted from the ore as concentrates.

The quality of the lead and zinc concentrates obtained in the scale-up experiments is shown in Table 7.12. Because the solubility of Cu, Cr, Fe, Mg, Al, and As oxides in alkaline solution are much lower than those of Zn and Pb, especially when

Table 7.11 Scale-up experiment for the production of lead and zinc concentrations in alkaline solution

Material	Composition of lead-zinc oxidized ore (%)	Zn 15.91, Pb 4.94, Al 0.16, As 0.11, Ca 3.09, Fe 6.2, Mg 0.09, Mn 1.83, Ba 0.05, Cu 0.02, Cd 0.01, Cr 0.01
	Mass of materiel	Ores 1200 g, NaOH 1440 g, water 6 L
Leaching	Leaching conditions	90 °C,120 min, initial concentration of NaOH 240 g/L
	Volume of leaching solution (L)	5.645
	Composition of leaching solution (g/L)	Zn 28.90, Pb 9.36, NaOH 164.34, As 0.161, Ca 0.082, Cu 0.004, Fe 0.005, Mg 0.01, Al 0.551, Cr 0.001, Cd 0.003
	Leaching rate of Zn (%)	85.45
	Leaching rate of Pb (%)	89.12
Lead precipitation	Precipitation conditions	90 °C, 30 min, Na_2S/Pb(w/w) 1.8
	Volume of the filter after precipitation (L)	5.580
	Composition of the filter (g/L)	Zn 28.80, Pb 0.011, NaOH 181.94, As 0.165, Ca 0.111, Cu 0.003, Fe 0.008, Mg 0.011, Al 0.580, Cr 0.001, Cd 0.004
	Mass of lead concentrations (g)	62.42
	Precipitation efficiency	99.88%
Zinc precipitation	Precipitation conditions	90 °C, 180 min, Na_2S/Pb (w/w) 2.4
	Volume of the filter after precipitation (L)	5.40
	Composition of the filter (g/L)	Zn 5.28, NaOH 232
	Mass of lead concentrations (g)	240.8
Total zinc extraction (%)		81.0
Total lead extraction (%)		83.01
Loss of NaOH (g)		187.2 g

Table 7.12 Quality and industry standard of lead and zinc concentrates from alkaline solution

Chemical composition (%)	Pb	Zn	As	Cu	MgO	Al_2O_3
Lead concentrate	78.84	1.8	0.06	0	0.03	0.27
Primary standard of lead concentrate (YS/T319–2007)	>65	≤4	≤0.2	≤1.2	≤1.0	≤2.0
Chemical composition	Zn	Pb	As	Cu	Fe	SiO_2
Zinc concentrate	52.38	0.20	0.04	0	0.06	0.08
Tertiary standard of zinc concentrate (YS/T320–2007)	>45	≤2.0	≤0.5	≤1.0	≤12	≤5.5

Table 7.13 Leaching toxicity extraction test of the leaching residue from alkaline solution

Elements (mg/L)	Zn	Pb	As	Hg	Cd	Ba	Bi	Cr	Cu	Ni
Leaching test	43.57	2.16	0.48	0	0	1.68	0.01	0.12	0.1	0.04
GB5085.3-2007	100	5	5	0.1	1	100	0.02	5	100	5

these oxides coexist in ores, the contents of impurity in the concentrates were much less than the index of the industrial requirements.

The loss of NaOH in the whole process was low. The initial concentration of NaOH in the leaching was 240 g/L and the concentration reduced to about 160 g/L. Fortunately, the alkaline solution will be regenerated both in the lead and zinc precipitation. The final NaOH concentration in the liquid of zinc precipitated returned to 232 g/L again. That is to say, the loss of NaOH in the leaching is equal to the caustic regenerated in the two-step precipitation. Hence, the theoretical alkali consumption of the whole process is zero. The natural filtration with filter cloth was carried out in the scale-up experiment. Except the volatilization loss in the heating process, the loss of NaOH comes mainly from the leakage of solution in the filtration and the entrainment loss of sulfide precipitate. If the pressure filter is used to solid-liquid separation, the loss of NaOH will be greatly reduced.

The leaching residue was washed with 5% NaOH solution and then with water. The results of the toxicity characteristic leaching procedure are lower than hazardous constituent limits of *Identification Standards for Hazardous Waste-Identification for Extraction Toxicity (China GB5085.3–2007, similar to USA TCLP)* (Table 7.13). It is shown that the leaching residue belongs to the nonhazardous solid waste. Such residue may be used to make bricks, to use as materials for road construction or as stock feed for steelmaking plants, or just to be placed in landfill if available, depending on the heavy metal contents and economic and technical feasibility study case by case.

Chapter 8
Spent Electrolyte Regeneration and Recovery of Associated Valuable Metals from Lean Leaching Solution

Abstract The impurities in the electrowinning solution can be accumulated during the cycle of alkaline solutions. The integrated process of regeneration and purification of spent electrolyte by causticization is developed. The addition of solid NaOH can be used to separate Na_2CO_3 in the spent electrolyte. The Na_2CO_3 is causticized using CaO to regenerate NaOH. During this separation and causticization process, the impurity ions in the spent electrolyte are removed simultaneously. The causticized alkaline solution is recycled to the next leaching step. Moreover, Mo, W, Nb, Ta, etc., will be also dissolved in alkaline solution under the leaching conditions of zinc and lead hazardous wastes. The Mo and W can be recovered with ion flotation and adsorbing colloid flotation process using primary amine dodecylamine (RNH_2) as collector and ferric ions or magnesium chloride as co-precipitants or depressant (modifier). Adsorbing colloid flotation (ACF) and ion flotation (IF) using ferric ions as the co-precipitant or sorbent and sodium dodecyl sulfate as the collector can be used for the effective removal of arsenic and molybdenum from aqueous solution. The synergistic extraction of phosphorus, arsenic, and silica from tungstate and molybdate solutions with primary amines and TBP can be realized. Mixed solutions of 12-molybdophosphoric acid and its reduced molybdenum blue solutions are found to be an effective scrubbing process for the combined removal of H_2S, SO_2, and NO_x from gas streams. When the molar ratio of $NO_2/(SO_2 + H_2S)$ in the waste gases is just equal to 1/2, H_2S, SO_2, and NO_x will be removed quantitatively, while the regeneration of scrubbing solution is not needed. H_2S and SO_2 react with molybdophosphoric acid solution to form elemental sulfur and H_2SO_4, respectively, while molybdophosphoric acid is reduced into blue species, molybdenum blue solution. The reactions of H_2S are rather quickly, while those of SO_2 are much more slowly. The resulting molybdenum blue solution is used further for the removal of NO_x, in which molybdenum blue is rapidly oxidized back into molybdophosphoric acid and NO_x is reduced into N_2. Porous white tungstic acid (white tungstic acid), whose composition is firstly revealed in this work, is used to produce new tungstoniobates and tungstotantalates and a series of heteropoly compounds. Moreover, a quick determination method of tungsten concentration is proposed according to the quantitative relationship between the precipitate of white tungstic acid and pH value of a certain range.

Y. Zhao, C. Zhang, *Pollution Control and Resource Reuse for Alkaline Hydrometallurgy of Amphoteric Metal Hazardous Wastes*, Handbook of Environmental Engineering 18, DOI 10.1007/978-3-319-55158-6_8

Keywords Regeneration of spent electrolyte • Molybdenum • Tungsten • Adsorbing colloid flotation (ACF) and ion flotation (IF) ion • Solvent extraction • 12-Molybdophosphoric acid • Combined removal of H_2S • SO_2 and NO_x from gas streams • Porous white tungstic acid (white tungstic acid) • Tungstoniobate • Tungstotantalate • Heteropoly compounds

8.1 Regeneration of Alkaline Spent Electrolyte Causticized by CaO

An increase in sodium hydroxide consumption and impurities accumulation will appear following the adoption of this technique of sodium hydroxide zinc metallurgy in the disposal of lean zinc-bearing ores and wastes, especially when the zinc grade is below 10%. In addition, it is found that arsenic impurities seriously affect the electrolysis of metallic zinc powders in alkaline solutions. Thus, reducing the NaOH requirements and removing impurity ions are important steps in the improvement of this technology.

Because the main forms of the lean zinc-bearing oxidized ores and wastes are ZnO, $ZnCO_3$, $ZnSiO_3$, and $ZnSO_4$, the spent electrolyte is, in fact, a mixed sodium carbonate (sodium sulfate, sodium silicate)/sodium hydroxide solution. The conversion of sodium carbonate into sodium hydroxide using lime, a process known as causticization, is one of the oldest processes used in the pulp and paper industry.

The solubility of Na_2CO_3 was observed to decrease as the NaOH concentration increased at 25 and 90 °C. When the NaOH concentration reached 350 g/L, the solubility of Na_2CO_3 decreased to less than 40 g/L. Thus, Na_2CO_3 may be separated from the alkaline solution by increasing the concentration of NaOH.

The effects of the initial NaOH and Na_2CO_3 concentration, CaO/ Na_2CO_3 weight ratios, and temperature on the NaOH regeneration in the causticization reaction were studied. The NaOH regeneration sharply decreased with increasing initial NaOH concentration. When the concentration reached 120 g/L, the value of the NaOH regeneration became negative. Meanwhile, the causticized alkaline solution became turbid, and the amount of caustic residue was observed to increase. The alkaline reduction may be caused by the loss of NaOH relating mainly to the leakage of the solution in the solid-liquid separation and the entrainment loss into the causticization precipitation, which had accrued in the high-concentration alkaline solution. The other reason may be that the NaOH, Na_2CO_3, $CaCO_3$, and $Ca(OH)_2$ had formed coprecipitations. However, caustic entrainment most likely existed in the Na_2CO_3 separated from the spent electrolyte. The excessive dilution will lead to a water imbalance in the sodium hydroxide zinc metallurgy process. Considering the above factors, the initial NaOH concentration of the causticization solution should be controlled in the range of 80–100 g/L.

The integrated process of regeneration and purification of spent electrolyte from sodium hydroxide zinc metallurgy by causticization is acted as a two-step process requiring a separation step and a causticization step. The first separation unit is a crystallization and precipitation unit for the separation of sodium carbonate from the spent electrolyte achieved by adding solid NaOH to make the alkaline concentration increase to over 350 g/L. After the mixture stands for hours, a solid-liquid separation is performed followed by the addition of a previously given amount of washing water from the leaching slag or zinc powder to prepare the causticization solution. In the second causticization stage, sodium carbonate, sodium silicate, and sodium sulfate are causticized using direct causticization with CaO to regenerate NaOH. The causticized alkaline solution mixing with the supernatant liquor of Na_2CO_3 separation is recycled to the next alkaline leaching operation in the process of sodium hydroxide zinc metallurgy. Simultaneously, the spent electrolyte is purified through the above operation because a portion of the impurity ions are precipitated into the caustic residue.

The spent electrolyte was purified and the impurity ions were partly wiped off in this process. The removal rates of, for instance, Fe, Cu, Cd, Cr, Mg, and Mn were 10–60%. In particular, the removal rate of As, an impurity with seriously detrimental effects on the electrolysis of metallic zinc powder in the alkaline solution, reached 62%. The spent electrolyte became clear and tawny, turning cloudy and gray-black after the scaled-up experiment. Furthermore, the main composition of the caustic residue is $CaCO_3$ alongside small black particles existing within the white caustic residue.

The raw materials that can be used in the process of sodium hydroxide zinc metallurgy include low-grade oxidized zinc ores and solid wastes containing zinc. The main compositions of these raw materials are ZnO, $ZnCO_3$, $ZnSiO_3$, and $ZnSO_4$. In the leaching and electrowinning operation, the following reactions occurred:

$$ZnO + 2NaOH = Na_2ZnO_2 + H_2O \tag{8.1}$$

$$ZnCO_3 + 4NaOH = Na_2ZnO_2 + Na_2CO_3 + H_2O \tag{8.2}$$

$$ZnSO_4 + 4NaOH = Na_2ZnO_2 + Na_2SO_4 + H_2O \tag{8.3}$$

$$ZnSiO_4 + 4NaOH = Na_2ZnO_2 + Na_2SiO_4 + H_2O \tag{8.4}$$

$$Na_2ZnO_2 + H_2O = Zn \downarrow + 2NaOH + \frac{1}{2}O_2 \tag{8.5}$$

When the composition of zinc in the raw materials is ZnO, two NaOH molecules are consumed in the alkaline leaching process, and the same numbers of NaOH molecules are regenerated in the electrowinning section. Thus, the theoretical alkali consumption of the process is zero. However, if $ZnCO_3$, $ZnSiO_3$, and $ZnSO_4$ are the main composition of the raw materials, four NaOH molecules are used, and only two are produced. The theoretical alkali consumption is 1.23 g NaOH/g Zn in this case.

In conventional chemical recovery, Na_2CO_3, Na_2SO_4, and Na_2SiO_3 are causticized with lime to form NaOH and a lime mud ($CaCO_3$), as shown in Eqs. (8.6, 8.7, and 8.8).

$$Na_2CO_3+CaO + H_2O = 2NaOH + CaCO_3 \tag{8.6}$$

$$Na_2SiO_4+CaO + H_2O = 2NaOH + CaSiO_4 \tag{8.7}$$

$$Na_2SO_4+CaO + H_2O = 2NaOH + CaSO_4 \tag{8.8}$$

From Eqs. (8.6, 8.7, and 8.8), one CO_3^{2-} group can regenerate two NaOH molecules. Thus, through the causticization process, the theoretical alkali consumption in the process of sodium hydroxide zinc metallurgy reduces to zero, regardless of the zinc composition in the raw materials. The conversion of Na_2CO_3 using lime is necessary to reduce the loss of NaOH in sodium hydroxide zinc metallurgy.

In addition to the regeneration of NaOH, the purification of the spent electrolyte is also realized by the process of regeneration and purification of spent electrolyte. In fact, the first sodium carbonate separation unit is the crystallization of sodium carbonate by the addition of solid NaOH to yield sodium carbonate supersaturated in the spent electrolyte. The supersaturation of Na_2CO_3 in the strong alkaline solution is an inherent impetus of crystal formation and growth. The precipitated Na_2CO_3 is an amorphous sediment with a latex-like appearance. During the crystallization of sodium carbonate, the impurity ions in the spent electrolyte, such as Fe, Cu, As, Mn, Mg, Cr, and Cd, also underwent a certain amount of precipitation. There may be two reasons for the precipitation of the impurity ions. The first reason is that the solubility of the impurity ions, particularly Fe, Cd, Mg, and Mn, decreases with increasing NaOH concentration. The high concentration of alkaline leads to supersaturation and precipitation of these impurity ions. However, the main reason is that the impurity ions are deposited via adsorption, coacervation, coprecipitation, and net-capture during the crystallization of sodium carbonate because the concentrations of impurity ions in the spent electrolyte are all milligram grade. In the end, the precipitation of the impurity ions enters the solid phase of the sodium carbonate separation unit and dissolves into the causticization solution.

In the second causticization unit, the direct causticization is achieved by the reaction of $Ca(OH)_2$ with sodium carbonate, sodium sulfate, and sodium silicate, forming NaOH and calcium carbonate, calcium sulfate, and calcium metasilicate precipitates. During this causticization operation, the removal of impurity ions, especially the removal of As, was observed by comparing the impurity ion concentrations in the causticization solution and causticized alkaline solution. The As removal may be caused by $Ca(OH)_2$ reacting with Na_3AsO_4 and Na_3AsO_3 to form $Ca_3(AsO_4)_2$ and $Ca_3(AsO_3)_2$ sediments, as shown in Eqs. (8.9 and 8.10):

$$3Ca(OH)_2+2Na_3AsO_4= Ca_3(AsO_4)_2 \downarrow +6NaOH \tag{8.9}$$

$$3Ca(OH)_2+2Na_3AsO_3= Ca_3(AsO_3)_2 \downarrow +6NaOH \tag{8.10}$$

At the same time, the other impurity ions formed insoluble metal hydroxides, carbonates, or subcarbonates and are removed into the caustic residue via coprecipitation, adsorption, coacervation, and net-capture.

Considering that the NaOH solution is constantly recycled in the sodium hydroxide zinc metallurgy process, and the chemical composition of the raw ore is notably complex, impurity ions such as As, Al, Cu, and Fe will accumulate in the recycled solution. These impurities will affect the quality of the metallic zinc powder. Thus, the impurity ions are partly removed, and the concentrations of the impurity ions in the recycled solution become stable in this process.

In summary, the addition of solid NaOH can be used to separate Na_2CO_3 in the spent electrolyte. The Na_2CO_3 is causticized using CaO to regenerate NaOH. During this separation and causticization process, the impurity ions in the spent electrolyte are removed simultaneously. The causticized alkaline solution is recycled to the next leaching step. The optimum causticization conditions were found to be an initial NaOH concentration of 80–100 g/L, 1.0 CaO/Na_2CO_3 weight ratio, 90 °C, and an initial Na_2CO_3 concentration above 40 g/L.

8.2 Flotation of Molybdate Oxyanions in Dilute Solutions Using Dodecylamine and Ferric Hydroxide

The metals (including molybdenum) are nonrenewable resources being consumed at an exponential rate as a result of population and capital growth. Thus, reclamation from industrial solid wastes and wastewaters, especially the rare or rare-earth metals, is becoming more and more attractive. Mo, W, and As are associated usually as an admixture with zinc and lead. The wastewater generated during hydrometallurgical treatment of these minerals contains generally both arsenic (V) and molybdenum(VI) as well as tungsten species.

The separation of arsenic from molybdate-containing waste streams and the subsequent recovery of molybdenum are apparently highly desirable in terms of resource reclamation, health, and environmental reasons. In view of the efficient operation of flotation for the treatment of industrial wastewaters, it seemed worthwhile to explore the possibility of using ion flotation for the quantitative removal and recovery of molybdenum as a saleable product from lean solutions or the separation of low concentrations of molybdenum.

Dodecylamine (RNH_2) in ethanol solution was used as collector for ion flotation, while RNH_2 and ferric hydroxide (formed in situ from ferric chloride) were used, respectively, as collector and co-precipitant for adsorbing colloid flotation. The recovery of molybdenum was strongly dependent on the pH value of solution (Fig. 8.1). The maximum recovery of molybdenum was obtained within the pH range 2.5–4.5, which was selected for the following experiments. At these conditions a white precipitate formed on the walls of glass column and no foam was produced on top of the feed solution. On the other hand, in separate experiments when arsenic was examined, it was found that it could not be removed from solution containing only arsenates at pH 2–6.

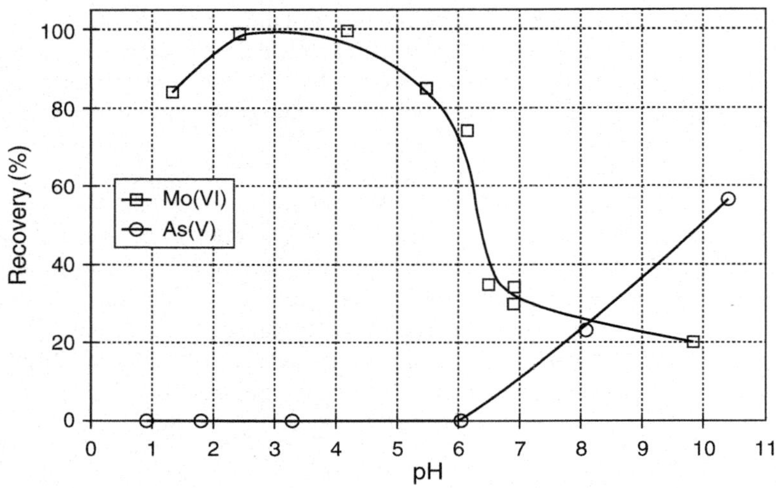

Fig. 8.1 Dependence of recovery of Mo(VI) and As(V) from aqueous solutions on pH by applying ion flotation

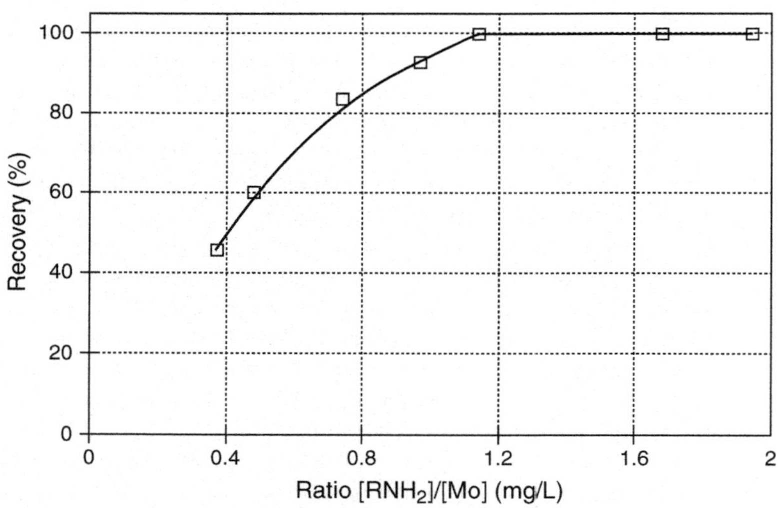

Fig. 8.2 Relationship between recovery of Mo(VI) from solutions by ion flotation and RNH_2/Mo concentration ratio

The recovery of molybdenum from aqueous solution increased almost linearly with the increase of dodecylamine content, and over 99% recovery was obtained when the RNH_2/Mo ratio exceeded 1.2 (Fig. 8.2). The addition of Cl^- or SO_4^{2-} ions was found to have little effect on the recovery of molybdenum from aqueous solutions by ion flotation, if their concentrations were lower than 10 g/L and 4 g/L,

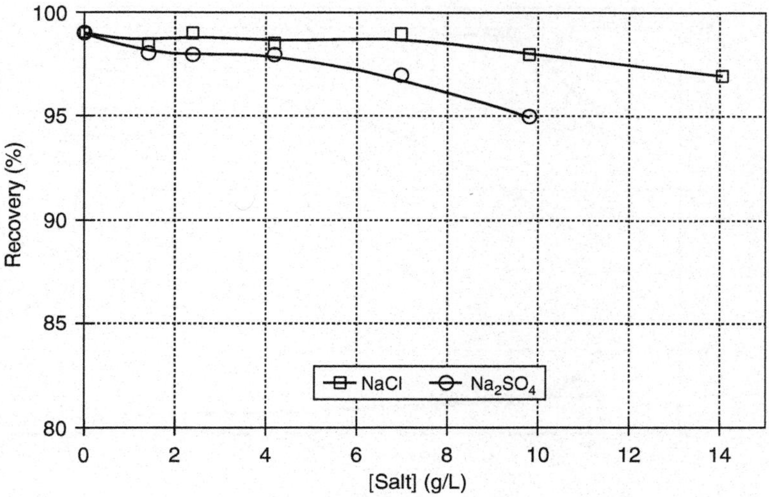

Fig. 8.3 Effects of NaCl and Na_2SO_4 on recovery of Mo(VI) by ion flotation from solutions

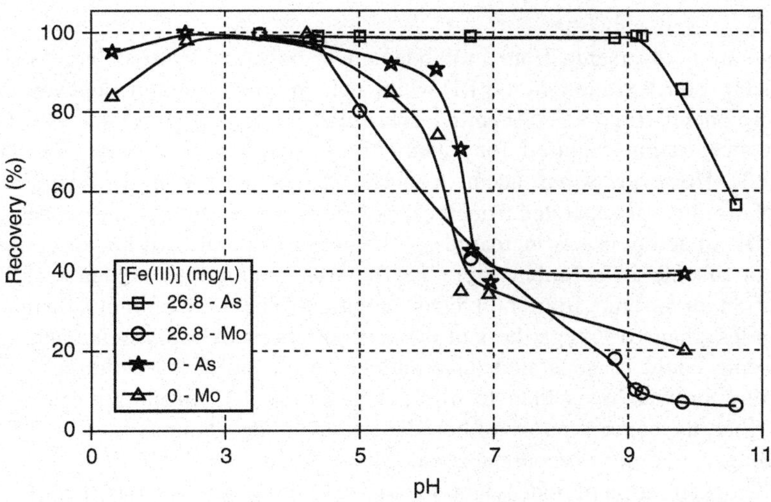

Fig. 8.4 Dependence of separation of arsenate anions from molybdate-containing solutions on pH with and without the presence of Fe(III)

respectively (Fig. 8.3). Since in most hydrometallurgical processes for molybdenum only HCl is used, the negative effects of SO_4^{2-} ions on flotation of molybdates could not be expected in actual industrial wastewater. If ion flotation is going to be applied for the treatment of leachates resulting from processing of molybdenum solid wastes, the use of sulfuric acid as a leaching agent could be simply avoided.

Figure 8.4 shows that Fe(III) existed in the solution can result in higher removal rate of arsenate ions, which closely reach to 100% at the pH of 5–9. The efficiency

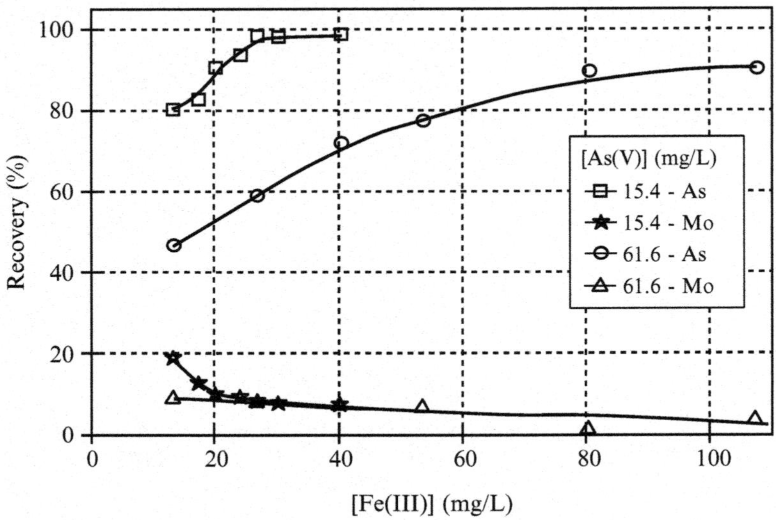

Fig. 8.5 Effects of Fe(III) concentration on the separation by adsorbing colloid flotation of As (V) from Mo(VI) solutions

of separation of arsenic from molybdenum in aqueous solution increased with increasing concentration of Fe(III),while the optimum amount depends on the arsenic content (Fig. 8.5). At initial concentration of arsenic 15.4 mg/L, the Fe (III) concentration required for quantitative separation of arsenic was around 27 mg/L. However, when arsenic concentration was 61.6 mg/L, no more than 90% of arsenic was separated from the molybdenum, even though the concentration of Fe(III) in solution was increased to 110 mg/L. Generally, at higher concentrations of arsenic in solution, more Fe(III) was required to obtain a sufficient separation of arsenic from molybdate solution. The loss of molybdenum was always less than 10%, regardless of the concentrations of both arsenic and Fe(III).

Arsenic could be separated quantitatively from molybdate solution when its concentration was lower than 15 mg/L (Fig. 8.6). If its concentration in solution was higher, the concentrations of both Fe(III) and dodecylamine must be increased accordingly, in order to separate arsenic (Figs. 8.6 and 8.7).

The concentration of dodecylamine required for maximum separation of arsenic from molybdate solutions also depends on the initial concentration of arsenic (Fig. 8.7). When the concentration of arsenic was 15.4 mg/L (with 26.8 mg/L Fe (III)), 354 mg/L of dodecylamine was required for quantitative separation of arsenic. Dodecylamine concentration of 700 mg/L was necessary to obtain the same separation efficiency, when the arsenic content was 61.6 mg/L (with 80.4 mg/L Fe(III)), resulting in increased loss of molybdenum. It was found by experiment that the least molar ratio of Fe(III)/As and RNH_2/As required for maximum separation of arsenic from molybdate solution increases (although not proportionally) with the increase of concentration of arsenic in this solution. The optimum concentrations of Fe(III) and RNH_2 vary with the initial arsenic

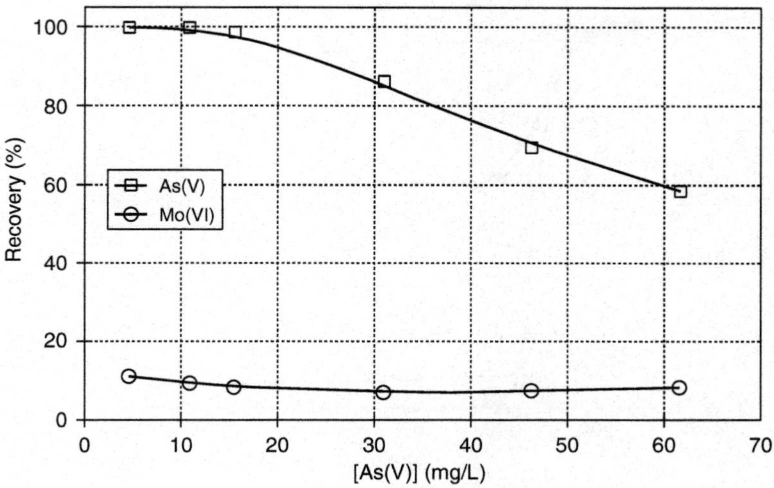

Fig. 8.6 Effects of As(V) content on the separation of As(V) from Mo(VI) solution by adsorbing colloid flotation

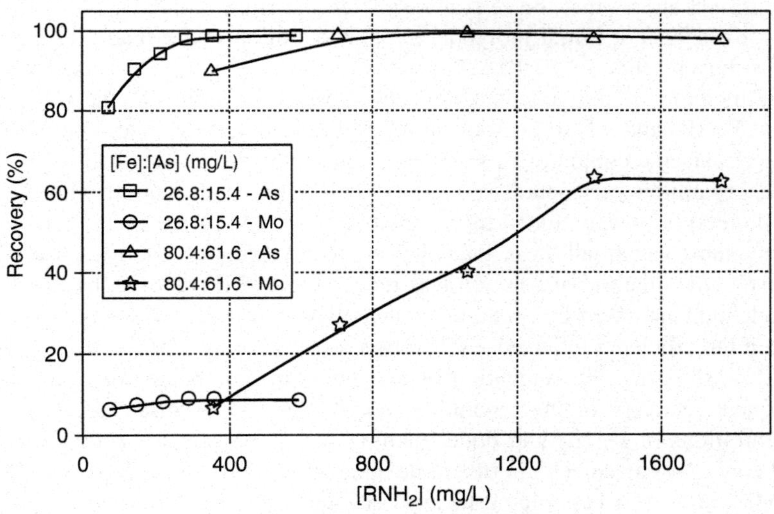

Fig. 8.7 Effects of RNH_2 content on the separation of As(V) from Mo(VI) solutions by adsorbing colloid flotation

concentration in the solution. Practically, somewhat higher Fe(III)/As molar ratio was preferable to a higher RNH_2/As molar ratio since the increase of dodecylamine concentration leads to higher loss of molybdenum, while an increase of Fe(III)/As molar ratio reduces this loss.

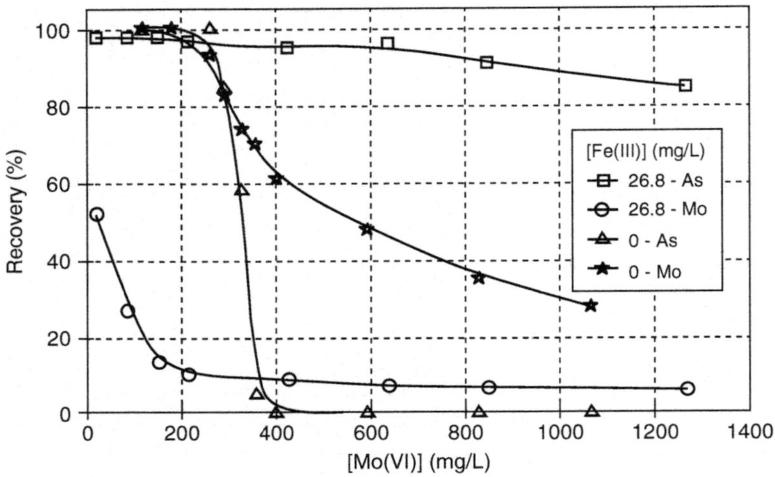

Fig. 8.8 Effects of Mo(VI) concentration on the separation of As(V) from Mo(VI) solutions by adsorbing colloid flotation

Figure 8.8 shows that the separation of arsenic from molybdate solutions was always over 97% at initial molybdenum concentration lower than 600 mg/L but only in the presence of Fe(III). The separation was slightly decreased with the further increase of molybdenum concentration in solution. On the other hand, without Fe(III) and when the concentration of molybdenum in solution exceeded 400 mg/L, arsenic could not be separated at all.

The separation of arsenic from molybdate solution decreased when sodium chloride or sulfate were added to the solution (Fig. 8.9). Nevertheless, the loss of molybdenum was small (less than 3–5%). The separation of arsenic could be improved when the pH of solution was reduced to 6–7 (Fig. 8.10), but the loss of molybdenum increased considerably. Obviously, a compromise should be made between the efficiency of arsenic separation and loss of molybdenum.

A flow sheet for the separation of arsenic from molybdate solutions and the subsequent recovery of the remaining molybdates can be proposed (Fig. 8.11). Concentrations of Fe(III) and dodecylamine should be selected according to the initial content of arsenic(V) in the dilute aqueous solution of molybdates.

Dodecylamine, a cationic surfactant, is often used for ion flotation in acidic media. In the present work, it has been found that dodecylamine could also be used for separation of arsenic from molybdate solutions containing Fe(III), even at pH from 9 to 9.2. A quaternary ammonium salt, such as cetyltrimethylammonium chloride (CTMA-Cl), can also be used for the separation of arsenic, but it is much less effective, giving 60–65% As removal, compared with 80–90% obtained with RNH_2. When the concentration of CTMA-Cl exceeds 134 mg/L, large amounts of foam are generated and overflow the top of the flotation column; hence, the concentration factor is reduced due to the substantial co-removal of the solution. For the same reason, ethanol (used alone), Aerofroth (a commercial

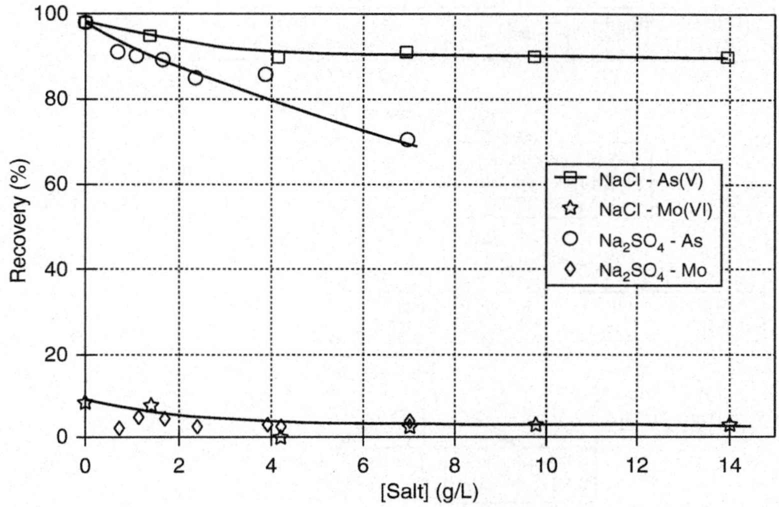

Fig. 8.9 Effects of NaCl or Na_2SO_4 addition on the separation of As(V) from Mo(VI) solutions by adsorbing colloid flotation

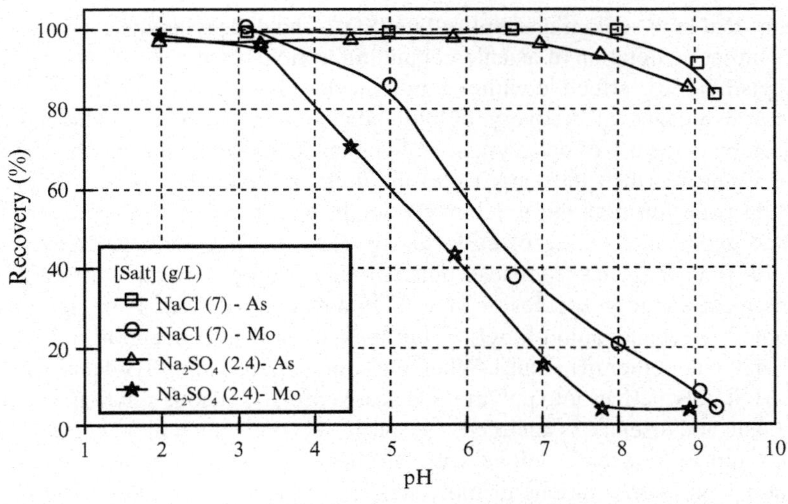

Fig. 8.10 Effects of solution pH on the separation of As(V) from Mo(VI) solutions in the presence of NaCl (7 g/L) and Na_2SO_4 (2.4 g/L)

frother), or sodium dodecyl sulfate (a common anionic surfactant) were not found effective when tested at the same pH value of 9. It was found that quantitative separation (and removal) of arsenic from molybdate solution containing Fe(III) can be always obtained with dodecylamine at pH of this solution ranging from 2 to 9.2.

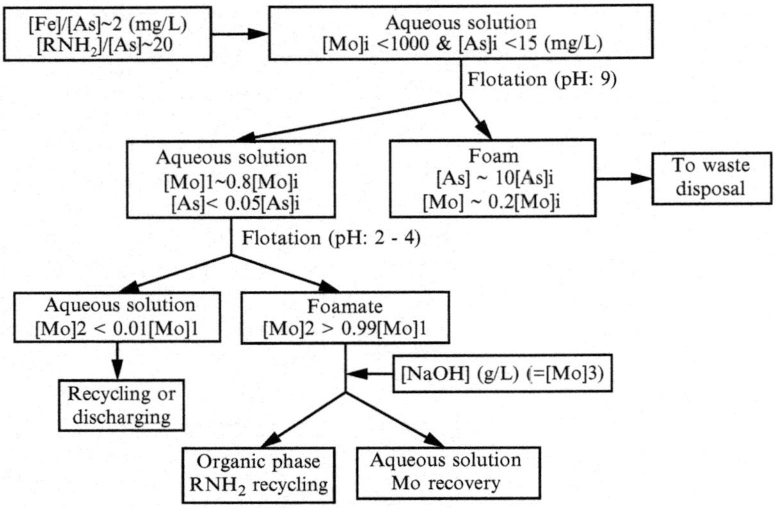

Fig. 8.11 Proposed flow sheet for the separation of As/Mo mixture of anions in aqueous solution and the subsequent recovery of molybdenum

It seems that both sodium dodecyl sulfate (SDS) and dodecylamine can be used for the quantitative flotation of arsenic-containing ferric flocs at pH 4–5. Nevertheless, dodecylamine was effective within a much wider pH range of 2–9.2.

The advantages of recovery of molybdates from aqueous solutions by ion flotation lie in the use of dodecylamine. In practice, dodecylamine can be dissolved in hot water and then introduced into the flotation unit. The volume of foamate resulting from ion flotation is relatively small, which makes further treatment of foamate and the recycling of collector simpler. It was found that recycling of collector from foamates from adsorbing colloid flotation is much more difficult.

The most common species of Mo(VI) in aqueous solutions is the tetrahedral, monomer molybdate anion MoO_4^{2-}, but it exists only at concentrations less than 10^{-4} M and at higher pH values. When the concentration of molybdenum exceeds 10^{-4} M, it polymerizes into polymolybdates. When an aqueous solution containing molybdate and arsenate is acidified, the species become much more complex since various molybdo-arsenic anions with different Mo/As molar ratio are formed depending on concentrations of molybdenum and arsenic, as well as on the pH. Molybdo-arsenic acid, $H_3AsMo_{12}O_{40}$, and its anion are the most stable in aqueous solution of pH less than 3 and have been studied extensively because of their importance for colorimetric determination of arsenic(V).

Both arsenic and molybdenum anions could be removed from solutions containing no Fe(III) by applying ion flotation, and the removal of arsenic was linearly related to that of molybdenum. Therefore, arsenic presumably floats as molybdo-arsenic anions. At lower pH the main arsenate species seems to be H_2AsO_4.

Recycling and subsequent reuse of the primary amine (RNH_2) during the recovery of molybdenum from lean solution by ion flotation was confirmed to be possible and practical, which can make the process cost-effective. In conclusion, adsorbing colloid flotation (with ferric ions as co-precipitants and dodecylamine as collector) can be used for the selective separation of arsenate anions from aqueous solutions containing molybdates at pH around 9.

8.3 Separation of Tungstates from Leaching Solution Using Ion Flotation

W in the zinc and lead wastes will be dissolved in alkaline solution. Although tungsten is considered as a relatively rare metal on the earth and its resources are rather limited, this metal and its compounds have found wide applications in the manufacture of steel alloys and in catalysts, pigments, dyestuffs, and electronics. All metals, including tungsten, are nonrenewable resources in general, but at present they are being consumed at an exponential rate. Therefore, the reclamation of metals, especially, the rare ones, from industrial solid wastes or from waste waters, is becoming increasingly attractive.

Nearly all tungstate compounds are currently produced by hydrometallurgical processes. The concentrates containing tungsten, which are produced mainly by the application of froth flotation, are subsequently pressure-leached by sodium or ammonium hydroxide solutions. Undissolved solids are discharged (after neutralization) as wastes, while the aqueous solutions produced are purified by removing initially the high content of silicates by partial acidification of the produced liquor at pH 8–9. The removal of arsenates and phosphates, as well as of the remaining silicates, is followed by precipitation of the respective insoluble compounds, after addition of magnesium chloride. The purified solutions, which are considered free from the presence of impurities, are further concentrated, and the products of sodium or ammonium tungstate are finally obtained (usually by crystallization). During the precipitation stage, a toxic solid waste containing the removed impurities (arsenates, phosphates, and silicates among others), as well as some tungsten, is generated.

Wastewaters produced during the mineral processing of tungsten ores and zinc and lead wastes or after the leaching of the resulting precipitates may contain 100–1000 mg/L W and 5–40 mg/L of As, P, or Si, although in certain cases the content of silicates may be rather higher. Moreover, tungsten or tungstate-laden spent catalysts, used pigments, dyestuffs, and other industrial products, generate annually large quantities of both solid wastes and waste waters. The solid wastes should be properly detoxified, which is generally realized by redissolving (leaching) the contained tungstate and other toxic metals and then removing them (preferably selectively) from the obtained solution. On the other hand, wastewaters should also be treated properly in order to meet environmental constraints, by recovering the valuable tungstate anions and removing the other toxic components (mainly arsenic).

The separation either of tungstates or of the other impurities from liquid waste streams can be carried out by applying several processes, such as precipitation with various precipitants, solvent extraction (with amines, tributyl phosphate, long-chain ketones), and ion exchange. However, when these methods are used, the arsenate, phosphate, and silicate content of the aqueous solutions is usually co-removed with tungsten, resulting in poor selectivity and economic impracticability. As the concentrations of tungstates in these solutions is usually relatively low (lower than 1 g/L), it seems that novel, cost-effective, and environmentally sound methods should be developed to recover tungsten selectively.

Various flotation methods (ion, precipitate, and adsorbing colloid flotations) have been studied, with the aim of effectively separating metallic ions from waters and wastewaters. Surfactants, which have been used for tungsten recovery by flotation, include colophony amine hydrochloride, cetylamine, and hexadecyl-trimethyl ammonium bromide. Owing to the efficient operation of flotation process for the treatment of industrial wastewaters, it was considered worthwhile to examine the possibility of using a common flotation technique for the selective separation of tungsten from aqueous mixtures containing the most commonly found impurities, such as arsenates, phosphates, and silicates.

Tungstate-, arsenate-, phosphate-, and silicate-bearing stock solutions were prepared by dissolving predetermined amounts of $Na_3PO_4{\cdot}2H_2O$, $Na_2HAsO_4{\cdot}7H_2O$, $Na_3PO_4{\cdot}12H_2O$, and $Na_2SiO_3{\cdot}9H_2O$ in deionized water. In the ion flotation experiments, which aimed to separate selectively tungstate from aqueous mixtures containing arsenates, feed was prepared by mixing a given volume of tungstate and arsenate solutions (and other components, when appropriate), followed after dilution in a beaker, by the addition of magnesium or ferric chloride and finally of ethanolic dodecylamine solution, while stirring with a magnetic stirrer. For comparison reasons, the content of ethanol in the feed solution was always kept at 0.5% (v/v). A total volume of 100 mL solution was used for all experiments. The feed solution was stirred until the pH value remained stable at the set value (after about 10–15 min), and then it was introduced into the flotation column. At optimum conditions a white precipitate was observed to attach to the glass walls of the flotation micro-cell. When magnesium chloride was used as depressant, the foam production was relatively small. An air flow rate of 200 mL/min and a 10 min flotation time were found to be necessary for equilibrium conditions. The feed was originally turbid but became clear while floating. After the necessary flotation time, gas was turned off and an aliquot of the remaining solution was taken from the lower part of the column, in order to determine the remaining concentrations of the metals.

The feed solution to be tested was prepared by simulating the actual concentrations of the wastewaters or leaching solutions of solid wastes containing tungsten. The initial concentration of each element in the feed was (unless otherwise stated) 184 mg/L W (1×10^{-3} M); 15 mg/L As(2×10^{-4} M); 15 mg/L P (5×10^{-4} M); 15 mg/L Si (5.3×10^{-4} M); 354 mg/L RNH_2 (2×10^{-3} M), i.e., stoichiometric for divalent WO_4^{2-}; 171 mg/L M Mg (7×10^{-3} M); and 53 mg/L Fe(III) (around 1×10^{-3} M). The pH of the solution to be treated was, in most experiments, kept

Table 8.1 Linear relationship between absorbance (A) and concentration of arsenic or phosphorus in the presence of increased tungsten concentrations, applying the hydrazine-reduced molybdenum blue colorimetric method (wavelength: 843 nm)

W content mg/L	As (mg/L) = a + b·A	R^2
0.00	As = −0.0210 + 2.912A	0.9999
1.48	As = 0.0263 + 2.757A	0.9995
2.96	As = 0.0562 + 2.833A	0.9998
4.40	As = 0.0861 + 2.877A	0.9995
6.00	As = 0.1400 + 2.896A	0.9996
7.20	As = 0.1844 + 2.927A	0.9997
	a = −0.0228 + 0.0275 W	0.9966
	b = 2.7340 + 0.2785 W	0.9678
	P(mg/L) = a′ + b′·A	
0.00	P = 0.0010 + 1.291A	0.9999
1.48	P = 0.0230 + 1.231A	0.9999
2.96	P = 0.0648 + 1.232A	0.9999
4.40	P = 0.0912 + 1.245A	0.9999
6.00	P = 0.1240 + 1.268A	0.9999
	a = −0.0056 + 0.0219 W	0.9959
	b = 1.2130 + 0.0084 W	0.9344

constant at 3.8. The percentage removal (Re%) of each component from the solution was calculated in the usual way.

Tungsten was analyzed, employing the $SnCl_2$-NHSCN-HCl colorimetric method. Colorimetric analytical determinations for phosphorus, arsenic, and silicon were also carried out according to standard procedures. The interferences caused by the tungsten content during the arsenic and phosphorus analytical determinations were eliminated as follows: to a 25 mL flask containing 0–70 μg of arsenic or 0–30 μg of phosphorus were added 0–184 μg W, as well as the other required chemical reagents, and the absorbance was measured at 843 nm. It was found that for a given concentration of P or As, the absorbance decreased linearly with the increase in the tungsten concentration. Linear correlation expressions were calculated and the results are presented in Table 8.1. After performing the tungsten determination in the solutions, the resultant value was inserted into the expressions given in Table 8.1. Therefore, the concentrations of arsenic or phosphorus could be also obtained. No interference as a result of the tungsten content was detected in the analytical determination of silicate when applying the standard ascorbic acid-reduced, molybdenum blue method.

The separation and recovery of tungstate from arsenate solutions was found to be strongly dependent on the pH value of the solutions, as well as on the type of depressant or co-precipitant used (Fig. 8.12). It was shown that addition of ferric chloride, a common flocculant reagent, which is also acting as precipitant, led to poor flotation separation, as over 95% of both arsenate and tungstate would be simultaneously removed from aqueous solutions when the pH value was lower than 6 and the Fe(III) concentration was higher than 50 mg/L. Further increase of ferric ion concentration at pH = 9 did not substantially affect the results of co-removal (Fig. 8.13).

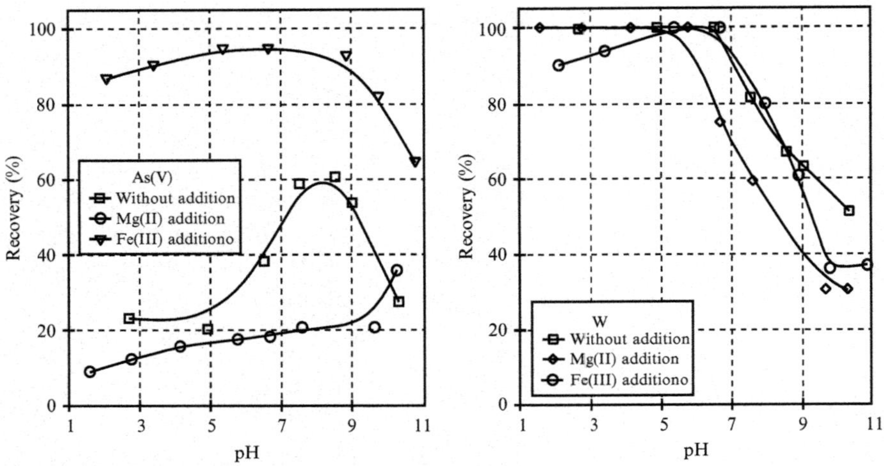

Fig. 8.12 Flotation recovery of tungstate from aqueous mixtures with arsenates by using dodecylamine: effects of pH in the presence and absence of ferric or magnesium chloride

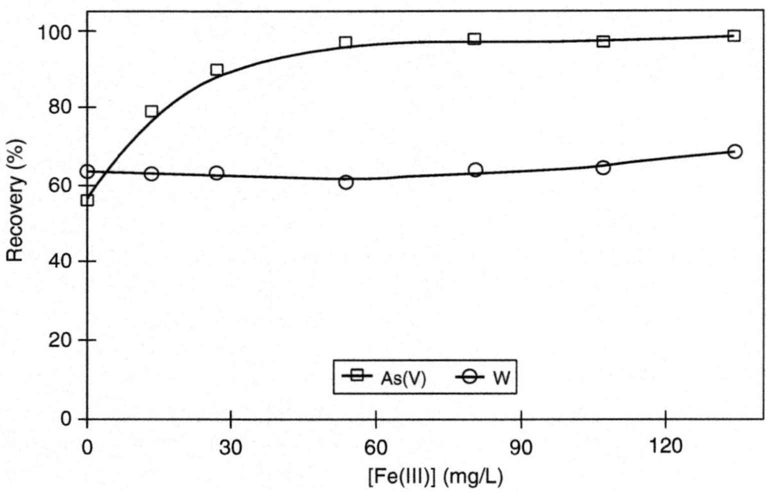

Fig. 8.13 Effects of ferric ions concentrations on the separation of tungstate from arsenate anions in aqueous mixtures at pH = 9

However, in cases other ions were not present, tungstates could be recovered quantitatively from aqueous mixtures with arsenates at pH values lower than 6 (Fig. 8.12). Simultaneously, the co-removal of arsenate was found to be around 25%, particularly when pH values were kept below 5. Further, when magnesium chloride was added to the solution, the co-removal of arsenate would be further depressed, and values lower than 17% were found in this case at pH values of the solution around 4. Therefore, it was decided to use magnesium ions as a flotation

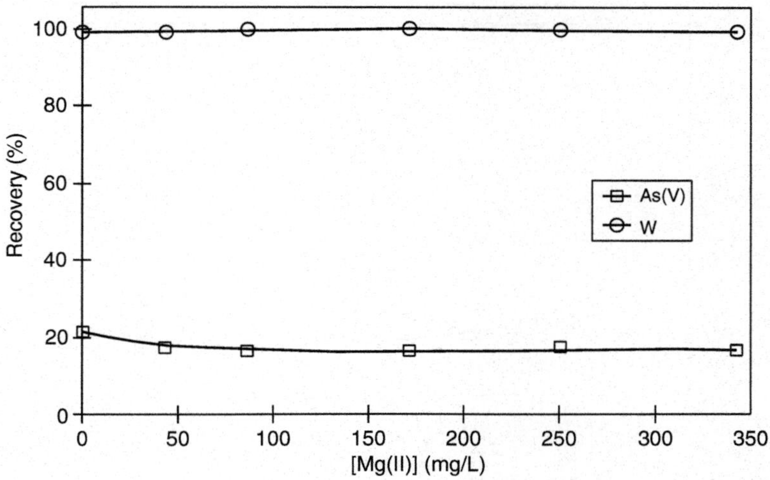

Fig. 8.14 Effects of magnesium ion addition on the flotation separation of tungstate from arsenate anions

modifying agent (depressant) at pH 3.8 in the following work, taking into consideration also that they may already exist in the actual leaching solutions of solid wastes which originated from the hydrometallurgical processing of tungsten-containing minerals.

The separation efficiency between tungstate and arsenate anions in aqueous solutions increased only slightly, when Mg(II) concentrations were increased up to 150 mg/L, decreasing the co-removal of arsenates (Fig. 8.14), while the quantitative recovery of tungstates was found to be unchanged (at pH 3.8). It should be stressed also that the effect of magnesium was more pronounced in subsequent experiments, when the presence of other impurities (phosphates, silicates) was also examined.

Dodecylamine, the surfactant used in this process, was acting as flotation collector for tungstates, influencing only slightly its separation. The required concentration for the maximum separation of tungstate and for the simultaneous lowest co-removal of arsenates was found to be about 350 mg/L, while the recovery of tungstates was always quantitative, even when as low as 60 mg/L of dodecylamine was used (Fig. 8.15).

The recovery of tungstate was found to be quantitative, irrespective of the initial arsenic concentration, while the percentage co-removal of arsenates was decreased as their initial concentration in solution was increased (Fig. 8.16). When the initial concentration of arsenic was kept at 15 mg/L (the concentration ratio of [W]/[As] was equal to 12 (in mg/L)), around 17% of arsenic was found to be co-removed with tungsten, i.e., the amount of arsenic co-floated was about 2.5 mg/L, which means that the [W]/[As] ratio in the foam was increased to 80.

It was proposed that the foam product could be redissolved and that the resultant aqueous solution could be further separated by the aforementioned flotation

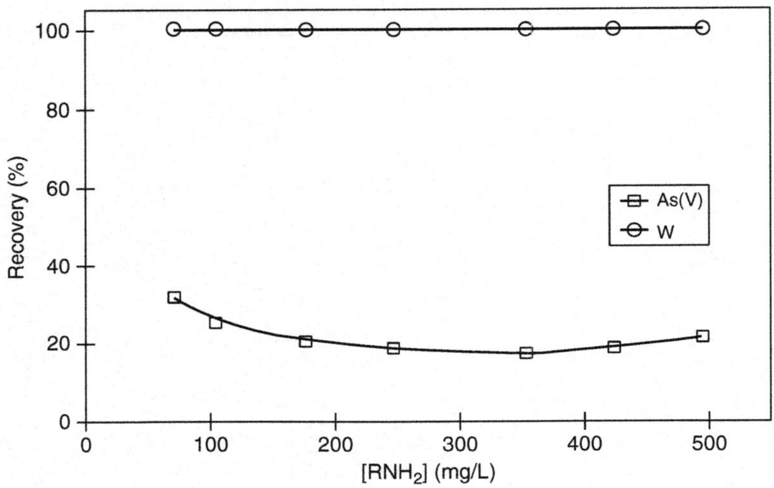

Fig. 8.15 Dependence of the separation of tungstate from arsenates on the concentration of dodecylamine

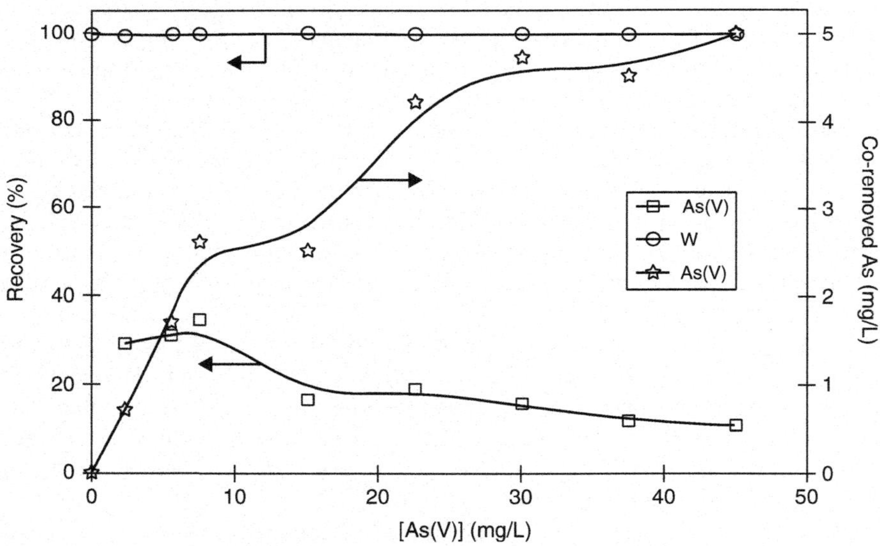

Fig. 8.16 Effects of arsenate concentration on its flotation separation from tungstate by dodecylamine (in the presence of Mg^{2+})

technique in a second stage. From Fig. 8.16, the amount of co-removed arsenic would be decreased to 0.7 mg/L when the initial concentration of arsenic is 2.5 mg/L. Consequently, the final (total) co-removal of arsenic would be around 5% of its initial concentration, after two subsequent flotation stages, and the

[W]/[As] ratio in the final foam product can be further increased up to 250. In this case, the concentration of the remaining arsenate, coexisting with tungstate, can be considered as acceptable, and tungsten would be recovered by subsequent processes, such as crystallization.

Around 90% of the primary amine could be recovered from the organic phase resulting from the dissolution of foam product with alkaline solutions, and the resultant aqueous solution can be used for the second flotation cycle by the same procedure as in the first flotation cycle. After the second re-dissolution of the foam product, 90% of the primary amine could be also recovered. Consequently, the total loss of primary amine in the two dissolution procedures may be estimated to be around 20%. In comparison with the common flotation processes, where the surfactants are usually not recovered, the total loss of primary amine was considered to be acceptable; additionally, the profit from tungsten recovery would also make up for some of the loss.

The co-removal of arsenates was increased considerably when the initial concentration of tungstates was increased to values higher than 200 mg/L (Fig. 8.17). Apparently, the flotation procedure developed is not suitable for the separation and recovery of tungstate from arsenate anions in aqueous mixtures containing relatively high tungsten concentrations. The recovery of tungstates was always quantitative when its initial concentration was lower than 450 mg/L at molar ratio of $[W]/[RNH_2]$ around 1:1.

The tungsten-bearing wastewaters, as well as the hydrometallurgical leaching liquors, may also contain impurities other than arsenates, such as phosphates and/or silicates, which should be separated, exactly as in the case of arsenate. The co-removal of arsenate in the foam product (or ion flotation concentrate) was

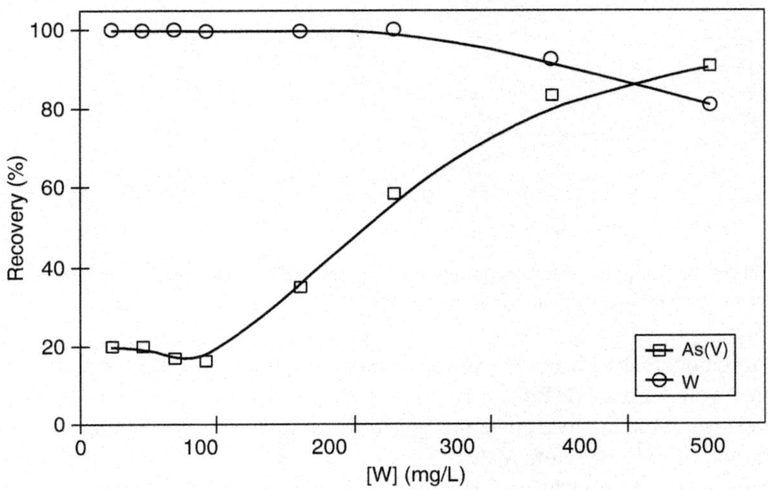

Fig. 8.17 Effects of tungstate concentration on its separation from arsenate in aqueous mixtures (in the presence of Mg^{2+})

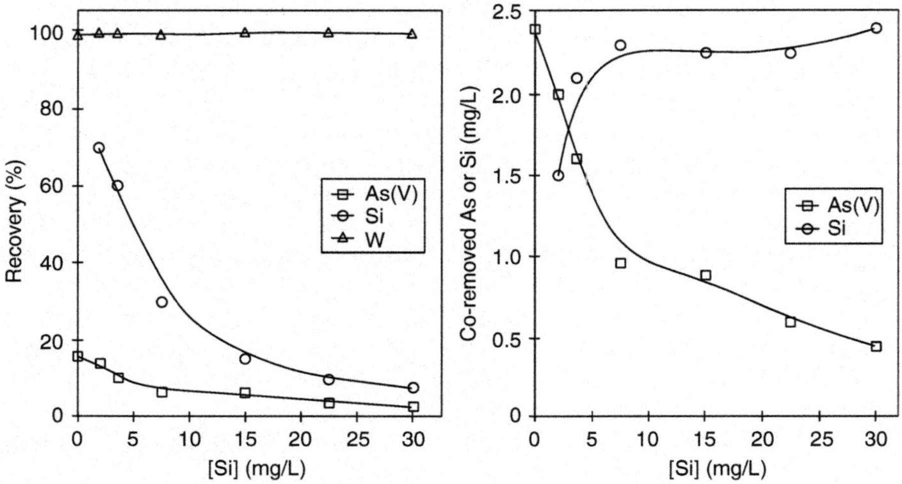

Fig. 8.18 Effects of silicate concentration on the separation of tungstate anions from tertiary aqueous mixtures also containing arsenates and silicates

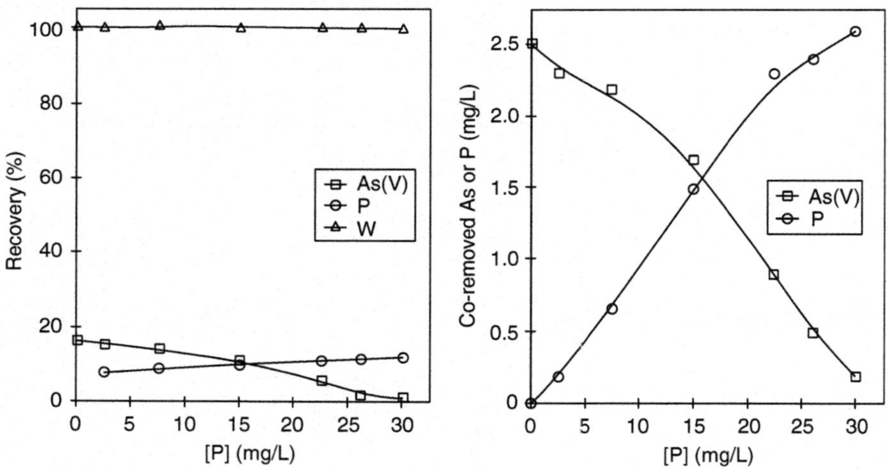

Fig. 8.19 Effects of phosphate concentration on the separation of tungstate from tertiary aqueous mixtures also containing arsenates and phosphates

found to be decreased when the concentrations of phosphates or silicates in tertiary mixtures were increased (Figs. 8.18 and 8.19). Nevertheless, the total concentration of impurities in the floated tungstates was found to be increased, owing to the partial co-removal of the other impurities.

When all these impurities (arsenates, phosphates, and silicates) exist together with tungsten in aqueous mixtures, the separation of them from tungsten was found to decrease as the presence of magnesium ions was increased (Fig. 8.20).

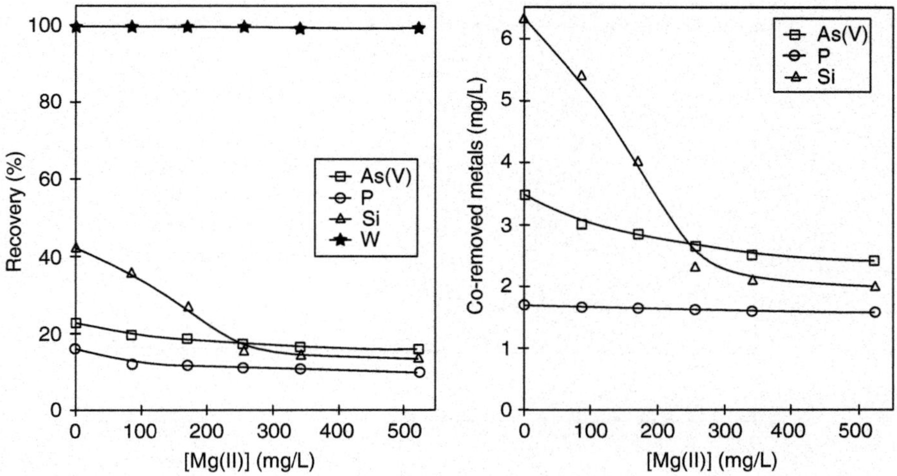

Fig. 8.20 Effects of Mg^{2+} ion concentration on the separation of tungstate from quaternary aqueous mixtures (containing also As, P, and Si)

Table 8.2 Simultaneous co-removal of low initial concentrations of impurities (As, P, and Si) with tungstate from quaternary aqueous mixtures; recovery of W: over 99.7% (in all cases)

No.	Initial concentration (mg/L)			Co-removed metal (mg/L)		
	P	As	Si	P	As	Si
1	1.5	1.5	1.5	0.2	0.5	0.9
2	3.0	3.0	3.0	0.5	1.0	1.8
3	4.5	4.5	4.5	0.9	1.2	2.7

The minimum amount of Mg(II) required for the lowest co-removal of impurities was found to be around 350 mg/L. The simultaneous removal of arsenic, phosphorus, and silicon with the floated tungsten was found to be 0.9 mg/L, 1.6 mg/L, and 2.1 mg/L, respectively, under the applied conditions (for one-stage flotation). According to the results given in Table 8.2, it may be proposed that the co-removed impurities with tungstates would be further separated by applying a second flotation stage to a substantially lower concentration. Therefore, acceptable levels for them could be achieved by common tungsten recovery processes, to be subsequently applied.

The presence of sodium chloride in the solution was not found to have any effect on the separation and recovery of tungstate anions from aqueous mixtures with arsenates, while the co-removal of arsenates was found to be increased quite considerably when the concentration of sodium sulfate was increased to values higher than 1 g/L (Fig. 8.21).

The aqueous chemistry of tungstates is very complex and similar to that of molybdates, forming a number of isopoly, as well as of heteropoly acids or the respective dissociated anions, such as $[W_6O_{24}]^{12-}$, $[W_{12}O_{48}]^{24-}$, $[W_{12}O_{39}]^{6-}$,

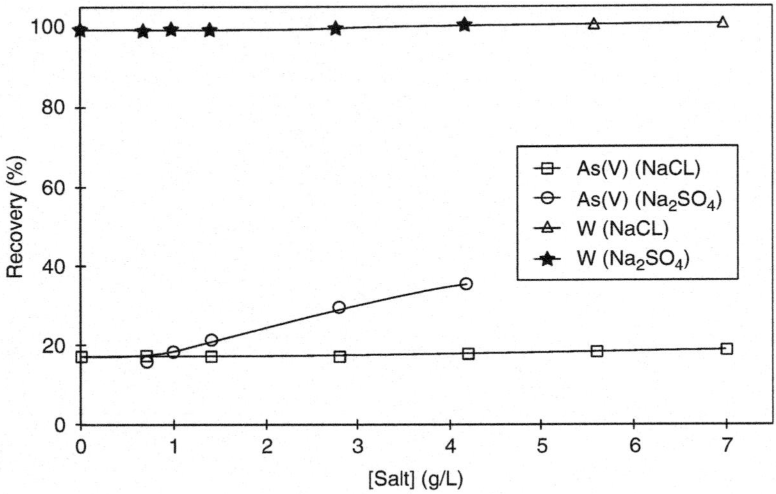

Fig. 8.21 Addition of salts (NaCl or Na_2SO_4) and their effects on the separation of tungstate from arsenates in dilute aqueous mixtures

$[W_{12}O_{41}]^{10-}$, $[PW_{12}O_{40}]^{3-}$, $[AsW_{12}O_{40}]^{3-}$, $[SiW_{12}O_{40}]^{4-}$, $[PAsW_{18}O_{63}]^{8-}$, $[PW_{11}O_{39}]^{7-}$, and $[SiW_{11}O_{39}]^{8-}$, when they exist in mixtures with the studied impurities (arsenates, phosphates, and silicates) at pH values lower than 10. It can be observed that all the aqueous species of tungsten, phosphorus, arsenic, and silicon, as well as their heteropoly-tungstic complexes in solutions having pH values higher than 2, where tungsten is precipitated as insoluble tungstic acid (H_2WO_4), may exist only as anions. These complex anions can be decomposed into the corresponding simpler ones only in stronger alkaline solutions, at solution pH values higher than 10. Consequently, the separation between these impurities and tungsten is considered a difficult task.

Among the components usually found in wastewaters or leaching solutions are certain cations, such as Ca^{2+}, Cu^{2+}, Zn^{2+}, Fe^{2+}, and Al^{2+}, the ferric ions especially contained in waste tailings of wolframite ores, as well as other cations, which may form precipitates with the studied impurities (arsenates, phosphates, and silicates), and these could affect the separation efficiency obtained by flotation, under certain pH conditions of the solution, mainly due to hydrolytic precipitation. In the presence of Fe^{3+} ions at pH = 3–5 (Fig. 8.12), but also at pH = 9 (Fig. 8.13) and when dodecylamine is used as surfactant, tungstates and arsenates were found to be floated together.

Better separation was achieved when foreign ions, such as Fe(III), were not present in the W/As aqueous mixture. The addition of Mg(II) ions in the same mixture, instead of Fe(III) ions, led to further depression of the co-removal of arsenates, especially in the pH range of 1.5–6, thus increasing the separation efficiency, although its presence was not found to be critical at this pH value and for the W/As mixture. However, there was still about 17% of the initial concentration of arsenate anions which would co-float with tungstates, probably in the form of un-dissociated complexes. Magnesium chloride, a common precipitant used in tungsten hydrometallurgy for the purification of concentrated W-containing solutions from the commonly coexisting impurities, is known to form insoluble precipitates with arsenate phosphate and silicate ions in neutral or weakly alkaline solutions, but not with tungstates.

For practical applications also, the recycling of used surfactant was considered to be important, in order to reduce the operating costs, a fact to which more and more attention has been paid. The foam products may be treated with caustic soda solution, causing the saponification and easy separation of the used surfactant, therefore permitting its recycle, while the resultant (more concentrated) tungsten solution can be further processed for the production of pure products.

It has shown that ion flotation, using dodecylamine as collector and magnesium chloride as depressant (modifier), can be successfully applied for the selective recovery of tungstate from other anions considered as impurities such as arsenates, phosphates, and silicates, at a solution pH value of 3.8. When the initial concentration of arsenates was kept at 15 mg/L, its co-removal was found to be lower than 17%. The concentration ratio of [W]/[As] could be increased from 12 to a value of 80, by applying one-stage flotation and to over 250 by two-stage flotation. A proposed flow sheet for the separation and recovery of tungstates in dilute aqueous mixtures with arsenates was suggested.

8.4 Removal of Molybdate and Arsenate from Aqueous Solutions by Flotation

The wastewater generated during the alkaline hydrometallurgical treatment of molybdenum minerals and zinc and lead wastes usually contains both arsenic (V) and molybdenum (VI) species. In addition, the impurities of arsenic in molybdenum leaching solutions are currently removed by precipitation by adding magnesium chloride. This results in a precipitate containing arsenic (V) and molybdenum, which should be further treated before it is environmentally safe for disposal. An alternative route for the detoxification of this precipitate is to retransfer any arsenic and molybdenum it contains back to the solution and then to remove them from the solution. In general, the content of molybdenum and arsenic in the wastewater generated during molybdenum processing or the leach solution of the resultant precipitate varies between 100–1000 mg/L and 5–40 mg/L, respectively. The removal of arsenic and molybdenum from waste streams is highly desirable, mainly for health and environmental reasons.

The removal of molybdenum (VI) and arsenic (V) depended strongly on the solution pH. Figure 8.22 shows that the maximum removal of both molybdenum and arsenic was obtained in the pH range of about 2.5–4.5, which was selected for the following experiments. The percent removal (denoted as Re) of arsenic was closely related to that of molybdenum. Arsenic could not be removed from a

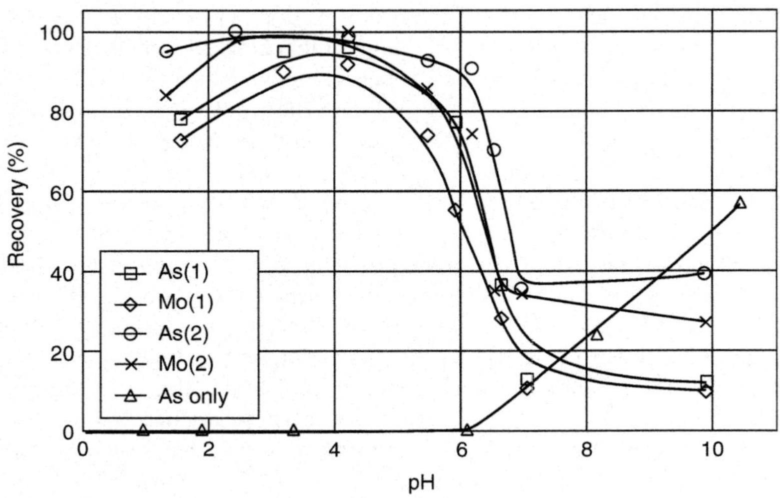

Fig. 8.22 Effects of solution pH on molybdenum and arsenic removal from solutions by ion flotation at different dodecylamine content: (1) 283.76, (2) 354.70 mg/L RNH_2

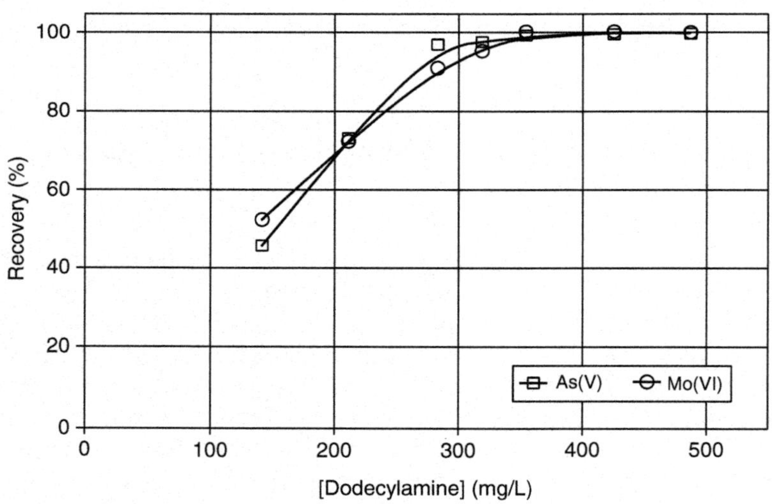

Fig. 8.23 Effects of dodecylamine addition during ion flotation at pH = 3.6

solution containing only As (V) at pH 2–6. The flotation behavior of molybdenum in solutions containing only molybdenum was found to be very similar to that in As-Mo systems (not shown in the Fig. 8.22).

The removal of molybdenum and arsenic from solution was found to increase linearly with an increase of dodecylamine content. Over 99% removal of both elements could be obtained at amine contents higher than 350 mg/L, as shown in Fig. 8.23. When the initial content of arsenic in solution was lower than 15 mg/L,

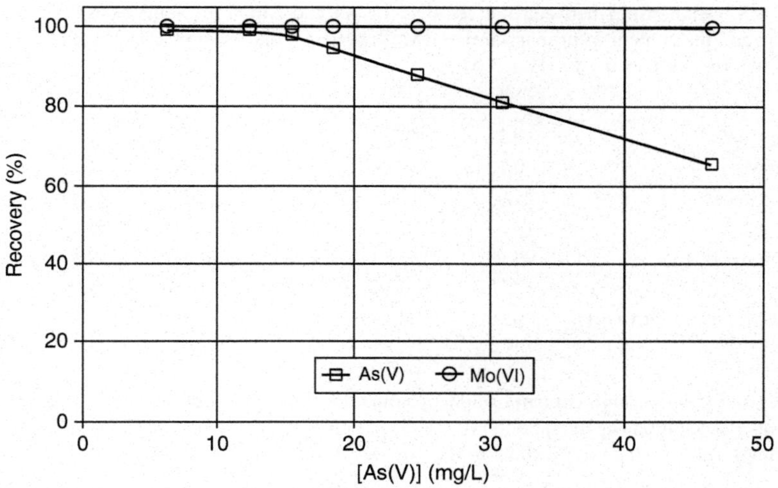

Fig. 8.24 Effects of arsenic(V) content in solution during ion flotation ($[RNH_2] = 354.70$ mg/L)

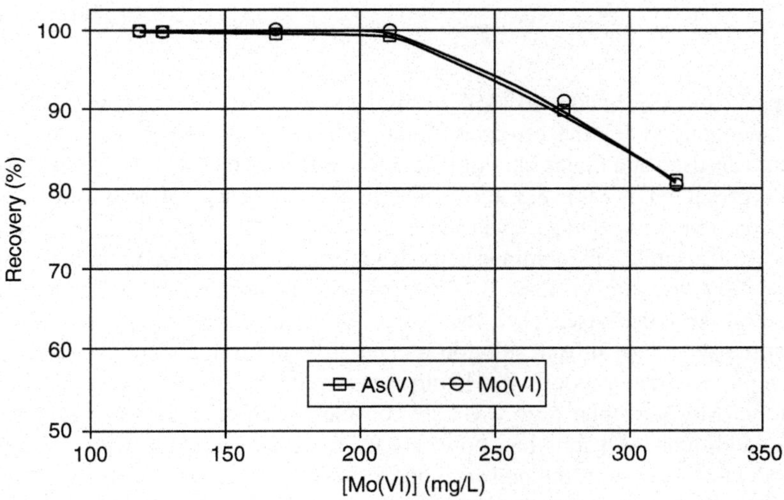

Fig. 8.25 Effects of initial molybdenum content in solution during ion flotation ([As] = 15.41 mg/L)

over 99.4% of arsenic could be removed, while the separation of molybdenum was always quantitative, regardless of its concentration, under the conditions of Fig. 8.24.

No arsenic was removed without the presence of molybdenum in solution; it seemed that arsenic floated as molybdo-arsenic anions. When the content of molybdenum was increased from 120 to 210 mg/L, its removal was found to be quantitative and then decreased linearly with any further increase of molybdenum, as shown in Fig. 8.25. When the concentration of molybdenum was lower than

Table 8.3 Relationship between the removal of As and Mo from aqueous solutions by ion flotation and the content of primary amine ([Mo]/[RNH_2] molar ratio of 1:1, [As] = 15.41 mg/L (0.205 M), [ethanol] = 0.5%, pH = 3.6)

No.	1	2	3	4	5	6
RNH_2 (mM)	0.11	0.22	0.44	0.77	1.1	1.32
Mo (mM)	0.11	0.22	0.44	0.77	1.1	1.32
RNH_2 (mg/L)	21.04	42.96	85.92	142.72	203.89	244.67
Mo (mg/L)	10.57	21.14	42.28	73.99	105.7	126.84
Mo removal (%)	99.99	99.89	99.99	99.87	99.87	99.76
As removal (%)	12.72	18.57	37.47	53.16	71.01	99.87
Initial Mo/As molar ratios	0.55	1.1	2.2	3.85	5.5	6.6

Table 8.4 Effects of high content of dodecylamine on the removal of arsenic and molybdenum from aqueous solutions by ion flotation ([Mo(VI)]/[As(V)] molar ratio of 3.6, [Mo] = 211.41 mg/L (2.20 mM), [As] = 46.23 mg/L (0.62 mM), pH = 3.6)

No.	RNH_2 (mg/L)	Mo removal (%)	As removal (%)
1	354.7	99.87	65.45
2	496.17	99.89	67.73
3	637.94	99.98	72.46

120 mg/L, the RNH_2 content (354.70 mg/L) was in great excess, and difficulties were encountered for the effective flotation of both elements. When the RNH_2 content was used at a molar ratio of [Mo]/[RNH_2] equal to 1:1,the respective results were as presented in Table 8.3. It can be noticed that the separation of molybdenum from solution was always quantitative. However, the simultaneous complete removal of arsenic was obtained only when the initial [Mo]/[As] molar ratio in the solution was higher than 6 if dodecylamine was present at the necessary concentration. Otherwise, when the initial [Mo]/[As] molar ratio was lower than 6, quantitative removal from solution was possible for molybdenum but impossible for arsenic. An increase of dodecylamine improved the separation of arsenic a little when the [Mo/As] molar ratio was kept constant at 3.6, as shown in Table 8.4.

The addition of Cl ions had no effect on the efficient removal of arsenic and molybdenum. However, the presence of SO_4^{2-} ions depressed the removal of both elements remarkably (Fig. 8.26). It is worth noting that in most molybdenum hydrometallurgical processes only HCI acid is used. Therefore, the negative effects of SO_4^{2-} ions on flotation are not expected. If ion flotation is going to be applied for the treatment of leachates resulting from molybdenum solid wastes, the use of sulfuric acid must be avoided.

When the initial [Mo]/[As] molar ratio in solution was lower than 6, it was impossible to remove arsenic completely by using dodecylamine as the surfactant. In this case, an alternative method should be applied. Figure 8.27 shows the relationship between the removal of both arsenic and molybdenum species and the pH of the solution when using Adsorbing colloid flotation (ACF) with ferric

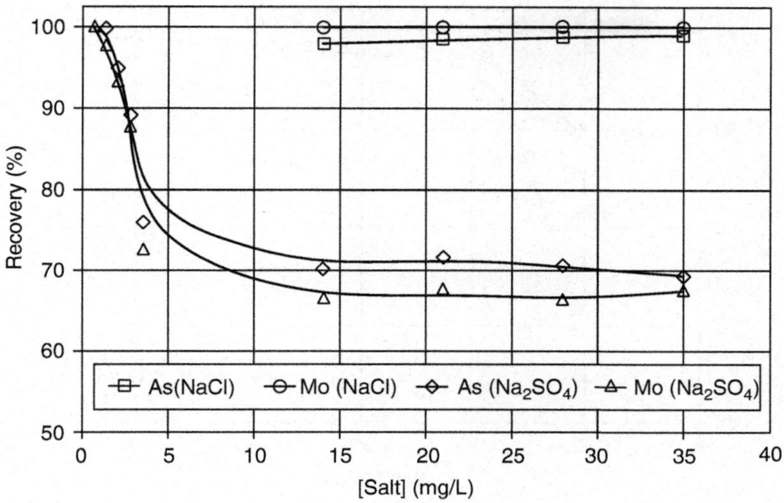

Fig. 8.26 Effects of impurities ions present as sodium salts during ion flotation at pH = 3.3

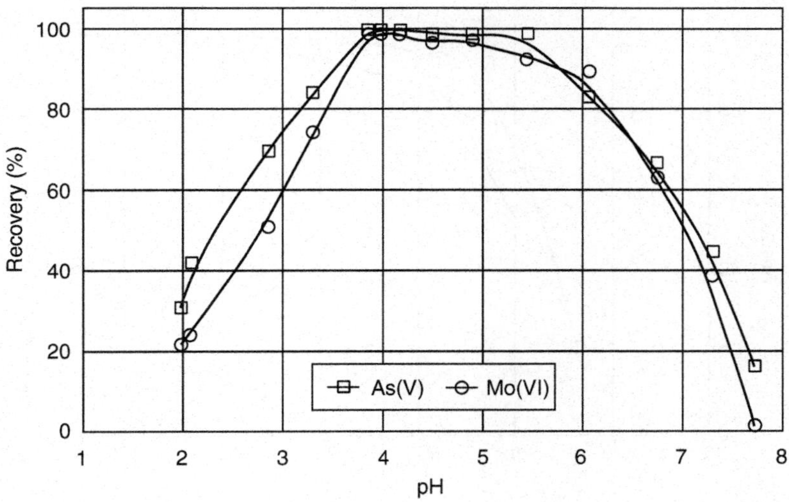

Fig. 8.27 Effects of solution pH on molybdenum and arsenic removal from solutions by adsorbing colloid flotation ([Mo] = 48 mg/L, [As] = 15.41 mg/L, [SDS] = 54.13 mg/L, [Fe (III)] = 134.89 mg/L)

hydroxide as the co-precipitant and sodium dodecyl sulfate as the collector. It can be seen that quantitative removal was obtained in the pH range 3.5–4.5, which was used for the following experiments. The results show that no (or little) flocs were formed when the solution pH was lower than 3.5, and the flocs formed at pH values over 4.5 could be floated effectively under the given conditions. However, the floated flocs settled rather quickly.

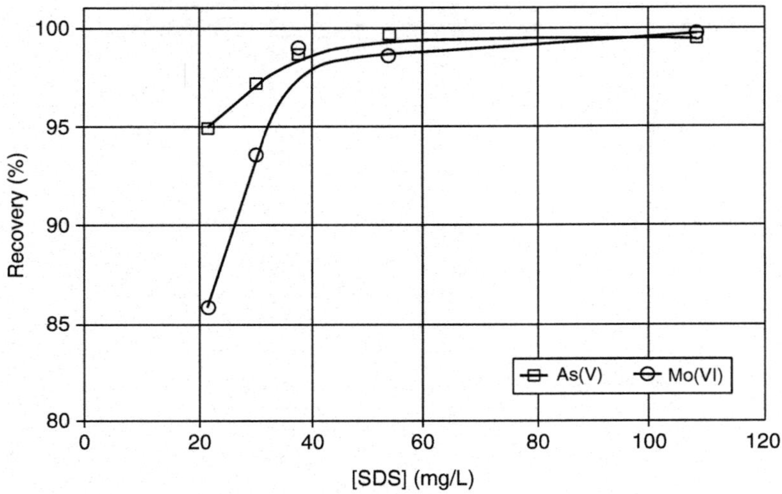

Fig. 8.28 Effects of sodium dodecyl sulfate addition during adsorbing colloid flotation ([Mo] = 48 mg/L, [As] = 15.41 mg/L, [SDS] = 54.13 mg/L, [Fe(III)] = 134.89 mg/L, pH 4.0)

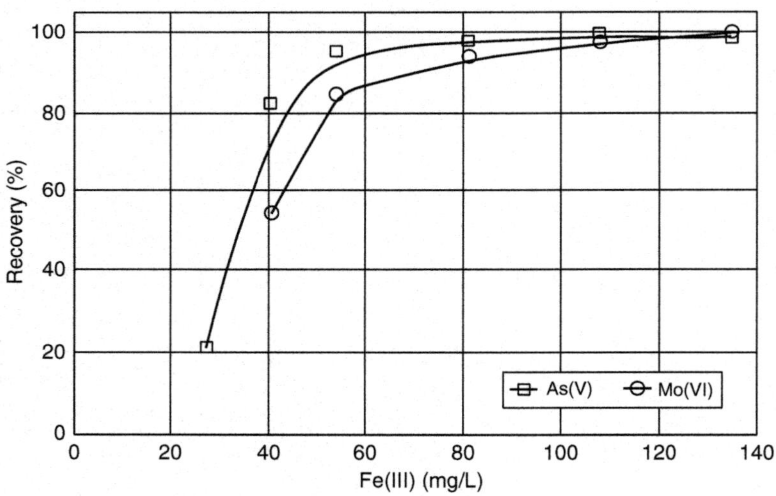

Fig. 8.29 Effects of ferric ion addition during adsorbing colloid flotation ([SDS] = 54.13 mg/L)

When the surfactant content in a solution was higher than 50 mg/L, 98% separation efficiency of arsenic and molybdenum was obtained (Fig. 8.28). Nevertheless, the presence of excessive amount of SDS led to the production of a large amount of foam, which removed a large part of the solution.

The removal of arsenic and molybdenum anions from an aqueous solution increased as the Fe^{3+} concentration was increased; an over 98% removal efficiency of molybdenum and arsenic was obtained when the ferric content was higher than 130 mg/L (Fig. 8.29).

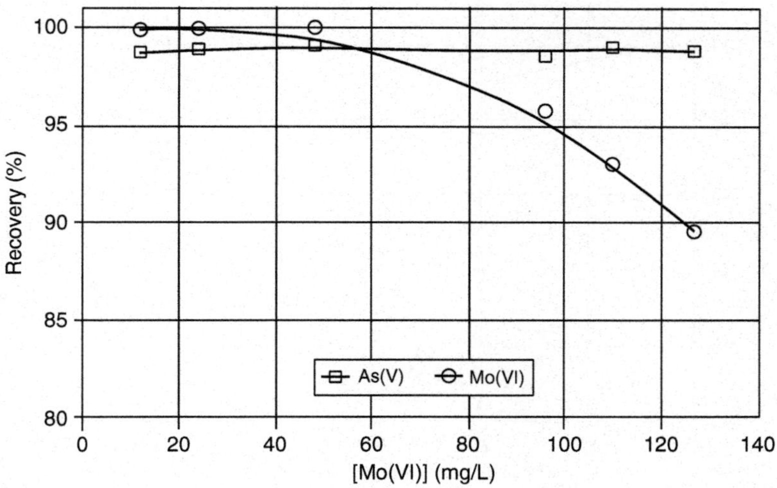

Fig. 8.30 Effects of molybdenum content in solution during adsorbing colloid flotation; [Fe] = 134.89 mg/L

The removal of arsenic was always over 95% regardless of the molybdenum content in the solution. However, the quantitative removal of molybdenum could only be obtained when its initial content in a solution was lower than 60 mg/L under the conditions shown in Fig. 8.30. It was observed in the experiments that a white precipitate was formed, so the flotation system described is not suitable for the removal of a high molybdenum content from aqueous solutions.

When the arsenic content in a solution was lower than about 60 mg/L, over 98% removal efficiency of arsenic could be obtained, as shown in Fig. 8.31. The yellow precipitate appeared in the experiments when the arsenic concentration in the solution was higher than 70 mg/L, and although this precipitate settled quickly, it was non-floatable by SDS.

Little effect by the addition to the solution of NaCl and Na_2SO_4 salts on flotation was noticed when the contents of these salts were lower than 4 g/L and 0.6 g/L, respectively (Fig. 8.32). However, the removals were decreased when the salt contents were further increased. Actual molybdenum hydrometallurgical wastewaters usually contain chloride ion, but little or no sulfates.

One of the advantages for the separation of arsenic and molybdenum ions from aqueous solution by ion flotation is that only dodecylamine as a collector needs to be added. In practical applications, dodecylamine can be dissolved in hot water and then introduced into the solution used for flotation, thus avoiding the consumption of ethanol, which had relatively little effect on flotation. Ion flotation has some disadvantages; it cannot be used for quantitative removal of arsenic from solution when the initial [Mo]/[As] molar ratio is lower than 6, in this case, although molybdenum can be removed completely regardless of the [Mo]/[As] ratio. Adsorbing colloid flotation (ACF) should be used for the complete removal of

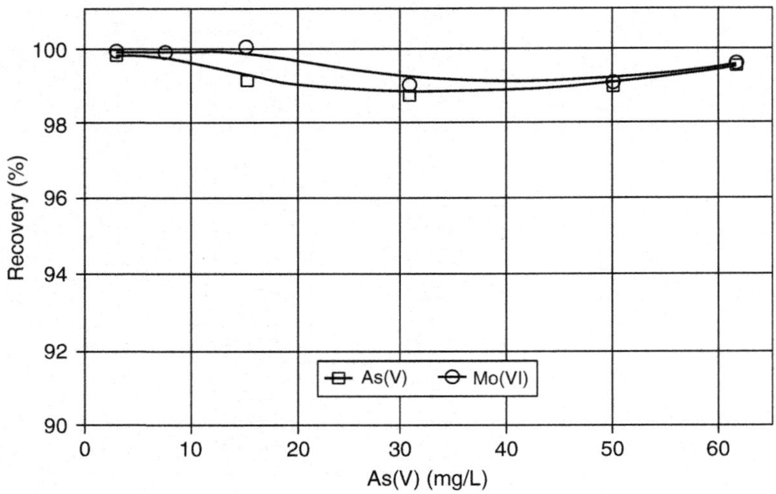

Fig. 8.31 Effects of arsenic content in solution during adsorbing colloid flotation ([Mo] = 48.00 mg/L)

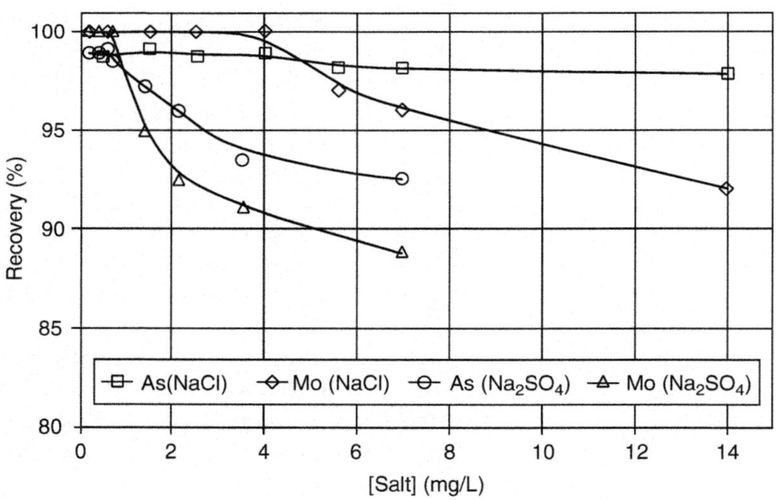

Fig. 8.32 Effects of foreign ions present as sodium salts during adsorbing colloid flotation

arsenic and molybdenum species from solution. This method is only suitable for the treatment of wastewaters containing arsenic and molybdenum at concentrations lower than 70 mg/L and 60 mg/L, respectively (regardless of the initial [Mo]/[As]] molar ratio).

According to the results presented in this work, it seems more favorable to use ion flotation for the removal of arsenic and molybdenum species from solution when the initial [Mo]/[As] molar ratio is higher than 6. The volume of foamate formed during ion flotation is relatively small. This is important to the further treatment of foamate and recycling of the surfactant when the recovered arsenic and molybdenum species are possibly salable products. It was also observed to be much more difficult to recycle the surfactant from the foamate during adsorbing colloid flotation (ACF).

Thus, both arsenic and molybdenum in aqueous solution can be effectively removed by ion flotation, using dodecylamine as the collector, when the initial [Mo]/[As] molar ratio is higher than 6, in the optimum pH range of 2.5–4.5, regardless of the initial content of arsenic and molybdenum. The float can be collected rather easily for further possible treatment and recycling of surfactant, as well as for recovery of the arsenic and molybdenum. The arsenate ion is considered to be floated as molybdo-arsenic foamate, and its removal is directly related to that of molybdenum.

Adsorbing colloid flotation (ACF) using ferric ions as the co-precipitant or sorbent and sodium dodecyl sulfate as the collector can be used for the effective removal of arsenic and molybdenum from aqueous solution at an optimum pH value of 3.5–4.5 when the contents of these two elements are lower than 70 mg/L and 60 mg/L, respectively, regardless of the initial [Mo]/[As] molar ratio in the solution. No direct relationship between the removal of arsenic and molybdenum from solution was observed. In practical applications, the ion flotation process is preferred because of its advantages over adsorbing colloid flotation, such as the easy collection of foamate. Therefore, the applied surfactant can be recycled, and the arsenic and molybdenum can be recovered.

8.5 Extraction of Phosphorus, Arsenic, and/or Silica from Sodium Tungstate and Molybdate Solutions with Primary Amine and Tributyl Phosphate as Solvents

8.5.1 Parameter Optimization for the Extraction of Phosphorus, Arsenic, and/or Silica from Sodium Tungstate and Molybdate Solutions with Primary Amine and Tributyl Phosphate as Solvents

The removal of phosphorus, arsenic, and silica as impurities from alkaline leaching solutions is an important step in the hydrometallurgy of tungsten and molybdenum. The traditional method used in industry is by precipitation with the addition of magnesium chloride to the leaching solution. This process has many problems, such as low recovery of metals, difficulty in disposing of the precipitate, and pollution arising from its storage. Solvent extraction can be considered as a better method for the removal of impurities in comparison with the precipitation method if an effective solvent system can be developed. The extractants for the recovery and separation of tungsten and molybdenum include Alamine 336, MIBK, D2EHPA and TBP, and tertiary amines, which is developed for the removal of the impurities by the extraction of tungsten or molybdenum and can be carried out only in acidic media. This extraction is accompanied by a number of problems, such as the formation and extraction of heteropoly acids of phosphorus, arsenic, and silica, resulting in a considerable decrease in the separation efficiency between tungsten or molybdenum and the impurities. The further removal of impurities co-extracted with tungsten or molybdenum is, in general, carried out by a chemical method, such as precipitation, with appreciable loss of tungsten or molybdenum. In order to increase the separation efficiency and prevent the formation of heteropoly acids, some reagents, such as hydrogen peroxide, are added to the aqueous solution before extraction. However, this procedure has not proved to be very effective.

Another approach for the removal of impurities by solvent extraction is by the extraction of phosphorus, arsenic, and silica as heteropoly acids. The solvent systems developed so far for the removal of impurities by solvent extraction have not been considered to be of practical significance and without any possibility of industrial application because of the low separation efficiency with high solution acidity required for extraction.

In this work, the synergistic extraction of phosphorus, arsenic, and silica from tungstate and molybdate solutions by a mixture of primary amine and TBP was presented systematically. A mechanical shaker was used to mix the aqueous and organic solutions. The aqueous phase was prepared from sodium salts of arsenic, phosphorus, silica, tungsten, and molybdenum. The concentrations of primary amine and TBP in the organic phase were expressed as volume percent: 1% primary amine is equal to 30.43 mM. Primary amine with a total carbon number of about 19–21 (R_1R_2-$CHNH_2$ or simplified as RNH_2 for brevity in the following expressions) was supplied by the Institute of Organic Chemistry, Academia Sinica, Shanghai, and was purified by distillation. Commercial kerosene was used as diluent. The other chemicals were of analytical grade purity. Colorimetric determination of phosphorus, arsenic, silica, tungsten, and molybdenum was carried out in 1 cm cells. The experimental results show that 2 min or 3 min of shaking were enough to approach extraction equilibrium in all experiments, while less 1 min was required for the stripping of the loaded organic phase. The phase ratio for extraction and stripping was 10:10 (mL), except for the study of the effect of phase ratio.

Methods for the colorimetric determination of phosphorus, arsenic, and silica in solution and their mixtures containing tungsten or molybdenum without preliminary separation were developed for this work. For mixtures containing molybdenum, phosphorus was determined by ascorbic acid-Sb reduced heteromolybdate blue in the presence of sodium sulfide. The total amount of phosphorus and arsenic was analyzed by hydrazine sulfate reduced heteromolybdate blue. For mixtures containing tungsten, phosphorus was determined by the addition of ammonium molybdate and ammonium vanadate to form molybdophosphovandatic acid, while the total amount of phosphorus and arsenic can be obtained by the acetone-sodium molybdate-perchloric acid method. Silica can be analyzed by oxalic acid-ascorbic

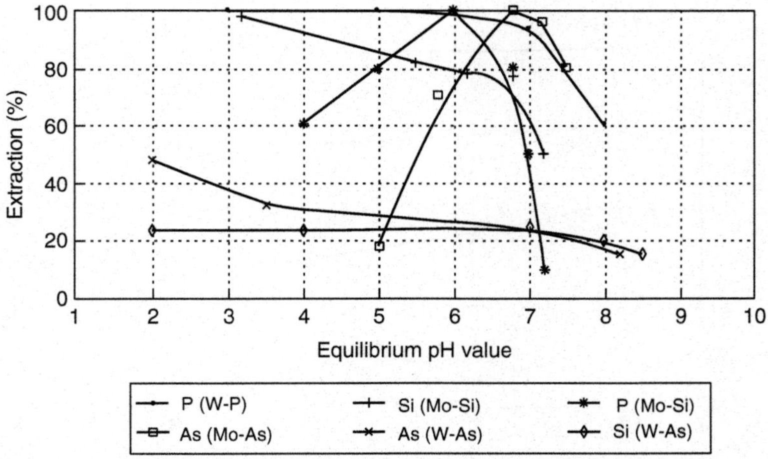

Fig. 8.33 Effects of pH on the single-impurity extraction from tungstate or molybdate solutions with 1% RNH_2 and 10% TBP in kerosene at 16 °C. Initial W or Mo: 40 g/L; initial P, As or Si in tungstate system: with As 1380 mg/L or P 120 mg/L or Si 151 mg/L; in molybdate systems: with As 121 mg/L or P 123 mg/L or Si 121 mg/L

acid reduced heteromolybdate blue regardless of the presence of tungsten or molybdenum.

The alkaline crude leaching solutions of sodium tungstate and ammonium molybdate were kindly supplied by the Zigong Cemented Carbide Company, Zigong, Sichuan Province of China. The composition of the leaching solutions is referred to in the text. The extraction and separation of individual elements of phosphorus, arsenic, and silica from tungstate and molybdate solutions were strongly dependent on the acidity of the aqueous phase, shown in Fig. 8.33. The separation of phosphorus, arsenic, and silica from tungstate solution and silica from molybdate solution increases with decreasing pH, while the maximum extraction and separation of phosphorus and arsenic from molybdate solutions can only be attained at an equilibrium pH of 6–7. In the following experiments, the equilibrium pH values of pH 7–8 for tungsten systems and pH 6–7 for molybdenum systems were adjusted according to the acidity required for separation. For the simultaneous extraction of phosphorus, arsenic, and silica, the results are shown in Figs. 8.34 and 8.35, which are in agreement with those given in Fig. 8.33.

Phosphorus, arsenic, and silica cannot be extracted without the presence of tungsten or molybdenum in the aqueous phase, which is one of the reasons for the suggestion that phosphorus, arsenic, and silica are extracted into the organic phase as heterotungstates or heteromolybdates. Figures 8.36 and 8.37 give the relationship between the simultaneous extraction of arsenic, phosphorus, and silica from tungsten and molybdenum solutions and the initial concentration of tungsten or molybdenum in the aqueous phase. It can be seen that the maximum extraction

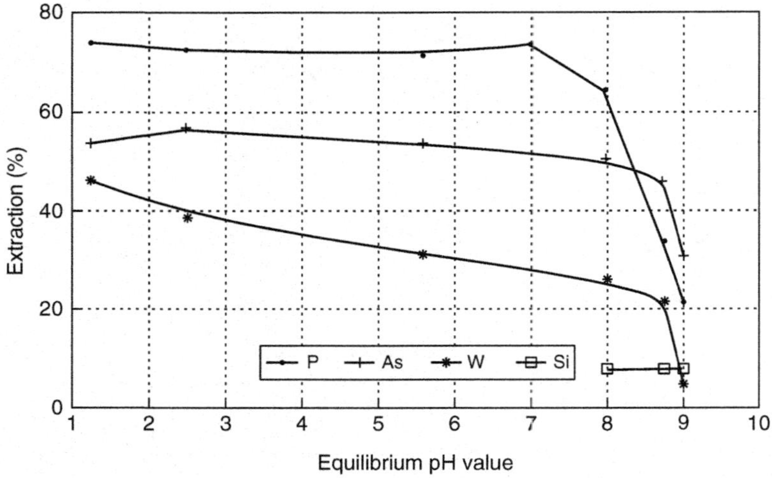

Fig. 8.34 Effects of pH on the simultaneous extraction of phosphorus, arsenic, and silica from tungstate solutions with 1% RNH_2 and 10% TBP in kerosene at 10 °C. Composition of initial aqueous solutions: 40 g/L W; 204 mg/L P; 245 mg/L As; 150 mg/L Si

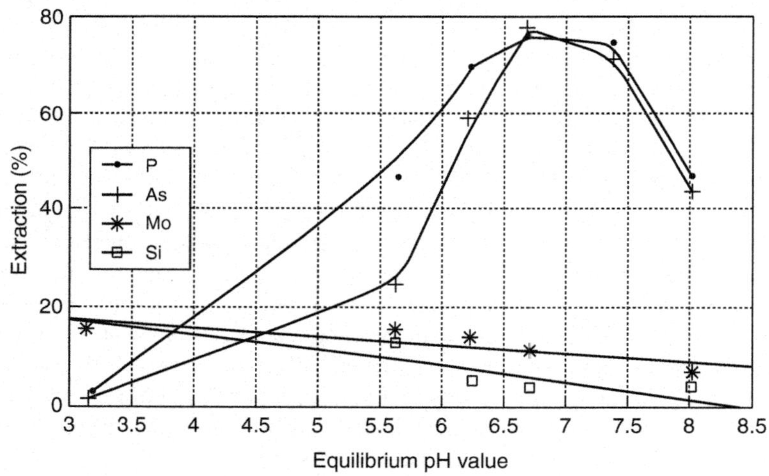

Fig. 8.35 Effects of pH on the simultaneous extraction of phosphorus, arsenic, and silica from molybdate solutions with 1% RNH_2 and 10% TBP in kerosene at 17 °C. Composition of initial aqueous solutions: 40 g/L Mo; 263 mg/L P; 242 mg/L As; 247 mg/L Si

can be obtained with an initial W concentration of over 30–40 g/L or Mo of over 10 g/L in the aqueous phase with pH 6–8.

The extraction of molybdenum or tungsten is dependent on the content of impurities in the aqueous solution. In many cases, such as when the contents of

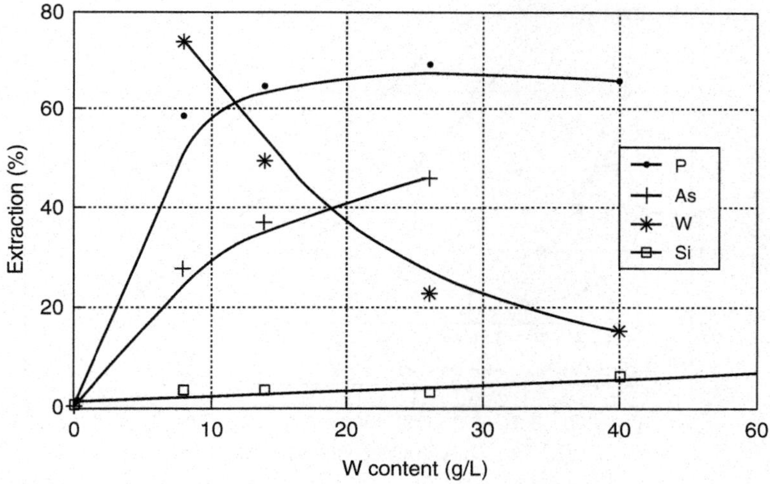

Fig. 8.36 Effects of initial tungsten content in aqueous solution on the simultaneous extraction of phosphorus, arsenic, and silica with 1% RNH_2 and 10% TBP in kerosene at 13 °C. Composition of initial aqueous solutions: 204 g/L P; 245 mg/L As; 150 mg/L Si; equilibrium pH for extraction: 7.1–7.6

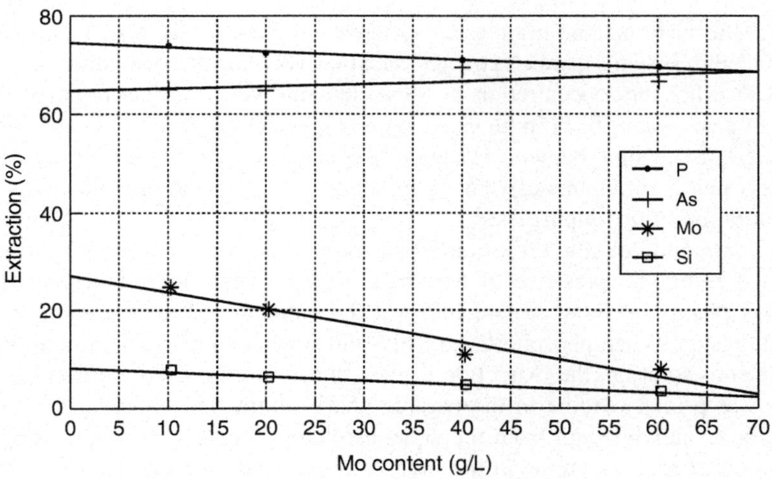

Fig. 8.37 Effects of initial molybdenum content in aqueous solutions on the simultaneous extraction of phosphorus, arsenic, and silica with 1% RNH_2 and 10%TBP in kerosene at 13 °C. Composition of initial aqueous solutions: 263 g/L P; 242 mg/L As; 249 mg/L Si; equilibrium pH for extraction: 6.3–6.7

phosphorus, arsenic, and silica are high enough, the molar ratios of tungsten or molybdenum to the impurities remain unchanged, based on the chemical stoichiometry of the extracted species.

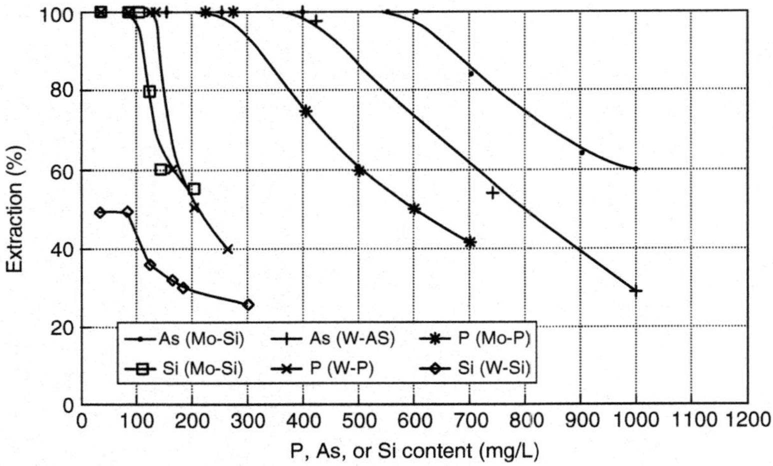

Fig. 8.38 The relationship between the initial content of phosphorus, arsenic, or silica in aqueous solution and the corresponding single-impurity extraction with 1% RNH_2 and 10% TBP in kerosene at 13–20 °C. Equilibrium pH for extraction: molybdate systems pH 6.3–6.7; tungstate systems pH 7.1–7.7; initial content in aqueous solution: 40 g/L W, or 40 g/L Mo

Phosphorus or arsenic in tungsten systems and phosphorus, arsenic, or silica in molybdenum systems can be extracted quantitatively under the conditions given in Fig. 8.38, when the concentrations of these elements were less than 100 mg/L P, Mo mg/L As, and 20 mg/L Si in tungsten systems and 300 mg/L P, 600 mg/L As, and 100 mg/L Si in molybdenum systems. When these elements are present in the aqueous phase simultaneously, the percentage extraction is dependent upon the concentration of the impurities.

The extraction of silica from tungstate solution is found to be low and ineffective. However, the presence of hundreds of milligrams of molybdenum in the aqueous phase can increase and improve the extraction and separation of silica from tungsten. When phosphorus, arsenic, and silica are present simultaneously in the aqueous solution, the extraction of these elements is decreased, under a given content of primary amine in the organic phase. Extraction of impurities as well tungsten or molybdenum from the aqueous phase increases with increasing concentration of primary amine in the organic phase (Figs. 8.39 and 8.40). It can be seen that the extraction increases rapidly as the content of RNH_2 increases. TBP alone cannot be used for the extraction of all elements studied, but its addition to the organic phase can improve the extraction of some of these elements considerably and prevent the formation of a precipitate in the organic phase during extraction. Figure 8.41 gives the relationship between the TBP content and the simultaneous extraction of phosphorus, arsenic, and silica from tungstate and molybdate solutions. It can be seen that no effect of TBP content is found for the extraction of silica, molybdenum, tungsten, and arsenic from tungstate. However, when the TBP

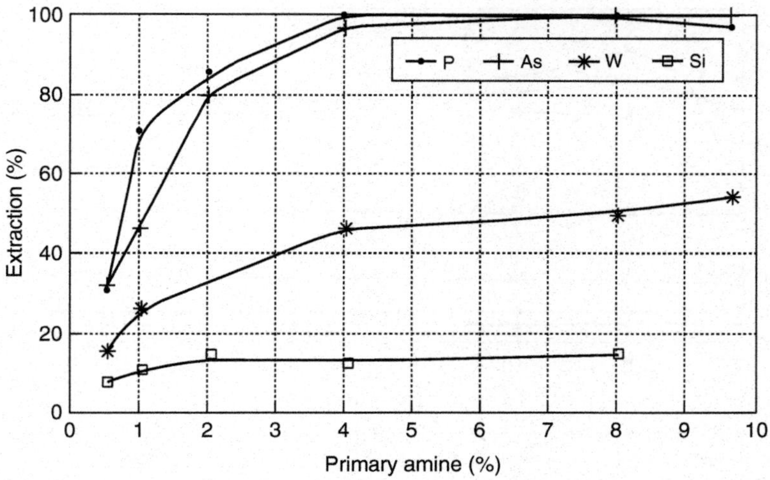

Fig. 8.39 Effects of primary amine content on the simultaneous extraction of phosphorus, arsenic and silica from tungstate solutions with 10% TBP in kerosene at 13 °C. Composition of initial aqueous solutions: 40 g/L W, 204 g/L P, 247 mg/L As, 151 mg/L Si; equilibrium pH for extraction: 7.3–8.1

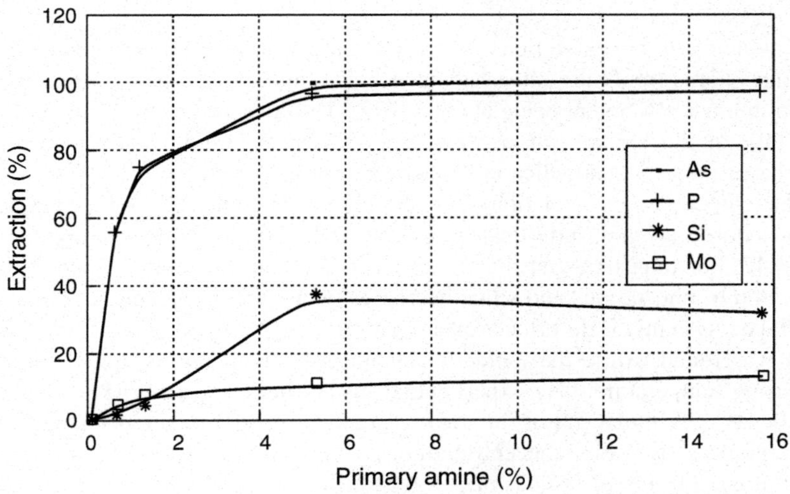

Fig. 8.40 Effects of primary amine content on the simultaneous extraction of phosphorus, arsenic, and silica from molybdate solutions with 10% TBP in kerosene at 12 °C. Composition of initial aqueous solutions: 40 g/L Mo, 262 mg/L P, 242 mg/L As, 151 mg/L Si, initial pH 6.5; equilibrium pH: 6.5–7.8. Note: the equilibrium pH value increases linearly from 6.5 to 7.8 as the RNH_2 content increases from 0 to 15%

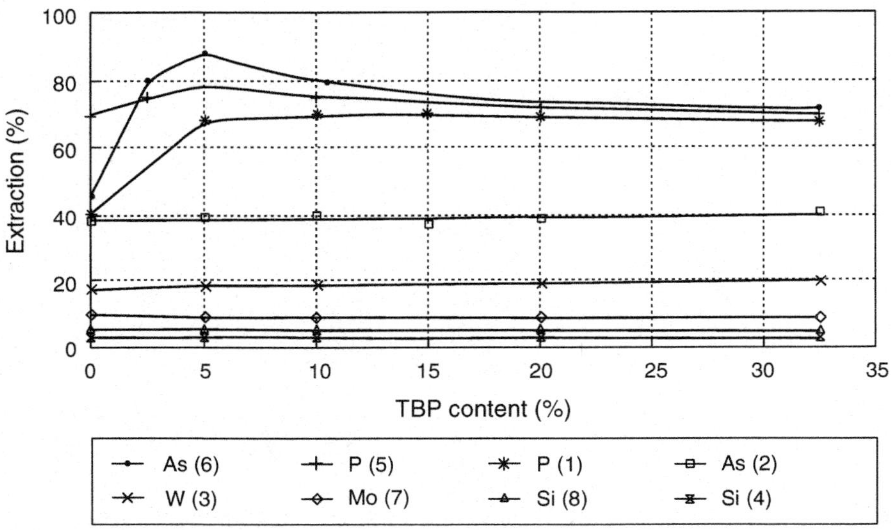

Fig. 8.41 Effects of TBP content on the simultaneous extraction of phosphorus, arsenic, and silica from tungstate and molybdate solutions with 1% RNH_2 in kerosene at 12 °C. Composition of initial aqueous solutions: Tungstate system, Lines 1–4: 40 g/L W, 262 mg/L P, 242 mg/L As, 151 mg/L Si, equilibrium pH 7.1–7.5. Molybdate system, Lines 5–8: 40 g/L Mo, 262 mg/L P, 242 mg/L As, 151 mg/L Si, equilibrium pH 6.5–6.8

content increases, the extraction of arsenic and phosphorus from molybdate systems and phosphorus from tungstate systems increases initially and then decreases, with the optimum TBP content being about 5–10%. The concentration of primary amine and TBP in the organic phase should be chosen according to the contents of phosphorus, arsenic, and silica in the aqueous phase.

The temperature of extraction can exert quite a different effect on the extraction of phosphorus, arsenic, and silica in tungstate and molybdate solutions, as shown in Fig. 8.42. For tungstate systems, when the extraction temperature increases, the extraction of phosphorus and silica decreases, while that of arsenic increases. For molybdate systems, little effects of temperature on the extraction of arsenic and silica was found, while the extraction of phosphorus increases remarkably with increasing temperature. The experimental results indicate the W/(P, As, Si) and Mo/(P, As, Si) molar ratios in loaded organic phase remain unchanged with temperature, phase ratio, concentration of primary amine, and TBP in the organic phase at equilibrium pH values of 7–8 for tungsten systems and 6–7 for molybdenum systems. An increase in the extraction phase ratio (O/A) can improve the extraction efficiency. However, the formation of a third phase may occur. It was found that an increase of the primary amine in the organic phase was preferable to an increase of the phase ratio, if the extraction efficiency was to be improved. The optimum phase ratio was found to be 1:1.

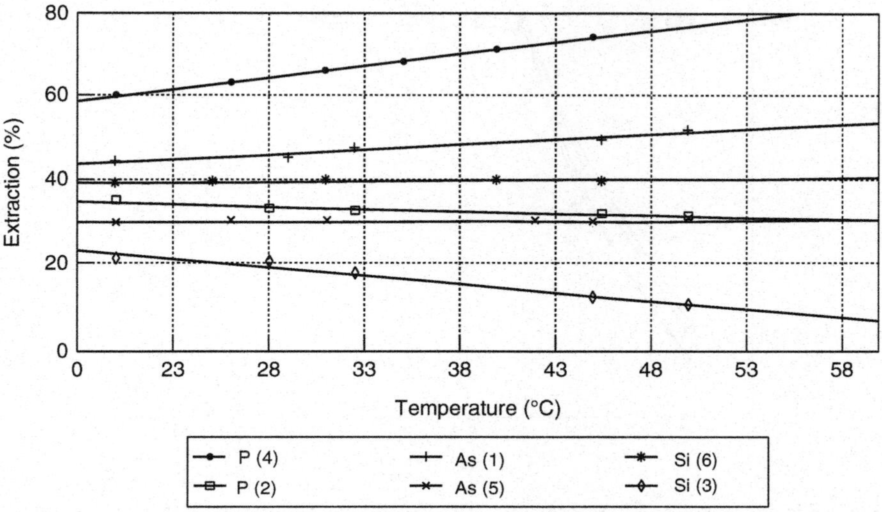

Fig. 8.42 Effects of temperature on the individual extractions of phosphorus, arsenic, and silica from tungstate and molybdate solutions with 1% RNH_2 and 10% TBP in kerosene. Composition of initial aqueous solutions: Tungstate system, Lines 1–3: 40g/L W, 689 mg/L As for W-As system; 40 g/L W, 408 mg/L P for W-P system; 40 g/L W, 301 mg/L Si for W-Si system. Molybdate system, Lines 4–6: 40 g/L Mo, 2047 mg/L As for Mo-As system; 40 g/L Mo, 126 mg/L P for Mo-P system; 40 g/L Mo, 151 mg/L Si for Mo-Si system

The loaded organic phase can be stripped quantitatively and easily with dilute NaOH solution. Figure 8.43 gives the results of studies on molybdate systems. It can be seen that practically complete stripping is obtained when the equilibrium pH value is higher than 8. The case of tungstate systems is very similar to the molybdate systems, except that the stripping pH should be kept over 9. The stripping may be carried out with a high phase ratio (O/A). Thus, the phosphorus, arsenic, and silica, as well as tungsten and molybdenum, are concentrated simultaneously. For example, for an aqueous solution containing 120 g/L Mo, 50 mg/L P, 70 mg/L As, and 40 mg/L Si, extracts with 5% primary amine and 10% TBP in kerosene and with a phase ratio of 1:1. The resulting loaded organic phase contains 4 g/L Mo, 50 mg/L P, 70 mg/L As, and 40 mg/L Si. Then, if the organic phase is stripped with NaOH solution with a phase ratio of 20:1, the contents of Mo, P, As, and Si are increased by 19 times, with a concentration of 80 g/L Mo, 1 g/L P, 1.4 g/L As, and 0.8 g/L Si in the stripped aqueous solution. The contents of Mo, As, P, and Si in this solution may be considered to be high enough for the preparation of heteropoly acids or their salts.

Besides phosphorus, arsenic, and silica, the crude leach solutions also contained other impurities, such as Ca, Cu, and Fe. The concentrations of phosphorus, arsenic, silica, and tungsten, as well as the extraction and separation of these three elements from the leaching solution, are given in Table 8.5. The concentrations of phosphorus, arsenic, and silica in the leach solution of ammonium molybdate were very low

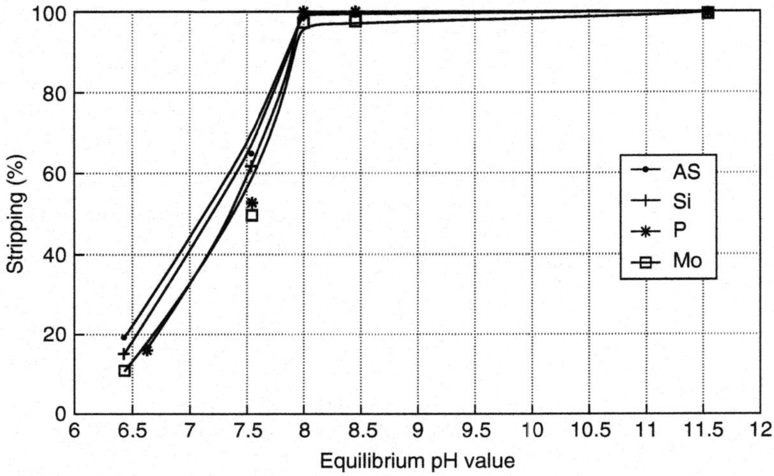

Fig. 8.43 The stripping of loaded organic phase of Mo-P-As-Si extraction system with 0.1 mol/L NaOH solution at 20 °C. Composition of loaded organic phase: 3.8 g/L Mo, 279 mg/L Si, 163 mg/L P, 1% primary amine and 10% TBP in kerosene, stripping phase ratio (O/A) = 1:1

Table 8.5 Extraction and separation of P, Si, and As from a crude leaching solution of sodium tungstate by 1% primary amine and 10% TBP in kerosene

No.	Initial pH	Equilibrium pH	W extr. (%)	P extr. (%)	Si extr. (%)	As extr. (%)
1	8.17	8.92	0.3	1.1	2.3	2.5
2	7.48	8.37	3.6	17.6	5.3	13.0
3	5.17	7.40	5.4	18.9	24.5	17.0

Solution contained 200 g/L W and other impurities such as ~1 g/L Ca,~0.05 g/L Cu, and 0.1 g/L Fe. P, As, and Si concentrations in the leach solution were 531 mg/L P, 1900 mg/L As, and 200 mg/L Si

because the molybdenite concentrates used for leaching had been pretreated. Therefore, these three impurities were added to the leach solution and the results are given in Table 8.6. It can be seen that the phosphorus, arsenic, and silica can be extracted and separated very effectively from the leach solutions of tungsten and molybdenum just as from the solutions prepared with pure sodium salts of all elements involved. The experimental results show that other impurities, such as Ca, Cu, Fe, and Al, present in the leaching solution cannot be extracted by the solvents used.

The extraction of heteropoly acids of phosphorus, arsenic, and silica as 12-heterotungstic and 12-heteromolybdic acids or their anions by neutral donor or acceptor extractants has been investigated extensively in strongly acidic medium to understand the extractability of these heteropoly acids. For the removal of impurities from alkaline leach solutions of tungsten and molybdenum, the extraction of

Table 8.6 Extraction and separation of P, As, and Si from a crude leach solution of ammonium molybdate containing about 120 g/L Mo with 1% primary amine and 10% TBP in kerosene

Effects of pH value in aqueous phase

No.	Equilibrium pH	Mo extr. (%)	Si extr. (%)	P extr. (%)	As extr. (%)
1	1.38	1.9	–	–	52.9
2	6.09	2.9	–	–	84.9
3	5.87	4.0	–	–	100.0

Effects of As added to the aqueous phase

No.	Equilibrium pH	As added (mg/L)	Total As (mg/L)	Si extr. (%)	P extr. (%)	As extr. (%)
1	7.20	0.0	17.2	–	–	90.7
2	7.13	168.0	185.2	–	–	100.9
3	7.10	389.2	406.4	–	–	82.7

Effects of P, As, and Si added to the aqueous phase

No.	Equilibrium pH	P added (mg/L)	As added (mg/L)	Total As (mg/L)	Si added (mg/L)	P extr. (%)	As extr. (%)	Si extr. (%)	Mo extr. (%)
1	7.02	99.0	0	17.2	0	100.0	0	0	3.3
2	6.96	208.5	0	17.2	0	70.0	0	0	3.1
3	6.93	208.2	206.8	224.0	0	44.3	67.4	0	3.3
4	6.86	208.5	206.8	224.0	83.8	37.9	59.4	28.0	3.3
5	7.04	0	0	17.2	60.0	–	0	54.0	3.1
6	6.94	0	0	17.2	133.4	–	0	42.1	3.3

Concentrations of P, As, and Si in leach solution: 0.0 mg/L P, 0.0 mg/L Si, and 17.2 mg/L As. Other impurities: ~1 g/L Ca, ~0.5 g/L Fe, ~0.05 g/L Zn, and 0.005 g/L Cu

these impurities as heteropoly acids or their anions is of great interest and of practical significance, considering that in the leaching solution, the contents of impurities are much lower than the contents of tungsten and molybdenum. Many problems remain to be solved for the removal of impurities from leach solutions of tungsten and molybdenum by either solvent extraction or precipitation. For example, low-quality concentrates of tungsten or molybdenum cannot be handled economically by the present process of precipitation because the leach solutions contain low concentrations of tungsten or molybdenum and high concentrations of impurities. It is predicted that the solvent system developed may be used for the removal of impurities from leach solutions of either high- or low-quality concentrates. However, further tests on a larger scale and developing a more effective solvent for the removal of silica may be necessary.

Tungsten or molybdenum can be stripped quantitatively from the loaded organic phase with dilute sodium hydroxide solution. The tungsten or molybdenum in the aqueous strip solution may be recovered by acidifying the solution to form heteropoly acids or their salts as mixtures. It is essential to find uses for these mixtures.

The extractability of phosphorus, arsenic, and silica from tungsten or molybdenum solutions by primary amine and TBP decreases in the order Mo-As > W-

As > Mo-P > W-P > Mo-Si > W-Si and As > P > Si for both tungsten and molybdenum systems.

The synergistic extraction of phosphorus, arsenic, and silica from tungstate and molybdate solutions with primary amines and TBP was investigated. The optimum operating conditions for the extraction and separation were studied. The effective extraction and separation of phosphorus, arsenic, and silica can be carried out at an equilibrium pH of 7–8 for tungsten systems and 6–7 for molybdenum systems with this solvent system.

8.5.2 Mechanism of Extraction of Phosphorus, Arsenic, and Silica from Tungstate and Molybdate Solutions

The extraction reaction of phosphorus, arsenic, and silica from tungsten or molybdenum solutions can be expressed as:

$$m\mathrm{H}^{+} + (L)_a + n(M)_a + x\mathrm{TBP} + y\mathrm{RNH_2} \xrightarrow{K} (LMn \cdot x\mathrm{TBP} \cdot y\mathrm{RNH_2})_o \quad (8.11)$$

where L represents phosphorus, arsenic, or silica; M represents tungsten or molybdenum; K represents the equilibrium constant of extraction; o and a represent organic and aqueous phase, respectively.

From Eqs. (8.11 and 8.12):

$$\log K = m\mathrm{pH} + \log(L)_o - \log(L)_a - n\log(M)_a - x\log(\mathrm{TBP})_o - y\log(\mathrm{RNH_2})_o \quad (8.12)$$

or:

$$\log(L)_o + m\mathrm{pH} - \log(L)_a = \log K + n\log(M)_a + x\log(\mathrm{TBP})_o + y\log(\mathrm{RNH_2})_o \quad (8.13)$$

Assuming $m = 1, 2, 3, \ldots$, then straight lines can be obtained by plotting [log $(L)_o-\log(L)_a + m\mathrm{pH}$] versus $\log(M)_a$. The slope and correlation coefficient can then be calculated. For a value of m, the corresponding value of n can be obtained. The value of m is chosen when the corresponding value of n has the highest correlation coefficient and was used in the next expression.

From Eq. (8.13), the following expression can be obtained:

$$\log K + n\log(M)_a - \log(L)_o + \log(L)_a = m\mathrm{pH} - x\log(\mathrm{TBP})_o - y\log(\mathrm{RNH_2})_o \quad (8.14)$$

Table 8.7 The values of m, n, x, and y of the extraction reaction expressed in Eq. (8.13) by linear correlation analysis

System	Number of data	n	m	x	y	Equilibrium pH	$x + y$
W-P	6	2	3	0.3	3	7–8	3.3
W-As	6	3	2	<0.5	3	7–8	3.5
W-Si	6	1	3	1	2	7–8	3
W	5	–	2	–	0.5	7–8	0.5
Mo-P	6	2	3	1	2	6–7	3
Mo-As	6	2	2	0.5	2	6–7	2.5
Mo-Si	6	4	5	<0.4	7	6–7	7
Mo	5	–	1	–	2	6–7	2

Note: Initial (W) or (Mo) 15–20 mM; initial P, As, or Si, 2–3 mM. The value of x for W or Mo systems is approximately equal to zero

The value of m can be calculated from the plot of $[n\log(M)_a - \log(L)_o + \log(L)_a]$ versus pH, which must be consistent with the value obtained in Eq. (8.14). Similarly, the values of x and y can be obtained and are given in Table 8.7. Thus, the extraction reactions can be given as follows:

$$3H^+ + H_2PO_4^- + 2HWO_4^- + 0.3TBP + 3RNH_2 \rightarrow H_3PW_2O_{10} \cdot 0.3TBP \cdot 3RNH_2 + 2H_2O \quad (8.15)$$

$$2H^+ + HAsO_4^{2-} + 3H_2WO_4 + xTBP + 3RNH_2 \rightarrow H_3AsW_3O_{13} \cdot xTBP \cdot 3RNH_2 + 3H_2O \quad (8.16)$$

$$3H^+ + SiO_3^{2-} + HWO_4^- + TBP + 2RNH_2 \rightarrow H_4SiWO_7 \cdot TBP \cdot 2RNH_2 \quad (8.17)$$

$$2H^+ + WO_4^{2-} + xTBP + 0.5RNH_2 \rightarrow H_2WO_4 \cdot xTBP \cdot 0.5RNH_2 \quad (8.18)$$

$$3H^+ + H_2PO_4^- + 2HMoO_4^- + TBP + 2RNH_2 \rightarrow H_3PMo_2O_{10} \cdot TBP \cdot 2RNH_2 + 2H_2O \quad (8.19)$$

$$2H^+ + HAsO_4^{2-} + 2H_2MoO_4 + 0.5TBP + 2RNH_2 \rightarrow H_3AsMo_2O_{10} \cdot 0.5TBP \cdot 2RNH_2 + 2H_2O \quad (8.20)$$

$$5H^+ + HSiO_3^- + 4HMoO_4^- + xTBP + 7RNH_2 \rightarrow H_8SiMo_4O_{18} \cdot xTBP \cdot 7RNH_2 + H_2O \quad (8.21)$$

$$H^+ + HMoO_4^- + xTBP + 2RNH_2 \rightarrow H_2MoO_4 \cdot xTBP \cdot 2RNH_2 \quad (8.22)$$

When the content of tungsten or molybdenum in the aqueous phase increases, the composition of the extracted species may vary. The relationship between W/P and RNH_2/P for the W-P system, W/As and RNH_2/As for the W-As system, Mo/P and RNH_2/P for the Mo-P system, Mo/As and RNH_2/As for the Mo-As system, Mo/Si, RNH_2/Si for the Mo-Si system, and the concentration of phosphorus, arsenic, and silica in the presence of relatively high concentrations of tungsten or

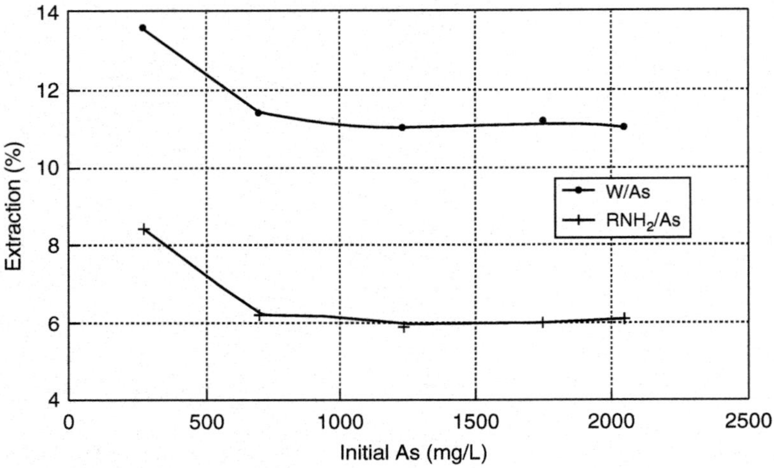

Fig. 8.44 Relationship between W/As or RNH_2/As molar ratios in the loaded organic phase and initial concentration of arsenic in the aqueous phase at equilibrium pH 7–7.8 extracted with 1% RNH_2 plus 10% TBP in n-heptane as solvent. Initial W 40 g/L

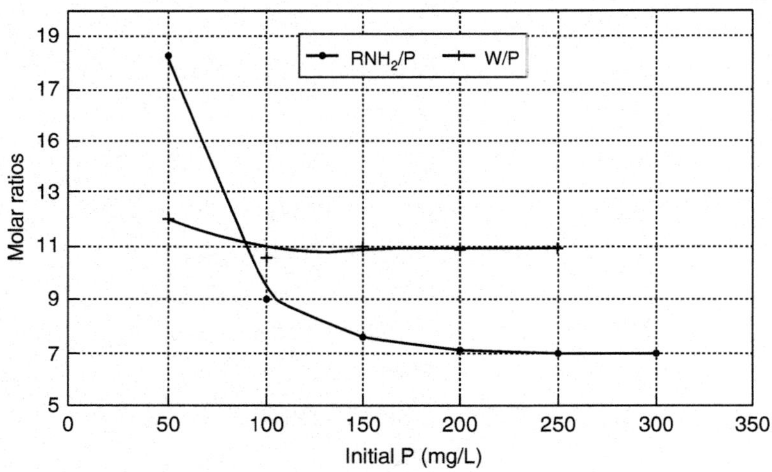

Fig. 8.45 Relationship between W/P or RNH_2/P molar ratios in the loaded organic phase and the concentration of phosphorus in the aqueous phase at equilibrium pH 7–7.9 extracted with 1% RNH_2 plus 10% TBP in n-heptane as solvent. Initial W 40 g/L

molybdenum in the aqueous phase are shown in Figs. 8.44, 8.45, 8.46, and 8.47 and Table 8.8. It can be seen that the W/As, W/P, Mo/As, Mo/P, and Mo/Si molar ratios approach 11, 11, 2.5, 3, and 9, while the corresponding RNH_2/As, RNH_2/P, RNH_2/As, RNH_2/P, and RNH_2/Si molar ratios approach 6, 7, 3, 3, and 7, respectively. The

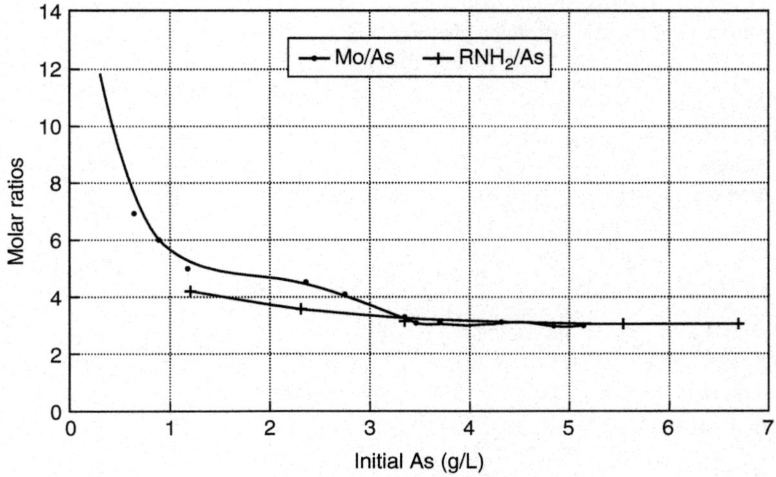

Fig. 8.46 Dependence of Mo/As or RNH_2/As molar ratios in the loaded organic phase on the initial concentration of arsenic in the aqueous phase at equilibrium pH 6–6.5 extracted with 1% RNH_2 plus 10% TBP in n-heptane as solvent. Initial Mo 40 g/L

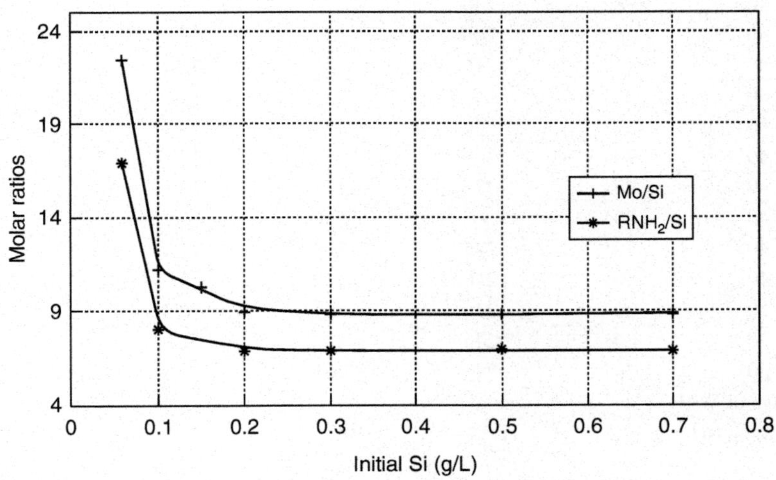

Fig. 8.47 Relationship between Mo/Si or RNH_2/Si molar ratios in the loaded organic phase and the initial concentration of silicon in the aqueous phase at equilibrium pH 6–6.8 extracted with 1% RNH_2 plus 10% TBP in n-heptane as solvent. Initial Mo 40 g/L

ratios then remain unchanged as the concentrations of phosphorus, arsenic, and silica in aqueous solution increases.

In the W-Si system, the molar ratios of W/Si in the loaded organic phase are all higher than 12 (Table 8.9). For the mixed systems containing either phosphorus, arsenic, silica, and tungsten or phosphorus, arsenic, silica, and molybdenum, the

Table 8.8 Mo/P and RNH_2/P molar ratios in the organic phase extracted with high Mo content in aqueous phase with 1% primary amine plus 10% TBP

No.	1	2	3	4	5	
Initial Mo/P in aqueous phase	4	6	9	12	15	Av.
Initial P(g/L)	6.9	6	5	4.2	3.7	
Mo/P in organic phase	2.58	2.66	2.51	2.38	2.67	2.56
RNH_2/P in organic phase	3.0	3.2	3.0	2.9	3.1	3.0

Equilibrium pH 6–7, initial Mo 40 g/L

Table 8.9 W/Si molar ratio in organic phase extracted with 1% primary amine plus 10% TBP

Initial W (g/L)	20	40	60	80	40	40	40	40
Initial Si (mg/L)	151	151	151	151	30	91	181	301
W/Si in organic phase	74.4	18.9	24	24	72	18.9	16.6	18.6

Equilibrium pH 7–8

Table 8.10 W/(P + As +Si) molar ratio in organic phase extracted with 1% primary amine plus 10% TBP

No.	Initial W (g/L)	Initial P (mg/L)	Initial As (mg/L)	Initial Si (mg/L)	W/(P + As + Si) in organic phase
1	40	82	246	151	16.8
2	40	409	246	151	11
3	40	818	246	151	11
4	40	204	49	151	11.5
5	40	204	246	151	11.1
6	40	204	393	151	12
7	40	204	738	151	11.8
8	40	204	246	60	11.2
9	40	204	246	301	11.3
10	40	204	246	542	12
11	10	204	246	151	10.4
12	20	204	246	151	11.1
13	60	204	246	151	12.8
Average of W/(P + As + Si) in organic phase					11.4

Equilibrium pH 7–8

molar ratio of W/(P + As + Si) is about 11 (Table 8.10), while the Mo/(P + As + Si) molar ratio varies (Table 8.11). It may be deduced that the extracted species of silica in the Si-W system is ($H_8SiW_{11}O_{39}$).

It can be seen that the composition of the extracted species is dependent on the initial concentration of tungsten and molybdenum. The main UV and IR absorbance peaks for the extracted species are given in Table 8.12. There are two maximum absorption peaks at 245 and 310 nm in the UV spectra of the loaded organic phase of tungsten or molybdenum only. The UV spectra of extracted species of the loaded

Table 8.11 Mo/(P + As + Si) molar ratios in the loaded organic phase extracted with 1% primary amine plus 10% TBP

No.	Initial Mo (g/L)	Initial P (mg/L)	Initial As (mg/L)	Initial Si (mg/L)	Mo/(P + As + Si) in organic phase
1	40	52	243	253	8.4
2	40	158	243	253	5.8
3	40	262	243	253	4.5
4	40	420	243	253	4.7
5	40	262	49	253	7.1
6	40	262	146	253	5.2
7	40	262	243	253	4.5
8	40	262	389	253	4.3
9	40	262	243	76	5.2
10	40	262	243	177	4.9
11	40	262	243	252	4.9
12	40	262	243	404	4.5
13	10	262	243	252	2.9
14	20	262	243	252	5.1
15	40	262	243	252	5.3
16	60	262	243	252	4.5

[a]Equilibrium pH 6–7

Table 8.12 IR and UVs spectra of loaded organic phase

System	Frequency or wavelength	Assignment
IR		
1% RNH_2 + 10% TBP in n-heptane	3150 cm^{-1}, w 1240 cm^{-1}, m	RNH_2, N–H TBP, P = O
Loaded organic phase (Mo/P = 2.5 or Mo/As = 3 or Mo/Si = 9)	3000 cm^{-1}, m 1240 cm^{-1}, s	RNH_3^+, N–H TBP, P = O
UV		
1% RNH_2 + 10% TBP in n-heptane solution	230 nm, s 270 nm, m	
Loaded organic phase of only Mo or W extracted species	310 nm, s 245 nm, m	Absorption of Mo
Loaded organic phase (Mo/P = 2.5 or Mo/As = 3 or Mo/Si = 9)	310 nm, s	No absorption at 245 nm

s = strong, *m* = medium, *w* = weak

organic phase with W/(P, As) a molar ratio of 11 and Mo/P, Mo/ As, and Mo/Si molar ratios of 2.5, 3, and 9, respectively, are completely identical with the only maximum absorption peak being at 310 nm. It seems that UV absorption of heteropoly acids or their anions is strongly dependent on the chemical environment such as solution composition and solvent. So far, the UV spectra of heteropoly acids or their anions cannot be explained satisfactorily.

The IR spectra of all extracted species were almost identical. It can be seen from Table 8.12 that the vibration absorption of N-H was shifted from 3150 cm^{-1} in RNH_2 to about 2980 cm^{-1} in the loaded organic phase, which was considered to be the absorption of N-H in cationic RNH_3^+. It can be concluded that primary amine in the loaded organic phase exists in cationic form, and the extracted species of all systems studied in fact exist as ion-associated species.

The vibration absorption intensity of the P = O group in TBP changes considerably after being contacted with aqueous solutions, while the absorption frequency is shifted only slightly, which contributes to the formation of hydrogen bonding between the O atom in the P = O group of TBP and acidic hydrogen atoms in the heteropoly acids (Table 8.12). It may be concluded from the UV and IR spectra that phosphorus, arsenic, and silica in the loaded organic phase are combined into the corresponding heteropoly acids.

It is proposed that phosphorus, arsenic, and silica are extracted into the organic phase as heteropoly acids by a solvation mechanism based on the generally accepted consideration that a primary amine exists as a neutral molecule in a medium of pH 6–8. Because heteropoly acids are strong acids and a primary amine is a weak base, the transfer of a proton from heteropoly acids to a primary amine occurs intramolecularly to form ion-associated species, which can be dissociated into free ions in a polar solvent to behave as a typical electrolyte. The mechanism of proton transfer may be expressed as follows (Fig. 8.48):where *M* represents W or Mo and only part of the structure of the extracted species is given for brevity. Due to the higher extraction power of the primary amine

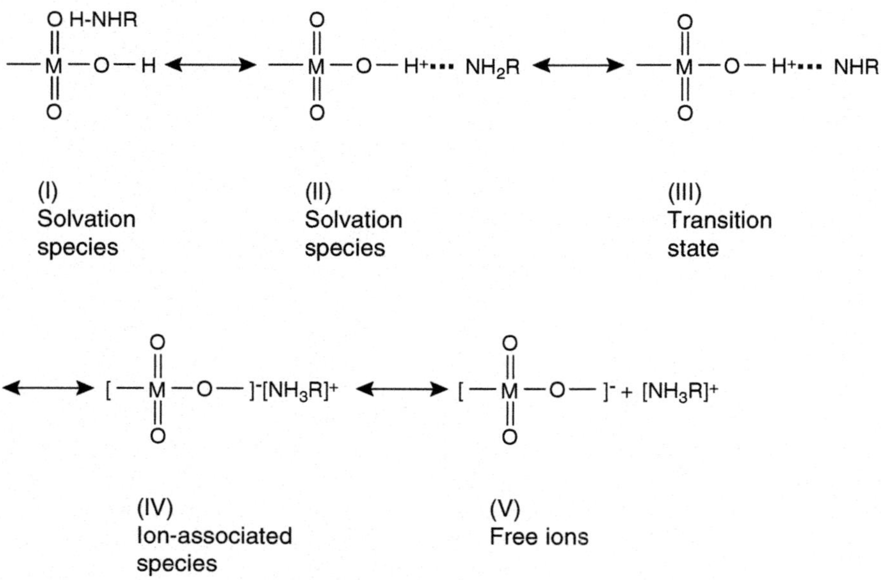

Fig. 8.48 The mechanism of proton transfer

compared with those of secondary and tertiary amines for the present system, it can be considered that the active hydrogen atoms in amines play the most important role during extraction, even though the spatial effect of amines should also be considered.

8.6 Combined Removal of SO_2, H_2S, and NO_x from Gas Streams by Chemical Absorption with Aqueous Solution of 12-Molybdophosphoric Acid and Its Reduced Species

SO_2, H_2S, and NO_x are main inorganic gaseous pollutants in various flue and stack gases and industrial waste gases. Their treatment is an important step in controlling and preventing atmospheric pollution. NO_x and SO_2 are generally found in effluents from combustion sources. Also, they exist in the waste gases emitted from chemical industries. The contents of NOx (as NO) and SO_2 separately in waste gas streams may range from 10 to 1000 mg/m^3 in flue gases or 100–5000 mg/m^3 in effluents of chemical industries. H_2S is generally found in raw town gas generated from the gasification of coal-containing sulfur and in natural gas. Moreover, H_2S is one of the main components responsible for odors, and it may also result from the incomplete combustion of organic wastes. For this reason sweetening of natural gases is an important process in chemical and fuel industries.

8.6.1 Solution Chemistry of Heteropoly Acids and Their Anions

The general formulas of heteropolyoxoanions may be represented by

$$[X_xM_mO_y]^{q-}(x \leq m) \tag{8.23}$$

where M is usually Mo and W, less frequently V, Nb, or Ta, or mixtures of these elements, in their highest (d^0, d^1) oxidation states. Such polyoxometalate anions form a structurally distinct class of complexes based predominantly, although not exclusively, upon quasi-octahedrally coordinated metal atoms. X, as heteroatoms, is found in more than 65 elements from all groups of the periodic table (except, at present, the rare gases).

However, only 12-molybdoheteropolyanions or 12-tungstoheteropolyanions of phosphorus, silica, arsenic, and germanium, such as $H_3[PMo_{12}O_{40}]$, $H_3[AsMo_{12}O_{40}]$, $H_4[SiMo_{12}O_{40}]$, $H_4[GeMo_{12}O_{40}]$, and their anions, as well as the corresponding tungstoheteropoly complexes, where the Mo and W exist in their oxidation state (VI), are considered to be important and used widely in industry as catalysts in petroleum refinery, corrosion inhibitor, dyestuff, etc.

These acids and their potassium and sodium salts are quite stable in aqueous solution and usually appear yellow color. When reductants, such as reduced Fe powder, ascorbic acid, H_2S (aqueous or gaseous), and SO_2, are added to the aqueous solutions, the Mo(VI) and W(VI) will be reduced to Mo(V) and W(V) or lower state, and a blue solution, usually named as molybdenum blue for molybdoheteropolyanions or less case, as tungsten blue for tungstoheteropolyanions, can be obtained readily. The composition of blue complexes is not yet well characterized. The stability of blue solution depends on the reductants used. When Fe(II) is used as reductant, the blue color of aqueous solution can be oxidized back into yellow upon heating in the air. However, when H_2S is used as reductant, the blue solution is found to be quite stable, even upon heating for several days in the air. In this case, relatively strong oxidants (e.g., MnO_4^{2-}, NO_2) should be used, in order to oxidize the Mo(V) into Mo(VI). The reactions may be expressed chemically as follows (represented by $H_3PMo_{12}O_{40}$):

Reduction reaction:

$$3H_3PMo_{12}(VI)O_{40} + Fe + 3H^+ \rightarrow 3H_4PMo_{11}(VI)Mo(V)O_{40} + Fe^{3+} \quad (8.24)$$

Molybdenum blue oxidation reaction:

$$4H_4PMo_{11}(VI)Mo(V)O_{40} + MnO_4^{2-} + 4H^+ \rightarrow 4H_3PMo_{12}(VI)O_{40} + Mn^{2+} + 4H_2O \quad (8.25)$$

Molybdenum blue has been used colorimetrically for the determination of heteroatoms (e.g., phosphorus, arsenic, silicon, germanium) for over 100 years. Generally speaking, all of the Mo(VI) in $[PMo_{12}O_{40}]^{3-}$ anion may be reduced from Mo(VI) to Mo(V), Mo(IV), and even Mo(III) by electro- and chemical reduction, while the Keggin structure of the anion remains unchanged. However, only the reduced species with one to six Mo(V) in the anion are stable in aqueous solution. Usually, only one to two Mo(VI) in the anions can be reduced to Mo(V) by moderate reductants. 12-Heteropoly anions and their corresponding reduced blue species will be decomposed into lower species, such as P_2Mo_5, P_2Mo_{18}, As_2Mo_6, PW_{11}, SiW_{11}, and AsW_{11} complexes, when the pH of aqueous solutions are increased to a value higher than 3. In these cases, several different anions may coexist in aqueous solutions.

12-Heteropoly acids or their anions can readily be prepared by acidifying the aqueous solutions containing the elements concerned with concentrated HCl to higher than 6 mol/L HCl and then extracted by ether. The ether is separated out, dissolves the heteropoly anions with water, and recrystallizes, and corresponding pure heteropoly acids can thus be obtained. Recently, a lot of work has been done on the recovery and preparation of heteropoly anions from industrial solid wastes containing molybdenum and tungsten by solvent extraction and flotation, from which a simple process that produced much cheaper but less purity product of the mixtures of heteropoly anions of phosphorus, arsenic, and silicon was developed.

8.6.2 Setup from Combined Removal of SO_2, H_2S, and NO_x from Gas Streams by Chemical Absorption with Aqueous Solution of 12-Molybdophosphoric Acid and Its Reduced Species

Yellow 12-molybdophosphoric acid is commercially available, with a purity of 99%. Molybdenum blue solution is not commercially available and should be prepared in laboratory. It is readily prepared qualitatively by the reaction of reductants such as ascorbic acid, reduced Fe powder, $FeSO_4$, and yellow 12-molybdosilicic acid solution. However, the quantitative preparation of molybdenum blue is much more difficult by chemical methods. In this work, it has found that H_2S can be quantitatively absorbed by 12-molybdophosphoric acid, and the chemical reaction is stoichiometric (refer to the text), which had been used for the quantitative preparation of the molybdenum blue based on the amount of H_2S absorbed. The ratios of yellow 12-molybdophosphoric acid and molybdenum blue in the mixed solution can be adjusted by controlling the amount of H_2S (aqueous solution) added.

NO, NO_2, SO_2, and H_2S used in the studies were commercially available as compressed gas which were diluted by compressed air (for NO_2, SO_2, and H_2S) or town gas (only for H_2S and NO) to the desired contents (see the text) and then passed through the mixed solutions.

All the chemical reagents used were of analytical grade unless indicated in the text. The composition of town gas used is found to be very complex, including around 19% CH_4, 46% H_2, 12% CO, 4% CO_2, 14% N_2, 0.5% O_2, and 3% unsaturated alkanes. The contents of SO_2 and H_2S in town gas were determined to be lower than 0.01 mg/m^3 and 0.5 mg/m^3, respectively. The experiments were conducted in a laboratory bench-scale dispersed-gas scrubbing reactor. It was constructed by a 20 mm i.d. and 500 mm long glass tubing, with a fritted glass disk as the bottom in the microcell, which had D4 nominal porosity (i.e., 10–16 μm). The schematic diagram is shown in Fig. 8.49.

NO_x: The sum of NO and NO_2 in the exit gases was measured colorimetrically with CrO_3-N-(1-naphthyl)-ethylenediamine dihydrochloride-sulfanilate solutions. SO_2 and H_2S were analyzed colorimetrically by K_2HgCl_4-pararosaniline and Cd $(OH)_2$-p-amino dimethylaniline dihydrochloride standard methods, respectively. The results were counter-tested with NOx, SO_2, and H_2S analyzers. Small differences were found for the instrumental and chemical analysis methods. However, the results of chemical analysis were reported in the paper, considering that the chemical methods used are of international standard, reliable, and much more convenient and accurate for determination of gas mixtures.

Average contacting time (t) for gas absorption was calculated as the expression of R/L, where R is the flow rates (cm/s) and L the height of solutions used for absorption. The change in contacting time can be obtained by adjusting the height of scrubbing solution used. The residence time of NO on oxidation tank shown in Fig. 8.49 was obtained by the formula of V/R, where V is the total volume of the oxidation tank (L)

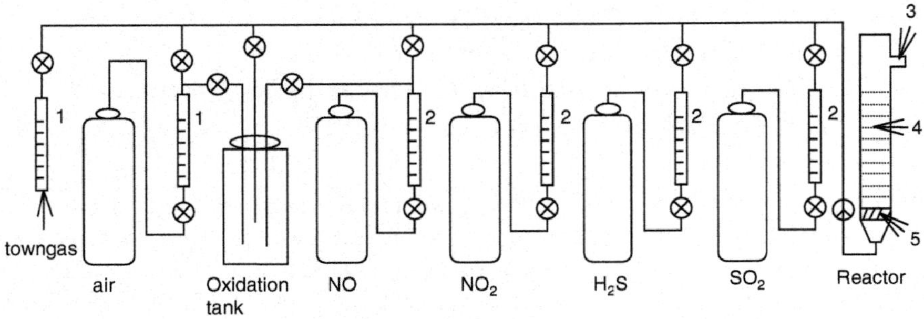

Fig. 8.49 Schematic diagram for the combined removal of SO_2, H_2S, and NO_x from gas streams. 1. Common flowmeters; 2. Microflowmeters; 3. Exit gas (for analysis); 4. Scrubbing solution; 5. D4 fritted glass disk

and *R* the flow rates of gas streams (L/min or mL/s). In order to change the residence time (oxidation strength) of NO in air streams, various tanks with different volume were used. The following experimental parameters were used, unless indicated otherwise: flow rate of gas streams (*R*), 2 L/min or 660 cm/min (11 cm/s); height of scrubbing solutions used (*L*), 40 cm; inner diameter of reactor, 2 cm; and contacting time for all waste gases (*t*), 40 cm/(11 cm/s) = 3.6 s. Contents of NO, NO_2, SO_2, and H_2S each (regardless of single components or multicomponents waste gases systems) in gas streams is 960 mg/m^3. $H_3PMo_{12}O_{40}$ content used for the absorption of SO_2 and H_2S is 0.001 mol/L. $H_3PMo_{11}(VI)Mo(V)O_{40}$ (molybdenum blue) used for the absorption of NO_x is 0.001 mol/L (prepared by the quantitative absorption of H_2S). No other acid was added to molybdophosphoric acid or molybdenum blue solutions, except in the case of testing of sulfuric acid effects on the absorption. The temperature is 20–25 °C.

8.6.3 *Removal of SO_2, H_2S, and NO_x from Gas Streams by Chemical Absorption with Aqueous Solution of 12-Molybdophosphoric Acid Solution*

Removal of H_2S or SO_2 was found to be a function of concentrations of H_2S or SO_2 and $H_3PMo_{12}O_{40}$, temperature, residence time, diameters of gas bubble, acidity, and the type of gas streams used (compressed air or town gas), as shown in Fig. 8.50. It can be seen that the removal efficiencies of H_2S were closely related to the types of gas streams used, and the efficiency in air was much higher than in town gas. For SO_2, the same removal efficiencies were found, and the curves are overlapped in the Fig. 8.50, regardless of types of gas streams used (air or town gas). Therefore, only air streams were used for the studies of SO_2 removal in the

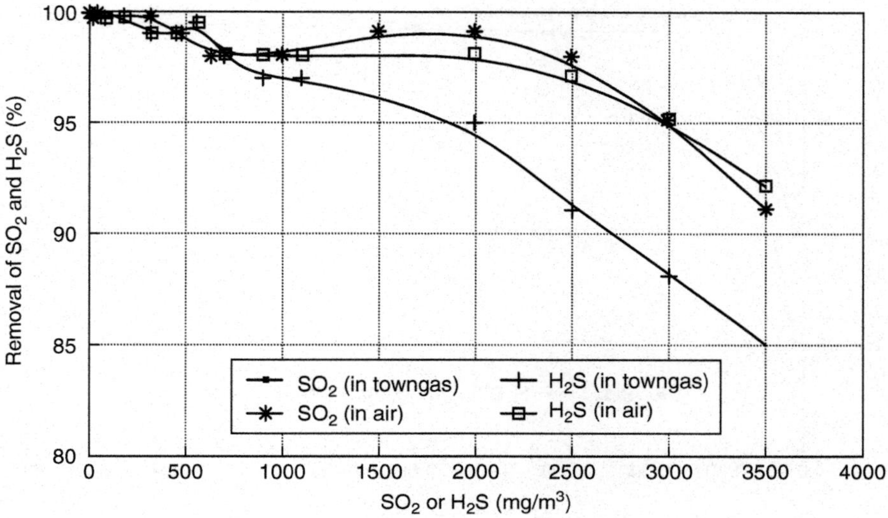

Fig. 8.50 Effects of SO_2 or H_2S concentrations on their removal from gas streams by molybdophosphoric acid solution

following experiments. Over 95% of SO_2 or H_2S can be removed, when their content in gas streams were lower than 3000 mg SO_2/m^3 (in both air and town gas) or 2000 mg H_2S/m^3 (in town gas) and 3000 mg H_2S/m^3 (in air).

Increase of molybdophosphoric acid concentration in scrubbing solutions can improve the removal efficiencies for SO_2 and H_2S (Fig. 8.51). It was seen that, when a concentration of 0.001 mol/L was used, both SO_2 and H_2S were removed nearly quantitatively from gas streams, regardless of the types of gas streams used. However, the removal efficiencies of H_2S in air were always higher than those in town gas, when concentration of molybdophosphoric acid was higher than 0.001 mol/L that was used in the following experiments.

The initial pH of 0.001 mol/L molybdophosphoric acid solution was determined to be around 3. When sulfuric and hydrochloric acids were added to the scrubbing solutions, the removal efficiencies of SO_2 and H_2S were decreased considerably (Fig. 8.52). In practical application, the addition of other mineral acids should be avoided. In this absorption process, the acidity of scrubbing solution was increased, which, however, was consumed completely in the regenerating process by the absorption of NO_2 or being oxidized by the air in the presence of catalysts.

The effect of contacting time between the waste gases removed and scrubbing solution on the removal efficiencies is shown in Fig. 8.53. Obviously, the longer the contacting time is, the higher removal efficiency of waste gases is obtained. When contacting time of around 3 s is used, the waste gases will be removed quantitatively from the gas streams.

The color of 12-molybdophosphoric acid solution is yellow. However, the color was changed to blue, when it was contacted with SO_2 or H_2S gas streams. For the

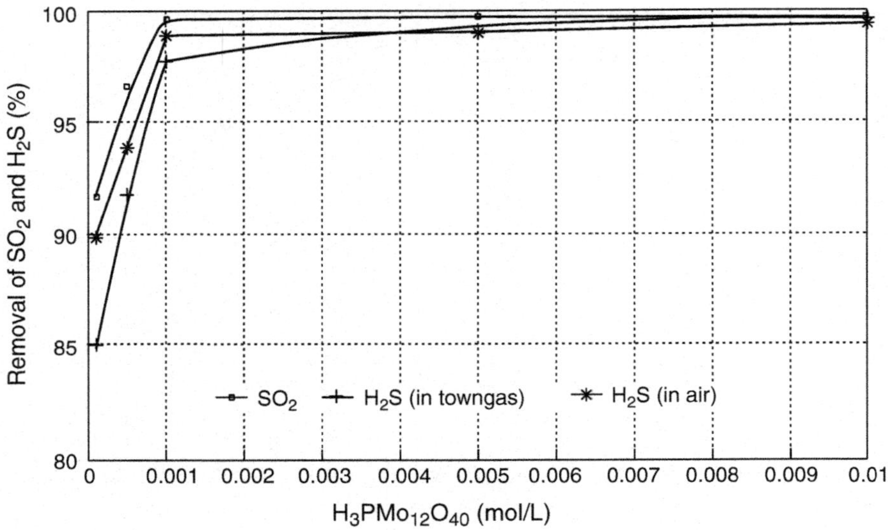

Fig. 8.51 Effects of concentration of 12-molybdophosphoric acid on the removal of SO_2 and H_2S from gas streams

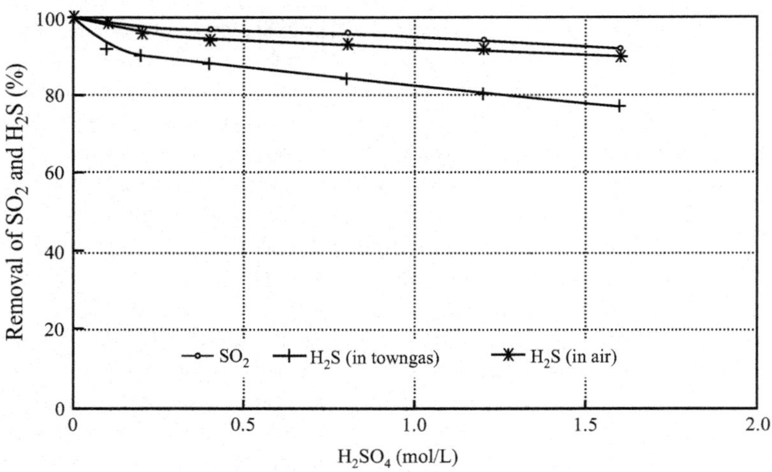

Fig. 8.52 Effects of H_2SO_4 concentrations on the removal of SO_2 and H_2S from gas streams by 12-molybdophosphoric acid solution

SO_2 absorption, the blue color appeared slowly. The violet-visible spectrophotometric determination shows that the maximum absorption wavelength of the resulting molybdenum blue solution was at 683 nm. The relationship between absorbance of the blue solution and residence time, after absorption of SO_2 by

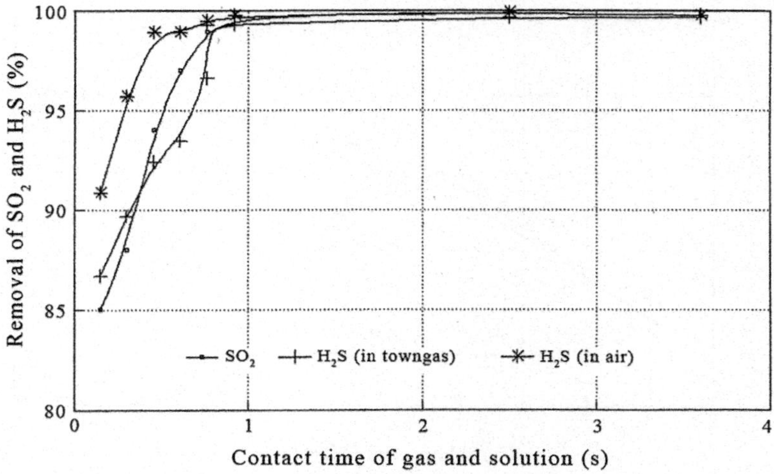

Fig. 8.53 Dependence of contact time of gas streams and scrubbing solutions of molybdophosphoric acid on the removal of SO_2 and H_2S

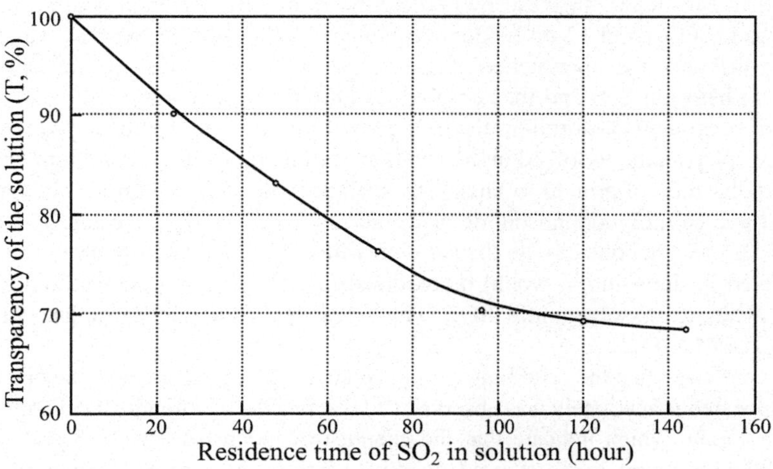

Fig. 8.54 Effects of residence time of SO_2 in molybdophosphoric acid solution being saturated with SO_2 on the spectrophotometric properties (absorbance) of the resulting blue solution

the scrubbing solution, is given in Fig. 8.54. After leaving for 4 days (96 h), the absorbance reached the maximum value. Two contacting alternatives were used for the absorption of SO_2 by molybdophosphoric acid; SO_2 was contacted with the solution continuously, without intervals, and in the other case waited for 24 h after the first absorption experiment was finished (the adjacent experimental points

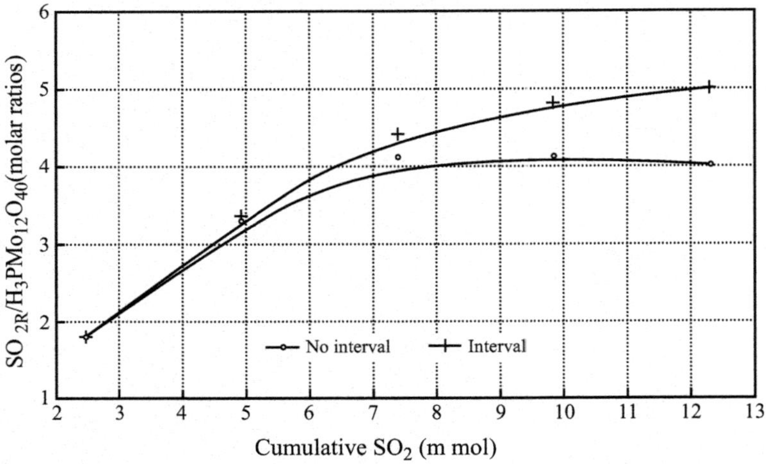

Fig. 8.55 Effects of contact arrangements on the cumulative absorption of SO_2 by molybdophosphoric acid solution

shown in Fig. 8.55). In these two cases, the ratios of maximum amount of SO_2 absorbed (SO_{2R}) and 12-molybdophosphoric acid, $SO_{2R}/H_3PMo_{12}O_{40}$, was found to be different and approached 4 and 5, respectively. Obviously, the chemical reaction between SO_2 and the scrubbing solution is quite complex. Firstly, in the case of continuous absorption, the SO_2 passed through the solution was in excess and was physically dissolved in the solution until it approached saturation, with the final molar ratio of around 4. Then, SO_2 reacted with water to form H_2SO_3, which would reduce molybdophosphoric acid into molybdenum blue species. Therefore, when SO_2 was in contact with the solution with an interval of 24 h, part of the SO_2 dissolved in the solution would be oxidized into H_2SO_4, and the solution can be used again to dissolve more SO_2, which results in the higher molar ratios of $SO_{2R}/H_3PMo_{12}O_{40}$.

In contrast, when the scrubbing solution contacted the H_2S streams, the blue color that appeared immediately was observed and reached the maximum absorbance value within 1 min, which indicate that the chemical reaction between H_2S and yellow molybdophosphoric acid occurred quickly. Like the case of SO_2 absorption, the maximum absorption wavelength of the resulting molybdenum blue solution was determined to be 683 nm, and the relationship between the absorbance and the cumulative amount of H_2S entered (inlet H_2S) or absorbed (H_2S_R) is given in Fig. 8.56. The more amount of H_2S entered into the solution, the lower the transparency of resulting blue solution was obtained. The minimum molar ratio of molybdophosphoric acid and the H_2S absorbed ($H_3PMo_{12}O_{40}/H_2S_R$) was found to be 2. It was supposed that in this case all the molybdophosphoric acid had been reacted with H_2S absorbed, and the concentrations of molybdophosphoric blue can be calculated by the H_2S absorbed, which has been used as a basis for the quantitative preparation of molybdophosphoric blue solutions (refer to the experimental section).

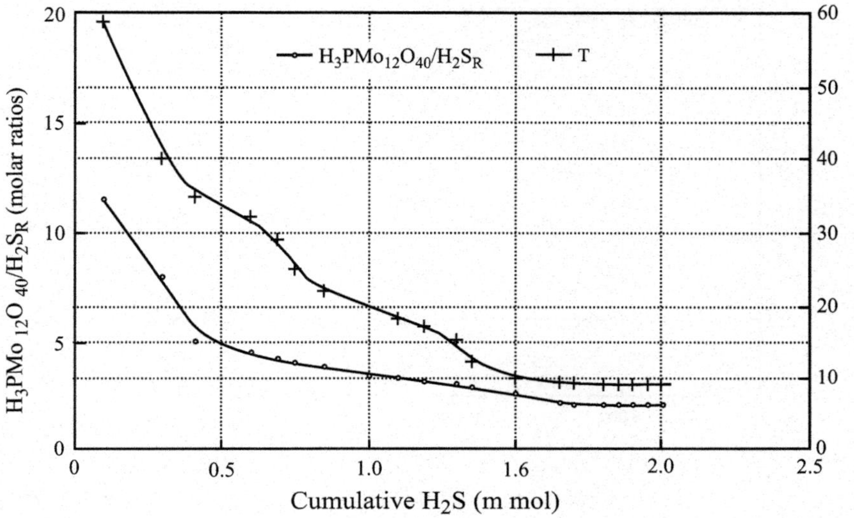

Fig. 8.56 Relationship between cumulative H_2S and $H_3PMo_{12}O_{40}/H_2S_R$

For the case of SO_2 absorption, the content of SO_4^{2-} in the resulting absorption solution was chemically determined to be nearly equal to that of SO_2 absorbed, after 15-day standing at room temperature, which indicated that it was reasonable to propose that all the SO_2 absorbed by the scrubbing solution was oxidized into H_2SO_4 by molybdophosphoric acid.

When molybdophosphoric acid solution was used for the absorption of H_2S, it was found that the solution became turbid as the absorption proceeded. After a two-day standing, the solution became clear, and a precipitate was settled down onto the bottom of reactor. The precipitate was collected, washed with water for five times, dried at room temperature, and then analyzed with thermogravimetric analyzer (TGA 51) and differential scanning calorimeter (DSC 910S), both made by TA Instruments, USA. The results are shown in Figs. 8.57 and 8.58. The corresponding result of pure sulfur is also given in the same figures. It can be seen that the TGA and DSC curves of both precipitate and pure sulfur are quite similar. In addition to this, it can be seen that the impurity (residue) of the pure sulfur was in the range of 1–2%, while that of the precipitate was around 5% (the error of TGA was about 1%). The impurity in precipitate may be the entrained salts of Mo and P, which remain in the residues and do not affect the purification of precipitate by sublimation usually used for the purification of raw sulfur.

In the DSC spectra, it can be found readily that the melting points of pure sulfur and the precipitate appeared at 106.33 °C (γ-sulfur) or 119.5 °C (β-sulfur) and 108.10 °C (γ-sulfur) or 120.48 °C (β-sulfur), and the corresponding boiling points appeared at 443.10 °C and 443.90 °C, respectively. Chemical analysis of the precipitate was also conducted, and the results were in good agreement with

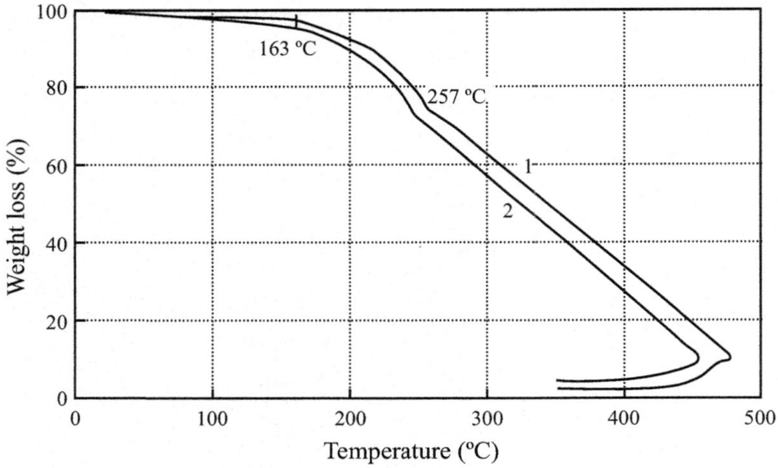

Fig. 8.57 TGA spectra of pure and the precipitated sulfur at temperature rate: 30 °C/min. 1, pure sulfur. 2, precipitated sulfur

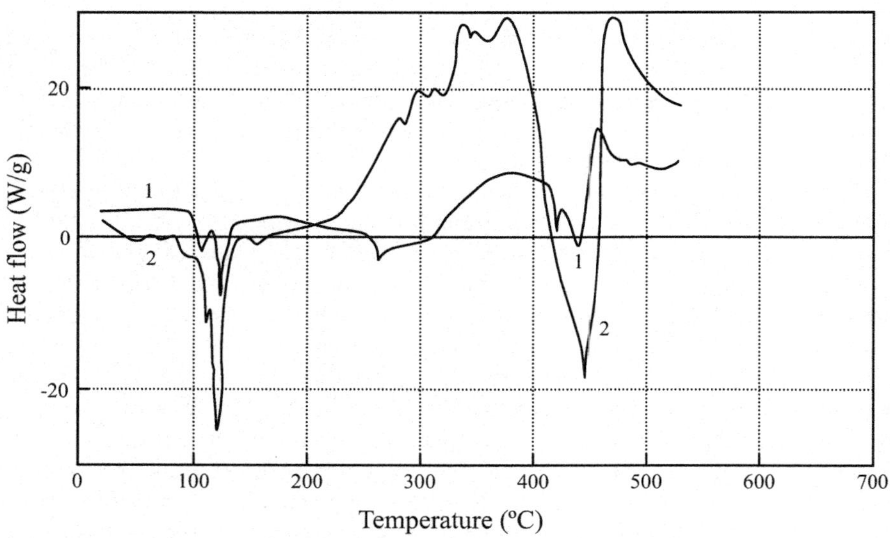

Fig. 8.58 DSC spectra of pure and the precipitated sulfur at temperature rate: 30 °C/min. 1, pure sulfur. 2, precipitated sulfur

those of TGA. The content of sulfur in precipitate was around 95%. The recovery of sulfur may be over 95%, based on the stoichiometry of the H_2S absorbed and careful collection of the precipitate.

8.6.4 Absorption of NO_2 and NO in Molybdenum Blue Solution

It is difficult to prepare molybdenum blue quantitatively by the reduction of yellow 12-molybdophosphoric acid solution with common reductants, such ascorbic acid, Fe powder, and Na_2SO_3. Nevertheless, according to the experimental results obtained above, it is possible to produce quantitative molybdenum blue by the reaction of H_2S and molybdophosphoric acid solution. Hence, a molybdenum blue solution with various concentrations, by controlling the absorbed H_2S, can be made. In the following experiments, 0.001 mol/L molybdenum blue solution was used.

The resulting molybdenum blue solution mentioned above can further be used for the effective removal of NO_2 from gas streams. When NO_2 gas streams was passed through the molybdenum blue solution, the intensity of blue color would be decreased gradually and finally became the original yellow color of $H_3PMo_{12}(VI)O_{40}$. The experimental results are shown in Figs. 8.59 and 8.60. It can be seen that the NO_2 in gas streams can be removed quantitatively, if the concentration of molybdenum blue solution is higher than 0.001 mol/L (Fig. 8.59) or NO_2 content in the streams was lower than around 500 mg/m^3 (Fig. 8.60). The removal efficiencies were found to be independent of the reaction temperature (from 20 to 90 °C) and the acidity of solution (from 0.0 to 0.3 mol/L H_2SO_4) in the tested range but dependent on the contacting time of NO_2 and molybdenum blue solution (Fig. 8.61).

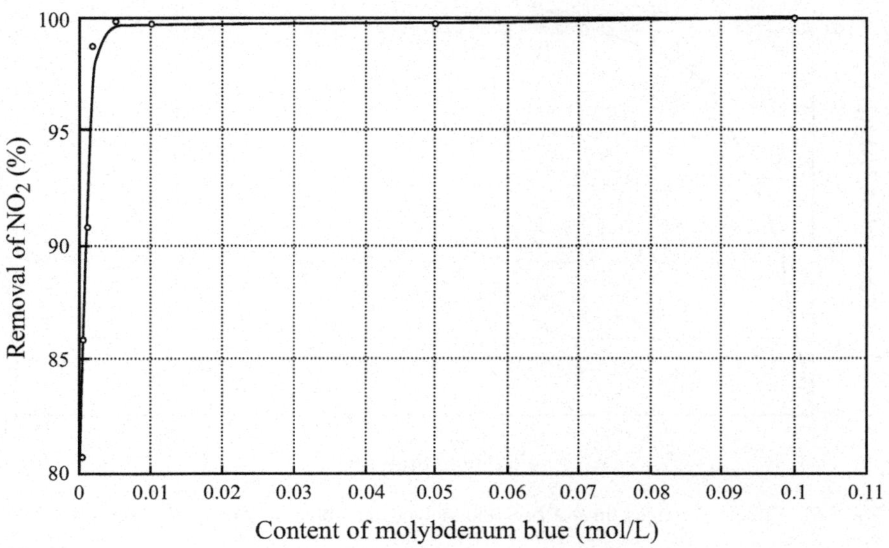

Fig. 8.59 Effects of molybdenum concentration of molybdenum blue solution on the removal of NO_2 from gas streams

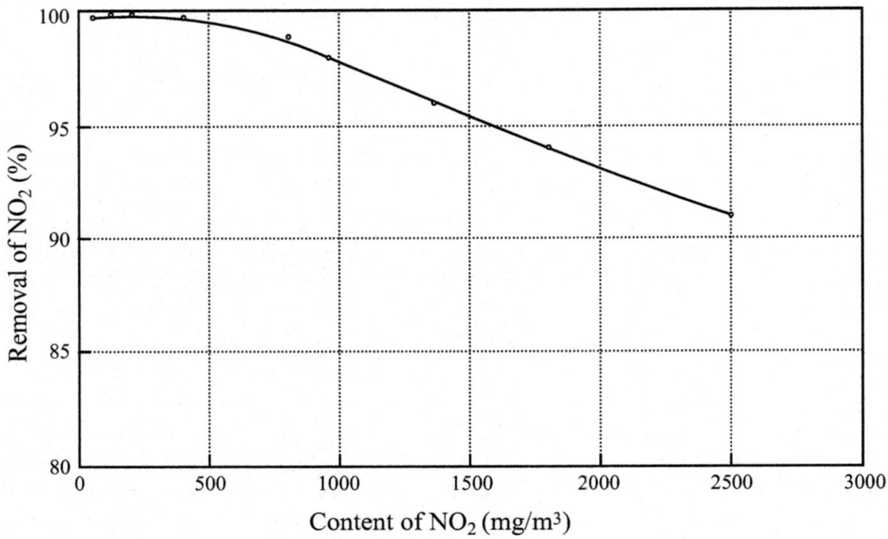

Fig. 8.60 Effects of NO_2 concentration on the removal of NO_2 from gas streams by molybdenum blue solution

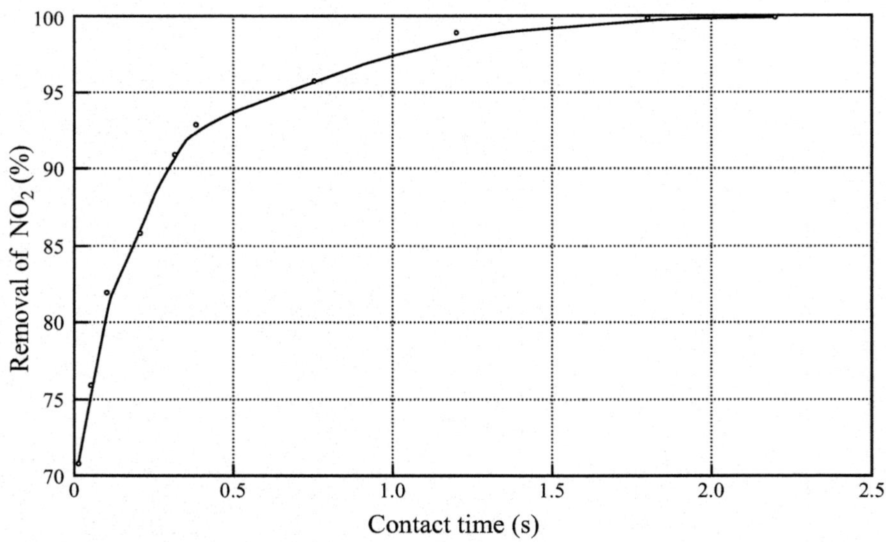

Fig. 8.61 Effects of contact time of NO_2 and molybdenum blue solution on the removal of NO_2 from gas streams

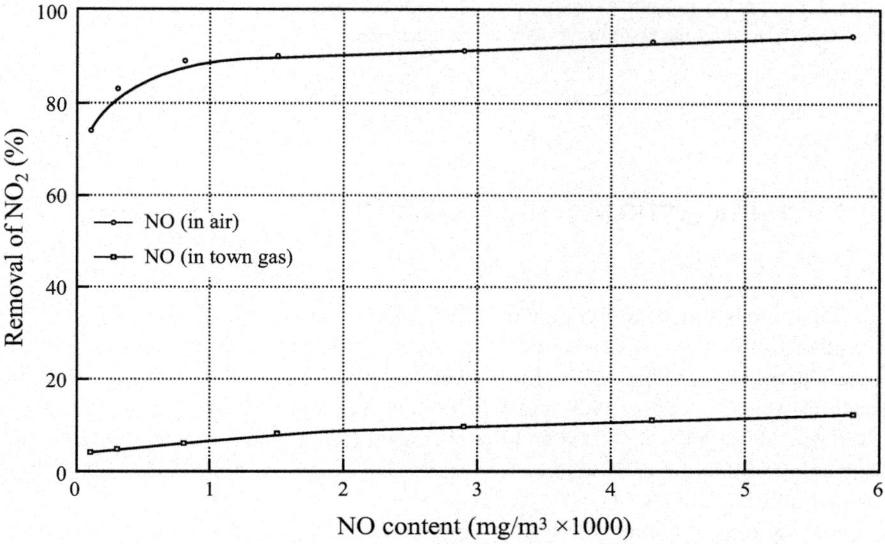

Fig. 8.62 Effects of NO concentration in air streams on its removal by molybdenum blue solution

Table 8.13 Effects of residence time of NO in air streams on the removal of NO by molybdenum blue solution

Residence time of NO in air(s)	0[a]	17	23	45	60
Removal of NO(%)	80.0	90.1	93.5	94.7	95.8

[a]Oxidation tank for NO shown in Fig. 8.48 was not used, and the air streams containing NO was passed into the scrubbing solution directly

Less than 10% of NO in town gas stream can be absorbed by molybdenum blue solution. However, when NO was oxidized by the air stream at the given residence time of 23 s as shown in Fig. 8.62, it could be removed effectively. Therefore, the absorption behavior of NO in air streams is similar to that of NO_2 and its removal was dependent on the oxidation strength (residence time) of NO in air streams. It can be seen that NO removal by molybdenum blue solution was affected remarkably by residence time in the oxidation tank and the results are shown in Table 8.13.

According to the results obtained, the chemical reactions of SO_2 or H_2S and molybdophosphoric acid could be expressed as follows:

$$SO_2 + H_2O \rightarrow H_2SO_3 \tag{8.26}$$

$$xH_2SO_3 + 2xH_3PMo_{12}(VI)O_{40} \rightarrow 2xH_4PMo_{11}(VI)Mo(V)O_{40}(\textit{Molybdenum blue}) + xH_2SO_4 \tag{8.27}$$

Table 8.14 The relationship between the H_2S absorbed by molybdophosphoric acid solution and the NO_2 removed by the resulting molybdenum blue solution

H_2S absorbed (m mol)	0.2	0.5	0.8	1.2	
NO_2 Removed (m mol)	0.11	0.24	0.41	0.65	
H_2S/NO_2	1.82	2.08	1.95	1.85	Average: 1.93

$$H_2S + 2H_3PMo_{12}(VI)O_{40} \rightarrow 2H_4PMo_{11}(VI)Mo(V)O_{40}(\textit{Molybdenum blue}) + S \tag{8.28}$$

The composition of molybdenum blue is very complex. Equations 8.27 and 8.28 are just to show the lowest chemical stoichiometry of absorption processes, not the exact chemical reactions. Therefore, it seems impossible to directly determine the reaction stoichiometry of NO_2 and Mo(V) in molybdenum blue solution. However, the ratios of the NO_2 absorbed and the H_2S used for the formation of molybdenum blue solution could be found, and the results show that the molar ratio of NO_2/H_2S was around 2 (Table 8.14).

The reaction products of NO_2 can only be NO, N_2O, N_2, and NH_4^+. It was the total amount of NO and NO_2 that was determined colorimetrically by CrO_3-N-(1-naphthyl)-ethylenediamine dihydrochloride-sulfanilate solutions. As mentioned above, the removal of NO_2 or NO_x from gas streams was nearly quantitatively, which indicates that little or no NO exist in the reaction products. Comparing the amount of NO_2 or NOx removed, the amount of NH_4^+ and N_2O was found to be negligible, based on the results obtained by chemical analysis. Thus, the predominant product should be N_2, and the chemical reactions could be represented as follows:

$$NO_2 + 4H_4PMo_{11}(VI)MO(V)O_{40} \rightarrow N_2 + 4H_3PMo_{12}(VI)O_{40} + H_2O \tag{8.29}$$

Combining Eqs. (8.26 – 8.29), it results:

$$2NO_2 + 4H_2SO_3 \rightarrow N_2 + 4H_2SO_4 \tag{8.30}$$

$$2NO_2 + 4H_2S \rightarrow N_2 + 4S + 4H_2O \tag{8.31}$$

8.6.5 Combined Removal of SO_2 and H_2S

SO_2 can react with H_2S to form elemental sulfur in gaseous environment. This unique reaction has been used for the sweetening of natural gas and coal gas for a long time, naming as clause process. In this work, it was found that the removal efficiencies can be improved considerably, when SO_2 and H_2S coexisted in gas streams (Table 8.15). By comparing with Fig. 8.50, it can be seen that a much higher concentration of H_2S in gas stream that would be removed quantitatively was allowed.

Table 8.15 Combined removal of SO_2 and H_2S in air streams by molybdophosphoric acid solution

SO_2 content (mg/m^3)	0.0	500.1	1560.0	3000.2
SO_2 removal (%)	–	99.1	99.7	99.0
H_2S content (mg/m^3)	3000.4	3000.4	3000.4	3000.4
H_2S removal (%)	95.1	96.2	97.5	99.5

Table 8.16 Combined removal of NO_x, SO_2, and H_2S in air streams with the mixed solutions of molybdophosphoric acid and its reduced molybdenum blue solution (50:50 in molar percentage)

NO_x[a] content (mg NO/m^3)	100.0	230.2	400.9	750.0	960.0
NO_x removal (%)	73.0	85.1	90.5	91.3	92.9
SO_2 content (mg/m^3)	960.0	960.0	960.0	960.0	960.0
SO_2 removal (%)	99.9	98.9	98.7	99.8	99.1
H_2S content (mg/m^3)	960.0	960.0	960.0	960.0	960.0
H_2S removal (%)	99.7	98.6	98.4	98.9	99.7

[a]The initial gas was NO and was oxidized in oxidation tank for 23 s (i.e., residence time)

8.6.6 Combined Removal of NO_x, SO_2, and H_2S

The results are shown in Table 8.16. Comparing with Figs. 8.50 and 8.51, it seems that the presence of NO_x in gas streams did not affect the removal efficiencies of SO_2 and H_2S and vice versa. It has been confirmed that all gases (SO_2, H_2S, and NO_2) can be removed from gas streams effectively.

8.6.7 Regeneration of Scrubbing Solution

In the case of H_2S and SO_2 removal, yellow molybdophosphoric acid was used, and the resulting molybdenum blue solution was found to be quite stable. The blue solution cannot be regenerated in the air or pure O_2, even upon heating. However, the blue color will fade and become yellow, when catalysts such as Cu(II) salts and NaCl are added to the solution and heated to 40 °C while the air was passed. The yellow solution can be reused for H_2S removal and the observed efficiencies were unchanged, even if over 100 times of recycling are carried out.

For NO_x absorption, the situation was just reverse. The molybdenum blue solution was used as scrubbing solution, and finally yellow molybdophosphoric acid was obtained. The regeneration of molybdenum blue solution may be realized by the addition of Fe powder, though it was not considered to be a good method.

It was possible that the regeneration of scrubbing solution is not required, when NO_x, SO_2, and/or H_2S coexisted in waste gases and the molar ratios of NO_2/(SO_2 + H_2S) was suitable. According to Eqs. 8.30 and 8.31, this ratio should be 1/2. Nevertheless, in most practical waste gases, the ratios of NO_x, SO_2, and/or H_2S

are not always equal to the value. Hence, regeneration of scrubbing solution is usually necessary.

The particles of elemental sulfur generated during the oxidation of H_2S by molybdophosphoric acid solution are observed to be quite small and cannot settle quickly. Nevertheless, the particles become bigger and settle gradually after 2–3 days standing. Some difficulties may be encountered when the process is applied industrially. A relatively large amount of scrubbing solution may be needed because the regeneration time would be quite long, owing to the longer time required for the collection of sulfur. In addition, the sulfur generated is too small and may be attached to the inner walls of scrubbing tower and the surface of filling materials in the tower, resulting to the maloperation of the process. Hence, the scrubbing tower should be specially designed and be favorable to collection of the sulfur and the washing of tower.

In order to observe the absorption behavior of heteropoly acids and their reduced species, pure 12-molybdophosphoric acids was used in this work. Nevertheless, in practical application, it is predicted that a large loss of scrubbing solution would occur and result to the increase of operation cost, if the entrained solution in the effluent gases was not recovered carefully. One of the solutions is to use a less purity molybdophosphoric acid such as that produced from solid wastes containing molybdenum and phosphorus. It is found that the presence of some ions such as K, Na, and common anions such as sulfate and chloride in the scrubbing solutions will not affect the absorption. Obviously, a mixture other than pure molybdophosphoric acid (e.g., industrial purity) can also be used as scrubbing agent, leading to the remarkable reduction in operating cost of the process. In petroleum and chemical industries, a large quantity of molybdophosphoric acid is used as catalysts, dyestuff, etc., resulting to the production of nearly equal amount of solid wastes containing the heteropoly acids. In hydrometallurgical industries of molybdenum and tungsten, hundreds and thousands of tons of solid wastes containing molybdenum or tungsten and phosphorus, arsenic, and silica are being generated. These wastes must be treated properly. These could be reclaimed effectively, and recovered as the original, but with less purity, heteropoly acids, which can be used as scrubbing agents.

Molybdophosphoric acid is a relatively strong oxidant. When some odorous compounds, such as organic sulfur and organic nitrogen, generating from the degradation of organic wastes, are passed through the scrubbing solutions, these odorous compounds will be absorbed chemically and removed to a great extent, from the gas streams. As a result, the strength of odor will be decreased considerably, and molybdophosphoric acid is reduced into molybdenum blue species. Based on the preliminary experimental results, it is believed that molybdophosphoric acid is suitable for the effective treatment of odor. Hence, the wet treatment process developed in this work can be used for the removal of NO_x, SO_2, and H_2S, as well as odor, individually or simultaneously, from gas streams, including the sweetening of raw town gas, natural gas, flue, or stack gases from power plants or municipal refuse incineration plants, effluents from chemical industries, and even the waste gases from wastewaters treatment plants.

H_2S and most of the odorous compounds are relatively strong reductants. These can react faster with molybdophosphoric acid solution. However, SO_2 is a moderate reductant and reacts slowly with the scrubbing solution. Similarly, NO_2 is a strong oxidant and can oxidize molybdenum blue solution quite fast. NO, in most cases, shows inert properties and cannot be removed by common wet scrubbing process, including molybdophosphoric acid solution. However, when it is oxidized by air, it is found that NO can also be removed effectively by the scrubbing process developed in this work.

In addition, 12-molybdosilicic acid and its reduced species can also be used for the effective removal of the waste gases from gas streams, but the regeneration of scrubbing solutions was found to be more difficult. Mixed solutions of 12-molybdophosphoric acid and its reduced molybdenum blue solutions was found to be an effective scrubbing process for the combined removal of H_2S, SO_2, and NO_x from gas streams. The affecting parameters such as the contents of the waste gases studied, acidity and concentrations of scrubbing solutions, contacting time of solutions and waste gases, were investigated. When the molar ratio of $NO_2/(SO_2 + H_2S)$ in the waste gases is just equal to 1/2, H_2S, SO_2, and NO_x will be removed quantitatively, while the regeneration of scrubbing solution is not needed. H_2S and SO_2 react with molybdophosphoric acid solution to form elemental sulfur and H_2SO_4, respectively, while molybdophosphoric acid is reduced into blue species, molybdenum blue solution. The reactions of H_2S are rather quickly, while those of SO_2 are much more slowly. The resulting molybdenum blue solution was used further for the removal of NO_x, in which molybdenum blue is rapidly oxidized back into molybdophosphoric acid and NO_x is reduced into N_2.

8.7 Recovery and Synthesis of Tungstotantalate and Tungstoniobate Using White Tungstic Acid

The heteropoly oxometalates have important applications in petroleum and pigment industry and other fields. W, Mo, V, Nb, and Ta can easily form heteropoly and isopoly oxometalates. Compared with the great variety of heteropoly oxometalates of tungsten, molybdenum, and vanadium, those of Nb and Ta are quite limited, such as tungstoniobates $[Nb_nW_{6\text{-}n}O_{19}]^{(2+n)\text{-}}$ ($n = 1$–4) and tungstotantalate $[TaW_5O_{19}]^{3-}$. It may be considered that other tungstoniobates and tungstotantalates with different compositions can be synthesized if a suitable method can be developed according to the similarity of the chemistry of W, Mo, V, Nb, and Ta.

As the main raw materials for preparing tungstoniobate and tungstotantalate, tungstic acid can be mainly classified as white tungstic acid and yellow tungstic acid. In this work, the quick determination method of tungsten concentration is proposed according to the quantitative relationship between the precipitate of white tungstic acid and pH value of a certain range. P, As, and Si can be simultaneously removed while white tungstic acid is prepared, which is used to produce new types of tungstoniobates and tungstotantalates and their organic compounds, such as $K_9[TaW_{11}O_{40}]$, $[(C_4H_9)_4N]_5K_2[TaW_{11}O_{40}H_2]$, $K_9[NbW_{11}O_{40}]\cdot 2H_2O$, $[(C_4H_9)_4N]_6K[NbW_{11}O_{40}H_2]$, $[(C_4H_9)_4N]_5K_2[NbW_{11}O_{40}H_2]$, and $[C(NH_2)_3]_{6.3}K[Nb_{1.3}W_{10.7}O_{40}H_2]\cdot H_2O$.

8.7.1 *Synthesis of Tungstotantalate*

Preparation of tungstotantalate anion and tungstoniobate anion is similar. Here takes the preparation of $[(C_4H_9)_4N]K_2TaW_{11}O_{40}H_2$ as an example.

1. Preparation of potassium tantalate

   Potassium tantalate was prepared by the reaction of tantalum pentoxide and potassium carbonate at 950 °C and then hydrolyzed in water and crystallized. The composition of potassium tantalate prepared was determined as $K_8Ta_6O_{19}\cdot(13\sim20)H_2O$, which was then dissolved in water with a concentration of 0.6 M Ta.

2. Preparation of mixed solution of potassium tantalate and potassium carbonate

   Mix potassium tantalate with Ta 0.6 M and 0.6 M potassium carbonate in a given proportion.

3. Preparation of white tungstic acid

   An aliquot of 100 mL 0.52 M sodium tungstate solution is added drop by drop to 200 mL 0.9 M nitric acid solution; the white tungstic acid is thus obtained (48 mmol as tungsten trioxide). The white tungstic acid is filtered and air-dried for use.

4. Synthesis of tungstotantalate anion

   (a) The white tungstic acid prepared (48 mmol as tungsten trioxide) is added to 400 mL water while stirred strongly with magnetic stirrer.
   (b) Potassium tantalate solution prepared from Step 2 is added to white tungstic acid at Step (a) drop by drop with a rate of about 0.2 mL/min till the pH value of the solution increases from its initial pH 2.0–2.3 to pH 5.0–5.3 at which all the white tungstic acid in solution is dissolved completely and a clear solution is thus obtained.
   (c) The solution is heated for about 10 min, filtered, and evaporated to 170 ml, and 15 mL tetrabutylammonium chloride solution with 50% concentration is added to produce organic compounds as required.

$K_9TaW_{11}O_{40}\cdot 15H_2O$ solution thus obtained had a composition of K_2O 11.8, WO_3 74.0, Ta_2O_5, 6.0, H_2O 7.9%, W/Ta = 11.1, and K/Ta = 9.1 (Cal. K_2O 12.2, WO_3 73.6, Ta_2O_5, 6.4, H_2O 7.8%). Nevertheless, yellow tungstic acid cannot react

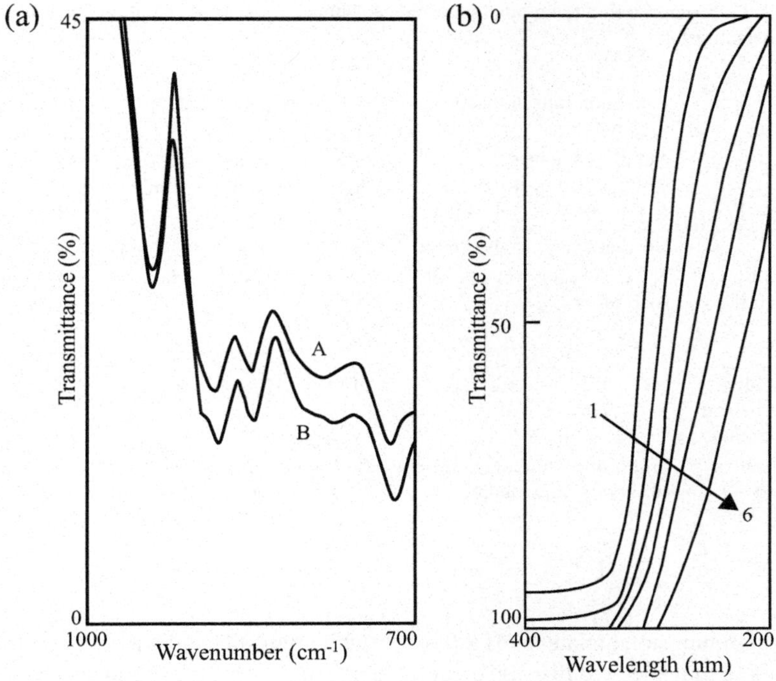

Fig. 8.63 (**a**) IR spectra of paratungstate (*A*) and potassium tantalate (*B*), (**b**) UV spectra of potassium tungstotantalate solution with different concentrations (mol/L). (1, 15×10^{-5}; 2, 6.0×10^{-5}; 3, 3.0×10^{-5}; 4, 1.8×10^{-5}; 5, 1.2×10^{-5}; 6, 0.6×10^{-5})

with potassium tantalate solution, showing the great difference of white and yellow tungstic acids.

During the reaction of white tungstic acid and potassium carbonate, the white tungstic acid could dissolve completely. The composition of formed crystal was determined as $K_{10}W_{12}O_{42}H_2{\cdot}2H_2O$. The IR spectrum and UV spectrum of potassium tungstotantalate were very similar with ones of potassium paratungstate (as shown in Fig. 8.63 and Table 8.17). The solubility of potassium tungstotantalate, in water at 25 °C, was moderate and similar with that of potassium paratungstate (as shown in Table 8.17).

While the potassium carbonate solution was added to potassium tungstotantalate solution, the pH increased from the initial value of around 5.3 and the UV spectrum of potassium tungstotantalate solution gradually approached that of the mixed solution of potassium tungstate and potassium tantalate and even overlapped at pH $>$ 13 (Fig. 8.64a). It was considered that potassium tungstotantalate disassociated in alkali medium, while no change of UV spectrum was found in acid medium, perhaps due to the protonation and non-changeable structure of tungstotantalate anion (Fig. 8.64b). In fact, the range of acidity for synthetizing

Table 8.17 Comparison among potassium tungstotantalate, potassium paratungstate, and potassium metatungstate

	Potassium tungstotantalate		Potassium paratungstate	Potassium metatungstate
	W—O	Ta—O	W—O	W—O
IR (cm^{-1})	934.4, 875, 837.5, 775.0, 706.3, 621.9, 531.3, 496.9, 437.5	887.5	934.4, 875.0, 837.5, 771.9, 703.1, 621.9, 531.3, 496.9, 437.5	960.0, 933.1, 886.2, 760.0, 580.1, 400.1
UV	No characteristic absorption		No characteristic absorption	256 nm
Solubility g/100 mL H_2O	25 °C, 3.1, hot water, very soluble		25 °C, 2.15; 100 °C, 6.6	25 °C, soluble; hot water, very soluble
The acidity of synthesis and stabilization	pH < 5.5		pH 5 ~ 8	pH < 4
Synthesis method	(a) Porous white tungstic acid (b) Potassium tantalate		(a) Porous white tungstic acid, yellow tungstic acid, or colloidal white tungstic acid (b) Potassium carbonate	The same as potassium paratungstate

potassium tungstotantalate is pH 2.0 ~ 2.3 (white tungstic acid aqueous solution) to 5.5 (white tungstic acid dissolved completely), in which the tungstotantalate anion keeps stable. Hence, it may be concluded that $[TaW_{11}O_{40}]^{9-}$ probably had a similar structure with dodecaparatungstate anion. There is no tetrahedral compound of TaO_4 reported, while the tetrahedron of VO_4 is certainly found. As V, Nb, and Ta are present in the same subgroup for similar properties, it is probably that $[TaW_{11}O_{40}]^{9-}$ is formed via the monosubstituted paratungstate anion with WO_6 replaced by TaO_4.

8.7.2 Synthesis of $[(C_4H_9)_4N]_5K_2TaW_{11}O_{40}H_2$

$[(C_4H_9)_4N]_5K_2TaW_{11}O_{40}H_2$ can be obtained when tetrabutylammonium chloride is added into $K_9TaW_{11}O_{40}{\cdot}15H_2O$ solution. Figure 8.65 presents the IR and Raman of $[(C_4H_9)_4N]K_2TaW_{11}O_{40}H_2$, respectively, which are quite similar with $[W_{12}O_{40}H_2]^{6-}$ and $[VW_{11}O_{40}H_2]^{7-}$. The IR data of relevant compounds are shown in Table 8.18. It can be seen that $[TaW_{11}O_{40}H_2]^{7-}$ almost has the same IR spectra with $[W_{12}O_{40}H_2]^{6-}$ and $[VW_{11}O_{40}H_2]^{7-}$, while there is a big gap between $[TaW_5O_{19}]^{3-}$ and $[NbW_5O_{19}]^{3-}$. In the Raman adsorption peaks of $[TaW_{11}O_{40}H_2]^{7-}$, 1092, 1010, 974, and 954 cm^{-1} belong to W-O adsorption peak, 900 cm^{-1} to Ta-O adsorption peak which is speculated from 918 cm^{-1}, and Nb-O adsorption peak in $[NbW_5O_{19}]^{3-}$.

In fact, there are a lot of heteropolyanions which have the same structure with $[TaW_{11}O_{40}H_2]^{7-}$ like $[(V)V_nW_{12-n}O_{40}]^{(3+n)-}$ ($n = 2,3,4$), $[MoW_{11}O_{40}H_2]^{6-}$, and so

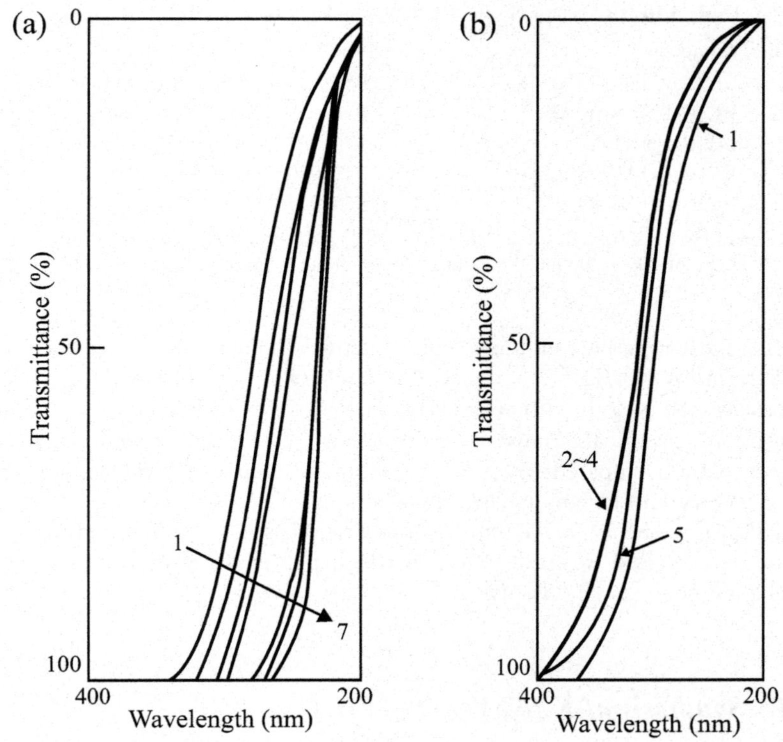

Fig. 8.64 (**a**) Relationship between UV spectra and pH value of potassium tungstotantalate with K_2CO_3 or KOH. Potassium tungstotantalate of 3.0×10^{-5} mol/L; 1, pH 5.31(initial); 2 to 6, pH 6.81, 11.0, 11.8, 12.7, 13.4; 7, the mixed solution of potassium tantalate (3.0×10^{-5} mol/L Ta) and potassium tungstate (33.0×10^{-5} mol/L) (**b**) Relationship between UV spectra and pH value of potassium tungstotantalate with HCl. Potassium tungstotantalate of 3.0×10^{-5} mol/L; 1, pH 5.31 (initial); 2 to 4, pH 2.65, 1.90, 1.36 (overlapped); 5, pH 0.93

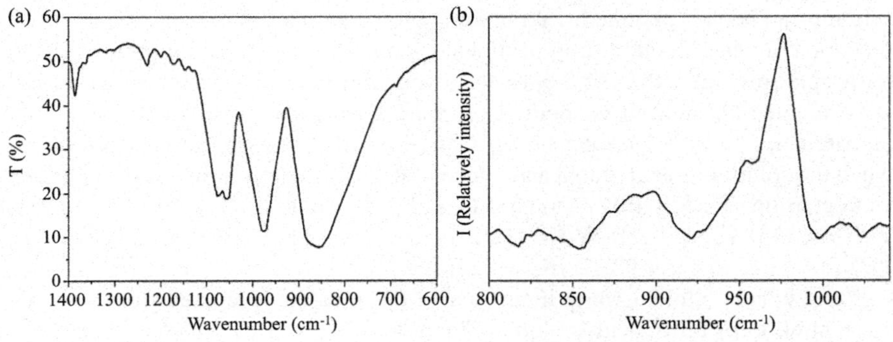

Fig. 8.65 IR spectra (**a**) and Raman spectra (**b**) of $[(C_4H_9)_4N]K_2TaW_{11}O_{40}H_2$

Table 8.18 IR spectra of relevant compounds of Ta and Nb

Compounds	ν/cm^{-1}
$[(C_4H_9)_4N]_5K_2TaW_{11}O_{40}H_2$	953, 944, 881, 791, 669, 587, 422
$[(C_4H_9)_4N]_3NbW_5O_{19}$ 2)	974, 957, 913, 884, 802, 733
$[(C_4H_9)_4N]_3TaW_5O_{19}$ 2)	975, 957, 907, 884, 805, 735
$[(CH_3)_4N]_6HVW_{11}O_{40}H_2$	942, 930, 867, 768, 580, 400
$[(CH_3)_4N]_6W_{12}O_{40}H_2$	960, 933, 886, 760, 600, 400

(1) ν ranges from 400 cm^{-1} to 1000 cm^{-1}. (2) Absorption peak within the scope of 400 ~ 700 cm^{-1} cannot be found in the literature

on. All of these negative ions are Keggin structure which can be regarded as a result that WO_6 in $[W_{12}O_{40}H_2]^{6-}$ is replaced by VO_6 or MoO_6. $[TaW_{11}O_{40}H_2]^{7-}$ may also possess the Keggin structure (WO_6 in $[W_{12}O_{40}H_2]^{6-}$ is replaced by TaO_6) based on its chemical composition, behavior of chemical reactions, IR, and Raman spectra. Compared with corresponding V compounds, the chemical properties are very similar among heteropolyanions of V, Nb, and Ta. The pH for $[MW_{11}O_{40}H_2]^{7-}$(M = V, Nb and Ta) preparation is 4.0 ~ 5.5 and porous white tungstic acid is used in the reaction in which potassium niobate and potassium tantalate are both easy to dissolve.

8.7.3 Synthesis of $K_9[NbW_{11}O_{40}]\cdot 2H_2O$

Niobic acid will be precipitated from soluble potassium niobate solution with pH value lower than 10–11 by the addition of mineral acids such as nitric acid to the solution when the concentration of niobate is higher than 0.01 M Nb. However, no precipitate of niobic acid is formed when aqueous solution of potassium niobate is added to the "solution" of white tungstic acid with initial pH 2–2.3 although the concentration of niobate solution is as high as 0.6 M and pH value up to about 5–5.3, in which all the white tungstic acid is dissolved completely and a clear solution can be thus obtained. Taking into account the facts that niobic acid will be precipitated when the first drop of niobate solution is added to the highly diluted nitric acid solution with pH 5.2 other than white tungstic acid, it may be considered that soluble tungstoniobate anion is formed during the addition of potassium niobate solution to white tungstic acid, and its potassium salt can be crystallized when the solution is evaporated and concentrated. The composition of the complex was determined as $K_9[NbW_{11}O_{40}]\cdot 2H_2O$. Cal. Nb 2.96, K 11.2, W 64.4, H_2O 1.1%, W/Nb = 11.0, K/Nb = 9.0, W/K = 1.22, Found Nb 3.0, K 11.9, W 65.4, H_2O 1.0%, W/Nb = 11.2, K/Nb = 9.2, W/K = 1.24.

Yellow and colloidal white tungstic acid (such as that prepared from the high concentration of nitric acid solution) cannot be dissolved by aqueous solution of potassium niobate to form tungstoniobate anion, which may be contributed to their much weaker acidity of aqueous solution in comparison with that of white tungstic

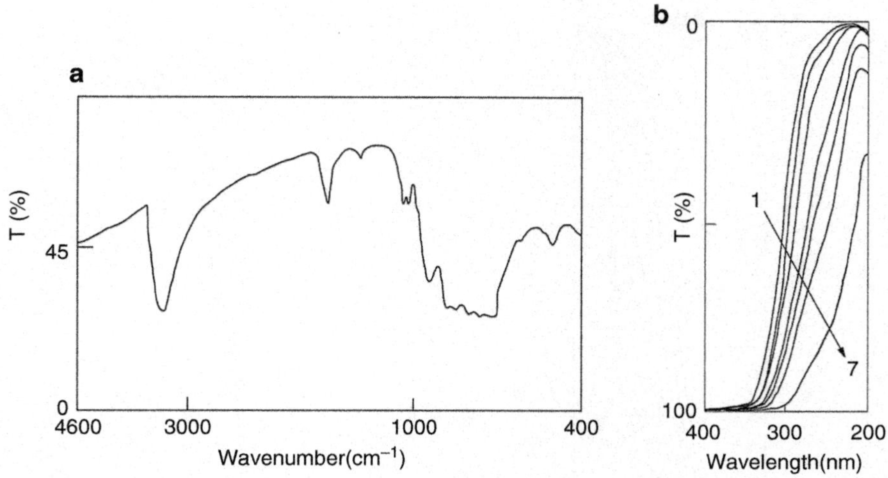

Fig. 8.66 (**a**) IR spectra of potassium 11-tungstoniobate. (**b**) The relationship between UV absorption shift and molar of aqueous solution of potassium 11-tungstoniobate. (1, 4.9×10^{-5}; 2, 3.9×10^{-5}; 3, 3.2×10^{-5}; 4, 1.9×10^{-5}; 5, 1.5×10^{-5}; 6, 1.0×10^{-5}; 7, 0.5×10^{-5}mol/L)

acid. IR spectra (Fig. 8.66a) of potassium 11-tungstoniobate are extremely similar to that of dodecameric anion of paratungstate, in which the absorption peaks of $\nu_{\text{W-O}}$ (940.37 cm^{-1}), $\nu_{\text{W-O-W}}$ (871.9 cm^{-1}), and $\nu_{\text{Nb-O}}$ (890.0 cm^{-1}) can be found as shown in Fig. 8.66a. The other absorption peaks are at 3427.0, 1637.5, 1384.4, 1075.0, 1043.7, 1028.1, 840.6, 769.0, 756.3, 709.4, 621.9, and 496.9 cm^{-1}.

UV spectra of aqueous solution of potassium tungstoniobate are also similar to that of dodecameric anion of paratungstate and no absorption peaks can be found in the UV region as given in Fig. 8.66b. It may be proposed reasonably that the structure of 11-tungstoniobate anion is similar to that of dodecameric anion of paratungstate based on the similarity of IR and UV spectra of these two different anions as mentioned above, resulting from the substituting of one WO_6 octahedron in the latter anion by one NbO_4 tetrahedron to form the former anion.

The stability of aqueous solution of potassium 11-tungstoniobate in alkaline or acidic media was studied spectrophotometrically. It is showed that the anion can be decomposed in alkaline solution of potassium carbonate or hydroxide, while the structure of the anion may be considered to keep unchanged in acidic medium as shown in Fig. 8.67, respectively. It can be seen that the UV absorption curve was shifted downward considerably when the pH value of aqueous solution was higher than the initial pH 5.4 by the addition of potassium carbonate solution, while only slight shift can be found in the curve by the addition of HCl solution even though the pH value of the solution was lower than 1.

It is the first time to report the synthesis of 11-tungstoniobate anion by the reaction of white tungstic acid and soluble potassium niobate solution. It is of great significance for the chemistry of heteropoly of niobium and tantalum which can

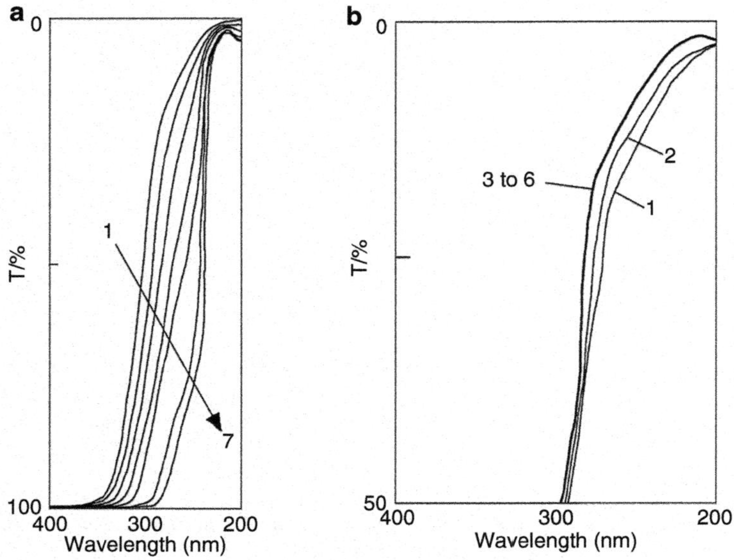

Fig. 8.67 (**a**) The effect of pH value on the UV absorption shift by the addition of potassium carbonate or hydroxide solution to aqueous solution of potassium 11-tungstoniobate with initial concentration of 3.9×10^{-5} M. 1, pH 5.4 (initial pH); 2, pH 7.2; 3, pH 10.2; 4, pH 10.8; 5, pH 11.1; 6, pH 13.0; 7, pH 13.6; (**b**) The dependency of UV absorption shift on pH value by the addition of HCl solution to aqueous solution of potassium 11-tungstoniobate with initial concentration of 3.9×10^{-5} M. 1, pH 5.4 (initial pH); 2, pH 7.2; 3 to 6, pH 1.4, 1.2, 1.0 and 0.9 (overlapped in the Figure).

also form tungstotantalate anion. Effective separation of niobium and tantalum may be carried out based on the great different properties of tungstoniobate and tungstotantalate anions.

There were many substitution derivatives of metatungstate, such as $[VW_{11}O_{40}H_2]^{7-}$ and $[MoW_{11}O_{40}H_2]^{6-}$, which all have Keggin structure, and most of them are substitution derivatives of V. Based on the chemical similarity of V, Nb, and Ta, the existence of tungstoniobate is reasonable. Moreover, the acidity of $[VW_{11}O_{40}H_2]^{7-}$ synthetized is pH 5, which is nearly consistent with $[NbW_{11}O_{40}H_2]^{7-}$.

8.7.4 *Synthesis of $[(C_4H_9)_4N]_6K[NbW_{11}O_{40}H_2]$ and $[(C_4H_9)_4N]_5K_2[NbW_{11}O_{40}H_2]$*

When tetrabutylammonium chloride is added to $K_9[NbW_{11}O_{40}]$ solution, two complexes can be obtained as determined: $[(C_4H_9)_4N]_6K[N_bW_{11}O_{40}H_2]$ (complex A), Cal. C 27.11, H 5.13, N 1.97, K_2O 1.10, Nb_2O_5 3.12, WO_3 60.03, W/Nb = 11.0, weight loss 35.75%, Found C 23.80, H 4.40, N 1.73, K_2O 2.42, Nb_2O_5 3.28, WO_3

Table 8.19 Relationship between K_2CO_3/Nb and the composition of complex (molar ratio)

K_2CO_3/Nb	Composition	Productivity (W)	Productivity (Nb)
Only Nb	–	30	15
1	$[(C_4H_9)_4N]_6K[N_bW_{11}O_{40}H_2]$ (complex A)	42	29
2	A + B, A mainly	23	30
4	$[(C_4H_9)_4N]_5K_2[N_bW_{11}O_{40}H_2]$ (complex B)	28	35

Note: 0.9 M Ni, 0,9 M potassium carbonate, initial pH 2.1 (white tungstic acid solution), terminal pH 5.3 ~ 5.7 (white tungstic acid dissolved completely)

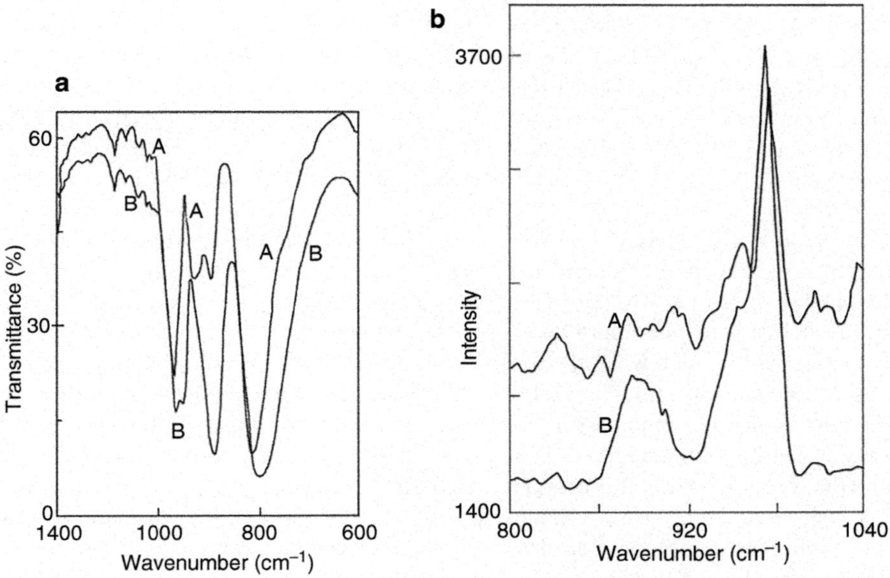

Fig. 8.68 (**a**) IR spectra and (**b**) Raman spectra of tetrabutylammonium chloride-11-tungstoniobate

63.47, W/Nb = 11.0, weight loss 35.73%, and $[(C_4H_9)_4N]_5K_2[N_bW_{11}O_{40}H_2]$ (complex B), Cal. C 23.80, H 4.40, N 1.73, K_2O 2.42, Nb_2O_5 3.28, WO_3 63.47, W/Nb = 11.0, weight loss 31.26%, Found C 23.90, H 4.30, N 1.80, K_2O 2.13, Nb_2O_5 3.10, WO_3 63.47, W/Nb = 11.1, weight loss 31.14%. The relationship between K_2CO_3/Nb ratio and composition of complexes is presented in Table 8.19.

The IR and Roman spectra are very similar with that of $[W_{12}O_{40}H_2]^{6-}$ and $[VW_{11}O_{40}H_2]$ as shown in Fig. 8.68. The $v_{Nb\text{-}O}$ peak (918.8 cm^{-1}) was extremely similar to that of quaternary] ammonium of $[NbW_5O_{19}]^{3-}$, while the $v_{Nb\text{-}O}$ peak of complex B was at 938 cm^{-1}. The Raman absorption peaks of 976, 958 cm^{-1} (W-O), and 918 cm^{-1}(Nb-O) were for complex A, while 974, 954 cm^{-1}(W-O), and 905 cm^{-1} for complex B. Other absorption peaks were 881, 890, and 896 cm^{-1} (common for complex A and B.)

Based on the chemical formula, IR, Raman, and chemical behavior of complex A and B, it was considered that they had same framework of $[W_{12}O_{40}H_2]^{6-}$ and structure of $[VW_{11}O_{40}H_2]^{7-}$, resulting from the substituting of one WO_6 octahedron in $[W_{12}O_{40}H_2]^{6-}$ by one NbO_4 tetrahedron to form $[NbW_{11}O_{40}H_2]^{7-}$. Complex A and B were found to be both soluble in acetonitrile, acetone, and methanol, while non-soluble in ether, ethanol, and water.

8.7.5 Synthesis of $[C(NH_2)_3]_{6.3}K[Nb_{1.3}W_{10.7}O_{40}H_2]\cdot H_2O$

When guanidine is added to $K_9[NbW_{11}O_{40}]$ solution, $[C(NH_2)_3]_{6.3}K[Nb_{1.3}W_{10.7}O_{40}H_2]\cdot H_2O$ can be obtained. The IR spectrum (Fig. 8.69a) of $[C(NH_2)_3]_{6.3}K[Nb_{1.3}W_{10.7}O_{40}H_2]\cdot H_2O$ was very similar to that of $[W_{12}O_{40}H_2]^{6-}$ and $[VW_{11}O_{40}H_2]^{7-}$, with characteristic absorption peaks of the prepared compound at 933.2, 875.8, 777.3 cm^{-1}, and $[C(NH_2)_3]_6[W_{12}O_{40}H_2]$ at 930.1, 876.2, 768.0 cm^{-1}, and $K_7[VW_{11}O_{40}H_2]\cdot 14H_2O$ at 925.0, 868.2, and 765.1 cm^{-1}. The absorption peaks of $\nu_{Nb\text{-}O}$ were found at 872 cm^{-1} in the Roman spectrum (Fig. 8.69b) of $[C(NH_2)_3]_{6.3}K[Nb_{1.3}W_{10.7}O_{40}H_2]\cdot H_2O$ and other absorption peaks at 526, 602, 618, 642, 952, and 1012 cm^{-1}.

The UV spectra of $[C(NH_2)_3]_{6.3}K[Nb_{1.3}W_{10.7}O_{40}H_2]\cdot H_2O$ solution are presented in Fig. 8.70a. The maximum absorption wavelength was 257 nm ($\varepsilon = 5.5 \times 10^4$ $l\cdot cm^{-1}\cdot mol^{-1}$), which was very similar with that of metatungstate solution, and the absorption intensity at 257 nm followed the Beer Law (Fig. 8.70b). When the pH of the tungstoniobate complex solution increased from the initial 5.37, the absorption intensity at 257 nm decreased. When the concentration of tungstoniobate solution and the mixed solution of potassium tungstic and potassium niobate was equal, two UV spectra of them were overlapped at pH > 13 (Fig. 8.71a). It confirmed that the tungstoniobate anion could decompose in alkali medium, completely at pH > 13. The results showed that the UV absorption intensity at 257 nm of the completely decomposed $[C(NH_2)_3]_{6.3}K[Nb_{1.3}W_{10.7}O_{40}H_2]\cdot H_2O$ aqueous solution also

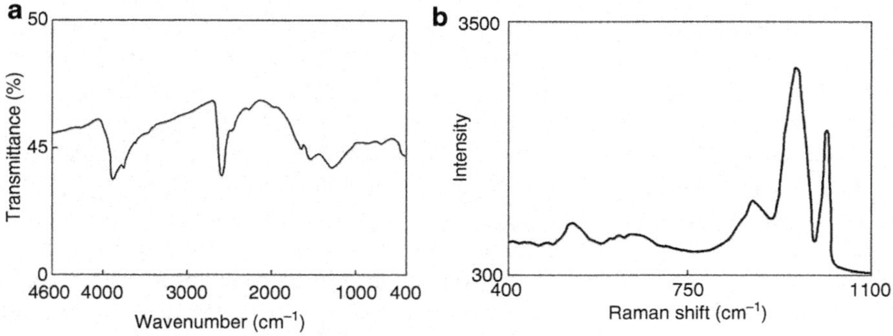

Fig. 8.69 (**a**) IR spectra and (**b**) Raman solid spectra of $[C(NH_2)_3]_{6.3}K[Nb_{1.3}W_{10.7}O_{40}H_2]\cdot H_2O$

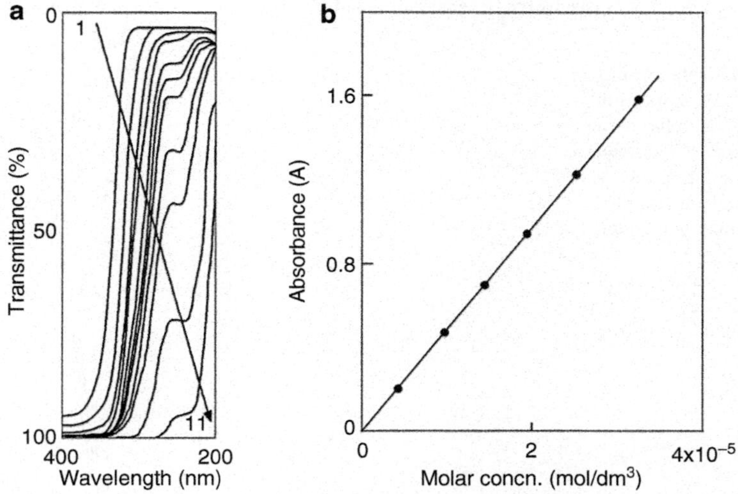

Fig. 8.70 (**a**) Relationship between UV absorption shift and concentration (mol/L) of aqueous solution of the tungstoniobate complex. (1–18×10^{-5}; 2–7.3×10^{-5}; 3–3.5×10^{-5}; 4–2.9×10^{-5}; 5–2.2×10^{-5}; 6–1.8×10^{-5}; 7–1.5×10^{-5}; 8–1.1×10^{-5}; 9–0.73×10^{-5}; 10–0.37×10^{-5}; 11–0.15×10^{-5}); (**b**) Liner relationship between concentration (mol/L) and absorbance of aqueous solution of tungstoniobate complex at 257 nm

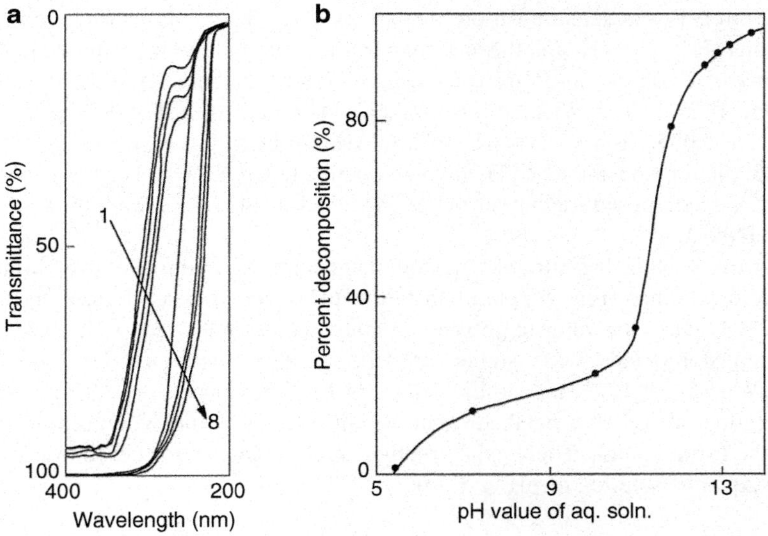

Fig. 8.71 (**a**) Effects of pH value on UV absorption shift by the addition of potassium carbonate or hydroxide solution to the tungstoniobate complex with initial concentration of 3.7 × 10^{-5} mol/L. (**b**) Relationship between percent decomposition and pH value by the addition of potassium carbonate or hydroxide solution to aqueous solution of tungstoniobate complex

No.	1	2	3	4	5	6	7
pH	5.4	7.5	10.0	11.0	12.9	13.1	13.7
B	1.01	0.86	0.78	0.66	0.09	0.08	0.07

No. 1 initial solution of the tungstoniobate complex (pH 5.4)
No. 8 mixed solution of potassium niobate (4.8 × 10^{-5} mol/L Nb) and potassium tungstate (39.59 × 10^{-5} mol/L W)

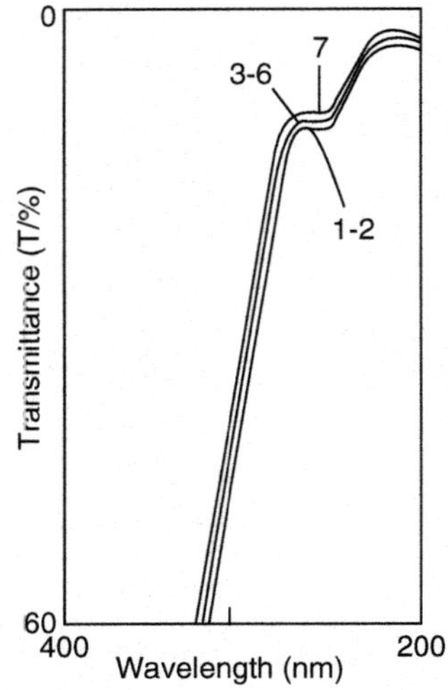

Fig. 8.72 Relationship between UV absorption shift and pH value by the addition of HCl solution to aqueous solution of tungstoniobate complex with initial concentration of 3.7×10^{-5} mol/L. 1, pH 5.4 (initial pH value); 2, pH 4.8; 3–6 pH 3.5, 2.9, 2.4, 1.5 (overlapped in Figure); 7, pH 0.9

followed Beer Law. Therefore, the relationship of decomposition rate and pH was established as follow.

Assuming the tungstoniobate complex at a certain pH (pH < 13), the absorbance at 257 nm is B, the decomposition rate is x ($0 < x < 1$), the absorbance of original solution (pH 5.37) is H, and the absorbance after completely decomposed is A. For the invariant volume and total concentration, and the characteristic absorption at 257 nm, $1-x = (B\text{-}Ax)/H$ or $x = (H\text{-}B)/(H\text{-}A)$ is obtained. In Fig. 8.70a, $H = 1.01$ and $A = 0.075$, so $x = (1.01\text{-}B)/0.935$. According to the given relationship of pH ~ B, the relationship of x-pH could be obtained (Fig. 8.71b). The tungstoniobate complex decomposed quickly at pH > 11, which is also the minimum acidity for stable $[Nb_6O_{19}]^{6-}$.

However, the UV spectra of tungstoniobate complex solution showed little shift in acidic medium (Fig. 8.72). In fact, the pH value increased from the initial 2.0 ~ 2.3 when the tungstoniobate complex was synthesized. It indicated that the tungstoniobate was stable in the acidic medium. In conclusion, $[Nb_{1.3}W_{10.7}O_{40}H_2]^{7.3-}$ has similar structure to paratungstate, resulting from the substituting of one WO_6 octahedron in the latter anion by one NbO_4 tetrahedron to form the former anion. There exist no other soluble niobate, except for niobic acid, in the aqueous solution of pH < 5 ~ 6.

8.7.6 Quick Determination of Tungsten Based on White Tungstic Acid: Gravimetric Method

The optimal acidity for white tungstic acid preparation is 1 mol/L HNO_3 and 0.5 mol/L Na_2WO_4 solution, with 95% W recovery. The effect of acidity of

Table 8.20 Effects of H_2SO_4 acidity on W precipitation percentage

No.	H_2SO_4, mol/L Initial point	Terminal point	H_2SO_4 consumption (C), mmol	Yield of WO_3, mmol	C/ WO_3	Precipitation percentage/%
1	5.0	2.0	15.0	8.96	1.67	108
2	4.5	1.6	12.4	8.46	1.46	102
3	4.0	1.4	12.7	8.53	1.49	103
4	3.5	1.2	10.6	8.71	1.22	105
5	3.0	0.99	11.1	8.58	1.29	104
6	2.5	0.70	10.9	8.36	1.30	101
7	2.0	0.52	9.5	8.62	1.10	100
8	1.5	0.32	8.6	8.23	1.04	99.5
9	0.95	0.045	8.6	8.31	1.04	100
10	0.90	0.035	8.3	8.33	1.00	100
11	0.85	0.027	8.0	8.30	0.94	100
12	0.80	0.022	7.6	8.03	0.95	97.1
13	0.75	0.015	7.2	7.45	0.97	90.0
14	0.65	0.016	6.0	6.38	0.98	76.9

Notes: 1. 0.827 mol/L Na_2WO_4 10 mL; reactants are strongly stirred by electromagnetic stirrer at 20 °C
2. No. 1 ~ 5, white tungstic acid is in colloidal state; no. 13, 14, white tungstic acid is dissolved in Na_2WO_4

H_2SO_4 or HNO_3 on W recovery is shown in Tables 8.20 and 8.21. White tungstic acid can be much more easily prepared in HNO_3 than in H_2SO_4, as the latter can make white tungstic acid colloidal and unfavorable precipitation. However, for a practical application, H_2SO_4 is preferred as SO_4^{2-} in the resultant wastewater can be discharged freely without regulated control, while the regulated TN is very low.

From Table 8.20, it can be seen that when $H_2SO_4 > 2$ mol/L, the precipitation rate of white tungstic acid exceeds 100%. When the terminal concentration of H_2SO_4 is below 0.02 mol/L (pH = 1.5 ~ 2), the precipitation rate will be under 100% because white tungstic acid dissolves into excessive Na_2WO_4 solution. Therefore, W can be quantitatively precipitated at initial H_2SO_4 concentration <2 mol/L and terminal H_2SO_4 concentration >0.02 mol/L (or pH < 1.5).

Similarly for HNO_3, the optimal initial HNO_3 concentration should be 1 ~ 2 mol/L and terminal concentration needs to be greater than 0.2 mol/L. Impurities like Cr^{2+}, Al^{3+}, and Fe^{3+} will not influence the determination.

Tungsten exists as monomeric WO_4^{2-} in Na_2WO_4 solution and then converts into tungstic acid as soon as dilute acid is dropped into the solution. Apparently, tungstic acid newly generated belongs to monomeric acid. White tungstic acid is acidic which contains the bond W-OH known from IR, while yellow species have none. The yellow tungstic acid can be expressed as $WO_3{\cdot}nH_2O$ ($n = 1 \sim 2$).

H^+ in white tungstic acid is easy to exchange with positive ion like Na^+ and Cs^+, which has the highest exchange capacity, and then get released. From Tables 8.22 and 8.23, it can be found that white tungstic acid has a high exchange capacity. However, it would decrease upon aging, proving that polymerization happens.

Table 8.21 Effects of HNO_3 acidity on W precipitation percentage

No	HNO_3, mol/L Initial point	Terminal point	HNO_3 consumption (C), mmol	Yield of WO_3, mmol	C/WO_3	Precipitation percentage/%	States
1	7.1	2.6	19.6	6.00	3.3	101	Yellow, colloidal
2	6.4	2.1	21.72	5.93	3.6	100	
3	5.7	1.9	18.24	5.94	3.0	100	
4	4.9	1.7	16.36	5.92	2.7	100	
5	4.2	1.3	15.66	5.95	2.6	100	
6	3.5	1.0	15.40	5.93	2.6	100	
7	2.1	0.48	11.64	6.08	1.9	102	Pale yellow, porous
8	1.4	0.16	10.36	5.54	1.9	93	White, porous
9	1.0	0.12	8.2	4.41	1.9	74	

Notes: 1. 0.595 mol/L Na_2WO_4 10 mL; HNO_3 10 mL; reactants are strongly stirred by electromagnetic stirrer at 20 °C
2. No. 1 ~ 6 are hard to be filtered and washed; only under the situation that $HNO_3 < 2$ mol/L, could the white tungstic acid be generated

Table 8.22 Ion exchange between H^+ from white tungstic acid and Na^+

No.	Aging time of white tungstic acid (h)	Exchange capacity (H^+, mmol/g)	WO_3/H^+
W_1	2	0.443	8
W_2	12	0.299	12
W_3	Yellow tungstic acid	0.040	90

Notes: 3 mol/L NaCl; 20 °C; white tungstic acid, $WO_3 \cdot 2.5H_2O$

Table 8.23 Ion exchange between H^+ from white tungstic acid and cations

Cations	Cs^+	NH_4^+	K^+	Na^+	Li^+	Ca^{2+}	Zn^{2+}
Exchange capacity (H^+, mmol/g)	0.839	0.644	0.639	0.443	0.368	0.385	0
W/H^+	4.3	5.6	5.6	8	9.8	4.7	

Notes: concentration of cations are all 3 mol/L except Cs_2SO_4 which is 1 mol/L

White tungstic acid belongs to monomeric acid initially with the theoretical value of W/H^+ 0.5. Then it would transform to $NaHWO_4$ or Na_2WO_4 due to exchanging with Na^+ over time if it is not filtered in time. Finally, products would polymerize into colloid.

Reaction processes are as follows:

$$2H^+ + WO_4^{2-} \rightleftharpoons H_2WO_4 \downarrow \text{ (white tungstic acid)} \tag{8.32}$$

$$H_2WO_4 + Na^+ \rightleftharpoons NaHWO_4 + H^+ \tag{8.33}$$

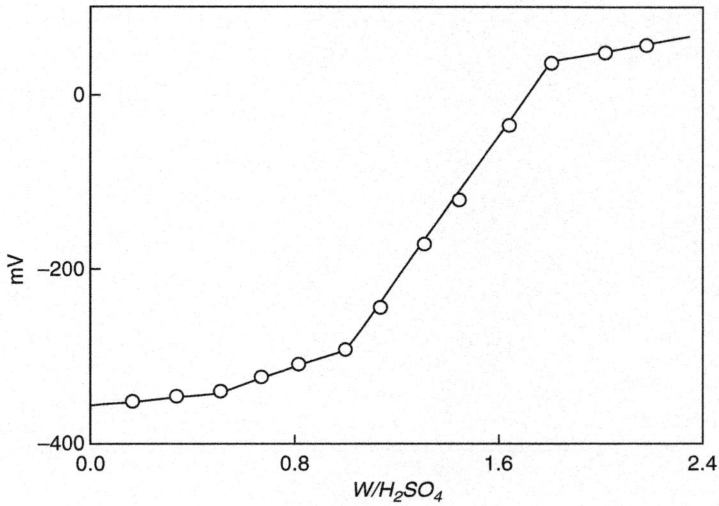

Fig. 8.73 The potential curve for adding Na_2WO_4 solution to H_2SO_4

$$NaHWO_4 + Na^+ \rightleftharpoons Na_2WO_4 + H^+ \tag{8.34}$$

$$2mH^+ + mNa_2WO_4 + (n - m)H_2WO_4 \rightleftharpoons H_{2x}W_nO_{3n+x} + 2mNa^+ + (n - x)H_2O \tag{8.35}$$

The value of n keeps growing up with the increase of aging time. Facts proved that white tungstic acid would transform to colloidal and yellow species after days aging in the original fluid. From the IR spectra, it can also get that the absorption peak at 1400 cm^{-1} is reduced progressively to zero with the increasing aging time. From Tables 8.22 and 8.23 and Figs. 8.73 and 8.74, it can be seen that white tungstic acid is separated from the original fluid at the initial time; its exchange capacity with positive ion will increase to the ideal value $H^+/W = 0.5$. The activity of white tungstic acid would reduce upon aging due to the decrease of monomeric acid. In another word, as long as white tungstic acid is separated from the original fluid, its activity is able to remain the same for a long time.

8.7.7 Removal of P, As, and Si During White Tungstic Acid Preparation

In alkaline solutions, P, As, and Si will coexist with tungstate or molybdate and must be removed, which can be realized when tungstic acid is prepared through generation of white tungstic acid by reverse titration method, i.e., addition of tungstate solution to dilute acidic solution. White tungstic acid was prepared under 0.5 mol/L

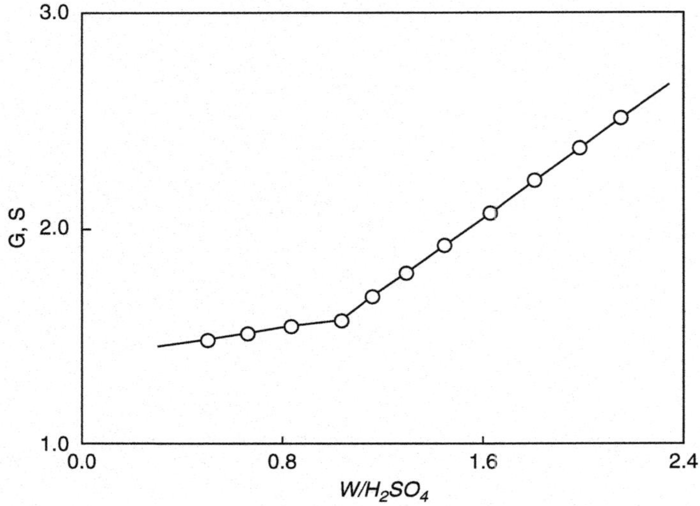

Fig. 8.74 The conductivity curve for adding Na_2WO_4 solution to H_2SO_4. (0.83 M Na_2WO_4, initial H_2SO_4 0.5M)

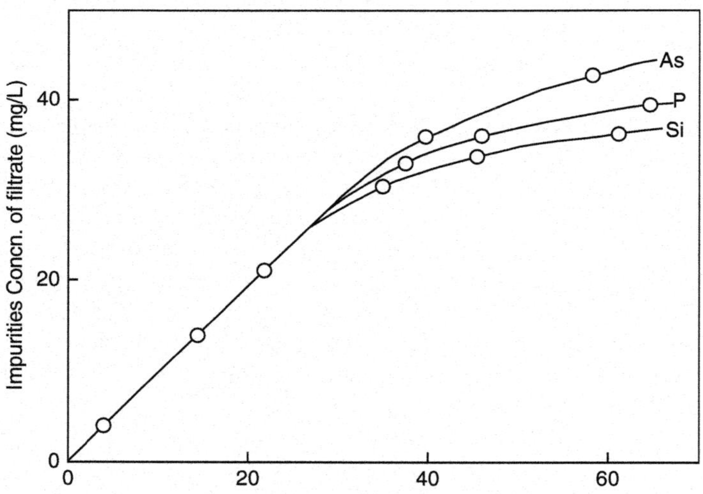

Fig. 8.75 Removal of P, As, and Si from tungstate solution through generation of white tungstic acid by reverse titration method. (0.5 mol/L Na_2WO_4 20 mL; 1 mol/L HNO_3 10 mL; terminal pH 1)

Na_2WO_4 in the presence of P, As, and Si sodium salts, and in 1 mol/L HNO_3, the result is presented in Fig. 8.75. The white tungstic acid product thus obtained contained negligible contents of these impurities at the concentrations of P, As, or Si below 30 mg/L, respectively, or the total concentration under 30 mg/L in the tungstate solutions.

The white tungstic acid can be filtered, water-washed, and dried. The filtrate contains impurities and soluble tungsten presents as heterotungstates, which depend on the impurity concentrations in the initial solution. Filtrate can be directly discharged when the concentration of heteropolytungstate is quite low. For example, in the initial solution with (P + As + Si) < 0.2 mmol/L, $Na_2WO_4 = 0.5$ mol/L, the filtrate will contain 2.4 mmol/L W and nearly the same initial impurity concentrations with a loss of 0.48% W, when the white tungstic acid is recovered. Filtrate should be recycled when the concentration of heteropolytungstate is at a higher level. For example, when (P + As + Si) =1 mmol/L, the loss of W could reach to 2.4% in the filtrate and is needed to be recycled by evaporating and concentrating filtrate and then separating out heteropolytungstate.

8.7.8 Practical Significance of White Tungstic Acid in Tungsten Metallurgy Industry

It is white tungstic acid instead of yellow ones being produced by reverse titration resulting to excellent removal of impurities. The differences between white and yellow tungstic acid is shown in Table 8.24. It is easy to know that white tungstic acid has much more remarkable performance on the isopolytungstate preparation with the advantages of high yield which is almost 100%, high speed, mild reaction, and so on. White tungstic acid produced under a higher acidity could be calcined to produce yellow WO_3 which has the same character with yellow tungstic acid. The white tungstic acid preparation method can be used in gravimetric method for quick determination of W and the removal of P, As, and Si.

Table 8.24 Characteristics comparison between white and yellow tungstic acid

Items	White tungstic acid	Yellow tungstic acid
Acidity	Acidic	Nonacidic
Solubility in ammonia	Soluble	Soluble
Reactivity with potassium niobate and potassium tantalate	Reactive	Nonreactive
Color (after burning)	Yellow or white[a]	Yellow
Be heated in dilute HCl	Transforming to yellow tungstic acid	
Capacity of producing metatungstate potassium or heteropoly tungstate	Strong capacity, high productivity	Low capacity; cannot be used to produce heteropoly tungstate directly

[a]White tungstic acid is still white after burning and that state can be maintained for a long time in a dryer under the production conditions with initial $HNO_3 < 0.5$ mol/L and terminal value of pH 1 ~ 2.5

Chapter 9
Industrial-Scale Production of Zinc Powder Using Alkaline Leaching-Electrowinning Processes

Abstract An integrated process consisted of raw materials milling and grinding, leaching, purification, electrowinning, regeneration of the spent electrolytes, vacuum drying, analysis, etc. is developed, and the industrial application flow sheet is designed, with annual yield 1500–2000 tons metallic zinc powder. The process developed has been successfully applied at several plants with a scale of 1000, 1500, and 2000 tons of metallic zinc powder annually. The long-term industrial operations have been shown that the process can use any raw materials containing Cl, F, Si, Fe, As, Cu, etc. The total Zn content of zinc powder produced is over 98% and the metallic zinc content over 94%, which reaches the Class 2 of the national standards of zinc powder quality. Hence, the process is quite different to the acidic process and can apply the raw materials unacceptable for the acidic process. The life cycle assessment method is used to investigate the environmental impacts of the new alkaline processes of production of Zn powder. The results indicate that the environmental impacts of alkaline leaching-electrowinning technology are less than the conventional acid leaching hydrometallurgical processes and pyrometallurgical processes. The energy consumption, greenhouse effect, and acidification in the new processes are 87.76%, 47.80%, and 0.08% of those in the conventional acid leaching hydrometallurgical processes and are 93.94%, 9.25%, and 0.14% of those in the pyrometallurgical processes.

Keywords Industrial scale • Process design • Alkaline leaching • Electrowinning • Life cycle assessment

9.1 Industrial Production of Zn Powder by Alkaline Process Using Brass Smelting Ash as an Example

The industrialization processes for the production of zinc powder using alkaline leaching-purification-electrowinning with an annual output of 1500 tons from wastes containing 20–40% zinc are described. The initial raw materials may include the steelmaking dust, slags from zinc refinery plants, lower-grade zinc oxidized ores, zinc and lead wastes containing chlorides, dust from heavy metal volatilization factories, etc.

Y. Zhao, C. Zhang, *Pollution Control and Resource Reuse for Alkaline Hydrometallurgy of Amphoteric Metal Hazardous Wastes*, Handbook of Environmental Engineering 18, DOI 10.1007/978-3-319-55158-6_9

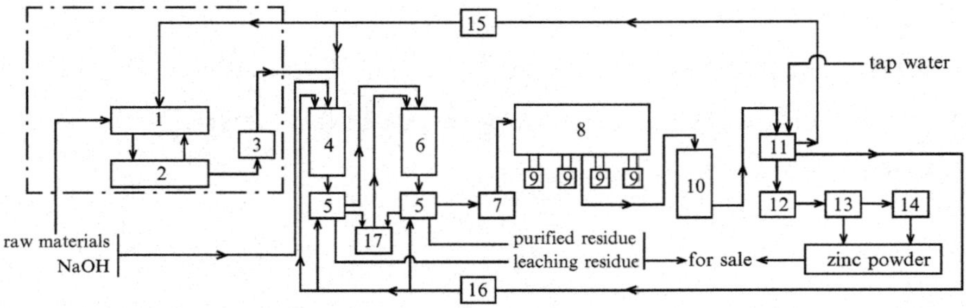

1 - Ball mill	7 - aging bunker	13 - sizing machinery
2 - spiral classifier	8 - electrolytic cell	14 - milling
3 - storing cistern	9 - circulatory pool	15 - electrolyte tank
4 - leaching kettle	10 - transfer kettle	16 - water pool
5 - filter press	11 - centrifuge	17 - water tank
6 - purifying kettle	12 - vacuum drier	

Fig. 9.1 Flow sheet of alkaline leaching, purification, and electrowinning process for recovery of Zn powder from brass smelting ash

A kind of brass smelting ash is used as an example of raw material in the plant, with a composition of (wt.%) Zn 55.21–59.47, Pb 8.07–12.40, Cu<0.38, Fe<0.56, As<0.004, Mn<0.22, Cd<0.025, Co<0.004, Mg<0.15, Al<1.65, little W, Mo, etc. The flowchart of the integrated alkaline process is shown in Fig. 9.1. Due to its small particle size, the brass smelting ash is leached directly without the milling and grinding pretreatment. After the raw material is leached by caustic sodium solution in a leaching tank, the slurry is filtered and the filtrate treated to remove impurities in a purification tank. The filtration operations of the leach and the purified slurry are completed, respectively, in a plate- and frame-type filter press.

After electrowinning, using stainless steel electrodes as anode and titanium alloy as cathode, the Zn powder is removed from cathode and separated from the spent electrolyte in a centrifuge, in which the zinc powder can be washed thoroughly with tap water. After dried in a double-cone rotary vacuum dryer, particle size is graded in an ultrasonic vibrating screen and milled for the oversize product in a self-designed machine. The metallic Zn powder is then analyzed, weighed, matched, and packed for sale.

A normal and good running of the process will not be carried out without the laboratory guidance in the plant. Based on the results of routine analysis, the required quantity of brass smelting ash, NaOH product, sodium sulfide hydrate, etc. is calculated by the laboratory for each case. The main calculations in laboratory and the corresponding formulas are as follows:

$$M_1 = V_1 \times (38 - V_2 \times C_0/V_1)/(m_0 \times \eta) \tag{9.1}$$

where M_1 (kg) the required quantity of the raw material, V_1 (m^3) the given volume of leaching solution, V_2 (m^3) the given volume of spent electrolyte, 38 (g/L) the expected zinc concentration in leach solution, C_0 (g/L) the zinc concentrations in spent electrolyte, m_0 (%) the percent of zinc content in the raw material, and η (%) the optimal leaching efficiency of Zn obtained from the laboratory experiments

$$M_2 = V_1 \times (C_2 - V_2 \times C_1/V_1)/m_1 \quad (9.2)$$

where M_2 (kg) the required quantity of NaOH product, C_2 (g/L) the given initial NaOH concentration in leaching solution, C_1 (g/L) the NaOH concentration in spent electrolyte, and m_1 (%) the percent of NaOH content in NaOH product.

$$M_3 = V_{3\times}w \quad (9.3)$$

where M_3 (kg) the required quantity of sodium sulfide hydrate, V_3 (m^3) the real volume of purifying solution, and w (g/L) the optimal addition of $Na_2S{\cdot}xH_2O$ ($7 \leq x \leq 9$) per liter of purifying solution obtained from laboratory experiments. In addition, when the composition of raw material changed, the operating conditions of leaching and purification are firstly gained and optimized by laboratory experiments and then adjusted in the production process.

9.1.1 Industrial-Scale Leaching

The maximum extraction of Zn and Pb from the brass smelting ash can be reached at 96–99% and 90–95%, respectively, when the leaching operation was conducted with 230 g/L NaOH at 80 °C for 90 min at the liquid-solid ratio (v/w) of 15. The concentrations of elements, such as Fe, Cu, Cr, Cd, Co, Ni, etc., were always lower than 10 mg/L in the leach solution. A proportion of Al and Mg can dissolve together with Zn and Pb, but their dissolution was depressed by the presence of Zn and Pb, and their concentrations were always lower than 2 g/L, which had no effect on the electrowinning of Zn. Therefore, the dissolution of these impurities can be negligible.

9.1.2 Industrial-Scale Purification

The purifying operation is conducted at 85 °C by adding sodium sulfide hydrate. Since the precipitates of sulfides (with PbS as the predominant compound) are colloidal and quite difficult to filter, the calcium oxide is added with a weight ratio of 0.8 (CaO to Na_2S). The filtration operation is also improved by allowing the solution to stand for 2–3 h to settle through gravity and heating it at 70–75 °C. By sodium sulfide precipitation and calcium oxide hydrolyzation, the lead concentration in the alkaline solution can be lower than 10 mg/L and arsenic lower than 30 mg/L, while the loss of Zn (about 35–40 g/L Zn in the purifying solution) was lower than 1.5 g/L. In addition, a portion of other impurities, such as Fe, Cu, Al, Mg, Mn, Cd, Co, Ni, etc., can also be removed by coprecipitation.

Table 9.1 Economic and technical norms of the electrolysis process

Operation parameter	Number range
Initial zinc concentration, g/L	35–40
Initial NaOH concentration, g/L	185–200
Ultimate zinc concentration, g/L	<6
Ultimate NaOH concentration, g/L	220–240
Electrolyte temperature, °C	40–50
Current density, A/m^2	800–1000
Cell potential, V	2.7–3.0

Fig. 9.2 Scanning electron microscope micrographs showing the surfaces of the zinc powder deposited at different temperatures. Zinc deposits obtained at (**a**) 40 °C and (**b**) 50 °C (1000 A/m^2, 60×)

9.1.3 Industrial-Scale Electrolysis

The economic and technical norms of the electrolysis process are shown in Table 9.1. The zinc powder deposited on cathode is gray with a slightly bright metallic luster. Although the zinc deposits, formed within the first minutes of electrolysis, are in compact form, they become level but unfirm during the following minutes of the operation and can autostrip or be peeled off easily from the substrate. The surface photographs of the deposited zinc powder at different temperature are shown in Fig. 9.2.

9.1.4 Industrial-Scale Zn Powder Filtration, Washing, and Desiccation

The operations of zinc powder filtration, washing, and desiccation, especially the first two steps, should be completed as soon as possible in case of metallic Zn oxidation. It is proposed the first two steps be finished within 40 min and the washing operation be terminated when the pH value of the washing water is lower than 8. After dried thoroughly and rapidly, the zinc powder is cooled down with recirculated cooling water in a double-cone rotary vacuum dryer. The permitted temperature of zinc powder unloading is at 50 °C in case of fire.

9.1.5 Industrial-Scale Sieving and Milling

The particle size range of the dried zinc powder is often from 0.25 to 0.028 mm or even smaller. By the process of sieving and milling, the zinc powder is particle size graded, and the oversize zinc powder is milled and rescreened. Then, the Zn powder is matched in accordance with user's demand. The loss of NaOH in whole process is dependent mostly on the type and composition of feed material. Additional loss of NaOH is caused by entrainments in the leach residue and the lead sulfide precipitate as well as by an accumulation of sodium salts in the leach solution. The average loss of NaOH is about 0.1 kg per kg of Zn powder when the brass smelting ash is used as raw material.

The main cost item of the process is the energy consumed by electrolysis (around 2.7–3.1 kWh/kg Zn powder). Additional energy required for mechanical raking, filtration, pumping, sieving, milling, etc. is about 1.3 kWh/kg Zn powder and for heating about 1.5 kWh/kg Zn powder. Moreover, the profit of selling the lead sulfides and the leach residues may offset the cost of purificants consumed in the purification process.

Hence, operational costs include electric energy consumed in the whole plant (about 4.0–4.4 kWh/kg Zn powder), heating of leaching, purification and zinc powder desiccation (about 1.5 kWh/kg Zn powder), and the loss of NaOH.

Table 9.2 compares different processes for the production of metallic Zn powder by treating oxidized zinc materials. About 10–15% of energy is reduced in this process in comparison to the classical air pulverization process, in which the Zn metal used as the feed material is produced by the traditional acidic process. The purification process in the alkaline process is simpler than the acidic-air pulverization process due to the undesired elements such as Fe, Cu, Cr, Cd, Co, Ni, etc. which can hardly be leached out in alkaline solution. The energy for electrolysis of Zn in alkaline solution is 2.7–3.1 kWh/kg Zn, which is less than in the acidic medium with 3.3 kWh/kg Zn. Moreover, the alkaline process is cleaner and environmentally friendly as no wastewater and hazardous wastes are excluded from the process.

The alkaline process is practicable for recovering Zn from the brass smelting ash at industrial scale, which has been confirmed with more than 1 year's good running in the plant. The production of Zn powder with a purity over 96% can be obtained finally, which can meet the Class 3 standard in comparison with the corresponding items of GB/T 6890–2000 (Chinese Industrial Standard, Zinc Powder). Compared with the classical acidic-air pulverization process, the alkaline process is considered as a simple, cost-effective, and environmentally friendly technology.

Table 9.2 Comparisons for different processes for the production of metallic Zn powder from oxidized zinc wastes (based on the production of 1 kg Zn)

Processes selected	The alkaline process	Acidic-air pulverization process
Raw material grade	Low to high	High
Energy for electrolysis	2.7–3.1 kWh	3.3 kWh
Energy for heating	1.5 kWh	1 kWh
Energy for evaporation	None	1 kWh
Unit consumption of total energy	5.5–5.9 kWh	6.0–6.4 kWh
Loss of leaching agents	Moderate	Large quantity
Purification of leach solution	Simple	Complex
Removal of Pb	By sulfide precipitation	By metallic Zn powder
Particle shape of zinc powder	Flake	Sphere
Feasibility	Small- to large-scale company	Small- to large-scale company
Capital costs	Small to large	Small to large

9.2 Overall Process for Production Zinc Powder Using Alkaline Leaching-Electrowinning Processes

Wastes containing zinc are introduced into the leaching tanks through a feeding runner after weighing, and then the waste electrolyte, residue-washing water, and said fresh sodium hydroxide are added to leach the wastes in the alkaline leaching units. Leaching tanks are heated by steam up to 85 °C for 90 min. The leaching residues are then filtered by plate and frame filter press and washed by low alkali washing solution from the zinc powder washing unit. Leach solutions and washing solutions are combined together, and then sodium sulfide, ferric sulfate, sodium silicate, and calcium oxide are introduced into the purification tanks so that the metal impurity can be separated. The metal precipitates contain over 20% Pb and can be sold to Pb smelters.

The purified solution containing zincate is electrolyzed in the electrolytic cells for hours depending on the zinc concentrations. The electrowinning operation stops when the zinc concentrations decrease to below 5 g/L. The zinc powder thus obtained is then separated by centrifugal equipment and dried in vacuum drier after washing using water. The dry zinc powder can be packaged directly and sold. The spent electrolyte is stored in a waste solution tank and recycled for the next leaching process. The whole process and device attachment are illustrated separately in Figs. 9.3 and 9.4, and the technical conditions of main operation procedures are offered in Table 9.3.

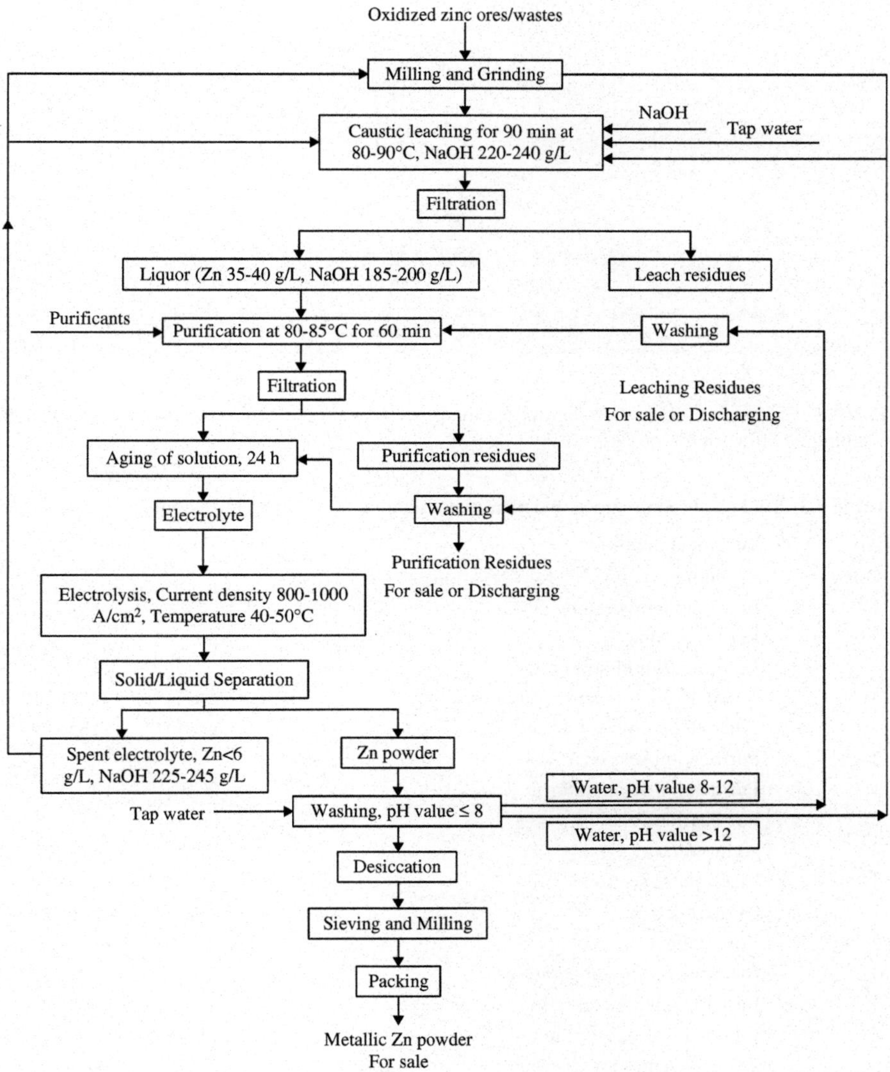

Fig. 9.3 Production flow sheet for Zn powder by alkaline leaching-electrowinning process

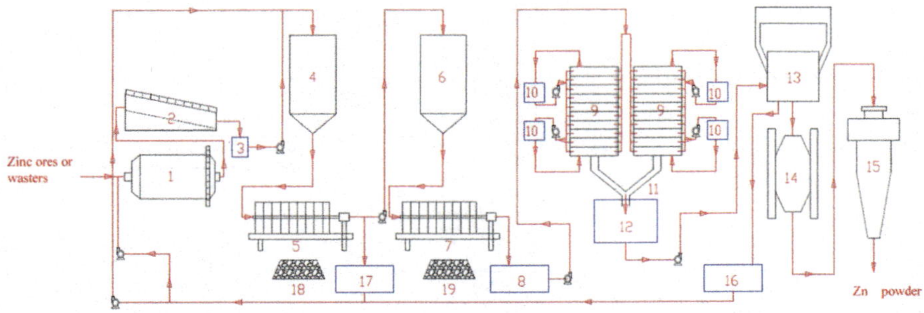

1—Ball mill; 2—classifier; 3—Slurry storage tank; 4—leaching tank; 5—Filter press of leaching; 6—purified-kettle; 7—Filter press of purifying; 8—Aging tank; 9—electrolytic bath; 10—circulation pool of electrolyte; 11—spent electrolyte chute; 12—washing-filter tank of zinc ; 13—centrifugal machine; 14—dryer; 15—aeropulverizer; 16—pool of spent electrolyte ; 17—pool of washing water; 18—leaching residue; 19—purifying residue

Fig. 9.4 Equipment schematic diagram for the process of production of Zn powder by alkaline leaching-electrowinning process

Table 9.3 Technical requirements for plant with alkaline leaching-electrowinning processes

No	Technical condition	Unit	Value
One	Leaching section		
1	Leaching temperature	°C	80–90
2	Leaching time	h	1–2
3	Leaching liquid-solid ratio		1:5–1:15
4	initial alkali concentration of leaching	g/L	220–240
5	Final alkali concentration of leaching	g/L	185–190
Two	Purification section		
1	Purification temperature	°C	70–80
2	Purification time	h	2–3 h
3	Aging time	h	48–72 h
Three	Electrolytic section		
1	Current density	A/m^2	800–1000
2	Bath voltage	V	2.7–2.9
3	Electrowinning temperature	°C	30–50
4	Initial alkali concentration of electrolyte	g/L	185–195
5	Initial zinc concentration of electrolyte	g/L	30–40
6	Zinc concentration of electrolyte after electrolyzing	g/L	8–10

9.3 Design for Production Equipment for Production of 1500 t/a Zinc Powder Using Alkaline Leaching-Electrowinning Processes

The number of leaching tanks is

$$N = \frac{Qt}{24V_0\eta} \tag{9.4}$$

where, N—number of leaching tanks
Q—daily amount of leaching solutions, m^3, $Q = 171$ m^3
t—interval time, h, $t = 8$ h
V_0—volume of leaching tank, m^3, $V_0 = 40$ m^3
η—utilization coefficient of the tank, $\eta = 0.85$

The number of leaching tanks is

$$N = \frac{Qt}{24V_0\eta} = \frac{171 \times 8}{24 \times 50 \times 0.85} = 1.34 \quad (9.5)$$

Three mechanical agitation tanks (Φ4000 × 4000 mm, V = 50 m^3) are thus chosen in this unit, two of them for daily uses and the other for backup supply. The body of leaching tank with conical bottom is constructed with reinforced concrete-lined antiseptic carbon steel. The spirally steel steam heating pipes, liquid level observation hole, long thermometer, thermowell, and liquid level scale are set in leaching tank. A turbo agitator, an antiseptic mixer, an antiseptic motor, and a speed reducer to drive the stirring process are installed. The underflow is derived from the bottom of leaching tank and then filtered through the filter. The waste electrolyte feed tube, washing water feed pipe, and alkali feed port are set at the upper portion of the leaching tank, while a removable cover is set at the top. The structure of the leaching tank is shown in Fig. 9.5.

After leaching and filtering operation, the leaching solutions are pumped to the purification tank and measure the impurity concentrations. Stir and heat the leaching solutions to above 70 °C, followed by the addition of Na_2S, sodium silicate, ferric sulfate, and calcium oxide, sequentially to precipitate lead, iron, arsenic, copper, etc. Filter the solutions after standing for 4 h, and pipe the purified solution to the aging tank for electrolysis.

Fig. 9.5 Structure of leaching tank for production of 1500 t/a zinc powder using alkaline leaching-electrowinning processes

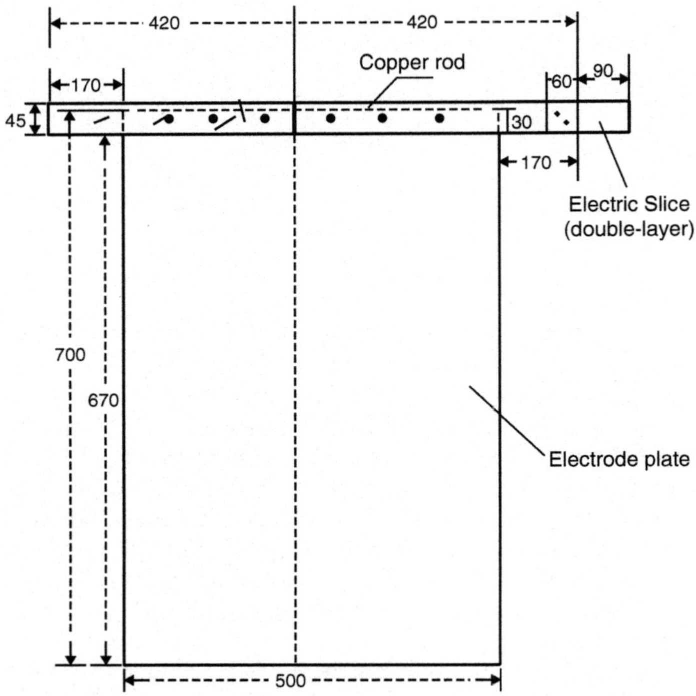

Fig. 9.6 Structure of cathode plate (mm) for production of 1500 t/a zinc powder using alkaline leaching-electrowinning processes

Two mechanical agitation tanks ($\Phi 4000 \times 4000$ mm, $V = 50$ m^3) for purification are coupled with the leaching tanks, made of reinforced concrete with conical bottom, and lined with carbon steel to avoid corrosion. Meanwhile, liquid observation well, long thermometer, thermowell, and liquid level scale are also installed.

The purified zincate solution is introduced into the electrowinning system to produce zinc powder. The cathodes consist of magnesium alloy electrode plate, copper conducting rod, electric slice, and insulating strip as shown in Fig. 9.6. Two edges of cathodes are insulated by polyethylene or rubber. The valid area of cathode plate is (700 mm–180 mm) $\times$ 500 mm $\times$ 2 = 0.52 m^2.

Anodes consist of stainless steel electrode plate, copper conducting rod, electric slice, and insulating strip as shown in Fig. 9.7, and two edges of electrode are insulated by polyethylene or rubber.

The daily amount of zinc powder produced in cathode is calculated as

$$Q_1 = \frac{Q \times b}{m\eta_1} = \frac{1500 \times 0.96}{330 \times 0.95} = 4.6 \text{ t} \tag{9.6}$$

where: Q_1—the daily amount of zinc in cathode, t

Q—the designed producing amount, t/a

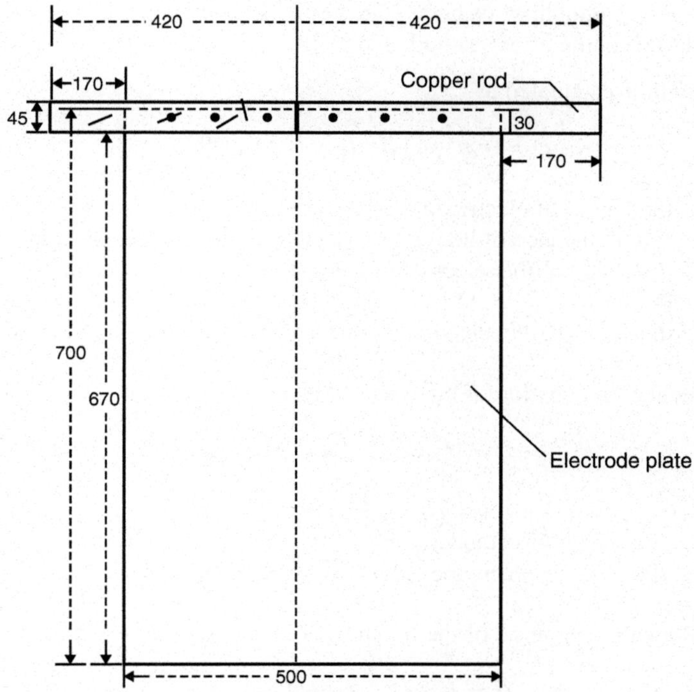

Fig. 9.7 Structure of anode plate (mm) for production of 1500 t/a zinc powder using alkaline leaching-electrowinning processes

b—the content of metallic zinc in zinc particle production, %
m—the annual amount of working day, 330 days generally, d
η_1—the direct yield of zinc powder after washing and drying, %, assuming 95%

The total valid area of cathode is calculated as

$$F = \frac{Q_1 \times 10^6}{D_i \eta_i \times 1.2195 \times t} = 232.75\ \mathrm{m}^2 \tag{9.7}$$

where: *F*—the total valid area of cathode, m^2
1.295—the electrochemical equivalent of zinc, g/(A·h)
D_i—the current density of cathode, A/m^2, $Di = 1000\ A/m^2$
η_i—current efficiency, %, $\eta_i = 90\%$
t—electrowinning time, h, $t = 6 \times 3$ h
Calculation for n_{Ca}—the amount of electric tanks—is

$$n_{tank} = \frac{F}{f_{ca} \times (n_{ca} - 0.5)} \approx 20 \tag{9.8}$$

where: n_{Ca}—the amount of cathode plate per tank, assuming 24
f_{Ca}—the valid area of per cathode plate, m^2

The length is calculated as

$$L = (n_{ca} - 1)d + l_1 + l_2 \sim 2200\,\mathrm{mm} \tag{9.9}$$

where: L—the length of electric tank, mm
d—the central distance of homopolarity plates, mm, assuming 70 mm
l_1—the distance of the electrode plate at inflow port and tank wall, mm, assuming 300 mm
l_2—the distance of the electrode plate at outflow port and tank wall, mm, assuming 300 mm

The calculation equation of the width of electrolysis bath is

$$B = b_{\mathrm{cathode}} + 2b_1 = 700\ \mathrm{mm} \tag{9.10}$$

where: B—the width of electrolysis bath, mm
b_{cathode}—the width of cathode, mm
b_1—the distance between the edge of cathode and side wall of bath, mm, 100 mm

The calculation equation of the depth of electrolysis bath is

$$H = h + h_1 + h_2 = 1200\ \mathrm{mm} \tag{9.11}$$

where: H—the depth of electrolysis bath, mm
h—the distance between the liquid level and the surface of bath, mm, 120 mm
h_1—the depth that the cathode plate being immersed in the electrolyte, mm, 500 mm
h_2—the distance between the underneath of cathode plate and the bottom of bath, mm, 580 mm

The calculation equation of the volume of electrolysis bath is

$$V_{\mathrm{bath}} = \mathrm{LBH} = 1.9\ \mathrm{m}^3 \tag{9.12}$$

where V_{bath}—the volume of electrolysis bath, m^3

The electrolysis bath is made of reinforced concrete. Hard PVC plate with 10 mm of thickness is adopted as the liner. The bottom of the electrolysis bath is a cone with an outlet for zinc powder. Its structure is showed in Fig. 9.8. The number of electrolysis bath is designed as 20, which is divided into four sections and five in each section. Each section is equipped with circulating pool and recycle pump for self-circulation so as to ensure that the solution concentration is homogeneous and has a better heat dissipation.

The calculation equation of the volume of circulating pool for electrolyte is

$$V_{\mathrm{circulating\ pool}} = \frac{V_1 n_{\mathrm{pool}} \times \eta_2 \times V_{\mathrm{pool}} \times n}{4 \times \eta_3 \times n} \approx 5.5\ \mathrm{m}^3 \tag{9.13}$$

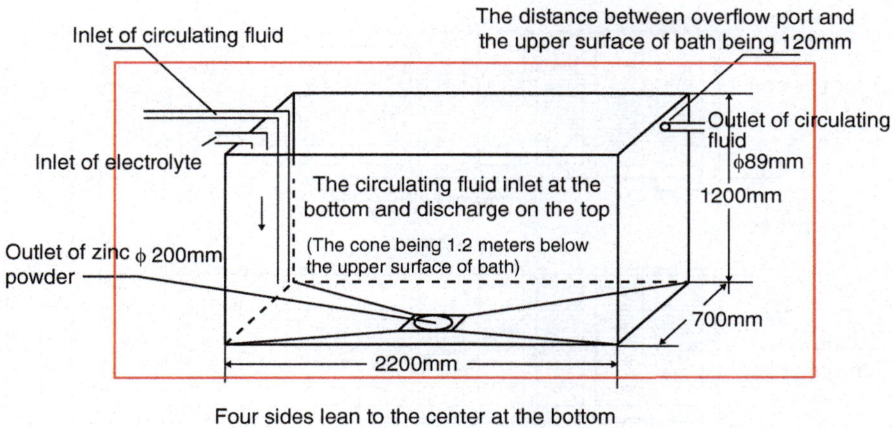

Fig. 9.8 Structure of electrolysis bath for production of 1500 t/a zinc powder using alkaline leaching-electrowinning processes

where: V_1—fresh electrolyte needed per day, m^3

n—the frequency of electrolysis per day, time, three times

η_2—the efficiency of utilization of electrolyte bath, 0.9

η_3—the efficiency of utilization of circulating pool for electrolyte, 0.95

The rectifier is determined by two parameters which are direct voltage and direct current. The specific calculation equations are as follows:

A. *Calculation of current intensity*

$$I = f_{\text{cathode}} \times (n_{\text{cathode}} n 0.5) \times D_{\text{i}} \approx 9800 \sim 12{,}500 \ \text{A/m}^2$$

B. *Direct voltage*

While electrolysis is carried out in three sections, namely, 15 baths simultaneously, the direct voltage needed is

$$15 \times 2.7 \sim 3.0 \ \text{V} = 40.5 \sim 43.5 \ \text{V}$$

While electrolysis is carried out in four sections, namely, 20 baths simultaneously, the direct voltage needed is

$$20 \times 2.7 \sim 2.9 \ \text{V} = 54 \sim 58 \ \text{V}$$

Thereby, the criteria to select rectifier are as follows: rated direct voltage being 40.5–58 V and rated direct current being 9800–12,500 A/m^2.

The configuration of electrowinning workshop is shown in Fig. 9.9. When the temperature is high, the efficiency will decrease heavily due to the dissolution of deposited zinc powder. During the process of electrolysis, the electroplating

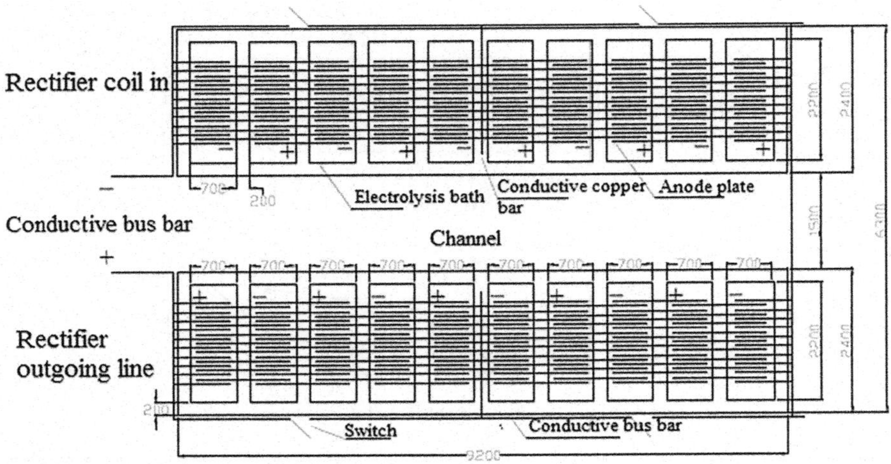

Fig. 9.9 Configuration of electrowinning workshop for production of 1500 t/a zinc powder using alkaline leaching-electrowinning processes

temperature should be maintained below 50 °C. In order to decrease the temperature, the electrowinning workshop should be ventilated well. Some space should be left on the bottom of electrolysis bath and circulating bath of electrolyte for ventilating. Considering the local climatic condition, cooling tower should be equipped if needed.

9.4 Industrial Design for Purging, Drying, and Crushing Working Section of Zinc Powder

After electrolysis, the zinc powder deposited from cathode is mixed with waste electrolyte and should be separated from waste electrolyte. The main facilities in this working section are filtration and purging equipment for zinc powder, drying facility for zinc powder, and grinding equipment for zinc powder. The deposited zinc powder is easy to be oxidized with high activity and also reacts with alkali during the filtration to dissolve again. Besides, if the zinc powder has not been purged thoroughly with water, they will react with the remained alkali and generate hydrogen which leads to spontaneous combustion. Thereby, the purging process of zinc powder should be thoroughly within a short time.

Type of XR1200-N top suspension manual uploading centrifuge is adopted in this technology. The rotary drum of centrifuge is driven by electric machinery to rotate. In feeding rotating speed phase, mixed solution of electrolyte and zinc powder are poured into the rotary drum through inlet pipe. Feeding is stopped when feedstock reaches the setting volume and they are raised to high-speed

Table 9.4 Model XR1200-N top-suspended manual discharge centrifuge

Model	Drum					Motor power (kW)	Size (mm)	Weight (kg)
	Diameter (mm)	Height (mm)	Maximum speed (rpm)	Maximum separation factor	Load limit (kg)			
XR1200-N	1200	1055	970	644	450	22	2210 × 1600 × 3414	4146

separation. Under the effect of centrifuge force, waste electrolyte is discharged through filter screen and rotating drum wall. The waste electrolyte is discharged into waste electrolysis bath through fluid discharge pipe, and zinc powder is intercepted in the rotary drum. The washing water for zinc powder is poured until the pH of water reaches neutral. After purging, stop the facility and dither the filter bag to make zinc powder loose. Then zinc powder is discharged from the bottom of the facility through the bottom cone of rotary drum. Under the bottom of facility connected with a dryer so as zinc powder would drop into it directly.

Adopting two top suspension manual uploading centrifuges with type of XR1200-N, the limit of feedstock is 400 kg each time. Three hundred eighty-five kilogram zinc powder is produced once in each electric tank. Table 9.4 shows the basic features of XR1200-N top-suspended manual discharge centrifuge.

The SZG double-taper rotary vacuum dryer is chosen for zinc powder drying, which is able to mix and dry the powder at the same time. The vacuum drying device is made up of the condenser, vacuum pumps, and drying machine. All the materials can be emitted and labor intensity can be reduced. Meanwhile, the rotation of both the container and materials increases the heat transfer coefficient and drying efficiency, which not only saves energy but fully and uniformly dries the powder.

The SZG double-taper rotary vacuum dryer has a double-taper rotary tank. Fill in the tube with steam or hot water to heat the wet zinc powder in vacuum. The steam evaporated by the wet powder is pumped by a vacuum pump. The vacuum environment and rotary process largely increase the drying efficiency, and thus a uniform drying is achieved.

Two SZG-1000 double-taper rotary vacuum dryers may be chosen based on the zinc powder production. They are installed just below the spout of the top-suspended manual discharge centrifuge. A vertical slide is equipped so that the clean zinc powder can directly be slipped into the dryer. Table 9.5 shows the basic features of the SZG-1000 double-taper rotary vacuum dryer. As zinc powder will be oxidized by water under high temperature, the steam should be pumped away in time during the drying process. In this case, the power of vacuum pump is amplified to 15–20 kW.

Selection of zinc powder crushing device plays an important role in the quality of product. Zinc powder will be oxidized, and zinc levels decrease by 1–2% points which may sacrifice the quality to secondary standard once using the common mill. For this reason, a set of zinc powder crushing device that comprises of feeder, smashing devices, motor, draught fan, decompress storehouse, storage compartment, and dust

Table 9.5 Model SZG-1000 double-taper rotary vacuum dryer

Model	Tank volume (L)	Load volume (L)	Weight (kg)	Rotating height (mm)	Motor power (kW)	Size (mm)
SZG -1000	1000	≤500	2800	2800	3	2860 × 1300
Model	Tank design pressure (MPa)	Tube design pressure (MPa)	Pump model and power		Operating temperature	
SZG -1000	0.1~0.15	≤0.09	SK-2.7B 15–20 kW		Tank ≤ 85 °C, Tube ≤ 140 °C	

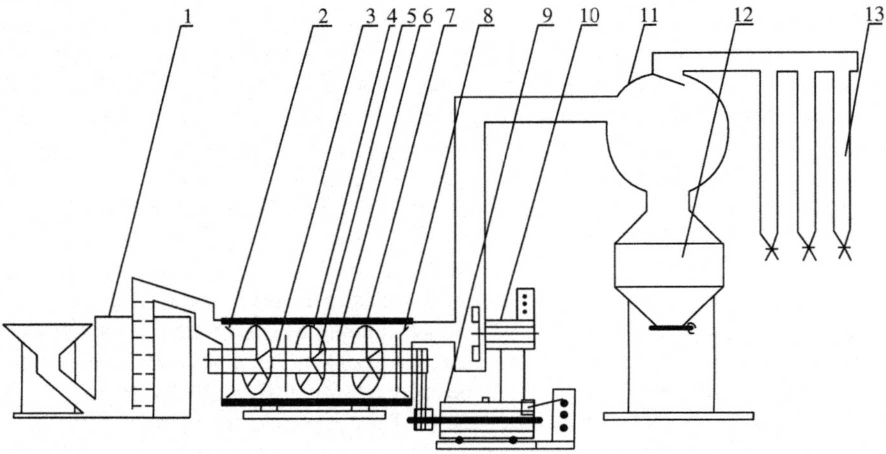

Fig. 9.10 Crushing device of zinc powder for production of 1500 t/a zinc powder using alkaline leaching-electrowinning processes. 1. feeder; 2. impeller; 3. spindle; 4. wheel; 5. tool; 6. swirling round plate; 7. cabinet; 8. discharge impeller; 9. motor; 10. fan; 11. decompression warehouse; 12. storage compartment; 13. dust collector

collector (Fig. 9.10) is designed. This device includes an externally casing, a spindle located in the axis line of whole device, several wheel disks fixed on the spindle at regular intervals, and placing cutting tools with the same angle. Between the wheel disks, there exist swirling round plates. Note that the impeller of feeder is situated at the inlet of crushing device as well as the impeller of material discharge located in the outlet of it. This device contains 1–15 cm gaps between casing and wheel disks. In this device, the inlet connects with feeder and the outlet coupled with fan which connected to the decompression warehouse by the pipe. As can be seen, the decompression warehouse has an exhaust outlet on the top connected with dust collector and an outlet for the product below lined with storage compartment as well as the motor nearby connected with the spindle of crushing device.

9.5 Industrial Operation Procedure in Leaching Process

1. Analyze the grade of the zinc hazardous wastes received and carry on the leaching test at the laboratory to determine the required concentrations of alkali, temperature, and leaching time.
2. Open the intake valve to make sure the waste electrolyte and washing water in leaching tank reach the required level.
3. Open the blender of leaching tank; analyze the concentrations of alkali and zinc in leaching solutions for every 5 minutes.
4. According to the results of Step 3, technician can calculate the amount of each ingredient according to the following formula:

$$\begin{aligned}\text{Adding amount of alkali} = {} & \text{Leaching volume of waste liquid} \\ & \times(\text{the needed concentration of alkali} \\ & -\text{the concentration of alkali in waste liquid})\end{aligned}$$

$$\begin{aligned}\text{Adding amount of raw material} = {} & \text{Leaching volume of waste liquid} \\ & \times(35 - \text{the content of zinc in waste liquid})/ \\ & (\text{leaching percent rate})\end{aligned}$$

5. Add alkali and raw material according to the results of Step 4.
6. Open steam valve and steam, and close it when the temperature reach the range of 80–90 °C (the exact value is determined by lab test).
7. If the raw materials contain much $ZnCO_3$ or $ZnSiO_3$, some CaO emulsions should be added after 30 min from the time achieving the highest temperature in Step 6.
8. When the temperature is in the range of 80–90 °C, the concentrations of alkali and zinc are analyzed, and the alkali should be added once the concentration of alkali is less than 180 g/L.
9. Leach for 1–2 h after achieving the highest temperature (the leaching time is determined by lab test).
10. Continue stirring. Press filter pressure button, push the pressure plate, and make all the filter plates pressured.
11. Start the feed pump to filter.
12. Take the leaching liquid for laboratory analysis in the outlet of filter press.
13. Open the valve of washing water so as to wash the leaching residues and take the washing solution for laboratory analysis when the filter completes.
14. Open the compressed air valve to make the residues pressured again with the purpose of reducing the moisture content.
15. Unload all the leaching residues and sample for laboratory analysis.

9.6 Industrial Operation Procedure in Purification

1. The leach solution is imported to purification tank after filtered by filter press, and the volume is measured, and then the concentrations of elements, such as Zn, NaOH, Pb, As, and Al, are analyzed.
2. The chemical agents for purifying operation should be established on the concentration of elements.
3. The purification can be directly operated if the temperature of filtrate is above 70 °C; otherwise, the temperature of filtrate should be heated.
4. The purifying operation is conducted according to Step 2 by adding 80% sodium sulfide at first. Sample 100 mL after stirring 30 min and then add some sodium sulfide to test if there are black precipitates of PbS. If so, the operation should be repeated unless no precipitate appears.
5. The mixture is stirred for 30 min after adding sodium silicate, and then ferric sulfate is added to the mixture and stirred for 30 min. Finally, the calcium oxide is added and stirred for 60 min.
6. Add 0.5 g sodium sulfide hydrate to purified solution of 50 mL after filtration; if there are no black precipitates appeared, then the purification is already completed. Else, sodium sulfide hydrate should be added and stirred for 20 min and all the steps should be repeated.
7. Stop stirring and settling for 4 h.
8. Send to filter press and guarantee the filter liquor is clear. Else, stop the filtration and start purifying again.
9. Sample the solution from filter press to the laboratory for analysis.
10. Send the purified solution to the aging tank for 48 h.
11. Unload the residues to the laboratory for analysis.

9.7 Industrial Operation Procedure in Electrolysis

1. The impurities of Pb, As, and Al in the purified solution should be less than 50 mg/L, 300 mg/L, and 2 g/L, respectively.
2. If not, the solution should be purified again and repeated the steps in Sect. 9.6.
3. Meanwhile, the purified solution should be sent into electrolysis bath and flowed circularly by circulating pump.
4. Check and close the outlet under the electrolysis bath. Open the inlet of liquid valve and let the electrolyte flow into electrolysis bath.
5. Start the circulating pump and the electrolyte flows circularly.
6. Start the rectifier and the rectifier cabinet. During the process, the rectifier alarms if there is any unusual situation, and then the high voltage should be shut down. Restart the machine after the fault is eliminated.
7. When electrolysis process is completed within the settled time, shut the bus bar switch control and stop power supply, and then open the outlet under the electrolysis bath and the waste electrolyte together with zinc powder flow

into zinc powder downspouting and go into the transient pool area of zinc powder washing. After that, let the electrolyte flow into electrolysis bath again and recover power supply.

8. During the electrolysis process, check the connector of negative plate, positive plate, and leading generatrix.
9. Sample the electrolyte to test the concentration of Zn every half hour.
10. Stop electrolysis when concentration of Zn is lower than 5–8 g/L, and shut down the rectifier.
11. Engineering measures should be taken to ensure the temperature of electrolyte is lower than 50 °C.

9.8 Industrial Operation Procedure in Zinc Powder Filtration, Washing, and Desiccation

1. Turn on power switch and set the speed regulation control in intermediate speed position and turn it on. When the frequency modulation of centrifugal machine achieves 25, start to feedstock.
2. Start the feed inlet pump of centrifugal machine and put the mixture of zinc powder and electrolyte into centrifugal machine.
3. The operator should focus on the status of the centrifugal machine. The machine and pump should be shut off if there is any unusual situation.
4. The load-carrying capacity should be no more than 450 kg for centrifugation at one time. Open the outlet of drain valve of Zn powder washing and close the outlet of electrolyte valve after 3 min. When the pH of Zn powder washing valve is neutral, close the outlet of drain valve.
5. Set the speed regulation control in neutral position and open the feed inlet pump. When the centrifugal machine stops, unload the materials.
6. After unloading, transfer the adhesive zinc powder into the vacuum drying tank and prepare to drying process.
7. Open the feed inlet port and put zinc powder of 400 kg into the equipment, and then close the inlet.
8. Start the vacuum system. Close the drain valve of vacuum pump and start the vacuum pump. When the vacuum pump operates normally, open the outlet of cooling water, open the outlet of buffer tank on the top of vacuum system which connects to drying equipment, and then open the outlet of buffer tank which connects to vacuum pump so the vacuum pump starts to run.
9. After running for a while, the outlet of heating steam of the drying equipment could be opened when the vacuum degree achieves 0.07 MPa.
10. Switch on the power of the drying equipment, start the forward button, and turn on the speed control switch. The drying equipment starts to work at 50 revolutions per hour.
11. During the drying process, the operator should focus on the changes of vacuum. The outlet of steam heating of the drying equipment should be closed when the

vacuum degree is lower than 0.05 MPa and opened when the vacuum degree is higher than 0.06 MPa.

12. Steam pressure should be controlled at 0.2~0.25 MPa and no more than 0.3 MPa.
13. The vacuum degree should be remained in 0.05 MPa and temperature remained in 105 °C.
14. As the drying process goes on, with the heating steam supplying continuously, the vacuum degree should be remained above 0.09 MPa; the temperature of vacuum exhaust tube decreases to room temperature. Meanwhile, the materials have dried completely, and then close outlet of steam equipment and enter into the cooling process.
15. Open the outlet of steam and drain the steam. Open the outlet of cooling water to cooling.
16. The drying process can be done when the temperature of materials decreases to 50 °C.
17. Close the outlet of buffer tank on the top of vacuum system which connects to drying equipment, and then close the outlet of buffer tank which connects to vacuum pump. The vacuum pump should be closed in the end.
18. Open the outlet of vacuum drying equipment slowly until the air pressure decreases to zero.
19. Send the dried zinc powder to the smash appropriate devices.
20. Send the particle size required of zinc powder to the laboratory for analysis.
21. Package the product of zinc powder according to GB/T6890-2000.

9.9 Analytical Requirements

Laboratory analysis plays a key role in the management and controlling of the production. On one hand, all the operational processes should be based on laboratory analysis. On the other hand, sampling inspection should be carried out to guarantee the product quality. The analysis of mass in alkaline leaching-electrowinning processes is listed in Table 9.6, and the analysis methods of samples are listed in Table 9.7.

1. Leaching section

   According to mass balance, the NaOH can be generated according to the reactions as follows:

$$PbO + NaOH + H_2O = NaPb(OH)_3$$
$$NaPb(OH)_3 + Na_2S = 3NaOH + PbS$$

   As can be seen from the above reactions, 2 mol of NaOH can be generated from 1 mol of Pb in raw materials. Because of the increase of NaOH with the rising of

Table 9.6 Analysis of mass in alkaline leaching-electrowinning processes

Components		Test parameters	Sample point
Raw material		Zn, Pb, and As content	Sampling with the car
Ball milling	Ore pulp	Zn and NaOH concentration	Slurry tank
Leaching	Leaching liquid	Zn, NaOH, and Na_2CO_3 concentration	Leaching autoclave
	Leached liquid	Zn, Pb, NaOH, and Na_2CO_3 concentration	The liquid outlet of leaching filter
	Leaching residue	Zn and Pb content	The spout of leaching filter
	Slag-washing liquid	Zn, Na_2CO_3 and Na_2CO_3 concentration	Slag-washing basin
Purification	Raw liquid of purification	NaOH, Na_2CO_3, Zn, Pb, As, and Al concentration	Purification kettle
	Purified slag	Zn, Pb, S, and moisture	The spout of purification filter
	Purified solution	NaOH, Na_2CO_3, Zn, Pb, As, and Al concentration	The liquid outlet of purification filter
Electrolyzation	Electrolysis liquid	NaOH, Na_2CO_3, Zn	Cell
	Electrolysis control system	Zn	Cell
Washing	Electrolysis wastewater	NaOH, Na_2CO_3, and Zn	Wastewater cell
	Zinc powder washing water	NaOH, Na_2CO_3, and Zn	Washing tank
Milling	Zinc powder	total zinc, metal zinc	The spout of aerofall mill

Pb contents, the high-quality lead ore should be used together with low-quality lead ore or zinc oxide ore.

2. Purification section

The loss is increasing because the ZnS may be generated through the reaction between Zn and Na_2S with a large amount Na_2S added. In order to solve this problem, the purified residue can be added to the leach tank mixed with the leach residue for the second leaching. The most of ZnS dissolved out using lead in waste materials according to the reactions as follows:

$$Na_2Pb(OH)_4(aq.) + ZnS(s) = Na_2Zn(OH)_4(aq.) + PbS(s)$$

The amount of purification residue is calculated according to the following:

$$\text{The addition} = \frac{\text{Lead grade in raw material} \times \text{the addition of raw material} \times 65.4}{\text{The percentage of zinc in the slag purification} \times 207.2}$$

Table 9.7 Analysis methods of samples

Samples	Test parameters	Analysis method
Leaching residue, raw materials, and purified slag	Zn	Determination of the amount of zinc content with EDTA titration method
	Pb	Determination of the amount of lead content with EDTA volumetric method
	As	Determination of arsenic content-molybdenum blue spectrophotometric method
	Moisture	Determination of volatile matter of pigments at 105 °C
Ore pulp, leaching liquid, purified solution, electrolysis liquid, slag-washing liquid, and electrolysis wastewater	NaOH	Neutralization titration method
	Zn, Pb and Al	Determination of zinc, lead, and aluminum with EDTA complexometric titration
	As	Determination of arsenic content-molybdenum blue spectrophotometric method
	S	Determination of sulfur content with the combustion-neutralization titration method
Zinc powder	Total zinc	Determination of total zinc content with Na_2EDTA titration method
	Metal zinc	Determination of metal zinc content with potassium permanganate titration method

3. Washing section of zinc powder

   Due to the zinc powder entrain alkali solution, the washing time will continue a long time, and the normal production cycle will be affected easily. The improvement method to solve this problem is to add the dilute acid to the cleaning liquid. When the pH of effluent reached 13, begin to switch to pH 5.5–6 dilute acid washing, and finally rinse 5 min with clean water. This way would reduce the amount of water significantly and time for washing.

9.10 "The Three-Waste" Emissions of Zinc Production Process by Alkaline Leaching-Electrolysis

"The three wastes" mean wastewater, waste gas, and solid wastes. Because the alkali leaching-electrolysis zinc powder production process is the wet process, the production process generates no emission except for the coal-fired boiler. The leach liquor produced through alkaline leaching-electrolysis zinc powder production process is completely reused. The water loss of leaching and electrolysis process is supplemented by slag-washing water and zinc-washing water, which can realize the balance of production of water and without generation of wastewater.

The waste slag produced through alkaline leaching-electrolysis process is mainly leaching residues. The yields of leaching residues are 3.5 t/t Zn from the raw material with 24% zinc, and the toxic extraction test of cleaned leaching residue meets the identification standards for hazardous identification for extraction toxicity (GB5085.3–1996). The cleaned leaching residues can be manufactured into tiles for the road together with the clay (to ensure the quality, the addition under 30%).

The quantity of leaching residues depends on the raw materials used. For the wastes containing over 70% zinc and lead, nearly no residue can be generated. In many cases, the raw wastes can contain Cu, In, and other metals, and the resultant residue can be sold at a high price.

9.11 Production Operations for Alkaline Leaching-Electrolysis Process

The program of producing 1500 tons zinc per year in Kunming city was operated officially in December 2006, 1000 tons zinc per year in Hangzhou city in July 2007, and 1000 tons zinc per year in Qujing City in November 2007. These factories mainly used zinc oxide and zinc slag as raw materials. The production was relatively stable, and the leaching rate of various raw material statistics is shown in Table 9.8. Leaching rate was generally above 90% and the lowest was 86%. The total Zn content of zinc powder was more than 98%, and the metallic zinc content

Table 9.8 Results of the alkaline leaching rate of materials used in industrial application

Item	The element content in raw material (%)			The average leaching rate of zinc (%)
	Zn	Pb	Fe	
NO. 1 (Zinc oxide ore)	17.32	2.25	8.05	90.27
NO. 2 (Zinc oxide ore)	22.63	1.96	4.67	93.32
NO. 3 (Ash)	24.45	2.44	5.31	95.7
NO. 4 (Ash)	48	4.69	2.22	92.15
NO. 5 (Ash)	30.19	10.62	0.77	86.02
NO. 6 (Ash)	18.42	8.34	4.67	90.20
NO. 7 (Ash)	48.71	0.51	0.96	91.6
NO. 8 (Zinc slag)	41.14	1.37	1.15	93.16
NO. 9 (The ash of lead ore)	59.47	12.4	0.56	99.37
NO. 10 (The ash of lead ore)	55.98	23.51	0.047	94.5
NO. 11 (The ash of steel ore)	30.07			88.05

was more than 94%, which reached to the Class 2 of the national standards of zinc powder. What's more, the operation is in good condition.

9.12 Life Cycle Assessment in the Process of Alkaline Hydrometallurgy for Zinc and Lead Hazardous Wastes

Life cycle assessment (LCA) is used to assess the potential environmental impacts associated with a product, process, or activity during its entire life cycle and is sometimes referred to as "cradle-to-grave" analysis. The new process of producing Zn powder by alkaline leaching and electrowinning is a reliable, cost-effective, and flexible process. The life cycle assessment method is used to investigate the environmental impacts of the new alkaline processes of Zn powder production and compared with traditional processes.

According to the ISO 14000 series (14041–14043), the technical framework for the LCA methodology as it is defined in ISO 14040 consists of four phases: (1) goal and scope definition, (2) inventory analysis, (3) impact assessment, and (4) interpretation. In order to investigate the environmental impacts of the new process producing Zn powder by alkaline leaching and electrowinning and compare the environmental profiles of the different processes of producing Zn powder, Figs. 9.11, 9.12, and 9.13 show the scope of models used for assessing the environmental impact of different processes during their whole life cycle. The functional unit is 1000 kg of zinc powder.

This involves collecting and analyzing the data concerning the relevant or main inputs (raw material and energy consumption) and outputs of emission such as SO_2, CO_2, NO_2, wastewater, and solid waste of a well-defined system. The initial data for the new alkaline process of production of metallic zinc powder are collected from the production line. The initial data for conventional acid leaching hydrometallurgical processes and pyrometallurgical processes are obtained from practicality

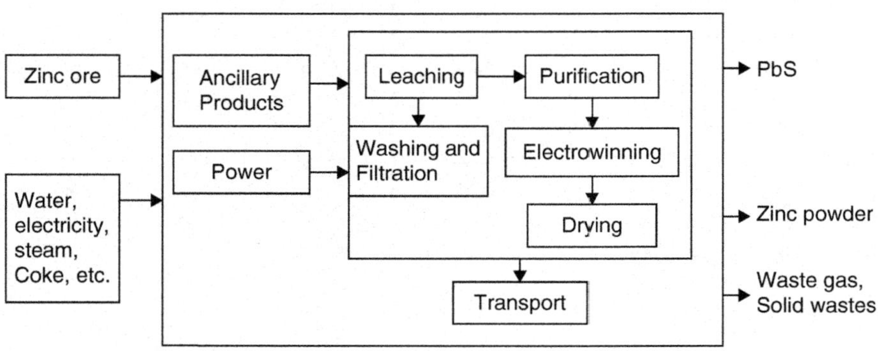

Fig. 9.11 Scope of LCA of alkaline leaching processes of Zn powder production

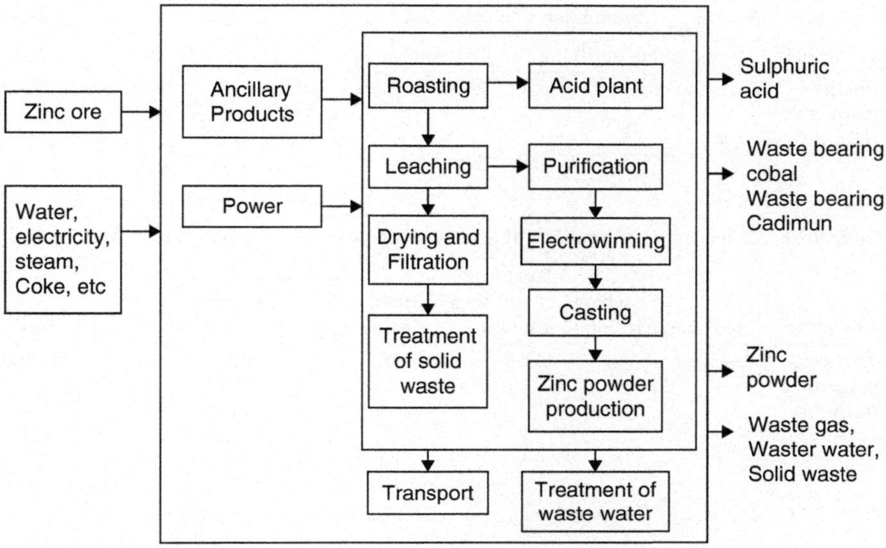

Fig. 9.12 Scope of LCA of acid leaching processes of production of Zn powder

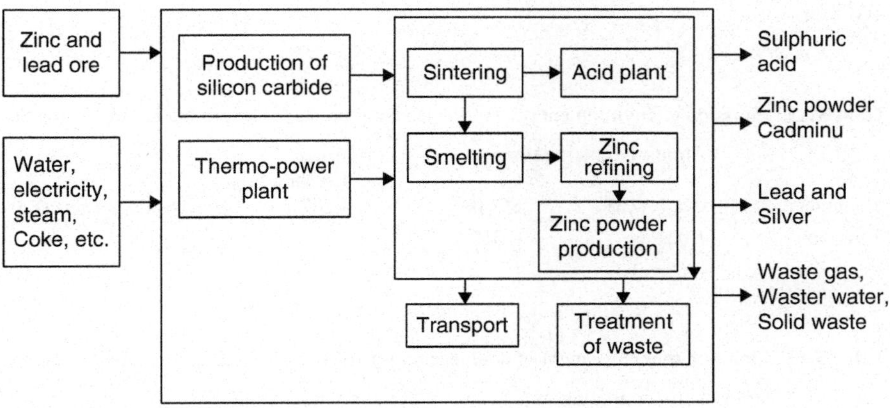

Fig. 9.13 Scope of LCA of pyrometallurgical processes of Zn powder production

manual and documents. All the data are treated by statistic method to get the average value as shown in Tables 9.9, 9.10, 9.11, 9.12, 9.13, 9.14, and 9.15.

The results indicate that the environmental impacts of alkaline leaching—electrowinning technology—are less than the conventional acid leaching hydrometallurgical processes and pyrometallurgical processes. The energy consumption, greenhouse effect, and acidification in the new processes are 87.76%, 47.80%, and 0.08% of those in the conventional acid leaching hydrometallurgical processes and are 93.94%, 9.25%, and 0.14% of those in the pyrometallurgical processes.

Table 9.9 Main energy consumption of alkaline leaching processes of production of Zn powder

Life cycle	Leaching purifying	Electrowinning	Drying	Auxiliary	Total
Energy consumption (GJ/t·Zn)	8.46	36.10	1.52	9.28	55.36

Table 9.10 Main energy consumption of acid leaching processes of production of Zn powder

Life cycle	Roasting	Leaching purifying	Treatment of leaching residues	Electro-winning	Casting	Zinc powder production	Auxiliary	Total
Energy consumption (GJ/t·Zn)	−4.27	3.44	11.58	39.46	1.60	1.38	9.89	63.08

Table 9.11 Main energy consumption of pyrometallurgical processes of production of Zn powder

Life cycle	Sintering smelting	Zinc refining	Zinc powder production	Auxiliary	Total
Energy consumption (GJ/t·Zn)	44.40	8.88	0.38	5.27	58.93

Table 9.12 Emission to environment of alkaline leaching processes of production of Zn powder

	Atmospheric emissions		Water discharges (kg/tZn)	
Pollutants	CO_2(t/tZn)	SO_2(kg)		Solid waste(t/tZn)
Emission	1.021	0.015	–	0.75

Table 9.13 Emission to environment of acid leaching processes of production of Zn powder

	Atmospheric emissions				Water discharges (kg/tZn)					Solid waste (t/tZn)
Pollutants	CO_2 (t/tZn)	SO_2 (kg/tZn)	As (kg/tZn)	Pb (kg/tZn)	Pb	Cd	Cu	As	Hg	
Emission	2.136	19.308	0.016	0.098	0.018	0.0026	0.005	0.001	4.34E-05	0.326

Table 9.14 Emission to environment of pyrometallurgical processes of production of Zn powder

	Atmospheric emissions			Water discharges (kg/tZn)				Solid waste (t/tZn)
Pollutants	CO_2(t/tZn)	SO_2(kg/tZn)	Hg (kg/tZn)	Pb	Cd	As	Hg	
Emission	11.035	11.104	0.035	0.139	0.008	0.0019	0.0007	0.588

Table 9.15 Eco-indicators of production of zinc powder in LCA

Process	GER (GJ/t·Zn)	GWP (t·CO_2 /tZn)	AP (t·SO_2·e/t Zn)	HME (kg·Pb/tZn)	SWB (t/tZn)
Alkaline leaching	55.36	1.021	0.015	0	0.75
Acid leaching	63.08	2.136	19.308	0.138	0.326
Pyrometallurgical (ISP)	58.93	11.035	11.104	0.204	0.588

Bibliography

Abdel EA (2000) Kinetics of sulfuric acid leaching of low-grade zinc silicate ore. Hydrometallurgy 55:247–254

Alan PB, Janet HM et al (1983) The alkaline electrolytic process for zinc production-a critical. Ind Eng Chem Prod Res Dev 22(2):263–272

Allen RJ, Foller PC, Vora RJ et al (1993) Full-scale hydrogen anodes for immersed-tank electrowinning. J Oper Manag 45(3):49–53

Amine AS (1995) Hydrometallurgical treatment of Egyptian zinc lead oxide ore, modeling. Meas Control C 49:1–12

Awakura Y, Kamei S, Majima H (1980) Kinetic study on non-oxidative dissolution of galena in aqueous acid solution. Metallurgical Transactions B (Process Metallurgy) 11B(3):377–381

Babu MN, Sahu KK, Pandey BD (2002) Zinc recovery from sphalerite concentrate by direct oxidative leaching with ammonium, sodium and potassium persulphates. Hydrometallurgy 64:119–129

Baik DS, Fray DI (2001) Electrodeposition of zinc from high acid zinc chloride solutions. J Appl Electrochem 31(10):1141–1147

Baldwin SA, Demopoulos George P, Papangelakis VG (1995) Mathematical modeling of the zinc pressure leach process. Metall Mater Trans B 26:1035–1047

Beckman O, Ostvold T (2000) Products formed during cobalt cementation on zinc in zinc sulfate electrolytes. Hydrometallurgy 54:65–78

Besecker CJ, Klemperer WG (1980) Polyoxoanion-supported metal carbonyls: syntheses of the $[(OC)_3M(Nb_2W_4O_{19})]^{3-}$anions (M = rhenium and manganese). J Am Chem Soc 102 (25):7598–7600

Besecker CJ, Klemperer WG, Maltbie DJ et al (1985) 17O nuclear magnetic resonance spectroscopy of polyoxometalates (2): heteronuclear decoupling of qnadrupolar nuele. Inorg Chem 24 (7):1027–1032

Besteti M, Ducati U et al (2001) Use of catalytic anodes for zinc electrowinning at high current densities from purified electrolytes. Can Metall Q 40:451–458

Bodas MG (1996) Hydrometallurgical treatment of zinc silicate ore from Thailand. Hydrometallurgy 40(1):37–49

Bonk S, Wicinski M et al (2004) Electrochemical characterizations of precipitates formed on zinc in alkaline sulphate solution with increasing pH values. Electrochem Commun 6:800–804

Calusaru R (1979) Electrodeposition of metal powders. Elsevier, Amsterdam

Y. Zhao, C. Zhang, *Pollution Control and Resource Reuse for Alkaline Hydrometallurgy of Amphoteric Metal Hazardous Wastes*, Handbook of Environmental Engineering 18, DOI 10.1007/978-3-319-55158-6

Caravaca C, Cobo A, Alguacil FJ (1994) Considerations about the recycling of EAF flue dusts as source for the recovery of valuable metals by hydrometallurgical processes. Resour Conserv Recycl 10(1–2):35–41

Carlos IA, Malaquias MA, Oizumi MM et al (2001) Study of the influence of glycerol on the cathodic process of lead electrodeposition and on its morphology. J Power Sources 92:56–64

Carlos IA, Siqueira J, Finazzi GA et al (2003) Voltammetric study of lead electrodeposition in the presence of sorbitol and morphological characterization. J Power Sources 117:179–186

Carlos IA, Matsuo TT, Siqueira J et al (2004) Voltammetric and morphological study of lead electrodeposition on copper substrate for application of a lead–acid batteries. J Power Sources 132:261–265

Cbalkley ME, Collins MJ, Ozberk EE (1993) Behavior of Sulphur in the sherrit zinc pressure leach process. Proceedings of the Intenraitonal Symposium on Zinc, p 325

Chaudhuri KB, Gluns L, Kawalec J (1984) KIVCET technology for complex lead concentrates smelting. 113th Annual Meeting – Metallurgical Society of AIME, Los Angeles

Chen AA, Dreisinger DB (1994) Ferric fluosilicate leaching of lead concentrates: part I. Kinetic studies. Metall Mater Trans B Process Metall Mater Process Sci 25(4):473–480

Clarence WB, Edger FS (1908) Preparation of potassium tantalate. J Am Chem Soc 30 (2):1637–1650

Collins MJ, Kofluk D (1998) Recovery of zinc from sulphuric concentrates. Hydrometallurgy 47:390

Collins MJ, Mc-Conaghy ET, Staufer RF et al (1994) Starting up the sherritt zinc pressure leach process at Hudson Bay. Hydrometallurgy 46:51–58

Copham PM, Fray DJ (1991) Selecting an optimum electrolyte for zinc chloride electrolysis. J Appl Electrochem 21(2):158

Corriose JP, Gely P, Viers P (1988) Thermodynamic and kinetic study of the pressure leaching of zinc sulfide in aqueous sulfuric acid. Hydrometallurgy 21:85–102

Dabbabi M (1976) Rasterization of $Nb_nW_{6\text{-}n}O_{19}{}^{(2+n)\text{-}}$. J Inorg Nucl Chem 38(3):1011–1014

Dabbabi M, Boyer M, Launay JP, Jeannin Y (1977) Propriétés électrochimiques de polyanions mixtes de composition $Nb_nW_{6-n}O_{19}{}^{(2+n)-}$(n=0, 1, 2, 3 et 4). J Electroanal Chem Interfacial Electrochem 76(2):153–164

Das SC, Subbaiah T, Sahoo PK et al (1989) Electrowinning of lead from its chloride bath containing organic additives. Hydrometallurgy 21(3):373–383

Dean AJ (2001) Langes handbook of chemistry, 14th edn. Plenum Press, New York

Donald JR, Pickles CA (1995) Kinetics of the reduction of the zinc oxide in zinc ferrite with iron. In: Queneau PB, Peterson RD (eds) 3rd Recycling of metals and engineered materials. International Symposium Minerals, Metals & materials Society, Warrendale, pp 603–621

Donald JR, Pickles CA (1996) Reduction of electric arc furnace dust with solid iron powder. Can Metall Q 35(3):255–267

Dutrizac JE (1980) Physical chemistry of iron precipitation in the zinc industry. IEEE International Conference on Acoustics, 1979 Metal Soc of AIME, pp 532–564

Dutrizac JE (1981) The dissolution of chalcopyrite in ferric sulphate and ferric chloride media. Trans Nonferrous Metals Soc China 12B:371–378

Dutrizac JE (1996) The effect of seeding on the rate of precipitation of ammonium jarosite and sodium jarosite. Hydrometallurgy 42:293–312

Elgersma F, Witkamp GJ, Van Rosmalen G (1993) Simultaneous dissolution of zinc ferrite and precipitation of ammonium jarosite. Hydrometallurgy 34:23–47

Feng L, Yang X, Shen Q et al (2007) Pelletizing and alkaline leaching of powdery low grade zinc oxide ores. Hydrometallurgy 89(3–4):305–310

Fleck DC, Sandberg RG, Wong MM (1983) Effects of impurities in electrolytes on electrowinning of lead from lead chloride. US Bureau of Mines, Avondale

Flynn CM, Pope MT (1971a) Tungstovanadate heterophy complexes (1): vanadium(V) complexes with the constitution $M_6O_{19}{}^{n-}$ and V:W $\leq$ 1:2. Inorg Chem 10(11):2524–2529

Flynn CM, Pope MT (1971b) Tungstovanadate heterophy complexes (2): products of acidification of $V_2W_4O_{19}{}^{4-}$. Inorg Chem 10(12):2745–2750
Frenay J, Elboudjian M (1990) Influence of the main parameters on zinc powder electrowinning in caustic soda medium. Proc Int Symp Electrometall Plant Pract 8:85–99
Frenay J, Ferlay S, Hissel J (1986) Zinc and lead recovery from EAF dusts by caustic soda process. In: Electric Furnace Proc., Treatment Options for Carbon Steel Electric Arc Furnace Dust, vol 43. Iron and Steel Society of Aime, USA, pp 171–175
Fugleberg, Sigmund P (1997) Method for leaching material containing zinc oxide and zinc silicate. Hydrometallurgy 46:338
Gabriel DS (2004) Kinetics and mechanism of the bacterial and ferric sulphate oxidation of galena. Hydrometallurgy 75:99–110
Gong YJ, Chen JY (1993) Kinetics of conversion of galena into lead carbonate in ammonium carbonate solution in the presence of cupric ion. Hydrometallurgy 33:177–195
Gord AR, Pickering RW (1975) Improved leaching technology in the electrolytic zinc industry. Metall Trans B (Process Metall) 6B:43–53
Gu YD (1985) Discrepancy between methods of preparing yellow and white tungstic oxide hydrates, and versatile uses of the latter. Chin J Appl Chem 2(4):9–15
Gu YD, Zhu SS (1982) The synthesis of porous white tungstic acid. Chem J Chin Univ 3 (1):138–139
Gürmen S, Emre M (2003) A laboratory-scale investigation of alkaline zinc electrowinning. Miner Eng 16(6):559–562
Harvey T, Tai YW, Paterson JG (1993) Kinetic investigation into the pressure oxidation of sphalerite from a complex concentrate. Miner Eng 6:949–967
Havlik T, Turzakova M, Stopic S et al (2005) Atmospheric leaching of EAF dust with diluted sulphuric acid. Hydrometallurgy 77(1–2):41–50
Hi SL, Dominique LP (1997) Determination of optimum leaching of roasted zinc concentrate in alkaline solution. J Chem Eng Jpn 30:59–63
Hideo S, Norio T, Hiroki K (1996) Recent operational improvements in the ISF plant. Metallurgical Rev Mining Metallurgical Instit Jpn 13:154–165
Hiroshi M, Ernest P, Yasuhiro A et al (1990) Fundamental studies on chlorine behavior as related to zinc electrowinning from aqueous chloride electrolytes. Metall Trans 21(2):251–258
Huot J, Boubour E et al (2004) Electrochemical performance of gelled zinc alloy powders in alkaline solutions. J Power Sources 128:343–349
Ivanov I (2004) Increased current efficiency of zinc electrowinning in the presence of metal impurities by addition of organic inhibitors. Hydrometallurgy 72:73–78
Jankola WA (1995) Zinc pressure leaching at Cominco. Hydrometallurgy 39:63–70
Jha MK, Kumar V, Singh RJ (2001) Review of hydrometallurgical recovery of zinc from industrial wastes. Resour Conserv Recycl 33(1):1–22
Kelebek S, Yörük S, Davis B (2004) Characterization of basic oxygen furnace dust and zinc removal by acid leaching. Miner Eng 17(2):285–291
Krysa BD (1995) Zinc pressure leaching at HB-MS. Hydrometallurgy 39:71–77
Kulenov AS, Sannikov YI, Slobodkin LV et al (1998) KIVCET process - a unique technology for the smelting of raw material containing lead, zinc and/or copper. Journal for Exploration, Mining and Metallurgy 51(4):273–279
Lachmann J, Pedlick M (1995) Kinetic anomalies of dissolution of sphalerite in ferric sulphate solution. Hydrometallurgy 37:89–96
Leclerc N, Meux E, Lecuire J (2002) Hydrometallurgical recovery of zinc and lead from electric arc furnace dust using mononitrilotriacetate anion and hexahydrate ferric chloride. J Hazard Mater 97:257–271
Leclerc N, Meux E, Lecuire JM (2003) Hydrometallurgical extraction of zinc from zinc ferrites. Hydrometallurgy 70(183):175
Li Q, Zhao YC (2014) Effects of cetyltrimethylammonium bromide and sodium lauryl sulfate on production of zinc powders by alkaline electrowinning. Russ J Non-Ferr Met 55(1):65–72

Li HG et al (2004) Hydrometallurgy. Middle and Southern University Press, Changsha
Liana M, Maurin G, Oniciu L et al (1996) Influence of metallic impurities on zinc electrowinning from sulphate electrolyte. Hydrometallurgy 43(1):345–354
Limp JL (1997) Oxidizing leaching of complex sulphides in ammonium chloride leaching kinetics. Rev Metal 33:258–270
Liu Q, Zhao YC, Zhao GD (2011) Production of zinc and lead concentrates from lean oxidized zinc ores by alkaline leaching followed by two-step precipitation using sulfides. Hydrometallurgy 110(1–4):79–84
Liu Q, Zhao GD, Zhao YC (2014) Regeneration and purification of spent electrolyte from sodium hydroxide zinc metallurgy using causticisation. Hydrometallurgy 144:107–113
Lu KY, Chen CY (1986) Conversion of galena to lead carbonate in ammonium carbonate solution-a new approach to lead hydrometallurgy. Hydrometallurgy 17:73–83
Lupi C, Pilone D (1997) New lead alloy anodes and organic depolarizer utilization in zinc electrowinning. Hydrometallurgy 44(3):347–358
Magdy G, Edward G (1987) Electrochemical phenomena in aqueous electrowinning of lead. J Appl Electrochem 17(6):1234–1245
Masahiro T, Harutoshi K (1995) Zinc-lead smelting and refining at Sumitomo Harima works. Metallurgical Review of Mining and Metallurgical Institute of Japan, 12:77–95.
Masakazu M (1995) Zinc-lead smelting at Hachinohe smelter. Metallurgical Review of Mining and Metallurgical Institute of Japan, 12:39–50.
Massacci P, Recinella M, Piga L (1998) Factorial experiments for selective leaching of zinc sulphide in ferric sulphate media. Int J Miner Process 53:213–224
Menad N, Ayala JN, Garcia-Carcedo F et al (2003) Study of the presence of fluorine in the recycled fractions during carbon-thermal treatment of EAF dust. Waste Manag 23(6):483–491
Morachevskii AG, Demidov AI, Vaisgant ZI et al (1996) Recovery of lead battery scrap using alkali– glycerol electrolyte. Russ J Appl Chem 69:412–414
Morin D, Gaunand A, Renon H (1985) Representation of the kinetics of leaching of galena by ferric chloride in concentrated sodium chloride solutions by a modified mixed kinetics model. Metall Trans B (Process Metallurgy) 16B(1):31–39
Muresan L, Maurin G, Oniciu L et al (1996) Influence of metallic impurities on zinc electrowinning from sulphate electrolyte. Hydrometallurgy 43:345–354
Murphy JE, Chambers MF (1991) Production of lead metal by molten-salt electrolysis with energy -efficient electrodes. RI 9335
Nathalie L, Eric M (2003) Hydrometallurgical extraction of zinc from zinc ferrites. Hydrometallurgy 70:175–183
Niederkorn JS (1985) Kinetic study on catalytic leaching of sphalerite. J Met 37:53–56
Nunez C, Espiell F, Garcia-Zayas J (1988) Kinetics of non-oxidative leaching of galena in perchloric, hydrobromic, and hydrochloric acid solutions. Metallurgical Transactions B (Process Metallurgy) 19(4):541–546
Nunez C, Espiell F, Garcia-Zayas J (1990) Kinetics of galena leaching in hydrochloric acid-chloride solutions. Metallurgical Transactions B (Process Metallurgy) 21(1):11–17
Olper M (1993) EZINEX process: a new and advanced way for electrowinning zinc from a chloride solution. Proceedings of the international symposium on zinc. Publish by Australasian Inst of Mining & Metallurgy, p 491
Olper M, Maccagni M (2000) Electrolytic zinc production from crude zinc oxides with the Ezincx Process. 2000 TMS Fall Extraciton & Processing, Pittsburgh
Orhan G (2005) Leaching and cementation of heavy metals from electric arc furnace dust in alkaline medium. Hydrometallurgy 78(3–4):236–245
Ozberk E, Jankola WA, Vecchiarelli M et al (1995a) Commercial operations of the sherritt zinc pressure leach process. Hydrometallurgy 39:49–52
Ozberk E, Colilns MJ, Makwana M et al (1995b) Zinc pressure leaching at the Ruhr-Zink refinery. Hydrometallurgy 39:53–62

Paramguru RK, Kuezeci E, Kammel R (1988) Direct electrowinning of lead from suspension galena concentrate anode in different electrolytes. Metall Trans B (Process Metallurgy) 19B (1):59–65

Peters E (1976) Direct leaching of sulphides: chemistry and applications. Trans Nonferrous Metals Soc China 7:505–517

Petrtova M, Noncheva Z, Dobrev T et al (1996) Investigation of the processes of obtaining plastic treatment and electrochemical behavior of lead alloys in their capacity as anodes during the electroextraction of zinc I. Behavior of Pb-Ag, Pb-Ca and Pb-Ag-Ca alloys. Hydrometallurgy 40:293–318

Phillips TA (1978) Economic evaluation of a leach-electrolysis process for recovering lead from galena concentrate, vol 08. US Bureau of Mines, Washington, p 733

Piron DL (1990) Recent improvements in zinc alkaline hydrometallurgy. In: Proceedings of the international symposium on Electrometallurgical Plant Practice, Canada, pp 129–140

Pope MT (1983) Heteropoly and isopoly oxometalates. Springer, Berlin/Heidelberg/New York/ Tokyo, pp 40–41

Rabah MA, El-Sayed AS (1995) Recovery of zinc and some of its valuable salts from secondary resources and wastes. Hydrometallurgy 37(1):23–32

Raghavan R, Mohanan PK, Verma SK (1999) Modified zinc sulphate solution purification technique to obtain low levels of cobalt for zinc electrowinning process. Hydrometallurgy 51:187–206

Ramachandran P, Venkateswaran KV, Srinivasan R (1985) Inhibition of PbO_2 formation during lead electrowinning. J Appl Electrochem 15(6):937–945

Rashkov S, Stefanov Y, Noncheva Z et al (1996) Investigation of the processes of obtaining plastic treatment and electrochemical behavior of lead alloys in their capacity as anodes during the electroextraction of zinc I. Electrochemical formation of phase layers on binary Pb-Ag and Pb-Ca and ternary Pb-Ag-Ca alloys in a sulphuric-acid electrolyte for zinc electroextraction. Hydrometallurgy 40:319–334

Rashkov S, Ts D et al (1999) Lead–cobalt anodes for electrowinning of zinc from sulphate electrolytes. Hydrometallurgy 52:223–230

Rocchiccioli-Deltcheff C, Thonvenot R, Dabbabi M (1977) Etude de la structure des niobotungstates $Nb_nW_{6-n}O_{19}^{(2+n)-}$. Spectrochim Acta A 33A(1):143–153.

Saba AE, Elsherief AE (2000) Continuous electrowinning of zinc. Hydrometallurgy 54:91–106

Salihoglu G, Pinarli V (2008) Steel foundry electric arc furnace dust management: stabilization by using lime and Portland cement. J Hazard Mater 153(3):1110–1116

Sato T, Lawson F (1983) Differential leaching of lead smelter slags with sulphurous acid and oxygen. Hydrometallurgy 11:371–388

Shu J (1999) V I Lakshmanan, J convey, sintering and ferrite formation during high temperature roasting of sulfide concentrates. Can Metall Q 10:215–225

Stanforth RR. Method for reduction of heavy metal leaching from hazardous waste under acidic and nonacidic conditions. U.S. Patent 5,037,479. 1991-8-6

St-Pierre J, Piron DL (1986) Electrowinning of zinc from alkaline solutions. J Appl Electrochem 16(3):447–456

St-Pierre J, Piron DL (1990) Electrowinning of zinc from alkaline solutions at high current densities. J Appl Electrochem 20(1):163–165

Swarnkar SR, Gupta BL, Sekharan RD (1996) Iron control in zinc plant residue leach solution. Hydrometallurgy 42:21–26

Tozawa K, Nishimura T, Akabori M (1992) Comparison between purification process for zinc leach solutions with arsenic and antimony trioxides. Hydrometallurgy 30:445–461

Trina MD, Amy N, George PD et al (2001) The kinetics of cobalt removal by cementation from an industrial zinc electrolyte in the presence of Cu, Cd, Pb, Sb and Sn additives. Hydrometallurgy 60(2):105–116

Tuffrey NE, Jiricny V, Evans JW (1985) Fluidized bed electrowinning of zinc from chloride electrolytes. Hydrometallurgy 15(1):33–54

Turan MD et al (2004) Recovery of zinc and lead from zinc plant residue. Hydrometallurgy 75:169–177

Vakil S (1996) Technological innovation in the zinc electrolyte purification process of a hydrometallurgical zinc plant through reduction in zinc dust consumption. Hydrometallurgy 40:247–262

Wend WJ (1963) Ammonium carbonate leaching of low grade of zinc ore. Eng Min J 154:84–90

Wong MM (1976) Leach-electrolysis method for producing lead. J Iron Steel Inst Jpn 2:603–621

Wong SM, Abrantes LM (2005) Lead electrodeposition from very alkaline media. Electrochim Acta 51:619–626

Xia DK, Pickles CA (1999) Caustic roasting and leaching of electric arc furnace dust. Can Metall Q 38(3):175–186

Zeng QH, Chou SZ (1985) New-type zinc powder for cobalt elimination in depth purification of $ZnSO_4$ solutions. Zinc 85 Proc Int Symp Extr Metall Zinc 127–140 Japan lead zinc development assoc, Jpn Mining & Metallurgical Inst of Japan

Zeydabadi BA, Mowla D, Shariat MH et al (1997) Zinc recovery from blast furnace flue dust. Hydrometallurgy 47(1):113–125

Zhang CL, Zhao YC (2009) Mechanochemical leaching of sphalerite in an alkaline solution containing lead carbonate. Hydrometallurgy 100(1):56–59

Zhang P, Thomas JO, Yu P (2001) Electrochemical characterization of the effects of impurities and organic additives in lead electrowinning from fluoborate electrolyte. Hydrometallurgy 61:207–221

Zhang CL, Zhao YC, Guo CX et al (2008) Leaching of zinc sulfide in alkaline solution via chemical conversion with lead carbonate. Hydrometallurgy 90(1):19–25

Zhang CL, Wang JW, Bai JF, Guan J (2013) Recovering lead from cathode ray tube funnel glass by mechano-chemical extraction in alkaline solution. Waste Manage Res 31:759–763

Zhao YC, Chen JY (1996) Extraction of phosphorus, arsenic and/or silica from sodium tungstate and molybdate solutions with primary amine and tributyl phosphate as solvents. I—synergistic extraction and separation of phosphorus, arsenic and/or silica from tungstate and molybdate solutions. Hydrometallurgy 42(3):313–324

Zhao YC, Gu YD (1991) Synthesis and properties of potassium 11-tungstoniobate. Chin Chem Lett 2(8):665–668

Zhao YC, Gu YD (1992) Synthesis and properties of potassium 11-tungstotantalate. Chin J Appl Chem 9(5):74–76

Zhao GW, Inman D (1995) Electrowinning zinc from chloride melts in bipolar cell. Trans Nonferrous Soc 5:45–48

Zhao YC, Li Q (2013) Production of ultrafine zinc powder from wastes containing zinc by electrowinning in alkaline solution. Braz J Chem Eng 30(4):857–864

Zhao YC, Stanforth R (2000a) Production of Zn powder by alkaline treatment of smithonite Zn–Pb ores. Hydrometallurgy 56(2):237–249

Zhao YC, Stanforth R (2000b) Integrated hydrometallurgical process for production of zinc from electric arc furnace dust in alkaline medium. J Hazard Mater 80(1):223–240

Zhao YC, Stanforth R (2001a) Selective separation of lead from alkaline zinc solution by sulfide precipitation. Sep Sci Technol 36(11):2561–2570

Zhao YC, Xie GY, Gu YD (1992a) Synthesis and properties of $[C(NH_2)_3]_{6.3}K[Nb_{1.3}W_{10.7}O_{40}H_2]\cdot H_2O$. Chinese J Inorg Chem 8(4):416–420

Zhao YC, Xie GY, Gu YD (1992b) The synthesis of $[(C_4H_9N)_4]_nK_{7-n}NbW_{11}O_{40}H_2$ complex (n = 5, 6). Acta Chim Sin 60:1178–1181

Zhao YC, Xie GY, Zhu SS, Gu YD (1993a) White tungstic acid and its applications. Chin J Rare Met 17(3):15–19

Zhao YC, Xie GY, Gu YD (1993b) The synthesis of $[(C_4H_9N)_4]_5K_2TaW_{11}O_{40}H_2$ complex. J Fudan Univ (Natural Science) 32(1):110–114

Zhao YC, Zouboulis AI, Matis KA (1996a) Removal of molybdate and arsenate from aqueous solutions by flotation. Sep Sci Technol 31(6):769–785

Zhao YC, Zouboulis AI, Matis KA (1996b) Flotation of molybdate oxyanions in dilute solutions. Part I. Selective separation from arsenate. Hydrometallurgy 43(1):143–154

Zhao YC, Zouboulis AI, Matis KA (1996c) Flotation of molybdate oxyanions in dilute solutions. Part II. Selective separation from phosphates, arsenates and silicates mixtures. Hydrometallurgy 43(1):155–167

Zhao YC, Zhu QS, Gu GW (1998) Combined removal of SO_2, H_2S and NOx from gas streams by chemical absorption with aqueous solution of 12-molybdophosphoric acid and its reduced species. Water Air Soil Pollut 102(2):157–176

ZhaoYC CJY (1996) Extraction of phosphorus, arsenic and/or silica from sodium tungstate or molybdate solutions with primary amine and tributyl phosphate as solvents. II—mechanism of extraction of phosphorus, arsenic or silica from tungstate or molybdate solutions. Hydrometallurgy 42(3):325–335

Zouboulis AI, Zhao YC, Matis KA (1996) Separation of tungstate from aqueous mixtures containing impurities (arsenate, phosphate and silicates anions) using ion flotation. J Chem Technol Biotechnol 67(2):204–212

Index

Y. Zhao, C. Zhang, *Pollution Control and Resource Reuse for Alkaline Hydrometallurgy of Amphoteric Metal Hazardous Wastes*, Handbook of Environmental Engineering 18, DOI 10.1007/978-3-319-55158-6

Printed by Printforce, the Netherlands